Buildings and
Schubert Schemes

Preface

This work first aims at introducing the minimal generalized galleries in the Tits building of a reductive group G over a field k, and at constructing a family of equivariant smooth resolutions for its Schubert varieties, which may be described by these galleries in terms of Tits geometry (cf. [47], Ch. 6–9). The second aim is to:

- introduce the Universal Schubert scheme and Schubert schemes of a reductive S-group G scheme, where S denotes an arbitrary base scheme, and its description by transposing the buildings terminology to this relative situation;

- construct a canonical equivariant Resolution of Singularities of the Universal Schubert scheme of G, which amounts to the construction of smooth Resolutions for each Schubert scheme;

- study the behaviour of these Resolutions under base extensions $S' \rightarrow S$.

The canonical Schubert scheme smooth resolution is obtained after a twisted constant finite extension of S. Under certain conditions these smooth resolutions are obtained as the fibers of the Universal Smooth Resolution of G. All these constructions rely on combinatorial data only. Techniques, definitions, and verifications of this part of the work are schematical and developed in the setting of S-reductive group schemes. This may be of interest as until now, few resolutions of singularities of algebraic varieties which are valid on fields of characteristic p are known.

The first seven chapters are written in the usual algebraic geometric style over a field language, as in Hartshorne's book [33]. They are relatively self-contained, and give an overview of the subsequent developments in the case of the linear group $Gl(k^{r+1})$ over a field k, or even over a base scheme S, thus they furnish both a guide and an example of the results of the rest of the book. The classical Schubert varieties of a linear group appear as subvarieties of the Flag varieties associated to k^{r+1} (resp. the projective space $\mathbb{P}(k^{r+1})$). They are indexed by couples formed by a matrix M (The relative position matrix) with non negative integers as entries, and by a flag D of k^{r+1}. A flag belongs to a Schubert variety if its subspaces satisfy intersection dimensional conditions defined by its corresponding couple.

The Schubert varieties of the Grassmann variety $\mathrm{Grass}_n(k^{r+1})$ of n-dimensional linear sub-spaces of k^{r+1} correspond to the classical Schubert varieties, for $k = \mathbb{R}$, or $\mathbb{C}$, which play an important role in Topology (cobordism theory), Algebraic Geometry, Representation Theory, determinantial ideals, and Singularity Theory, particularly in the definition of Boardman-Thom singularities. These varieties have been studied for a long time by classical authors in connection with ennumerative geometry problems. One of the main results was obtained by F. Severi and (his student) J. Todd, who first obtained a formula for the arithmetic genus of a subvariety of the projective space in terms of the intersection properties of Schubert varieties. Ehresmann proved that "the classes defined by the Schubert varieties of $\mathrm{Grass}_n(k^{r+1})$, give a basis of their $\mathbb{Z}$-homology (resp. cohomology)" [28]. Characteristic classes of a non-singular algebraic variety may be defined, by means of a classifying map, as the pull-back of special Schubert Varieties. More generally Chevalley-Demazure proved that the Chow ring of generalized Flag varieties (Varieties of parabolics) is given in terms of Schubert varieties. The cohomological point of view was further developed in A. Borel thesis in terms of the transgression homomorphism. Finally, invariant differential forms were obtained by A. Weil and S. Chern representing dual classes of Schubert varieties.

Equivariant smooth resolutions of singularities of Schubert varieties are obtained as **Configurations varieties**. The underlying sets of points of these varieties are subsets of finite products of Flag varieties (resp. Grassmann varieties) of k^{r+1}, or more generally subsets of finite products of the Flag complex of k^{r+1}, defined by the incidence relation between flags. The **Flag complex** of k^{r+1} is a simplicial complex whose vertices correspond with subspaces, and its simplices with flags of k^{r+1}. It is endowed with a natural incidence relation: Two subspaces of k^{r+1} are incident if one of them is contained in the other one (cf. [47], 1.2). A finite graph is naturally associated with a Configurations variety, such that to each one of its vertices is associated a type of a flag, and to each one of its edges a type of incidence of flags. Such a graph is called a **typical graph**.

A generalized gallery of the Flag complex is, by definition, the image of a linear typical graph by a mapping preserving types, and corresponds to a point of a Configurations variety defined by this graph. Among the linear typical graphs are the **minimal generalized galleries of types** characterized as linear typical graphs defining Configurations varieties **birationally equivalent** to Schubert varieties. To such a linear graph is thus associated the relative position matrix indexing the Schubert variety birationally equivalent to the Configurations variety defined by this linear graph. A **minimal generalized gallery** of the Flag complex may be seen as a generic point of such a Configurations variety. It is worth noting that minimal generalized galleries, which play a central role in this work, are characterized by a combinatorial property.

For a Schubert variety of $Gl(k^{r+1})$, a family of Configurations varieties, which are smooth resolutions of this variety, is thus obtained. To a relative position matrix is associated a typical graph (resp. a minimal generalized gallery of types), thus defining a canonical smooth resolution of a Schubert variety defined by this matrix. As an application of these canonical smooth resolutions a characterization of the Singular Locus of the Schubert varieties of $Gl(k^{r+1})$, which holds for all characteristics, is given. In fact, this was the first such characterization (Author Thesis 1983, cf. [16]).

This result is completed by the construction of Configurations varieties giving a Nash smooth resolution of singularities for each Schubert variety in a Grassmannian, i.e. the pull back of the tangent module to the corresponding Schubert cell admits a locally free extension to this smooth resolution.

The subcomplex of the Flag complex whose simplices are given by the flags adapted to the canonical basis of k^{r+1}(**The canonical Apartment**), plays the main role in the determination of minimal generalized galleries, and relate this determination to calculations in the symmetric group in $r + 1$ letters.

To generalize the above results for $Gl(k^{r+1})$ to a reductive k-group G observe that the Flag complex of k^{r+1} may be rendered solely in terms of the linear group by means of the bijective correspondence $D \mapsto P_D$, where P_D denotes the stabilizer of D in $Gl(k^{r+1})$ (as it follows from a result of A. Borel-Cl. Chevalley, cf. [6], Th. 11.16). Thus the stabilizers of flags correspond to the parabolic subgroups of the linear group, i.e. the smooth subgroups P such that the quotient space $Gl(k^{r+1})/P$ is a projective variety. Then the incidence relation between flags corresponds with the opposite of the inclusion relation between parabolics, and generalized galleries become configurations of parabolics defined by the incidence relation. This correspondence establishes a simplicial complex isomorphism between the Flag complex of k^{r+1} and the Tits Building $I(Gl(k^{r+1}))$. Recall that the **Tits Building** $I(G)$ of the k-reductive group G is the simplicial complex whose simplices are the parabolic subgroups of G, endowed with the incidence relation given by the opposite relation to the inclusion of parabolics (cf. [47], Ch. 5). All the definitions stated in the setting of Flag complex may be naturally transposed to that of general buildings. As a particular case a minimal generalized gallery (resp. Configurations varieties) in a building $I(G)$ of a reductive k-group G is defined following the same pattern of its corresponding definition in the Flag complex, i.e. as a generic point of a Configurations variety defined by a minimal generalized gallery of types (resp. as a set of images in $I(G)$ of a typical graph by typical graphs morphisms). The construction of a family of smooth resolutions of Schubert varieties of a reductive k-group G may be thus carried out by determinig the minimal generalized galleries in an apartment.

The Apartment of the Flag complex of k^{r+1} corresponds to the subcomplex of $I(G)$ defined by the Weyl group of G endowed with the canonical generating reflexions (**The Coxeter Complex** of G). The latter plays the same role in the determination of minimal generalized galleries in $I(G)$ that the former one in the Flag complex.

There is a threefold reason for introducing the relatively heavy machinery of building theory in the study of Schubert varieties and their smooth resolutions:

- Smooth resolutions appear as classes of linear subcomplexes of $I(G)$, i.e. families of generalized galleries whose corresponding galleries of types are defined by minimal generalized galleries in $I(G)$.

- Relations between these smooth resolutions may be easily stated in $I(G)$ (Adjacency, composition).

- By means of the **Incidence Geometries associated to buildings**, geometrical representations of smooth resolutions may be obtained in algebraic symmetric spaces, or in terms of well known combinatorics (The twenty seven lines on a cubic surface complex, the twenty eight bitangents to a plane quartic complex $\cdots$ etc.).

As an example the geometric realization of $I(G(l(k^{r+1})))$ is given by the flags of $\mathbb{P}^r(k)$, and that of an apartment by the barycentric subdivision of the r-dimensionnal simplex. Thus the geometric realization of $I(G)$ furnishes a natural frame to develop Schubert calculus and Configurations varieties, as in the case of the linear group (cf. Appendix).

Chapters 11 to 15 are written in the schematically setting. The technical results concerning the schematic closure needed in our developments are all included in the last two sections of Chapter 15. A Schubert scheme of G is defined as the **schematic closure of a Schubert cell** in the Parabolics scheme of G (cf. [23], Ch. XXV). The Schubert cells of a reductive S-group G over an arbitrary base scheme S are defined in terms of couples of parabolics in **standard position**, i.e. such that they contain locally in S a common maximal torus of G. These couples are classified modulo the adjoint action of G on the scheme of couples of parabolics in standard position, by the twisted constant finite **scheme of types of relative positions**. The above definition makes sens as Schubert cells are quasi-compact if S is affine, and the Parabolics scheme of G is projective and thus separated.

It is shown how all the building constructions above, in the case of a reductive group over an algebraically closed field, may be carried out in the relative case to produce resolutions of singularities of Schubert schemes associated to a reductive S-group scheme G.

The construction of galleries configurations smooth resolutions of Schubert schemes may be carried out once the basical building machinery is extended to reductive S-groups. The parabolic subgroups scheme $\mathcal{P}ar(G)$, endowed with the incidence relation, plays the role of a relative building. Remark that the only couples of parabolics that we consider in this setting of the relative building, are those in **standard position** (cf. loc. cit.). Actually, the relevance of the incidence relation is that it allows defining configurations subschemes of finite products of $\mathcal{P}ar(G)$. The complexes associated to buildings (Weyl complex, typical simplex, convex hulls, root subcomplexes,...etc.) become, in this setting, twisted constant finite schemes endowed with incidence relations. These schemes are all defined by etale descent from the case where G is splitted.

In order to make all constructions intrinsic the **Universal Schubert scheme** and its **Universal Smooth Resolution** are introduced. The former is naturally a scheme over that of types of relative positions, and the latter a scheme over that of minimal galleries of types. The main fact is that the Universal Smooth Resolution is a canonical smooth resolution of the Universal Schubert scheme, after the twisted constant finite extension of the scheme of types of relative positions by the scheme of minimal galleries of types.

The question arises naturally arises as to know whether the construction of the smooth resolutions of the Schubert schemes associated to a reductive S-group scheme G commutes with base extensions $S' \to S$. Chapter 16 is devoted to this question. The Universal smooth resolution of the Universal Schubert Scheme associated to a Chevalley group scheme G over $\mathbb{Z}$ defined by a $\mathbb{Z}$-root data $\mathcal{R}$ is introduced and it is shown that there exists an open not-empty subset U of $Spec(\mathbb{Z})$ with the following property. The Universal Smooth resolution of G_U commutes with the base extensions $S \to U$.

Roughly speaking that means that the above Global smooth resolution connects (almost) all the possible characteristic p smooth resolutions of all Schubert schemes corresponding to groups of type $\mathcal{R}$. It follows a condition on S so that the construction of the Universal smooth resolution of a reductive S-group scheme G commutes with base extensions $S' \to S$.

In the Appendix we establish a correspondence between the main theorems of [47] concerning automorphisms group of a building $I(G)$ of a reductive group on an algebraically closed field and the canonical set of generators and relations defining G as given in SGA3. This suggests how building calculations may be expressed into canonical generators and relations. It is also dicussed how the constructions of this work may be represented in the Tits incidence geometries corresponding to the reductive groups of differents types (Schubert Geometry), and how different smooth resolutions given by different galleries are related by braid relations.

One of the aims at the outset of this work was to find a common setting for the smooth resolutions given by configurations varieties associated to flag varieties (cf. [12] and [13]), and the smooth resolutions, associated to Schubert varieties in a quotient space G/B of a k-reductive group by a Borel subgroup (cf. [21]). The latter are given by the contracted products, corresponding to usual minimal galleries, first introduced by R. Bott and H. Samelson in [7], and generalizing a construction of M. Morse in [38].

The author thanks Professor P. Deligne that once remarked him that the above mentioned smooth resolutions by configurations varieties may be rendered by some kind of minimal gallery; and professor Lê Dũng Tráng, who first suggested that he should write this book.

Carlos Contou-Carrère

Contents

Chapter 1

Grassmannians and Flag Varieties

This chapter is an introduction to a building point of view for describing Grassmann varieties (resp. flags varieties) of k^{r+1} (where k denotes a field). The combinatorial grassmannians associated with a finite set are introduced. The elements of this set correspond to the subspaces of k^{r+1} adapted to the canonical basis. They index the canonical affine open sets of Grassmann varieties and allow a geometrical interpretation of their natural coordinates. Generally the set of combinatorial flags, corresponding to the flags of k^{r+1}, adapted to the canonical basis, indexes the affine canonical open sets of flags varieties. The algebraic variety structure of Grassmannians (resp. flag varieties) is defined in terms of local canonical coordinates in these open sets. The natural fibering in Grassmannians of flag varieties is obtained. Their projective variety structure over k is shown in terms of the Plücker and Segre's embeddings.

1.1 Grassmann variety and combinatorial grassmannian

Let k be a field. Given a finite set $E \neq \emptyset$ (resp. $F \neq \emptyset$), we denote by k^E the k-vector space with a basis $e_E = (e_i)_{i \in E}$ (**the canonical basis of k^E**) indexed by E, and $M_{E \times F}(k)$ stands for the set of k-matrices M with rows (resp. columns) indexed by E (resp. F), endowed with the canonical k-vector space structure. Write $k^\emptyset = \{0\}$.

For $n \in \mathbb{N}$, let $Grass_n(k^E)$ be the set of n-dimensional subspaces of k^E. Of course, if $|E| < n$, then $Grass_n(k^E)$ is an empty set, where given a finite set F we put $|F| = cardinal\ of\ F$. Write:

1

$$Grass(k^E) = \coprod_{n \in \mathbb{N}} Grass_n(k^E) \ .$$

With the aim of defining a k-variety structure on $Grass_n(k^E)$ we introduce the **combinatorial grassmannian** defined by the finite set E; given by the class of its subsets

$$Grass(E) = \coprod_{n \in \mathbb{N}} Grass_n(E) = \mathcal{P}(E) \ ,$$

where $Grass_n(E) = \mathcal{P}_n(E)$ denotes the class of subsets of E containing n elements (cf. [50]). To a not empty subset H of E is associated a direct sum decomposion

$$k^E = k^H \oplus k^{H^\perp} \ ,$$

with $H^\perp = E - H$ and k^H the k-vector subspace with canonical basis indexed by H $\subset$ E. Denote by $\pi_H : k^E \longrightarrow k^H$ (resp. $\pi_{H^\perp} : k^E \longrightarrow k^{H^\perp}$) the canonical projection, and define

$$U_H = \{S \in Grass_n(k^E) |\ rank\ (\pi_H)_S = n\} \ ,$$

where, given a k-subspace $S \subset k^E$, $(\pi_H)_S$ (resp. $(\pi_{H^\perp})_S$) is the restriction of π_H (resp. $\pi_{H^\perp}$) to S. Denote by e_H the canonical basis of k^H and given $S \in U_H$ let $(\pi)_H^{-1}(e_H)$ be the lifted basis of S.

Definition 1.1 *Let $S \in U_H$, and $M((\pi_H)_S^{-1}(e_H)) \in M_{E \times H}(k)$ be the matrix whose j-th column vector $(j \in H)$ is given by the e_E-coordinates of the j-th vector of the lifted basis of S:*

$$\tilde{e}_H(S) = (\pi_H)_S^{-1}(e_H) = ((\pi_H)_S^{-1}(e_j))_{j \in H}$$

of S, where $e_H = (e_j)_{j \in H}$ denotes the canonical basis of k^H. Denote by

$$M_H(S) = (\xi_{ij}^H(S))_{(i \in H^\perp, j \in H)} \in M_{H^\perp \times H}(k)$$

the $H^\perp \times H$-submatrix of $M((\pi_H)_S^{-1}(e_H))$ given by its $H^\perp$-row vectors. We call $(\xi_{ij}^H(S))$ U_H-coordinates of S, and $M_H(S)$ the U_H-coordinates matrix of S.

Fix a total order on E given by some bijection $E \simeq [\![1, |E|]\!]$. Given $H \in Grass_n(E)$ there are induced bijections: $H \simeq [\![1, |H|]\!]$ $(|H| = n)$ (resp. $H^\perp \simeq [\![1, |E| \quad |H|]\!]$). Observe that the lifted basis $(\pi_H)_S^{-1}(e_H)$ is naturally ordered by the order of H induced by that of E. Denote by: $(\beta_j)_{1 \leq j \leq |H|}$ (resp. $(\alpha_i)_{1 \leq i \leq (|E| - |H|)}$) the corresponding total order of H (resp. $H^\perp$). Write

$$e_H = (e_{\beta_j})_{1 \leq j \leq |H|} \ (\text{resp. } e_{H^\perp} = (e_{\alpha_i})_{1 \leq i \leq (|E| - |H|)}),$$

and given $S \in Grass_n(k^E)$, write $\tilde{e}_H(S) = (\tilde{e}_{\beta_j}(S))_{1 \leq j \leq |H|} = (\pi_H)_S^{-1}(e_H)$. The lifted basis $\tilde{e}_H(S)$ and the U_H-coordinates matrix $M_H(S) = (\xi_{ij}^H(S)) \in M_{|H^\perp| \times |H|}$ are related as follows:

$$\tilde{e}_{\beta_j}(S) = e_{\beta_j} + \sum_{1 \leq i \leq (|E| - |H|)} \xi_{ij}^H(S) e_{\alpha_i} \; .$$

Observe that the matrix $M((\pi_H)_S^{-1}(e_H))$ may be obtained from a basis $\tilde{e}_j = \sum_{1 \leq i \leq |E|} a_{ij} e_i$ $(1 \leq j \leq |H|)$ of S by **normalization**. Write $A((\tilde{e}_j)) = (a_{ij}) \in M_{|E| \times |H|}(k)$ (the $|E| \times |H|$-matrices with k-coefficients) and denote by $A((\tilde{e}_j))_H$ the $|H| \times |H|$-submatrix defined by the row vectors of $A((\tilde{e}_j))$ indexed by H. The condition $S \in U_H$ implies that $A((\tilde{e}_j))_H$ is an invertible matrix. The result is:

$$M((\pi_H)_S^{-1}(e_H)) = A((\tilde{e}_j)) \times A((\tilde{e}_j))_H^{-1} \; .$$

Thus the matrix $M_H(S)$ is the $|H^\perp| \times |H|$-submatrix of $A((\tilde{e}_j)) \times (A((\tilde{e}_j)))_H^{-1}$ whose row vectors are indexed by $H^\perp$. Observe that the image of the j-th column vector of the submatrix $A((\tilde{e}_j)) \times A((\tilde{e}_j))_H^{-1}$ by $(\pi_H)_S$ is the j-th vector of the basis e_H ordered by the order of E.

Definition 1.2 *Denote by $M_{E \times H}^*(k) \subset M_{E \times H}(k)$ the subset of rank $|H|$-matrices. We say that $M \in M_{E \times H}^*(k)$ is **normalized** if its $H \times H$-submatrix $(a_{\alpha\beta})$ satisfies $(a_{\alpha\beta}) = (\delta_{\alpha\beta})$.*

The particular case where E is totally ordered and $H \subset E$ is endowed with the induced order is of interest for us. The coefficients of M may be ordered accordingly. In this case every matrix $A \in M_{(|E|-n) \times n}(k)$ gives rise to a $E \times H$-normalized matrix in a canonical way. It may be noted that $M((\pi_H)_S^{-1}(e_H)) \in M_{E \times H}^*(k)$ is a normalized matrix.

On the other hand, the set $Grass_n(k^E)$ identifies with the set of equivalence classes: $M_{E \times [\![1,n]\!]}^*(k) / \sim$, defined by the equivalence relation:

$$M \sim M' \Longleftrightarrow M = M' \times \Lambda \; ,$$

where $\Lambda \in Gl(k^n)$. The equivalence relation " $\sim$ " is defined by the canonical mapping $M_{E \times [\![1,n]\!]}^*(k) \longrightarrow Grass_n(k^E)$, associating with $A = (a_{\alpha j})$ the k-subspace k^E, $S = Vect((\sum a_{\alpha j} e_\alpha))$. There is a canonical section of this mapping on U_H given by $S \mapsto M((\pi_H)_S^{-1}(e_H))$ (cf. definition 1.15). The fiber of this mapping over the subspace S identifies with the set of its ordered basis (cf. definition 1.15).

Given a subspace S and $H, H' \in Grass_n(E)$ we shall now compare its U_H-coordinates with its $U_{H'}$-coordinates. Let $H' \in Grass_n(E)$ such that $S \in U_H \cap U_{H'}$. It results from the above description of the coordinates matrix $(\xi_{ij}^H(S))$(resp. $(\xi_{ij}^{H'}(S))$) the following relation:

$$(\xi_{ij}^{H'}(S)) = \psi^{(H',H)}((\xi_{ij}^{H}(S))),$$

where $\psi^{(H',H)}$ denotes a rational function of the U_H-coordinates of S.

More precisely stated, the affine space $\mathbb{A}_H = \mathbb{A}(M_{H^\perp \times H}(k))$ defined by the k-vector space of $H^\perp \times H$-matrices with coefficients in k is the k-variety with coordinates ringing the polynomial algebra $k[X_{ij}^H]$ in the set of indeterminates $(X_{ij}^H)_{(i,j) \in H^\perp \times H}$, and $\psi^{(H',H)}$ is a rational function in the variables (X_{ij}^H). Let $\mathbb{A}_{HH'} \subset \mathbb{A}_H$ denote the domain of definition of $\psi^{(H',H)}$, which is an open Zariski subset of $\mathbb{A}_H$.

On the other hand, there is a bijective mapping $U_H \longrightarrow \mathbb{A}_H$ defined by the U_H- coordinate matrix $S \mapsto M_H(S)$. The intersection $U_H \cap U_{H'}$, corresponds by this mapping, to the set of k-points $\mathbb{A}_{HH'}(k)$ of $\mathbb{A}_{HH'}$, and, as it is easy to see, the rational function $\psi^{(H',H)}$ induces an isomorphism of k-varieties:

$$\psi^{(H',H)}|_{\mathbb{A}_{HH'}} : \mathbb{A}_{HH'} \longrightarrow \mathbb{A}_{H'H} ,$$

which for the sake of briefnnes is denoted by $\psi^{(H',H)}$. The set $(\psi^{(H',H)})_{(H',H) \in Grass_n(E) \times Grass_n(E)}$ satisfies the **cocycle condition**:

$$\psi^{(H'',H)} = \psi^{(H'',H')} \circ \psi^{(H',H)} .$$

Thus, the couple $((\mathbb{A}_H)_{H \in Grass_n(E)}, (\psi^{(H',H)}))$ defines a k-variety. It is immediate that, by construction, its underlying set of k-points corresponds to $Grass_n(k^E)$.

Definition 1.3 *If no confusion arises denote also by* $Grass_n(k^E)$ *this variety* (**the Grassmann variety of the n-th dimensional subspaces of** k^E), *and write:*

$$Grass(k^E) = \coprod_{n \in \mathbb{N}, n \leq |E|} Grass_n(k^E)$$

(**The Grassmann variety of** k^E). *By construction, the indexed set* $(U_H)_{H \in Grass_{\underline{n}}(E)}$ *defines an affine open covering*

$$Grass_{\underline{n}}(k^E) = \bigcup_{H \in Grass_{\underline{n}}(E)} U_H$$

(**the canonical open covering of** $Grass_{\underline{n}}(k^E)$)

Remark 1.4 *There is a natural action of the k-group* $Gl(k^E)$ *of k-automorphisms of* k^E *on* $Grass_n(k^E)$:

$$Gl(k^E) \times Grass_n(k^E) \longrightarrow Grass_n(k^E),$$

$(g, S) \mapsto g \cdot S$. *This action may be easily proved, in terms of U_H-coordinates, to be algebraic and* **homogeneous**. *We shall later prove that* $Grass_n(k^E)$ *is in fact isomorphic to an homogeneous space defined by a subgroup of* $Gl(k^E)$.

1.1.1 Grassmannian functor

The following section gives a brief functorial approach to the Grassmannian variety. We refer the reader to [24] for more details. Let $\mathscr{G}rass_n(k^E)$ be the k-functor, from k-algebras to Sets, associating to a k-algebra A the set of rank n projective A-submodules M of A^E.

It is well-known that $\mathscr{G}rass_n(k^E)$ is a sheaf for the Zariski topology. There is a functorial isomorphism

$$\iota : Grass_n(k^E) \longrightarrow \mathscr{G}rass_n(k^E)$$

defined by $\iota_A : s \mapsto \Gamma(Spec(A), s^*(\xi_n))$, where s denotes a section $s : Spec(A) \longrightarrow Grass_n(k^E)$, and ξ_n the tautological $\mathscr{O}_{Grass_n(k^E)}$-module (cf. 1.13).

On the other hand, if A is a local ring the set $\mathscr{G}rass_n(k^E)(A)$ identifies with the set of equivalence classes of rank n $E \times [\![1, n]\!]$-matrices with coefficients in A: $M^*_{E \times [\![1,n]\!]}(A)/ \sim$, defined by the equivalence relation: $M \sim M' \Longleftrightarrow M = M' \times \Lambda$, where $\Lambda \in Gl(A^n)$. By construction of $Grass_n(k^E)$ and by using the section of $M^*_{E \times [\![1,n]\!]}(k) \longrightarrow Grass_n(k^E)$ on U_H obtained by normalization of a matrix defining a basis of S, it follows that ι_A is a bijection, and finally, by a standard argument that ι is a functorial isomorphism. Equivalently $\mathscr{G}rass_n(k^E)$ may be described as associating with a k-algebra A the set of locally trivial $\mathscr{O}_{Spec(A)}$-submodules $\mathscr{M}$ of $\mathscr{O}^E_{Spec(A)}$ which are locally direct factors.

1.2 The Plücker embedding of the Grassmannian

A Grassmann variety may be canonically embedded in a projective space as a closed subvariety thus proving that it is a projective variety. Assume that the canonical basis e_E of k^E is an ordered basis. Denote by $e_E = (e_i)_{i \in E}$ the canonical basis of k^E, and fix a total ordering of E which gives an ordered basis of k^E. Let $H \in Grass_n(E)$. We suppose H endowed with the induced ordering, and we write $H = \{i_1, ..., i_n\}$ with $i_1 < ... < i_n$, and $i_1(H) = i_1, ..., i_n(H) = i_n$. The n-th exterior product $\bigwedge^n k^E$ of k^E is endowed with the canonical basis $\bigwedge e_E = (\wedge e_H)_{H \in Grass_n(E)} = (e_{i_1(H)} \wedge ... \wedge e_{i_n(H)})_{H \in Grass(E)}$.

There is a mapping

$$i_P = i_{P,n} : Grass_n(k^E) \longrightarrow \mathbb{P}(\bigwedge^n k^E) = Grass_1(\bigwedge^n k^E,$$

where $\mathbb{P}(\bigwedge^n k^E)$ denotes the projective space associated with $\bigwedge^n k^E$, defined as follows. Let $\tilde{e} = (\tilde{e}_j)_{1 \leq j \leq n}$ be a basis of S. Write: $\tilde{e}_j = \sum_{i=1}^{|E|} a_{ij} e_i$, and $A(\tilde{e}) = (a_{ij})$. The result is:

$$\tilde{e}_1 \wedge ... \wedge \tilde{e}_n = \sum_{H \in Grass(E)} \Delta_H(\tilde{e}) e_{i_1(H)} \wedge ... \wedge e_{i_n(H)} ,$$

where $\Delta_H(\tilde{e}) = \det A(\tilde{e})_H$, and $A(\tilde{e})_H$ denotes the $|H| \times |H|$-submatrix defined by the H-rows of $A(\tilde{e})$. The mapping $i_{\mathcal{P}}$ associates with $S \in Grass_n(k^E)$ the point of $\mathbb{P}(\overset{n}{\bigwedge}k^E)$ with homogeneous coordinates (resp. **Plücker coordinates**)$(\Delta_H(\tilde{e}))_{H \in Grass_n(E)}$. Let us see that this mapping is well defined, i.e. that the point in $\mathbb{P}(\overset{n}{\bigwedge}k^E)$ corresponding to S does not depend on the choice of the basis $\tilde{e}$ defining its homogeneous coordinates. If $\tilde{e}'$ is another basis of S, we might then write $A(\tilde{e}') = A(\tilde{e}) \times B$, where $B \in Gl(k^{|H|})$. Thus for $H \in Grass_n(E)$ the result is $(\Delta_H(\tilde{e}')) = (det(B)\Delta_H(\tilde{e}))$ proving that the image is independant of the choice of a basis of S.

In fact, the mapping $i_{\mathcal{P}}$ is induced by a morphism of k-varieties $Grass_n(k^E) \longrightarrow \mathbb{P}(\overset{n}{\bigwedge}k^E)$. So to see this we write $(x_H)_{H \in Grass_n(E)}$ for the homogeneous coordinates in the projective space $\mathbb{P}(\overset{n}{\bigwedge}k^E)$ defined by the basis $\overset{n}{\bigwedge}e_E$ of $\overset{n}{\bigwedge}k^E$ and given $H_0 \in Grass_n(E)$, let $(\frac{x_H}{x_{H_0}})_{H \in Grass_n(E)}$ be the affine coordinates of the open affine set:

$$\mathbb{P}_{(H_0)} = \{(x_H) \in \mathbb{P}(\overset{n}{\bigwedge}k^E) \mid x_{H_0} \neq 0\} .$$

Lemma 1.5 *Keep the above notation and terminology.*

1) The restriction mapping $i_{\mathcal{P},H_0} : U_{H_0} \longrightarrow \mathbb{P}_{(H_0)}$ is given by $i_{\mathcal{P},H_0} : S \longmapsto$ $(\frac{\Delta_H(\tilde{e})}{\Delta_{H_0}(\tilde{e})})_{H \in Grass_n(E)}$, where $\tilde{e}$ is a basis of S, and is induced by a morphism of k-varieties, which we denote also by $i_{\mathcal{P},H_0}$.

2) The morphism $i_{\mathcal{P},H_0}$ defines a closed embedding.

Proof *Given $S \subset k^E$ corresponding to a point $x \in U_{H_0}$, let $\tilde{e}$ be a basis of S, $\tilde{e}_{H_0}(S) = (\pi_{H_0})_S^{-1}(e_{H_0}) = (\tilde{e}_{\beta_j}(S))_{1 \leq j \leq n}$ the normalized basis of S defined as in 1.1 , and $M_{H_0}(S) = (\xi_{ij}^{H_0}(S))$ $(1 \leq i \leq |H_0^{\perp}|, 1 \leq j \leq |H_0|)$ the coordinate matrix. Thus:*

$$\tilde{e}_{\beta_j}(S) = e_{\beta j} + \sum_{1 \leq j \leq |H_0^{\perp}|} \xi_{ij}^{H_0}(S)e_{\alpha i} .$$

From $A(\tilde{e}) \times A(\tilde{e})_{H_0}^{-1} = A(\tilde{e}_{H_0}(S))$ one deduces:

$$(\forall H \in Grass_n(E)) \ \frac{\Delta_H(A(\tilde{e}))}{\Delta_{H_0}(A(\tilde{e}))} = \Delta_H(A(\tilde{e}_{H_0}(S))) .$$

The right member being a polynomial in the $(\xi_{ij}^{H_0}(S))$ one concludes that the $\mathbb{P}_{(H_0)}$-affine coordinates of S are polynomials in the $M_{H_0} = (\xi_{ij}^{H_0}(S))$. This proves our first assertion.

The second statement of the lemma follows from the following relation between the coordinates matrix $M_{H_0}(S)$ and the affine coordinates $i_{\mathcal{P},H_0}$:

$S \mapsto \Delta_H(A(\tilde{e}_{H_0}(S)))$ *of* S. *Observe that* $\Delta_{H_0}(A(\tilde{e}_{H_0}(S)))$. *Let* $H_0(i,j) = \{i\} \cup (H - \{j\})$ *for* $(i,j) \in H_0^\perp \times H_0$, *then:*

$$\frac{\Delta_{H_0(i,j)}(\tilde{e}_{H_0}(S))}{\Delta_{H_0}(\tilde{e}_{H_0}(S))} = \xi_{ij}^{H_0}(S).$$

From the proof of 1) *it can be deduced that there exists a family of polynomials* $(P_H((X_{ij})))_{H \in Grass_n(E)}$ *satisfying* $\frac{\Delta_H(\tilde{e}_{H_0}(S))}{\Delta_{H_0}(\tilde{e}_{H_0}(S))} = P_H((\frac{\Delta_{H_0(i,j)}(\tilde{e}_{H_0}(S))}{\Delta_{H_0}(\tilde{e}_{H_0}(S))}))$. *Thus the image of* $i_{\mathcal{P},H_0}$ *is the closed subvariety of* $\mathbb{P}_{(H_0)}$ *defined by the ideal given by* $(\frac{x_H}{x_{H_0}} - P_H(\frac{x_{H_0(ij)}}{x_{H_0}}))_{H \in Grass_n(E)}$. *This suffices to show that* $i_{\mathcal{P},H_0}$ *defines a closed embedding.*

Thus we have the following

Proposition 1.6 *The family of embeddings* $(i_{\mathcal{P},H})_{H \in Grass_n(E)}$ *defines an embedding of k-varieties* $i_{\mathcal{P}} : Grass_n(k^E) \longrightarrow \mathbb{P}(\bigwedge^n k^E)$ *called the Plücker embedding. More precisely, given* $H_0, H_1 \in Grass_n(E)$ *there is an isomorphism* $U_{H_0} \cap U_{H_1} \simeq U_{H_0} \times_{\mathbb{P}(\bigwedge^n k^E)} U_{H_1}$ *induced by* $i_{\mathcal{P},H_0} \times i_{\mathcal{P},H_1}$. *In fact,* $Grass_n(k^E)$ *is a proper k-variety, and thus* $i_{\mathcal{P}}$ *is a closed embedding (cf. [24], §9, [25]).*

Proof *The coordinates transformation cocycle, defined on* $\mathbb{P}_{(H_0)} \cap \mathbb{P}_{(H_1)}$, *from the affine coordinates* $(\frac{x_H}{x_{H_0}})$ *on* $\mathbb{P}_{(H_0)}$ *to affine coordinates* $(\frac{x_H}{x_{H_1}})$ *on* $\mathbb{P}_{(H_1)}$ *is given by* $\frac{x_H}{x_{H_1}} = \frac{x_{H_0}}{x_{H_1}} \frac{x_H}{x_{H_0}}$. *Accordingly the* H_1-*affine coordinates* $(\frac{\Delta_H(\tilde{e})}{\Delta_{H_1}(\tilde{e})})$ *of* S, *are related to the* H_0-*affine coordinates* $(\frac{\Delta_H(\tilde{e})}{\Delta_{H_0}(\tilde{e})})$ *of* S *by:* $\frac{\Delta_H(\tilde{e})}{\Delta_{H_1}(\tilde{e})} = \frac{\Delta_{H_0}(\tilde{e})}{\Delta_{H_1}(\tilde{e})} \frac{\Delta_H(\tilde{e})}{\Delta_{H_0}(\tilde{e})}$. *Thus the family of embeddings defines a morphism. We now proceed to show that this morphism is itself an embedding.*

Given a section (S, S') *of* $U_{H_0} \times_{\mathbb{P}(\bigwedge^n k^E)} U_{H_1}$, *the* $\mathbb{P}_{(U_0)}$-*affine coordinates (resp.* $\mathbb{P}_{(U_1)}$-*affine coordinates) of* $i_{\mathcal{P},H_0}(S)$ *(resp.* $i_{\mathcal{P},H_1}(S')$*) are given by the* $\bigwedge^n e_E$-*components of the exterior product* $\wedge \tilde{e}_{H_0}(S) = \tilde{e}_{H_0,1}(S) \wedge \cdots \wedge \tilde{e}_{H_0,n}(S)$ *(resp.* $\wedge \tilde{e}_{H_1}(S) = \tilde{e}_{H_1,1}(S) \wedge \cdots \wedge \tilde{e}_{H_1,n}(S)$*). As by hypothesis* $i_{\mathcal{P},H_0}(S) = i_{\mathcal{P},H_1}(S')$, *there exists a section* λ *of* $\mathcal{O}^*_{\mathbb{P}(\bigwedge^n k^E)}$ *satisfying* $\wedge \tilde{e}_{H_0}(S) = \lambda(\tilde{e}_{H_0,1}(S))$. *On the other hand, the set of sections* v *(resp.* v'*) of* S *(resp.* S'*) is characterized by* $\wedge \tilde{e}_{H_0}(S) \wedge v = 0$ *(resp.* $\wedge \tilde{e}_{H_1}(S) \wedge v' = 0$*), so we conclude that* $S = S'$.

Let us see that $Grass_n(k^E)$ *is a proper variety and then its image by* $i_{\mathcal{P}}$ *in* $\mathbb{P}(\bigwedge^n (k^E))$ *is also proper and thus, a closed subset of* $\mathbb{P}(\bigwedge^n (k^E))$. *Let* K *be a field extension of* k *endowed with a valuation* v, *and with integers subring* $A_K \subset K$. *Given a* K-*vector subspace* $S \subset K^E$ *it may be shown that there exists an* A_K-*submodule* $S_{A_K} \subset A_K^E$ *satisfying:* $S = K \otimes_{A_K} S_{A_K}$ *and giving a* A_K-*section of* $Grass_n(k^E)$. *Let* $(\Delta_H(S))$ *be the homogeneous coordinates of* S *calculated with respect to some basis of* S. *Choose* $H_0 \in Grass(E)$ *such* *that the valuation $v(\Delta_{H_0}(S))$ is minimal amongst the set $(v(\Delta_H(S)))$. Define $S_{A_K} = Vect((\tilde{e}_j)_{j \in H_0})$, where:*

$$\tilde{e}_j = e_j + \sum_{i \in H^\perp} \frac{\Delta_{H_0(i,j)}(S)}{\Delta_{H_0}(S)} e_i \ .$$

It is clear that the generic fiber $K \otimes_{A_K} S_{A_K}$ of S_{A_K} is equal to S, and that S_{A_K} is a direct factor of A_K^E. By the valuation criterium it may be concluded that $Grass_n(k^E)$ is a k-proper variety, and that $i_{\mathcal{P}}$ is a closed embedding. This proves the last assertion.

Remark 1.7 *The exterior product $\overset{n}{\wedge} k^E$ is a $Gl(k^E)$-representation, so there is a natural action of $Gl(k^E)$ on $\mathbb{P}(\overset{n}{\wedge} k^E)$. The Plücker morphism $i_{\mathcal{P}}$ is $Gl(k^E)$-equivariant.*

1.3 Flag varieties

Flag varieties are natural generalizations of Grassmannians. Let $\underline{n} = (n_1 < \ldots < n_l < n_{l+1})$ be an increasing sequence of positive integers with $n_{l+1} = |E|$. Define the set of flags of k^E of type $\underline{n}$ by:

$$Drap_{\underline{n}}(k^E) = \{(V_1 \subset \ldots \subset V_l \subset V_{l+1}) \mid V_i = n_i\text{-dimensional subspace of } k^E \ \}.$$

Thus $Drap(k^E)$ is the subset of the set theoretic product:

$$\prod_{i=1}^{l+1} Grass_{n_i}(k^E) = \prod_{i=1}^{l} Grass_{n_i}(k^E),$$

formed by the $(V_i)_{1 \leq i \leq l+1}$ satisfying the inclusion conditions $V_i \subset V_{i+1}$ for $1 \leq i \leq l$. More precisely stated, there is a subfunctor $\mathcal{D}rap_{\underline{n}}(k^E)$ of $\prod_{i=1}^{l} Grass_{n_i}(k^E)$, from k-algebras to sets, associating with A the set of flags of projective submodules of A^E of type $\underline{n}$ (cf. [24]). The set of sections of $\mathcal{D}rap_{\underline{n}}(k^E)$ over k is given by $Drap(k^E)$.

Let us show that $\mathcal{D}rap_{\underline{n}}(k^E)$ is the underlying set of a k-subvariety of the product of k-varieties $\prod_{i=1}^{l} Grass_{n_i}(k^E)$ (i.e. that it is representable by a k-scheme of finite type). Observe that there is a closed embedding:

$$\prod_{i=1}^{l+1} i_{\mathcal{P}}^{(n_i)} : \prod_{i=1}^{l+1} Grass_{n_i}(k^E) \longrightarrow \prod_{i=1}^{l+1} \mathbb{P}(\overset{n_i}{\wedge} k^E)$$

given by the product of the corresponding Plücker embeddings.

It suffices to see that there is a set of equations in the Plücker coordinates of the V_i's, characterizing the image of the composed morphism of functors:

$$\mathcal{D}rap_{\underline{n}}(k^E) \longrightarrow \prod_{i=1}^{l+1} \mathbb{P}(\bigwedge^{n_i} k^E),$$

for $l = 2$. Let V_1 (resp. V_2) be a m-dimensional (resp. n-dimensional) subspace of k^E. We suppose $m < n$. Let $e_1 = (e_{1i})_{1 \le i \le m}$ (resp. $e_2 = (e_{2j})_{1 \le j \le n}$) be a basis of V_1 (resp. V_2).

The Plücker coordinates of V_1 (resp. V_2) are thus given by $(\Delta_H(e_1))_{H \in Grass_m(E)}$ (resp. $(\Delta_J(e_2))_{J \in Grass_n(E)}$), with:

$$e_{11} \wedge \ldots \wedge e_{1m} = \sum_{H \in Grass_m(E)} \Delta_H(e_1) e_{i_1(H)} \wedge \ldots \wedge e_{i_m(H)}$$

(resp. $e_{21} \wedge \ldots \wedge e_{2n} = \sum_{J \in Grass_n(E)} \Delta_J(e_2) e_{i_1(J)} \wedge \ldots \wedge e_{i_n(J)}$).

Let $v = \sum_{1 \le i \le |E|} x_i e_i \in k^E$, then

$$e_{11} \wedge \ldots \wedge e_{1m} \wedge v = \sum_{K \in Grass_{m+1}(E)} \phi_K(e_1, v) e_{i_1(K)} \wedge \ldots \wedge e_{i_{m+1}(K)}$$

(resp.

$$e_{21} \wedge \ldots \wedge e_{2n} \wedge v = \sum_{L \in Grass_{n+1}(E)} \phi_L(e_2, v) e_{i_1(L)} \wedge \ldots \wedge e_{i_{n+1}(L)}) \ .$$

Here the $(\phi_K(e_1, v))_{K \in Grass_{m+1}(E)}$ (resp. $(\phi_L(e_2, v))_{L \in Grass_{n+1}(E)}$) are linear forms in the $(x_i)_{1 \le i \le |E|}$ with coefficients in the set of coordinates $(\Delta_H(e_1))_{H \in Grass_m(E)}$ (resp. $(\Delta_J(e_2))_{J \in Grass_n(E)}$). The following proposition is easily verified.

Proposition 1.8 *The kernel of the linear mapping:*

$$k^E \longrightarrow \bigwedge^{m+1} k^E \times \bigwedge^{n+1} k^E$$

defined by:

$$v \longmapsto ((e_{11} \wedge \ldots \wedge e_{1m}) \wedge v, (e_{21} \wedge \ldots \wedge e_{2n}) \wedge v)$$

is equal to $V_1 \cap V_2$. Hence, if we denote by r the rank of the linear system:

$$(\phi_H(e_1, v) = 0)_{H \in Grass_{m+1}(E)}; (\phi_J(e_2, v) = 0)_{J \in Grass_{n+1}(E)}$$

we have $r = |E| - dim\ V_1 \cap V_2$. Thus the inclusion condition $V_1 \subset V_2$ translates as "$r = |E| - m$", i.e. $r =$ the minimal possible rank of this linear system. It follows that the inclusion condition $V_1 \subset V_2$ is obtained as the simultaneous vanishing of the set:

$$(M_\alpha((\Delta_H(e_1)); (\Delta_J(e_2))))$$

of the $(|E| - m + 1)$-minors of this linear system.

It results from this the:

Proposition 1.9 *The subset $Drap_{\underline{n}}(k^E) \subset \prod_{i=1}^{l+1} Grass_{n_i}(k^E)$ is the underlying set of a k-subvariety which we denote also by $Drap_{\underline{n}}(k^E)$ if no confusion arises. The product action of $Gl(k^E)$ on $\prod_{i=1}^{l+1} Grass_{n_i}(k^E)$ stabilizes $Drap_{\underline{n}}(k^E)$ and induces a transitive left action (resp. an homogeneous action) of $Gl(k^E)$ on $Drap_{\underline{n}}(k^E)$.*

Remark 1.10 *The use of the term k-subvariety is rather abusive. The above equations characterize the subfunctor on k-algebras*

$$\mathcal{D}rap_{\underline{n}}(k^E) \subset \prod_{i=1}^{l+1} Grass_{n_i}(k^E)$$

associating with a k-algebra A the set of flags $\mathcal{D}$ of local direct summands of A^E of type $\underline{n} = (0 < n_1 < ... < n_l < n_{l+1} = |E|)$, i.e. the β-th module of $\mathcal{D}$ being of rank n_β. Thus it follows that $\mathcal{D}rap_{\underline{n}}(k^E)$ is representable by a finite type k-scheme. We shall later prove that $Drap_{\underline{n}}(k^E)$ is, in fact, a k-variety in Serre's sense from its decomposition in a sequence of locally trivial fibrations with grassmannians as basis and typical fiber. On the other hand, there is a functorial isomorphism:

$$\iota : Drap_{\underline{n}}(k^E) \longrightarrow \mathcal{D}rap_{\underline{n}}(k^E),$$

defined by $\iota_A : s \mapsto \Gamma(Spec(A), s^*(\xi_{\underline{n}}))$, where $s : Spec(A) \longrightarrow Drap_{\underline{n}}(k^E)$ denotes a section and $\xi_{\underline{n}}$ the canonical flag on $Drap_{\underline{n}}(k^E)$. (cf. 1.13)

Write:

$$Drap(k^E) = \coprod_{\underline{n} \in typ(E)} Drap_{\underline{n}}(k^E) \text{ (the flag variety of } k^E),$$

where:

$$typ(E) = \{\underline{n} \in \mathbb{N}^{*l+1} \mid \underline{n} = (0 < n_1 < ... < n_l < n_{l+1} = |E|)\},$$

i.e. the set of length $(l + 1)$ strictly increasing sequences of positive integers with $n_{l+1} = |E|$. Denote by $typ(E)$ the **set of types of combinatorial flags** of E. Consider that $Grass_n(k^E) = Drap_{(n<|E|)}(k^E)$.

Write $typ(\mathcal{D}) = \underline{n} = (0 < n_1 < ... < n_l < n_{l+1} = |E|)$, if $\mathcal{D} = (V_1 \subset ... \subset V_l \subset k^E)$, with $dim\, V_i = n_i$ $(1 \leq i \leq l + 1)$. One says that l is the **length** of $\mathcal{D}$, and that $\underline{n}$ is **the type** of $\mathcal{D}$. Write $l(\mathcal{D}) = l$. Define $typ : Drap(k^E) \longrightarrow typ(E)$ by $typ : \mathcal{D} \mapsto typ(\mathcal{D})$.

It is useful considering the combinatorial analogues of the above definitions for $Drap(E)$. Given $\underline{n} \in typ(E)$ let

$$Drap_{\underline{n}}(E) = \{(E_1 \subset ... \subset E_l \subset E) \mid E_i \in \mathcal{P}(E) \text{ and } |E_i| = n_i \},$$

and

$$Drap(E) = \coprod_{\underline{n} \in typ(E)} Drap_{\underline{n}}(E) \textbf{ (The combinatorial flags of } E\textbf{)}.$$

Given $D = (E_1 \subset \cdots \subset E_l \subset E_{l+1}) \in Drap(E)$ write $typ(D) = (|E_1| < \cdots < |E_l| < |E_{l+1}|)$, $l(D) = l$ (length of D), and $l(\underline{n}) = l$. Denote the empty flag by (E), and write $typ((E)) = (|E|)$ (resp. $l((E)) = 0$). Observe that if $E = I_{r+1} = \{1, \cdots, r+1\}$ the set $Drap(E)$ may be identified with the **first barycentrical subdivision** $\Delta^{(r)'}$ of the **combinatorial simplex** $\Delta^{(r)}$ defined by I_{r+1}.

Definition 1.11 *Let* $D = (J_1 \cdots \subset J_l \subset E) \in Drap_{\underline{n}}(E)$. *One associates with D the open affine subset of $Drap_{\underline{n}}(k^E)$:*

$$U_D = Drap_{\underline{n}}(k^E) \cap \prod_{1 \leq i \leq l(\underline{n})} U_{J_i},$$

where $\prod_{1 \leq i \leq l(\underline{n})} U_{J_i} \subset \prod_{i=1}^{l+1} Grass_{n_i}(k^E)$, *denotes the product of the open affine subvarieties* $U_{J_i} \subset Grass_{n_i}(k^E)$. *Remark that U_D is the pullback of the product of open affine sets* $\prod \mathbb{P}_{(J_i)} \subset \prod \mathbb{P}(\overset{n_i}{\wedge} k^E)$ *by the product embedding* $\prod_{i=1}^{l+1} i_{\mathcal{P}, n_i} : \prod_{i=1}^{l+1} Grass_{n_i}(k^E) \longrightarrow \prod_{i=1}^{l+1} \mathbb{P}(\overset{n_i}{\wedge} k^E)$, *with* $l = l(\underline{n})$, *and that* $(U_D)_{D \in Drap_{\underline{n}}(E)}$ *defines an open affine covering of* $Drap_{\underline{n}}(k^E)$:

$$Drap_{\underline{n}}(k^E) = \underset{D \in Drap_{\underline{n}}(E)}{\cup} U_D \textbf{ (The canonical open covering of }$$
$$Drap_{\underline{n}}(k^E)).$$

There is a canonical embedding of $Drap_{\underline{n}}(k^E)$ in a projective space

$$i_{\mathcal{S}} = i_{\mathcal{S}, \underline{n}} : Drap_{\underline{n}}(k^E) \longrightarrow \mathbb{P}(\underset{1 \leq i \leq l}{\otimes} \overset{n_i}{\wedge} k^E),$$

obtained by composing the embedding $\mathcal{D}rap_{\underline{n}}(k^E) \longrightarrow \prod_{i=1}^{l+1} \mathbb{P}(\overset{n_i}{\wedge} k^E)$ with **Segre's embedding** (cf. [24], §9,[25]) defined as follows. Assume that the basis e_E is ordered, and that $\underset{1 \leq i \leq l}{\otimes} \overset{n_i}{\wedge} k^E$ is endowed with the basis

$$\Big(\underset{1 \leq \beta \leq l}{\otimes} \wedge e_{H_\beta} \Big)_{H \in \prod_{i=1}^{l+1} Grass_{n_i}(E)},$$

where $\wedge e_{H_\beta}$ denotes the ordered product of the canonical basis of k^{H_β}, ordered by the lexicographical ordering induced by that of e_E. Let

$(x_{\underline{H}})_{\underline{H}\in \prod_{i=1}^{l+1} Grass_{n_i}(E)}$ be the homogeneous coordinates of a point in the pro-

jective space $\mathbb{P}(\underset{1\le i\le l}{\otimes} \overset{n_i}{\wedge} k^E)$ relatively to this basis. Given a flag $\mathscr{D} = (S_1 \cdots \subset S_l \subset k^E)$, and for all β a basis $e_{S_\beta} = (e_{S_\beta 1}, \cdots, e_{S_\beta n_\beta})$ of S_β, write $\wedge e_{S_\beta} = e_{S_\beta 1} \wedge \cdots \wedge e_{S_\beta n_\beta}$. Define

$$i_S : \mathscr{D} \mapsto (\Delta_{H_1}(S_1) \times \cdots \times \Delta_{H_l}(S_l))_{\underline{H}},$$

where $\underline{H} = (H_1, \cdots, H_l)$ runs on $\prod_{i=1}^{l+1} Grass_{n_i}(E)$, and

$$\wedge e_{S_\beta} = \sum_{H\in Grass_{n_\beta}(E)} \Delta_H(S_\beta) e_{i_1(H)} \wedge \ldots \wedge e_{i_{n_\beta}(H)} .$$

The coefficients $(\Delta_H(S_\beta))$ are the homogeneous coordinates of the image of S_β by the Plücker embedding. Thus the homogeneous coordinates of $i_S(\mathscr{D})$ are given by the coordinates of the tensor product $\underset{1\le\beta\le l}{\otimes} \wedge e_{S_\beta}$ relatively to the basis $(\underset{1\le\beta\le l}{\otimes} \wedge e_{H_\beta})_{\underline{H}}$.

Observe that, $\underset{1\le i\le l}{\otimes} \overset{n_i}{\wedge} k^E$ being a representation of $Gl(k^E)$, there is a natural action of $Gl(k^E)$ on $\mathbb{P}(\underset{1\le i\le l}{\otimes} \overset{n_i}{\wedge} k^E)$ and that the embedding i_S is $Gl(k^E)$-equivariant.

Definition 1.12 *We call $i_{S,\underline{n}}$ the **Segre's embedding** of $Drap_{\underline{n}}(k^E)$. It is obtained by composing the product of Plücker embeddings $\prod_{i=1}^{l+1} i_{\mathcal{P},n_i}$ with the Segre embedding of the product of projective spaces $\prod_{i=1}^{l+1} \mathbb{P}(\overset{n_i}{\wedge} k^E)$ in the projective space $\mathbb{P}(\underset{1\le i\le l}{\otimes} \overset{n_i}{\wedge} k^E)$.*

1.3.1 Flag varieties fiber decomposition

We give a k-variety structure, in Serre's sense, to $Drap_{\underline{n}}(k^E)$ based on its natural fibering.

Definition 1.13 *There is a locally free rank n module ξ_n over $X = Grass_n(k^E)$ defined as follows. Let $(U_H)_{H\in Grass_n(E)}$ be the canonical open covering of $Grass_n(k^E)$. Define ξ_n in terms of local data. Let $x \in U_H$ correspond to the subspace $S \subset k^E$, and $\tilde{e}_{H,x} = (\tilde{e}_{i,x})_{i\in H} = (\pi_H)_S^{-1}(e_H)$ be the lifted basis of S by $(\pi_H)_S$ obtained from the canonical basis of k^H. We assume E is totally ordered. A rank n morphism of $\mathscr{O}_{U_H}$- modules:*

$$\phi_H : \mathscr{O}_{U_H}^n \longrightarrow \mathscr{O}_{U_H}^E$$

may be defined by $\phi_{H,x} : (\lambda_i) \mapsto \sum\limits_{1 \leq i \leq n} \lambda_i \tilde{e}_{i,x}$. *The image* $\xi_{n,U_H} = Im\ \phi_H \subset$ $\mathscr{O}_{U_H}^E$ *is thus a rank n direct factor submodule satisfying:*

$$(\xi_{n,U_H})_x = S \ ,$$

and ϕ_H defines an isomorphism $\phi_H : \mathscr{O}_{U_H}^n \simeq \xi_{n,U_H}$. *Thus the family* $(\phi_{H'}^{-1} \circ \phi_H)$ *given by* (ϕ_H) *fulfills the cocycle condition and defines a rank n submodule* ξ_n *of* $\mathscr{O}_X^E$ *which we call* **the tautological module of** $Grass_n(k^E)$.

Let $V(k^n)$ denote the k-variety with coordinate ring $k[X_i]_{1 \leq i \leq n}$ (i.e. the k-variety corresponding to $Spec\ k[X_i]_{1 \leq i \leq n}$), and $V(\phi_{H'}^{-1} \circ \phi_H) : (U_H \cap U_{H'}) \times V(k^n) \longrightarrow (U_H \cap U_{H'}) \times V(k^n)$ the morphism of k-varieties associated to $\phi_{H'}^{-1} \circ \phi_H$. From the cocycle condition satisfied by $(\phi_{H'}^{-1} \circ \phi_H)$ we deduce that $(V(\phi_{H'}^{-1} \circ \phi_H))$ also satisfies the cocycle condition, and thus defines a vector fiber bundle on $Grass_n(k^E)$ with typical fiber $V(k^n)$ which we denote by $V(\xi_n)$ (The **associated vector bundle** $V(\xi_n)$ to ξ_n).

The fiber bundle $Grass_m(\xi_n)$ for $m \leq n$ (resp. $Drap_{\underline{m}}(k^n)$ for $\underline{m} \in typ(I_n)$) may be defined in a similar way following the same pattern of the definition of $V(\xi_n)$, by considering the left action on $(U_H \cap U_{H'}) \times Grass_m(k^n)$ (resp. $(U_H \cap U_{H'}) \times Drap_{\underline{m}}(k^n)$) induced by the cocycle $(V(\phi_{H'}^{-1} \circ \phi_H))$.

Remark 1.14 *The above constructions hold for any rank n locally free module η over a k-variety X giving rise to fiber bundles $V(\eta) \longrightarrow X$ (resp. $Grass_m(\eta) \longrightarrow X$, $Drap_{\underline{m}}(\eta) \longrightarrow X$). The fiber $V(\eta)_x$ (resp.$Grass_m(\eta)_x$, $Drap_{\underline{m}}(\eta)_x$) identifies with $V(\eta_x)$. (resp.$Grass_m(\eta_x)$, $Drap_{\underline{m}}(\eta_x)$). Moreover:*
"if X is a k-smooth variety (resp. a k-integral variety), it follows that $V(\eta)$ (resp. $Grass_m(\eta)$, $Drap_{\underline{m}}(\eta)$) is a k-smooth variety (resp.k-integral variety)".

It is easy from this remark to obtain another proof of proposition 1.9. Observe that the equations given in the proof of proposition 1.9 define the embedding of $Drap_{\underline{m}}(k^E)$ in a product of grassmannians in terms of the corresponding Plücker coordinates.

Denote by $p_\beta : Drap_{\underline{n}}(k^E) \longrightarrow Grass_{n_\beta}(k^E)$ the canonical projection, and by $p_\beta^*(\xi_{n_\beta})$ the pull-back of the tautological module ξ_{n_β}. Define the **tautological flag** $\xi_{\underline{n}}$ of submodules of $\mathscr{O}_{Drap_{\underline{n}}(k^E)}^E$ by:

$$\xi_{\underline{n}} = (p_1^*(\xi_{n_1}) \cdots \subset p_l^*(\xi_{n_l}) \subset \mathscr{O}_{Drap_{\underline{n}}(k^E)}^E).$$

The inclusions between these submodules are evident by definition. Write $\xi_{n_\beta} = p_\beta^*(\xi_{n_\beta})$ if no confusion arises.

Let $\underline{n} = (n_1 < \cdots < n_l < n_{l+1}) \in typ(I_{r+1})$. Write $\underline{n}_\beta = (n_1 < \cdots < n_{\beta-1} < n_\beta) \in typ(I_{n_\beta})$ (resp. $\underline{n}^\beta = (n_\beta < \cdots < n_l < n_{l+1}) \in typ(I_{n_{l+1}})$). For $2 \leq \beta \leq l$ a locally trivial fiber bundle is associated:

$$Drap_{\underline{n}}(k^E) = Drap_{\underline{n}_\beta}(\xi_{n_\beta}) \longrightarrow Drap_{\underline{n}^\beta}(k^E),$$

over $Drap_{\underline{n}^\beta}(k^E)$, where $\xi_{\underline{n}^\beta} = (\xi_{n_\beta} \subset \cdots \subset \xi_{n_l} \subset \mathscr{O}^E_{Drap_{\underline{n}^\beta}(k^E)})$ (**The canon-ical fiberings of $Drap_{\underline{n}}(k^E)$**). In schematic terminology this fibering gives the structure of $Drap_{\underline{n}}(k^E)$ as a $Drap_{\underline{n}^\beta}(k^E)$-scheme.

Definition 1.15 *Let $\mathcal{H}om(\mathscr{O}^m_X, \xi_{\underline{m}})$ (resp. $\mathcal{I}som(\mathscr{O}^m_X, \xi_{\underline{m}})$), where $X = Drap_{\underline{m}}(k^E)$, denotes the sheaf of germs of $\mathscr{O}_X$-homomorphisms (resp. iso-morphisms) $\nu : \mathscr{O}^E_X \longrightarrow \mathscr{O}^E_X$ satisfying $\nu(\mathscr{O}^{m_\beta}_X) \subset \xi_{m_\beta}$.*
*It is clear that $\mathcal{H}om(\mathscr{O}^m_X, \xi_{\underline{m}})$ is a locally free $\mathscr{O}_X$-module. Thus we may con-sider the vector fiber bundle $V(\mathcal{H}om(\mathscr{O}^m_X, \xi_{\underline{m}}))$. Define **the Stiefel variety** of $\xi_{\underline{m}}$ by*

$$Stief(\xi_{\underline{m}}) \subset V(\mathcal{H}om(\mathscr{O}^m_X, \xi_{\underline{m}})) \ ,$$

*the sub-bundle representing $\mathcal{I}som(\mathscr{O}^m_X, \xi_{\underline{m}})$, whose sections correspond to the sections of $\mathcal{I}som(\mathscr{O}^m_X, \xi_{\underline{m}})$. Let $D_{\underline{m}} = (k^{m_1} \cdots \subset k^{m_l} \subset k^{m_{l+1}}(= k^{|E|}))$. The typical fiber of $Stief(\xi_{\underline{m}}) \longrightarrow Drap_{\underline{m}}(k^E)$, is the Stiefel variety $Stief(D_{\underline{m}})$ whose points are given by the ordered basis of $k^{|E|}$ adapted to $D_{\underline{m}}$ (cf. defini-tion 2.1). There is a principal natural right action of the stabilizer subgroup $Stab\, D_{\underline{m}} \subset Gl(k^{|E|})$ on $Stief(\xi_{\underline{m}})$. We may thus consider $Stief(\xi_{\underline{m}})$ as the $Stab\, D_{\underline{m}}$-**principal bundle associated to** $V(\xi_{\underline{m}})$.*

For more details about the Stiefel variety we refer the reader to [24], 9.10.

1.3.2 The fiber decomposition of the Flag Varieties canonical open subvarieties

Keep the same notation of the preceding section. Given $D = (H_1 \cdots \subset H_l \subset H_{l+1}) \in Drap_{\underline{n}}(E)$ let

$$D^\beta = (H_\beta \cdots \subset H_l \subset H_{l+1}) \in Drap_{\underline{n}^\beta}(E)$$

(resp.

$$D_\beta = (H_1 \cdots \subset H_{\beta-1} \subset H_\beta) \in Drap_{\underline{n}_\beta}(H_\beta)) \ .$$

Definition 1.16 *By definition a flag $\mathscr{D} = (\mathscr{H}_1 \cdots \subset \mathscr{H}_l \subset \mathscr{H}_{l+1})$ of lo-cally free $\mathscr{O}_X$-modules over a k-variety X is **split** if there exists submodules $(\mathscr{H}^\perp_i)_{1 \leq i \leq l}$ with:*

$$\mathscr{H}_{i+1} = \mathscr{H}_i \oplus \mathscr{H}^\perp_i.$$

Let $\xi_{\underline{n}^\beta} = (\xi_{n_\beta} \subset \cdots \subset \xi_{n_l} \subset \xi_{n_{l+1}})$ be the canonical flag on $Drap_{\underline{n}^\beta}(k^E)$. The module $(\xi_{n_\beta})_{U_{D^\beta}}$ is endowed with a split flag $\mathscr{D}_\beta = (\mathscr{H}_1 \cdots \subset \mathscr{H}_\beta)$ characterized as follows. Let $(S_\beta \cdots \subset S_l \subset S_{l+1})$ correspond to the section x of $U_{D^\beta} \subset Drap_{\underline{n}^\beta}(k^E)$ on $X = Spec(A)$, with A a k-algebra. By definition of U_{D^β}, there is an isomorphism of $\mathscr{O}_X$-modules: $(\pi_{H_\beta})_{S_\beta} : S_\beta \longrightarrow \mathscr{O}^{H_\beta}_X$. Then $\mathscr{D}_\beta$ is given by:

$$(\mathscr{D}_\beta)_x = ((\pi_{H_\beta})_{S_\beta})^{-1}(\mathscr{O}^{D_\beta}_X) \ ,$$

where $\mathscr{O}_X^{D_\beta} = (\mathscr{O}_X^{H_1} \subset \cdots \subset \mathscr{O}_X^{H_\beta})$ and $\mathscr{O}_X^F$ denotes the free $\mathscr{O}_X$-module with a basis indexed by the finite set F; the supplementary submodules $(\mathscr{H}_{\beta'}^{\perp})_{1 \leq \beta' \leq \beta - 1}$ being given by

$$(\mathscr{H}_{\beta'}^{\perp})_x = ((\pi_{H_\beta})_{S_\beta})^{-1}(\mathscr{O}_X^{(H_{\beta'+1} - H_{\beta'})}).$$

Definition 1.17 *Define an open subvariety*

$$U_{\mathscr{D}_\beta} \subset Drap_{\underline{n}_\beta}(\xi_{n_\beta})$$

by the following condition on its sections: $(\mathscr{S}_1 \cdots \subset \mathscr{S}_\beta)$ is a section of $U_{\mathscr{D}_\beta}$ if and only if for $1 \leq \beta' \leq \beta$ the restrictions

$$(\pi_{\mathscr{H}_{\beta'}})_{\mathscr{S}_{\beta'}} : \mathscr{S}_{\beta'} \longrightarrow \mathscr{H}_{\beta'}$$

are isomorphisms. The projection $\pi_{\mathscr{H}_{\beta'}}$ is defined in terms of the direct sum decomposition given by the splitting of the flag $\mathscr{D}_\beta$.

By restriction of the base and of the fiber the locally trivial fiber bundle $Drap_{\underline{n}_\beta}(\xi_{n_\beta}) \longrightarrow Drap_{\underline{n}^\beta}(k^E)$ gives rise to a locally trivial fiber bundle:

$$U_{\mathscr{D}_\beta} \longrightarrow U_{D^\beta},$$

with $U_{\mathscr{D}_\beta} \subset Drap_{\underline{n}_\beta}(\xi_{n_\beta})$ (resp. $U_{D^\beta} \subset Drap_{\underline{n}^\beta}(k^E)$), and typical fiber $U_{D_\beta} \subset Drap_{\underline{n}_\beta}(k^{n_\beta})$.

Thus we have the following

Proposition 1.18 *The open affine subvariety $U_D \subset Drap_{\underline{n}}(k^E)$ (cf. Definition 1.11) is decomposed in a sequence of locally trivial fibrations $(U_{D^\beta} \rightarrow U_{D^{\beta+1}})_{1 \leq \beta \leq l}$ with typical fiber $U_{H_\beta} \subset Grass_{n_\beta}(k^E)$ induced by the canonical fibering decomposition $(Drap_{\underline{n}_\beta}(k^E) \rightarrow Drap_{\underline{n}_\beta}(k^E))_{1 \leq \beta \leq l}$.*

1.3.3 Coordinates for a Canonical open subvariety of a Flag variety

Assume that E is ordered totally by ω_E. Let $D = (H_1 \subset \cdots \subset H_l \subset H_{l+1} = E) \in Drap_{\underline{n}}(E)$ $(\underline{n} = (n_1 < \cdots < n_l < n_{l+1} = |E|))$. Write:

$$\mathbb{A}_D^E = \mathbb{A}(\prod_{1 \leq \beta \leq l} M_{H_\beta^{\perp} \times H_\beta}(k)) \text{ (resp. } \mathbb{A}_D = \mathbb{A}(\prod_{1 \leq \beta \leq l} M_{(H_{\beta+1} - H_\beta) \times H_\beta}(k))), \text{ and}$$
$$\mathbb{A}_{\underline{n}}^E = \mathbb{A}(\prod_{1 \leq \beta \leq l} M_{(|E| - n_\beta) \times n_\beta}(k)) \text{ (resp. } \mathbb{A}_{\underline{n}} = \mathbb{A}(\prod_{1 \leq \beta \leq l} M_{(n_{\beta+1} - n_\beta) \times n_\beta}(k))).$$

The order of E allows defining an isomorphism $\mathbb{A}_D \simeq \mathbb{A}_{\underline{n}}$. Let $H_\beta \simeq [\![1, n_\beta]\!]$ (resp. $(H_{\beta+1} - H_\beta) \simeq [\![1, n_{\beta+1} - n_\beta]\!]$) be the bijection induced by the order of E. Thus there are a k-vector space isomorphism $M_{(H_{\beta+1} - H_\beta) \times H_\beta}(k) \simeq M_{(n_{\beta+1} - n_\beta) \times n_\beta}(k)$, and a k-variety isomorphism $\mathbb{A}_D \simeq \mathbb{A}_{\underline{n}}$. Similarly one obtains an isomorphism: $\mathbb{A}_D^E \simeq \mathbb{A}_{\underline{n}}^E$.

Let $U_{H_\beta} \subset Grass_{|H_\beta|}(k^E)$ be the affine open subset as in Definition 1.1. One has an affine open subset $\prod_{1 \leq \beta \leq l} U_{H_\beta} \subset \prod_{1 \leq \beta \leq l} Grass_{|H_\beta|}(k^E)$ and a k-isomorphism $\prod_{1 \leq \beta \leq l} U_{H_\beta} \simeq \mathbb{A}_D^E$ defined by:

$$(S_\beta)_{1 \leq \beta \leq l} \mapsto M_D((S_\beta)_{1 \leq \beta \leq l}) = (M_D(S_\beta))_{1 \leq \beta \leq l}$$

where $M_D(S_\beta) = M_{H_\beta}(S_\beta)$ (cf. definition 1.1). The $\prod_{1 \leq \beta \leq l} U_{H_\beta}$-coordinates of (S_β).

Let $\mathscr{D} = (S_1 \subset \cdots \subset S_l \subset S_{l+1} = k^E)$ correspond to a point in

$$U_D = (\prod_{1 \leq \beta \leq l} U_{H_\beta}) \cap Drap_{\underline{n}}(k^E).$$

We consider now the U_D-coordinates of $\mathscr{D}$.

Write $M_D(\mathscr{D}) = M_D((S_\beta)_{1 \leq \beta \leq l})$. Define

$$m_D(S_\beta) = M_{H_\beta}(\pi_{H_\beta+1}(S_\beta)) ,$$

where the right member denotes the U'_{H_β}-coordinates matrix of $\pi_{H_\beta+1}(S_\beta) \subset k^{H_\beta+1}$, with $U'_{H_\beta} = \{S \in Grass_{|H_\beta|}(k^{H_\beta+1}) \mid rank\ (\pi'_{H_\beta})s = |H_\beta| \}$ and π'_{H_β} is the projection given by the direct sum decomposition: $k^{H_\beta+1} = k^{H_\beta} \oplus k^{(H_\beta+1-H_\beta)}$. In fact $\pi'_{H_\beta} \circ \pi_{H_\beta+1} = \pi_{H_\beta}$.

Notation 1.19 *Let $M_\beta \in M_{H_\beta^\perp \times H_\beta}(k)$ (resp. $m_\beta \in M_{(H_{\beta+1}-H_\beta) \times H_\beta}(k)$). Denote by $A(M_\beta) \in M_{E \times H_\beta}(k)$ (resp. $A(m_\beta) \in M_{H_{\beta+1} \times H_\beta}(k)$) the matrix obtained by completing M_β (resp. m_β), with "ones" and "zeros", in a H_β-normalized matrix. Write:*

$$\overline{A}_D(S_\beta) = A(M_D(S_\beta)) = A(M_{H_\beta}(S_\beta)) \ (resp.\ A_D(S_\beta) = A(m_D(S_\beta)))\ .$$

It is easy to check the following relations between $(\overline{A}_D(S_\beta))$ and $(A_D(S_\beta))$ hold for $1 \leq \beta \leq l$:

$$\overline{A}_D(S_\beta) = A_D(S_l) \times A_D(S_{l-1}) \cdots \times A_D(S_\beta) .$$

Thus the sequence $(m_D(S_\beta))$ determines the product coordinates $(M_D(S_\beta))$ of $(S_\beta) \in \prod_{1 \leq \beta \leq l} U_{H_\beta}$, where $\mathscr{D} = (S_1 \cdots \subset S_l \subset k^E)$.

PROPOSITION - DEFINITION 1.20 *The morphism $U_D \longrightarrow \mathbb{A}_D$ defined by $\mathscr{D} \mapsto m_D(\mathscr{D}) = (m_D(S_\beta))_{1 \leq \beta \leq l}$ is an isomorphism. The image of $\mathscr{D}$ by the isomorphism $U_D \simeq \mathbb{A}_D \simeq \mathbb{A}_{\underline{n}}$ (resp. $m_D(\mathscr{D})$) gives by definition the U_D-* **canonical coordinates** *of $\mathscr{D}$.*

Proof *Let $(m_\beta) \in \mathbb{A}_D$ for $1 \leq \beta \leq l$.*

Define $\overline{A}((m_\beta)_{1 \leq \beta' \leq \beta}) = A(m_l) \times A(m_{l-1}) \cdots \times A(m_\beta) \in M_{E \times H_\beta}(k)$.

It is easy to verify that $\overline{A}((m_\beta)_{1 \leq \beta' \leq \beta})$ is a H_β-normalized matrix. Denote by $(\overline{A}((m_\beta)_{1 \leq \beta' \leq \beta}))_{H_\beta^\perp}$ the $H_\beta^\perp \times H_\beta$-submatrix of $\overline{A}((m_\beta)_{1 \leq \beta' \leq \beta})$ given by the row vectors indexed by $H_\beta^\perp$. Define $\mathbb{A}_D \longrightarrow \mathbb{A}_D^E$ by $(m_{S_\beta}) \mapsto ((\overline{A}((m_\beta)_{1 \leq \beta' \leq \beta}))_{H_\beta^\perp})_{1 \leq \beta \leq l}$. From the above relation it results that this morphism is followed by the isomorphism $\mathbb{A}_D^E \simeq \prod_{1 \leq \beta \leq l} U_{H_\beta}$ factors through the embedding

$$U_D = (\prod_{1 \leq \beta \leq l} U_{H_\beta}) \cap Drap_{\underline{n}}(k^E) \hookrightarrow \prod_{1 \leq \beta \leq l} U_{H_\beta} \subset \prod_{1 \leq \beta \leq l} Grass_{|H_\beta|}(k^E),$$

and is the reciprocal isomorphism of $U_D \longrightarrow \mathbb{A}_D$. This proves our assertion.

Remark 1.21 *Definition 1.20 applied to $U_{(H \subset E)} \subset Drap_{(|H| \subset E)}(k^E) = Grass_{|H|}(k^E)$ gives $m_{(H \subset E)}((S \subset k^E)) = M_H(S)$. Thus the $U_{(H \subset E)}$-canonical coordinates of $(S \subset k^E)$ are precisely the coordinates of S as defined in 1.1.*

1.4 Chains

A chain of a finite set is informally speaking a redundant flag, given by an ascending sequence of subsets, with the inclusion relations not necessarily strict. Chains play a role in the construction of minimal galleries in $\Delta^{(r)'}$.

Definition 1.22 *Let E be a finite set with $|E| = r+1$. Endow $\mathrm{Grass}(E)$ with the natural order relation defined by the inclusion of subsets of E.*

  *a) Define a **chain of E of length l** as an increasing function $f : [\![1, l+1]\!] \longrightarrow \mathrm{Grass}(E)$ satisfying $f(l+1) = E$. Let*

$$\mathrm{In}\, f := \{i \in [\![1, l+1]\!] \mid f(i-1) \subset f(i),\ f(i-1) \neq f(i)\}$$

  $(f(0) = \emptyset)$ be the set of strictly increasing points of f, and f' be the restriction of f to $\mathrm{In}\, f$. Write

$$\varphi(f) = f'.$$

  *(**The flag defined by the chain f**). Given a chain f of E of length l, i.e. f is defined on $[\![1, l+1]\!]$, we associate to f the increasing sequence of integers*

$$\overline{\mathrm{typ}}\, f := (|f(i)|)_{1 \leqslant i \leqslant l+1}.$$

b) *Write*
$$\overline{\mathrm{typ}}(E) := \mathrm{Im}\ \overline{\mathrm{typ}},$$

the set of types of chains.

Given $\underline{n} \in \overline{\mathrm{typ}}(E)$ *such that* $\underline{n} = (n_1 \leqslant \cdots \leqslant n_l \leqslant r + 1)$, *we say that* $\underline{n}$ *is of length* l. *Denote by* $\mathrm{Chain}_{\underline{n}}(E)$ *the set of chains of* E *of type* $\underline{n}$. *Write*

$$\mathrm{In}\ \underline{n} := \{i \in [\![1, l+1]\!]|\ n_{i-1} < n_i \quad (n_0 = 0)\} = (i_1 < \cdots < i_{l'} < i_{l'+1}),$$

and $\underline{n}' = (n_{i_1} < \cdots < n_{i_{l'}} < r + 1)$. *Here* l' *denotes the length of* $\underline{n}' \in \mathrm{typ}(E)$. *Clearly we have the inclusion* $\mathrm{typ}(E) \subset \overline{\mathrm{typ}}(E)$ *as the subset of strictly increasing sequences of* $[\![1, r+1]\!]$ *with maximum equal to* $r + 1$.

c) *Given the couple of chains* $(f, f') \in \mathrm{Chain}_{\underline{m}}(E) \times \mathrm{Chain}_{\underline{n}}(E)$ $(\underline{m}, \underline{n} \in \overline{\mathrm{typ}}(E)$, *with length* $\underline{n} = \lambda$, *and length* $\underline{m} = l)$ *write*

$$\mathscr{M}(f, f') = (f(\alpha) \cap f'(\beta))$$

(resp.
$$M(f, f') = (|f(\alpha) \cap f'(\beta)|) \in \mathbb{N}^{(\lambda+1) \times (l+1)})$$

(The relative position matrix of the chains f **and** f'**).**

d) *Let* $\underline{n} \in \overline{\mathrm{typ}}(E)$. *Define a* **chain of subspaces** $\mathcal{D}$ *of* k^E *of type* $\underline{n}$ *following the pattern of the definition of a chain of* E. *Denote by* $\mathrm{Chain}_{\underline{n}}(k^E)$ *the set of chains of type* $\underline{n}$. *Write*

$$Chain(k^E) = \coprod_{\underline{n} \in \overline{\mathrm{typ}}(E)} Chain_{\underline{n}}(k^E)$$

(resp.

$$Chain(E) = \coprod_{\underline{n} \in \overline{\mathrm{typ}}(E)} Chain_{\underline{n}}(E)).$$

Given $\mathcal{D} \in Chain_{\underline{n}}(k^E)$ *let* $\varphi(\mathcal{D})$ *be the corresponding flag of type* $\underline{n}'$.

Define the following ordering on the set of chains $Chain(E)$ (resp. $Chain(k^E)$):

Definition 1.23 *Given chains* $\mathcal{D}$ *and* $\mathcal{D}'$ *of* E *defined respectively by*

$$f : [\![1, \lambda+1]\!] \longrightarrow \mathrm{Grass}(I_{r+1}) \quad and \quad f' : [\![1, \lambda'+1]\!] \longrightarrow \mathrm{Grass}(I_{r+1})$$

with $\lambda' \leqslant \lambda$, *write* $\mathcal{D}' \subset \mathcal{D}$ *if* f' *factors as* $f' = f \circ \varphi$, *with* $\varphi : [\![1, \lambda'+1]\!] \longrightarrow [\![1, \lambda+1]\!]$ *strictly increasing. Following the same pattern define* $\mathcal{D}' \subset \mathcal{D}$ *for* $\mathcal{D}', \mathcal{D} \in Chain(k^E)$. *Clearly "* $\subset$ *" is an order relation.*

We have $\mathcal{D}' \subset \mathcal{D} \Rightarrow \varphi(\mathcal{D}') \subset \varphi(\mathcal{D})$, i.e. if $\varphi(\mathcal{D}') = (H_1 \subset \cdots \subset H_\lambda \subset E)$ (resp. $\varphi(\mathcal{D}) = (J_1 \subset \cdots \subset J_l \subset E)$), then $\{H_1, \cdots, H_\lambda\} \subset \{J_1, \cdots, J_l\}$. The following definition plays a role in the contruction of minimal galleries giving rise to resolutions of singulaties of Schubert varieties.

Definition 1.24 *Let* $(D, H) \in Drap(E) \times Grass(E)$ *(resp.* $(\mathscr{D}, S) \in Drap(k^E) \times Grass(k^E)$*). Denote by* $D \cap H \in Chain(E)$ *(resp.* $\mathscr{D} \cap S \in Chain(k^E)$*) the chain given by the intersections of the elements of* D *(resp.* $\mathscr{D}$*) with* H *(resp.* S*).*

Chapter 2

Schubert Cell Decomposition of Grassmannians and Flag Varieties

The classical indexation of Schubert cells (resp. Schubert varieties) in Grassmannians, by increasing functions from an integral interval into another, is not easily adapted to Flag varieties (cf. [29], and [36]). Instead we introduce a more suitable general indexation by **Relative Position Matrices** for Schubert varieties in Flag varieties. These matrices are in bijection with classes of the symmetric group and thus give a geometric interpretation of the orbits in Bruhat decomposition of the linear group. On the other hand, we shall see that such a matrix summarizes the construction of a minimal generalized gallery and thus a block decomposition of the parametrizing subgroup of its corresponding Schubert cell. The corresponding Young diagram reflects this block decomposition.

From the construction of a basis adapted to a couple of flags (see below), one deduces that the quotient set of the couples of flags in k^{r+1} by the natural action of $Gl(k^{r+1})$ is given by the set of Relative Position Matrices. The parabolic subgroups of $Gl(k^{r+1})$ are defined as the stabilizers of flags in $Gl(k^{r+1})$. The set of these subgroups correspond bijectively to the set of flags of k^{r+1} and represent the points of certain $Gl(k^{r+1})$-homogeneous spaces. The Schubert cell decomposition of the Flag variety $Drap(k^E)$ defined by a fixed flag corresponds to the Bruhat decompositions of $Gl(k^E)$ given by a couple of parabolic subgroups. It is shown how a combinatorial flag is associated with a subset of the set of roots of $Gl(k^{r+1})$. The parabolic subgroups given by the flags adapted to the canonical basis correspond to these subsets.

2.1   The relative position matrix of a couple of flags

Definition 2.1 *Given a flag* $\mathcal{D} = (V_1 \subset \dots \subset V_l \subset k^E) \in Drap_{\underline{n}}(k^E)$ *(* $\underline{n} = (n_1 < \dots < n_l < n_{l+1} = |E|)$ *) and an ordered basis* $\tilde{e} = (\tilde{e}_i)_{1 \leq j \leq |E|}$ *of* k^E *, we say that* $\tilde{e}$ **is adapted to** $\mathcal{D}$ *(or that* $\mathcal{D}$ *is adapted to* $\tilde{e}$ *) if*

$$V_i = Vect((\tilde{e}_j)_{1 \leq j \leq n_i}).$$

The direct sum decomposition $k^E = \bigoplus_{\alpha \in A} F_\alpha$ **is adapted to** $\tilde{e}$ *(or that* $\tilde{e}$ *is adapted to this direct sum decomposition), if for each* F_α *there exists a subset* $\tilde{e}_\alpha \subset \tilde{e}$ *with* $F_\alpha = Vect(\tilde{e}_\alpha)$ *.*

A direct sum decomposition $k^E = \bigoplus_{\alpha \in A} F_\alpha$ **is adapted to** $\mathcal{D}$ *(or* $\mathcal{D}$ *is adapted to* $k^E = \bigoplus_{\alpha \in A} F_\alpha$ *), if for every* $1 \leq i \leq l$ *, there exists a subset* $A_i \subset A$ *satisfying* $V_i = \bigoplus_{\alpha \in A_i} F_\alpha$ *. A flag* $\mathcal{D}$ *is adapted to the basis* e_E *if there exists an ordering of* $\tilde{e}_E$ *of* e_E *such that* $\mathcal{D}$ *is adapted to* $\tilde{e}_E$ *.*

Given a couple of flags $(\mathcal{D}, d)$ of k^E we proceed to show that there exists a basis e_E of k^E adapted to both $\mathcal{D}$ and d. This result is a particular case of a general one stating that given a couple of parabolics in a reductive group there exists a maximal torus contained in both.

Lemma 2.2 *Let*

$$
\begin{array}{ccc}
V_1 & \longrightarrow & V_3 \\
\downarrow & & \downarrow \\
V_2 & \longrightarrow & V_4
\end{array}
$$

be a cartesian diagram of inclusions of subspaces of k^E *, i.e.* $V_1 = V_2 \cap V_3$ *, and* $V_2 = V_1 \oplus S_2$ *(resp.* $V_3 = V_1 \oplus S_3$ *) direct sum decompositions of* V_2 *and* V_3 *. Then there are the direct sum decompositions:*

$$V_2 + V_3 = V_1 \oplus S_2 \oplus S_3 \ (resp\ V_4 = V_2 \oplus S_3 = V_3 \oplus S_2).$$

Proof *Clearly* $V_2 + V_3 = V_1 + S_1 + S_3$ *. On the other hand,* $S_3 \cap (V_1 + S_2) = S_3 \cap V_2 \subset V_1 \cap S_3 = \{0\}$ *(resp.* $S_2 \cap (V_1 + S_3) = S_2 \cap V_3 \subset V_1 \cap S_2 = \{0\}$ *).* *If* $v = s_2 + s_3$ *with* $v \in V_1$ *,* $s_2 \in S_2$ *,* $s_3 \in S_3$ *, we deduce that* $s_3 = v - s_2 \in S_3 \cap V_2 \subset V_1 \cap S_3 = \{0\}$ *. Thus* $v = s_2$ *, and finally* $v = s_2 = 0$ *.*

Let $\mathcal{D} = (V_1 \subset \dots \subset V_l \subset k^E) \in Drap_{\underline{m}}(k^E)$ (resp. $d = W_1 \subset \dots \subset W_\lambda \subset k^E) \in Drap_{\underline{n}}(k^E)$), where $\underline{m} = (m_1 < \dots < m_l < |E|)$ (resp.$\underline{n} = (n_1 < \dots < n_\lambda < |E|)$). We introduce the $(l+1) \times (\lambda+1)$-matrix with coefficients in $Grass(k^E)$:

$$\mathcal{M}(\mathcal{D}, d) = (V_i \cap W_j).$$

Proposition 2.3 *There is a direct sum decomposition* $k^E =$
$\bigoplus_{(i,j)\in[\![1,l+1]\!]\times[\![1,\lambda+1]\!]} S_{ij}$ *of* k^E *adapted to* $\mathscr{D}$ *and* d.

Proof *We order lexicographically* $[\![1,l+1]\!]\times[\![1,\lambda+1]\!]$, *and we proceed by recursion to construct* (S_{ij}). *We begin by splitting the first row* $(V_1\cap W_j)_{j\in[\![1,\lambda+1]\!]}$ *in a direct sum:*

$$V_1 \cap W_j = \bigoplus_{1\leq\alpha\leq j} S_{1\alpha} \ (1\leq j \leq \lambda+1).$$

We conclude the proof by recursion applying 2.2 to the cartesian squares:

$$
\begin{array}{ccc}
V_i\cap W_{j+1} & \longrightarrow & V_{i+1}\cap W_{j+1} \\
\uparrow & & \uparrow \\
V_i\cap W_j & \longrightarrow & V_{i+1}\cap W_j
\end{array},
$$

following the lexicographical ordering. Define S_{i+1j+1} *as a subspace of* k^E *satisfying:*

$$V_{i+1}\cap W_{j+1} = (V_i\cap W_{j+1} + V_{i+1}\cap W_j) \oplus S_{i+1j+1}.$$

Corollary 2.4 *There is a basis* $\tilde{e} = \coprod_{(i,j)\in[\![1,l+1]\!]\times[\![1,\lambda+1]\!]} \tilde{e}_{ij}$ *adapted to the direct sum* $k^E = \bigoplus_{(i,j)\in[\![1,l+1]\!]\times[\![1,\lambda+1]\!]} S_{ij}$, *i.e.* $S_{ij} = Vect(\tilde{e}_{ij})$, *and giving rise, after a convenient re-ordering, to a basis adapted to* $\mathscr{D}$ *(resp. d).*

It may be noted that some of the S_{ij} may be reduced to the zero subspace $S_{ij} = \{0\}$, in which case we write $\tilde{e}_{ij} = \emptyset$.
There is a natural action of $Gl(k^E) = Aut_k(k^E)$ on $Drap(k^E)$:

$$Gl(k^E) \times Drap(k^E) \longrightarrow Drap(k^E)$$

defined by $(g,\mathscr{D}) \mapsto g(\mathscr{D})$, where $\mathscr{D} = (V_1 \subset ... \subset V_l \subset k^E)$, and $g(\mathscr{D}) = (g(V_1) \subset ... \subset g(V_l) \subset k^E)$. It is easy to see that this action is algebraic, i.e. that the above mapping is the underlying mapping of a morphism of k-varieties. We aim at giving a description of the quotient set $(Drap(k^E) \times Drap(k^E))/Gl(k^E)$ of $Drap(k^E) \times Drap(k^E))$ under the diagonal action of $Gl(k^E)$.

Definition 2.5 *There is a canonical mapping*

$$\psi_E : Drap(E) \longrightarrow Drap(k^E)$$

defined by : $\psi_E : D \mapsto k^D$, *where* $D = (E_1 \subset ... \subset E_l \subset E)$, $k^D = (k^{E_1} \subset ... \subset k^{E_l} \subset k^E)$, *and* $k^{E_1} = Vect((e_j)_{j\in E_i})$ *for* $1\leq i \leq l$. *Given* $H \in Grass(E)$ *write* $\psi_E(H) = k^H$.

Remark that the image of ψ_E is the set of flags adapted to the canonical basis e_E of k^E.

On the other hand, there is a combinatorial natural action of the symmetric group $\mathfrak{S}_E = Aut(E)$ on $Drap(E)$:

$$\mathfrak{S}_E \times Drap(E) \longrightarrow Drap(E)$$

defined by: $(D, g) \mapsto g(D)$, with $D = (E_1 \subset \ldots \subset E_l \subset E)$ and $g(D) = (g(E_1) \subset \ldots \subset g(E_l) \subset E)$, and a group homomorphism: $\alpha : \mathfrak{S}_E \longrightarrow Gl(k^E)$ associating with $g \in \mathfrak{S}_E$ the automorphism $\alpha_g : k^E \longrightarrow k^E$ of the vector space k^E defined by : $\alpha_g(e_i) = e_{g(i)}$. It is immediate that the canonical mapping $Drap(E) \longrightarrow Drap(k^E)$ is $(Gl(k^E), \mathfrak{S}_E)$-equivariant. We may thus state:

Proposition 2.6 *The induced canonical mapping*

$$(Drap(E) \times Drap(E))/\mathfrak{S}_E \longrightarrow (Drap(k^E) \times Drap(k^E))/Gl(k^E)$$

is a bijection.

The proof of this proposition is based on several definitions and a lemma.

Definition 2.7 *Let* $(\mathscr{D}, \mathscr{D}') \in Drap(k^E) \times Drap(k^E)$ *(resp.* $(D, D') \in Drap(E) \times Drap(E)$*). Write :*

$$M(\mathscr{D}, \mathscr{D}') = (dim(V_i \cap W_j)) \in \mathbb{N}^{(l+1)\times(\lambda+1)}$$
$$(resp. \quad M(D, D') = (|E_i \cap F_j|) \in \mathbb{N}^{(l+1)\times(\lambda+1)}),$$

where $\mathscr{D} = (V_1 \subset \ldots \subset V_l \subset k^E)$ *and* $\mathscr{D}' = (W_1 \subset \ldots \subset W_l \subset k^E)$ *(resp.* $D = (E_1 \subset \ldots \subset E_l \subset E)$ *and* $D' = (F_1 \subset \ldots \subset F_l \subset E)$*).*
Write $\mathcal{M}(D, D') = (E_i \cap F_j) \in Grass(E)^{(l+1)\times(\lambda+1)}$.

There are natural mappings

$$Drap(k^E) \times Drap(k^E) \longrightarrow \coprod_{l,\lambda \in \mathbb{N}} \mathbb{N}^{(l+1)\times\lambda+1)}$$
$$(resp. \ Drap(E) \times Drap(E) \longrightarrow \coprod_{l,\lambda \in \mathbb{N}} \mathbb{N}^{(l+1)\times\lambda+1)})$$

defined by $(\mathscr{D}, \mathscr{D}') \mapsto M(\mathscr{D}, \mathscr{D}')$ (resp. $(D, D') \mapsto M(D, D')$). Clearly there is a commutative diagram:

$$
\begin{array}{ccc}
Drap(k^E) \times Drap(k^E) & \longrightarrow & \coprod_{l,\lambda \in \mathbb{N}} \mathbb{N}^{(l+1)\times\lambda+1)} \\
\downarrow & \nearrow & \\
Drap(k^E) \times Drap(k^E)/\mathfrak{S}_E & &
\end{array}
\tag{2.1.1}
$$

where the vertical arrow represents the quotient mapping.

Definition 2.8 *Let* $(\underline{m}, \underline{n}) \in typ(E) \times typ(E)$. *Write*

$$Relpos_{(\underline{m},\underline{n})} = Relpos_{(\underline{m},\underline{n})}(E) = (Drap(E)_{\underline{m}} \times Drap(E)_{\underline{n}})/\mathfrak{S}_E$$

and

$$Relpos = Relpos(E) = \coprod_{(\underline{m},\underline{n}) \in typ(E) \times typ(E)} Relpos_{(\underline{m},\underline{n})}.$$

We call $Relpos(E)$ the set of types of relative position of $Drap(E)$. Let $\mathfrak{S}_m = \mathfrak{S}_{I_m}$ with $I_m = [\![1, m]\!]$. If E is totally ordered we may assume that $E = I_{r+1}$. There is a mapping

$$typ(I_{r+1}) \longrightarrow Drap(I_{r+1})$$

given by $\underline{m} = (m_1 < ... < m_l < r+1) \mapsto D_{\underline{m}}$ with $D_{\underline{m}} = (I_{m_1} \subset ... \subset I_{m_l} \subset I_{r+1})$. When we compose this mapping with $Drap(I_{r+1}) \longrightarrow Drap(k^{r+1})$ we obtain : $D_{\underline{m}} \mapsto k^{D_{\underline{m}}} = (k^{m_1} \subset ... \subset k^{m_1} \subset k^{r+1})$.

Denote by $\mathfrak{S}_D$ the stabilizer $Stab\, D$ of the combinatorial flag $D = (E_1 \subset ... \subset E_l \subset E)$ in $\mathfrak{S}_E$. We have :

$$\mathfrak{S}_D = \prod_{i=1}^{l+1} \mathfrak{S}_{(E_i - E_{i-1})},$$

where $E_0 = \emptyset$, i.e. $\mathfrak{S}_D$ may be seen as the subgroup of $\mathfrak{S}_E$ preserving the partition $E = \coprod_{1 \leq i \leq l+1} (E_i - E_{i-1})$. For $E = I_{r+1}$ and $D = D_{\underline{m}}$, write $\mathfrak{S}_{\underline{m}} = \mathfrak{S}_{D_{\underline{m}}}$. Thus there is the identification $\mathfrak{S}_{\underline{m}} = \prod_{i=1}^{l+1} \mathfrak{S}_{(m_{i+1} - m_i)}$.

Observe that the mapping $Drap_{\underline{m}}(E) \times Drap_{\underline{n}}(E) \longrightarrow \mathbb{N}^{(l+1) \times \lambda+1}$ factors as

$$Drap_{\underline{m}}(E) \times Drap_{\underline{n}}(E) \longrightarrow Relpos_{(\underline{m},\underline{n})}(E) \longrightarrow \mathbb{N}^{(l+1) \times (\lambda+1)}.$$

Lemma 2.9 *The mapping* $Relpos_{(\underline{m},\underline{n})}(E) \longrightarrow \mathbb{N}^{(l+1) \times \lambda+1}$ *induced by* $(D, D') \mapsto M(D, D')$ *is an injective mapping, i.e. the* $\mathfrak{S}_{r+1}$-*orbit of a couple* (D, D') ($D = (E_1 \subset ... \subset E_l \subset E), D' = (E_1 \subset ... \subset E_l \subset E)$, $typ(D) = \underline{m}$, $typ(D') = \underline{n}$) *is characterized by the relative position matrix* $M(D, D') = (|E_i \cap F_j|)$.

Proof *Let* $D_{\underline{m}} = (I_{m_1} \subset I_{m_2} \cdots \subset I_{m_l} \subset I_{r+1})$. *It is evident that the natural mapping*

$$(\{D_{\underline{m}}\} \times Drap_{\underline{n}}(E))/\mathfrak{S}_{\underline{m}} \longrightarrow (Drap_{\underline{m}}(E) \times Drap_{\underline{n}}(E))/\mathfrak{S}_{r+1}$$

is a bijection. Thus it suffices to prove that:

$$M(D_{\underline{m}}, D) = M(D_{\underline{m}}, D') \Rightarrow (\exists\, w \in \mathfrak{S}_{\underline{m}})\, w(D) = D'.$$

Let $D = (H_1 \subset \cdots \subset H_l \subset I_{r+1})$ *(resp.* $D' = (H'_1 \subset \cdots \subset H'_l \subset I_{r+1}))$, $I_{m_0} = \emptyset$, and $I_{m_{l+1}} = I_{r+1}$. As $M(D_{\underline{m}}, D) = M(D_{\underline{m}}, D')$. It follows that

$$|(I_{m_\alpha} - I_{m_{\alpha-1}}) \cap H_1| = |(I_{m_\alpha} - I_{m_{\alpha-1}}) \cap H'_1||$$

for $1 \leqslant \alpha \leqslant l + 1$. *From the decomposition*

$$H_1 = \coprod_{1 \leq \alpha \leq l+1} (I_{m_\alpha} - I_{m_\alpha-1}) \cap H_1 \ \text{(resp.} \ H'_1 = \coprod_{1 \leq \alpha \leq l+1} (I_{m_\alpha} - I_{m_\alpha-1}) \cap H'_1)$$

it results that there exists $w_1 \in \mathfrak{S}_{\underline{m}}$ *with* $w_1(H) = H'_1$. *Therefore, we have* $M(D_{\underline{m}}, w_1(D)) = M(D_{\underline{m}}, D')$, *and we may thus suppose* $H_1 = H'_1$. *We achieve the proof by induction. Let us suppose that* $H_1 = H'_1, \cdots, H_\beta = H'_\beta$ *(*$1 < \beta < l$*). Following the above reasoning with* $H_{\beta+1}$ *and* $H'_{\beta+1}$ *instead of* H_1 *and* H'_1 *we deduce that there exists* $w_{\beta+1} \in \mathfrak{S}_{\underline{m}}$ *with* $w_{\beta+1}(H_{\beta+1}) = H'_{\beta+1}$ *and* $w_{\beta+1|H_\beta} = 1_{H_\beta}$. *It follows from this that there exists* $w \in \mathfrak{S}_{\underline{m}}$ *satisfying* $w(D) = D'$.

Proof of Proposition 2.6

Proof *The image of* $Drap(E) \longrightarrow Drap(k^E)$ *defined by* $D \mapsto k^D$ *is given by the set of flags adapted to the canonical basis* e_E *of* k^E. *On the other hand, we know by 2.3, that for a couple of flags* $(\mathscr{D}, \mathscr{D}')$ *with* $typ(\mathscr{D}) = \underline{m}$ *and* $typ(\mathscr{D}') = \underline{n}$ *there exists a basis* $\tilde{e}_E$ *of* k^E *adapted both to* $\mathscr{D}$ *and* $\mathscr{D}'$. *Let* $\alpha \in Gl(k^E)$ *be the automorphism defined by* $\alpha(\tilde{e}_E) = e_E$. *Then* e_E *is adapted to the couple* $(\alpha(\mathscr{D}), \alpha(\mathscr{D}'))$, *i.e.* $\alpha(\mathscr{D})$ *and* $\alpha(\mathscr{D}')$ *are both adapted to* e_E. *It follows that the induced mapping* $Drap_{\underline{m}}(E) \times Drap_{\underline{n}}(E)/\mathfrak{S}_E \longrightarrow Drap_{\underline{m}}(k^E) \times Drap_{\underline{n}}(k^E)/Gl(k^E)$ *is surjective.*

The injectivity of this mapping results immediately both from Lemma 2.9 and the following commutative diagram

$$
\begin{array}{ccc}
Drap_{\underline{n}}(E \times Drap_{\underline{m}}(E)/\mathfrak{S}_E & \rightarrow & \coprod_{l,\lambda \in \mathbb{N}} \mathbb{N}^{(l+1)\times(\lambda+1)} \\
\downarrow & \nearrow & \\
Drap_{\underline{n}}(k^E) \times Drap_{\underline{m}}(k^E)/Gl(k^E). & &
\end{array}
\tag{2.1.2}
$$

Remark that the oblique arrow is the mapping $(\mathscr{D}, \mathscr{D}') \mapsto M(\mathscr{D}, \mathscr{D}')$, *that the horizontal arrow is injective and that the down arrow is surjective. This achieves the proof.*

2.2 Schubert cells and Schubert varieties

In what follows we identify the set $Relpos(E)$ of types of relative position of flags of E with its image in the set of matrices with integral coefficients $\coprod_{l,\lambda \in \mathbb{N}} \mathbb{N}^{(l+1)\times\lambda+1)}$. The $\mathfrak{S}_E$-orbits in $Drap(E) \times Drap(E)$ are thus represented by matrices with coefficients in $\mathbb{N}$.

From proposition 2.6 it results that the mapping $Drap(k^E) \times Drap(k^E) \longrightarrow \coprod_{l,\lambda \in \mathbb{N}} \mathbb{N}^{(l+1) \times \lambda+1)}$ corresponds to the quotient mapping of $Drap(k^E) \times Drap(k^E)$ under the action of $Gl(k^E)$. Actually it is induced by an algebraic morphism, as it will be seen later.

By the following definitions we introduce the main objects of interest in this chapter.

Definition 2.10 *Let $(M, \mathscr{D}) \in Relpos(E) \times Drap(k^E)$ (resp. $(M, D) \in Relpos(E) \times Drap(E)$). Define*

$$\Sigma(M) \subset Drap(k^E) \times Drap(k^E)$$

as the fiber over M of the quotient mapping, and

$$\Sigma(M, \mathscr{D}) \subset Drap(k^E)$$

as the fiber over $\mathscr{D}$ of the mapping $\Sigma(M) \longrightarrow Drap(k^E)$ induced by the second projection.
Write $\Sigma(M, D) = \Sigma(M, k^D)$. We call

$$\Sigma(M)$$

the Universal Schubert cell of type M, *and*

$$\Sigma(M, \mathscr{D})$$

the Schubert cell defined by the flag $\mathscr{D}$ and the type M.

Remark 2.11 *All the mappings (resp. group actions) above are induced by k-morphisms. This may be stated in scheme theory as follows. $\mathcal{D}rap(k^E)$ is a representable k-functor, and thus the product functor $\mathcal{D}rap(k^E) \times \mathcal{D}rap(k^E)$ is representable too. On the other hand, the diagonal action of $Gl(k^E)$ on this product is functorial and thus algebraic. It is easy to see that the quotient functor $\mathcal{D}rap(k^E) \times \mathcal{D}rap(k^E)/Gl(k^E)$ is representable by the variety of flags relative positions, and thus the quotient morphism is given by a k-morphism. All these assertions will be proved in the following chapters.*

From the remark one concludes

Proposition 2.12 $\Sigma(M)$ *(resp. $\Sigma(M, \mathscr{D})$) is the underlying set of a k-variety.*

The following is a direct proof of this proposition.

Proof *It suffices to prove the first assertion. Let $\mathscr{D} = (V_1 \subset ... \subset V_l \subset k^E)$, $\mathscr{D}' = (W_1 \subset ... \subset W_l \subset k^E)$, $typ(\mathscr{D}) = \underline{m}$, $typ(\mathscr{D}') = \underline{n}$, and $M(\mathscr{D}, \mathscr{D}') = (m_{ij})$. The condition $(\mathscr{D}, \mathscr{D}') \in \Sigma(M)$ splits in the set of conditions: $rk\ (V_i \cap$

$W_j) = m_{ij}$. *By the proof of Proposition 1.9 we know that this intersection rank condition may be written in terms of the vanishing of a set of homogeneous polynomials in the Plücker coordinates of V_i and W_j.*

On the other hand, observe that the product $Drap(k^E)_{\underline{m}} \times Drap(k^E)_{\underline{n}}$ is first embedded in a product of grassmannians, and secondly in a product of projective spaces, by taking the product of the corresponding Plücker embeddings. Thus the image of $\Sigma(M) \subset Drap(k^E)_{\underline{m}} \times Drap(k^E)_{\underline{n}}$ is characterized by a set of equations in terms of Plücker coordinates. We conclude that $\Sigma(M)$ is a k-variety (cf. remark 1.10).

Remark 2.13 *A more detailed description of the k-variety structure of $\Sigma(M)$ (resp. $\Sigma(M, \mathscr{D})$) will be given later.*

Proposition 2.14 *Let $M \in Relpos_{\underline{m}}(E) = \coprod\limits_{\underline{n} \in typ(E)} Relpos_{(\underline{m},\underline{n})}(E)$. The k-morphism induced by the first projection $\Sigma(M) \subset Drap(k^E)_{\underline{m}} \times Drap(k^E) \longrightarrow Drap(k^E)_{\underline{m}}$ defines a locally trivial fibration in the Zariski topology with typical fiber $\Sigma(M, D_{\underline{m}})$.*

Proof *There is a natural left action of $Stab\ D_{\underline{m}}$ on $\Sigma(M, D_{\underline{m}})$. One may thus define the contracted product:*

$$Stief(\xi_{\underline{m}}) \times_{Stab\ D_{\underline{m}}} \Sigma(M, D_{\underline{m}}),$$

and a canonical isomorphism:

$$Stief(\xi_{\underline{m}}) \times_{Stab\ D_{\underline{m}}} \Sigma(M, D_{\underline{m}}) \longrightarrow \Sigma(M),$$

showing that the universal Schubert cell defined by M:

$$\Sigma(M) \longrightarrow Drap_{\underline{m}}(k^E),$$

defines a locally trivial fibration over $Drap_{\underline{m}}(k^E)$.

We may now state the

Definition 2.15 *Let $\mathscr{D} \in Drap(k^E)_{\underline{m}}$, and $(\underline{n}, \underline{m}) \in typ(E) \times typ(E)$. Clearly we have the following cell decompositions:*

$$Drap(k^E)_{\underline{m}} \times Drap(k^E)_{\underline{n}} = \coprod\limits_{M \in Relpos_{(\underline{m},\underline{n})}} \Sigma(M)$$

(The Universal Schubert cell decomposition of $Drap(k^E)_{\underline{n}} \times Drap(k^E)_{\underline{m}}$),
(resp.

$$Drap(k^E) \times Drap(k^E) = \coprod\limits_{M \in Relpos} \Sigma(M)$$

(The Universal Schubert cell decomposition of $Drap(k^E) \times Drap(k^E)$),

$$Drap(k^E)_{\underline{n}} = \coprod_{M \in Relpos_{(\underline{m},\underline{n})}} \Sigma(M, \mathscr{D})$$

(The Schubert cell decomposition of $Drap(k^E)_{\underline{n}}$ defined by $\mathscr{D}$).

Let $E = I_{r+1}$, and $D_{\underline{r}} = (I_1 \subset ... \subset I_r \subset I_{r+1})$. One obtains the classical Schubert cell decomposition of the Grassmannian $Grass(k^{r+1}) = \coprod_{1 \leq n \leq r} Grass_n(k^{r+1})$:

$$Grass(k^{r+1}) = \coprod_{M \in Relpos_{(n<r+1)}} \Sigma(M, D_{\underline{r}}).$$

Definition 2.16 *Let $M \in \mathrm{Relpos}_{(\underline{n},\underline{m})}$, and $\mathscr{D} \in Drap_{\underline{m}}(k^{r+1})$. Denote by*

$$\overline{\Sigma}(M, \mathscr{D}) \ (resp. \ \overline{\Sigma}(M))$$

the Zariski closure of $\Sigma(M, \mathscr{D})$ (resp. $\Sigma(M)$) in the k-variety $Drap_{\underline{n}}(k^{r+1})$ (resp. $Drap_{\underline{n}}(k^{r+1}) \times Drap_{\underline{m}}(k^{r+1})$).
We call

$$\overline{\Sigma}(M)$$

the Universal Schubert variety of type M
    (resp.

$$\overline{\Sigma}(M, \mathscr{D})$$

the Schubert variety of type M defined by the flag $\mathscr{D}$).

2.3   The Schubert cell decomposition as an orbit decomposition

Remark 2.17 *From the isomorphism*

$$Drap(k^E) \times Drap(k^E)/Gl(k^E) \simeq Relpos(E)$$

it is deduced that the Universal Schubert cell decomposition $Drap(k^E) \times Drap(k^E) = \coprod_{M \in Relpos(E)} \Sigma(M)$ is the decomposition in $Gl(k^E)$-orbits of $Drap(k^E) \times Drap(k^E)$ under the diagonal action.
Fix $\mathscr{D} \in Drap_{\underline{m}}(k^E)$ and denote by $P(\mathscr{D}) = Stab \, \mathscr{D} \subset Gl(k^E)$ the stabilizer of $\mathscr{D}$. From the fact that $Drap_{\underline{m}}(k^E)$ is homogeneous under $Gl(k^E)$, one deduces the isomorphism

$$Drap_{\underline{n}}(k^E)/Stab \, \mathscr{D} \simeq Drap_{\underline{m}}(k^E) \times Drap_{\underline{n}}(k^E)/Gl(k^E),$$

and thus that $Drap(k^E)_{\underline{n}} = \coprod_{M \in Relpos_{(\underline{n},\underline{m})}} \Sigma(M, \mathscr{D})$ is the orbit decomposition of $Drap_{\underline{n}}(k^E)$ under $Stab \, \mathscr{D} \subset Gl(k^E)$.

Assume that E is identified with I_{r+1}, i.e. that E totally ordered and $|E| = r + 1$. Thus one can write I_{r+1} for E. Let $D_{\underline{r}} = (I_1 \subset I_2 \cdots \subset I_r \subset I_{r+1})$ $(resp. D_{\underline{n}} = (I_{n_1} \subset \cdots \subset I_{n_l} \subset I_{r+1}))$ be the maximal length flag defined by the order of I_{r+1} (resp. the flag given by the type $\underline{n}$). Write $\mathscr{D}_{\underline{n}} = k^{D_{\underline{n}}} = \mathscr{D}_{\underline{n}} = (k^{n_1} \subset k^{n_2} \cdots \subset k^{n_l} \subset k^{r+1})$ (resp. $\mathscr{H} = k^H$ for $H \in Grass(I_{r+1})$).

Proposition 2.18 *Given $\mathscr{D} \in Drap(k^{r+1})$ there exists $D' \in Drap(I_{r+1})$ and $\alpha \in Stab\ \mathscr{D}_{\underline{r}}$ with $\alpha \cdot k^{D'} = \mathscr{D}$. Such that D' is unique with this property.*

The proof of the proposition results immediately from 2.17 and the following lemma.

Lemma 2.19 *Let $M \in Relpos_{(\underline{r},\underline{n})}(I_{r+1})$. There exists one and only one combinatorial flag $D' \in Drap_{\underline{n}}(I_{r+1})$ with $M = M(D_{\underline{r}}, D')$.*

Proof *Write $M = (m_\alpha \beta) \in Relpos(I_{r+1})$ as a $1 \times (\lambda + 1)$-matrix consisting of the $(r + 1)$-uples $(\underline{m}_\beta)_{1 \le \beta \le \lambda + 1}$ given by the columns of M. Let In $\underline{m}_\beta = (i_1 < \cdots < i_{l(\beta)} < i_{l(\beta)+1})$ be the set of jump points of $\underline{m}_\beta$ and $\underline{m}'_\beta = (m_{i_1 \beta} < \cdots < m_{i_{l(\beta)} \beta} < m_{i_{l(\beta)+1} \beta})$ where $m_{i_{l(\beta)+1} \beta} = m_{r+1 \beta}$, and*

$$m_{i_1 \beta} = 1, \cdots, m_{i_{l(\beta)} \beta} = l(\beta), m_{i_{l(\beta)+1} \beta} = l(\beta) + 1.$$

Write $H_\beta = \{i_1, \cdots, i_{l(\beta)}, i_{l(\beta)+1}\}$ and

$$D' = (H_1 \cdots \subset H_\lambda \subset I_{r+1}).$$

It is easy to see that D' satisfies $M(D_{\underline{r}}, D') = M$, and that it is the only combinatorial flag satisfying this condition. In fact, it suffices to verify this equality for $\underline{n} = (n < r + 1)$. In this case the result is immediate.

It is important to associate the following combinatorial objects to the Schubert cells (resp. Schubert varieties) corresponding to definitions 2.10, 2.15, and 2.16. It should first be noted that $M(D, D') = M(k^D, k^{D'})$.

Definition 2.20 *Let*

$$\Sigma^{comb}(M) = (\psi_E \times \psi_E)^{-1}(\Sigma(M)) \subset Drap(E) \times Drap(E)$$

(resp.

$$\Sigma^{comb}(M, D) = \psi_E^{-1}(\Sigma(M, D)) \subset Drap(E)).$$

Thus one has

$$\Sigma^{comb}(M) = \{ (D, D') \in Drap(E) \times Drap(E) |\ M(D, D') = M \}$$

(resp.

$$\Sigma^{comb}(M, D) = \{ D' \in Drap(E) |\ M(D, D') = M \}).$$

In other words, $\Sigma^{comb}(M)$ is the fiber of $Drap(E) \times Drap(E) \longrightarrow Relpos(E)$ over M, and $\Sigma^{comb}(M, D)$ is the fiber of the mapping $\Sigma^{comb}(M) \longrightarrow Drap(E)$ induced by the first projection.

Let $D \in Drap_{\underline{m}}(E)$. From definitions 2.20 and 2.15 one obtains the following disjoint unions:

$$Drap(E) = \coprod_{M \in Relpos_{\underline{m}}} \Sigma^{comb}(M, D)$$

(resp.

$$Drap(E) \times Drap(E) = \coprod_{M \in Relpos} \Sigma^{comb}(M)).$$

Define the combinatorial closure of $\Sigma^{comb}(M)$ (resp. $\Sigma^{comb}(M, D)$) as

$$\overline{\Sigma}^{comb}(M) = (\psi_E \times \psi_E)^{-1}(\overline{\Sigma}(M))$$

(resp. $\overline{\Sigma}^{comb}(M, D) = \psi_E^{-1}(\overline{\Sigma}(M, D))$.

From 2.19 one obtains the fact:

PROPOSITION - DEFINITION 2.21 *Let* $E = I_{r+1}$ *and* $D = D_{\underline{r}}$ *be the canonical maximal length flag of* I_{r+1}; *Then* $\Sigma^{comb}(M, D)$ *consists of only one element:*

$$\Sigma^{comb}(M, D) = \{D_M\}.$$

We call D_M **the center of** $\Sigma^{comb}(M, D)$.

From lemma 2.9 it is deduced that given $D \in Drap_{\underline{m}}(E)$ (resp. $D', D'' \in Drap_{\underline{n}}(E)$, and $M \in Relpos_{(\underline{m}, \underline{n})})$ with $M = M(D, D') = M(D, D'')$, there exists $g \in \mathfrak{S}_D$ satisfying: $g(D') = D''$. One deduces that $\Sigma^{comb}(M, D)$ is homogeneous under $\mathfrak{S}_D$. In fact, $\Sigma^{comb}(M, D) \subset Drap_{\underline{n}}(E)$ is the $\mathfrak{S}_D$-orbit defined by the relative position matrix M.

2.4   The Permutation group action on the combinatorial flags set

This section serves as an example of the action of the Weyl group on the apartment of the building of a reductive group (cf. Chapter 8).

Definition 2.22 *Denote by* $Ord(E)$ *the* **set of total orderings** *of* E ($|E| = r + 1$, $\underline{r} = (1 < \cdots < r < r + 1)$). *There is a canonical bijection:*

$$Drap_{\underline{r}}(E) \simeq Ord(E),$$

between the set $Drap(E)_{\underline{r}}$ *of* **maximal length combinatorial flags** *of* E *and the set of total orderings of* E, *defined by* $D \mapsto w_D = (a_1 < ... a_r < a_{r+1})$ *where* $E_1 = \{a_1\}$, $E_{i+1} = E_i \cup \{a_{i+1}\}$, ($1 \leq i \leq r$). *Let* $H \in Drap_n(E)$.

*Denote by $\omega_{D \cap H}$ the **ordering induced** on H by ω_D. We make it explicit by writing $H = (i_1 < \cdots < i_n)$. Observe that the set $Ord(E)$, is left principal homogeneous under $\mathfrak{S}_E$.*

*Let $\tau_D : E \simeq [\![1, |E|]\!]$ be the **order preserving bijection induced by the ordering** ω_D of E, defined by $\tau_D(i) = \inf\{\alpha \in [\![1, |E|]\!] \mid i \in E_\alpha \}$.*

Let $D, D' \in Drap_{\underline{r}}(E)$. There is a unique strictly increasing mapping $w(D, D') : E_{\omega_D} \longrightarrow E_{\omega_{D'}}$ between E ordered respectively by ω_D and by $\omega_{D'}$ satisfying

$$w(D, D')(D) = D'.$$

It follows that $Drap_{\underline{r}}(E)$ is a principal homogeneous set under the natural action of $\mathfrak{S}_E$. A set of generators of the group of bijections $\mathfrak{S}_E$ of E is associated with the ordering $\omega_D = (i_1 < \cdots < i_{|E|})$ of E defined by τ_D, namely

$$S_D = \{(i_1, i_2), \cdots, (i_{|E|-1}, i_{|E|})\}.$$

Define the **length** $l_{S_D}(w)$ of an element $w \in \mathfrak{S}_E$ as the length of a minimal length word in S_D giving w.

Definition 2.23 *Let the notation be as in the proof of proposition 2.19. Let D be a maximal length flag and $(D, D') \in Drap_{\underline{r}}(E) \times Drap_{\underline{n}}(E)$, where $D' = (H_1 \subset \cdots \subset H_\lambda \subset H_{\lambda+1})$. Define a total ordering of E as follows. Write $E = \coprod_{0 \le \beta \le \lambda} (H_{\beta+1} - H_\beta)$, where $H_{\lambda+1} = E$ (resp. $H_0 = \emptyset$). Let $\omega_{(D,D')}$ be the ordering of E given by the above partition and the ordering induced by ω_D on the $(H_{\beta+1} - H_\beta)$'s. Thus by definition*

$$i \underset{\omega_{(D,D')}}{\le} j$$

*if $i \in (H_{\beta+1} - H_\beta)$, and $j \in (H_{\beta'+1} - H_{\beta'})$ for $\beta < \beta'$, or if $i, j \in (H_{\beta+1} - H_\beta)$ and $i \underset{\omega_D}{\le} j$. Denote by $proj_{D'} D$ the maximal length flag of E defined by $\omega_{(D,D')}$ (**the projection of** D **on** D'), according to the general notation to be later introduced in this work in the setting of buildings. Define $w(D, D') = w(D, proj_{D'} D)$ Observe that $D' \subset proj_{D'} D$ (cf. definition 1.23), and $w(D, proj_{D'} D)(D_{\underline{n}}) = D'$ if $\underline{n} = typ(D')$ and $D_{\underline{n}} \subset D$.*

From the general properties of Coxeter groups (cf. Chapter 8) it follows that $w(D, D')$ may be characterized as the minimal length element $w \in \mathfrak{S}_E$ relatively to S_D satisfying $D' \subset w(D)$.

Remark 2.24 *Let $E = I_{r+1}$, and $D = D_{\underline{r}}$. There is the following connection between the center D_M of $\Sigma(M, D) \subset Drap_{\underline{n}}(E)$ and $w(D, D_M)$:*

$$w(D, D_M)(D_{\underline{n}}) = D_M ,$$

where $D_{\underline{n}} = (I_{n_1} \cdots \subset I_{n_\lambda} \subset I_{r+1})$, the sub-flag of $D_{\underline{r}}$ of type $\underline{n}$, and $w(D, D_M)$ is the minimal length element of $\mathfrak{S}_{r+1}$ with this property with respect to the canonical set of generators of $\mathfrak{S}_{r+1}$ $S = \{(12), \cdots, (r\,r+1)\}$.

2.4.1   The Schubert and Bruhat cell decomposition

We may now state the following relation between Schubert and Bruhat cell decompositions (cf. [9], [23], Exp. XXVI)

Proposition 2.25 *Keep the assumptions and notation of 2.23.*

1. The set $Drap_{\underline{n}}(E)$ is homogeneous under the action of $\mathfrak{S}_E = \mathfrak{S}_{r+1}$ of E. Thus there is a canonical bijection:

$$\mathfrak{S}_{r+1}/\mathfrak{S}_{\underline{n}} \simeq Drap_{\underline{n}}(E)$$

(resp.

$$\coprod_{\underline{n} \in typ(E)} \mathfrak{S}_{r+1}/\mathfrak{S}_{\underline{n}} \simeq Drap(E)),$$

defined by $\overline{w} \mapsto w(D_{\underline{n}})$. The relation of inclusion between flags corresponds to the opposite of the relation of inclusion between classes.

2. There is a canonical bijection between the types of relative position between a flag of type $\underline{m}$ and another of type $\underline{m}$ and a set of double classes

$$\mathfrak{S}_{\underline{m}} \backslash \mathfrak{S}_E / \mathfrak{S}_{\underline{n}} \simeq Relpos_{(\underline{m},\underline{n})}(E) \ .$$

As a particular case of 1. It follows that $Drap_{\underline{r}}(E)$ is a principal homogeneous set under $\mathfrak{S}_E$.

Proof   *Let $D' \in Drap_{\underline{n}}(E)$ we have then $w(D_{\underline{r}}, D')(D_{\underline{n}}) = D'$, which proves the homogeneity of $Drap_{\underline{n}}(E)$ under $\mathfrak{S}_E$. As $\mathfrak{S}_{\underline{n}} = Stab\,D_{\underline{n}}$, there is a bijection $\mathfrak{S}_{r+1}/\mathfrak{S}_{\underline{n}} \simeq Drap_{\underline{n}}(E)$, defined by $\overline{w} \mapsto w(D_{\underline{n}})$. The decomposition*

$$Drap_{\underline{n}}(E) = \coprod_{M \in Relpos_{(\underline{r},\underline{n})}(E)} \Sigma^{comb}(M, D_{\underline{r}}) = \coprod_{M \in Relpos_{(\underline{r},\underline{n})}(E)} \{D_M\} \ ,$$

proves that there is a bijection $\mathfrak{S}_E/\mathfrak{S}_{\underline{n}} \simeq Relpos_{(\underline{r},\underline{n})}(E)$ defined by $\overline{w} \mapsto M(D_{\underline{r}}, w(D_{\underline{n}}))$.
On the other hand, the decomposition

$$Drap_{\underline{n}}(E) = \coprod_{M \in Relpos_{(\underline{m},\underline{n})}(E)} \Sigma^{comb}(M, D_{\underline{m}}) \ ,$$

is the $\mathfrak{S}_{\underline{m}}$-orbit decomposition of $\coprod_{M \in Relpos_{(\underline{r},\underline{n})}(E)} \{D_M\}$. So the mapping $\mathfrak{S}_{\underline{m}} \backslash \mathfrak{S}_E / \mathfrak{S}_{\underline{n}} \longrightarrow Relpos_{(\underline{m},\underline{n})}(E)$ defined by $\overline{w} \mapsto M(D_{\underline{m}}, w(D_{\underline{n}}))$ is a bijection. We have thus proved 1. and 2.

Remark 2.26 *Let $M \in Relpos_{(m,n)}$ and $\mathcal{C}_M \subset \mathfrak{S}_{\underline{m}} \backslash \mathfrak{S}_E / \mathfrak{S}_{\underline{n}}$ the double class given by M and $\mathcal{C}'_M \subset \mathfrak{S}_E / \mathfrak{S}_{\underline{n}}$ the corresponding subset of $\mathcal{C}_M$ in $\mathfrak{S}_E / \mathfrak{S}_{\underline{n}}$. There is a bijection $\mathcal{C}'_M \simeq \Sigma^{comb}(M, D)$ defined by $\overline{w} \mapsto \overline{w}(D_{\underline{n}})$.*

Write $\Sigma^{comb}(M, D)$ ("*par abus de langage*") for the image $\psi_E(\Sigma^{comb}(M, D)) \subset Drap(k^E)$.

Proposition 2.27 *The Stab $k^{D_{\underline{m}}}$-orbit decomposition of $Drap_{\underline{n}}(k^E)$ may be re-written as follows:*

$$Drap_{\underline{n}}(k^E) = \coprod_{M \in Relpos_{(\underline{m},\underline{n})}} Stab\ k^{D_{\underline{m}}} \cdot \Sigma^{comb}(M, D_{\underline{m}}) =$$

$$\coprod_{\overline{w} \in \mathfrak{S}_{\underline{m}} \backslash \mathfrak{S}_E / \mathfrak{S}_{\underline{n}}} Stab\ k^{D_{\underline{m}}} \cdot k^{w(D_{\underline{n}})}\ .$$

The natural action of $Gl(k^E)$ on $Drap_{\underline{n}}(k^E)$ allows defining a morphism

$$p_{\underline{n}} : Gl(k^E) \longrightarrow Drap_{\underline{n}}(k^E)$$

by $g \mapsto g \cdot D_{\underline{n}}$. Thus the orbit decomposition in proposition 2.27 lifts to a double class decomposition of $Gl(k^E)$.

Proposition 2.28 *With the above notation one has:*

$$Gl(k^E) = \coprod_{\overline{w} \in \mathfrak{S}_{\underline{m}} \backslash \mathfrak{S}_E / \mathfrak{S}_{\underline{n}}} Stab\ k^{D_{\underline{m}}} \cdot w \cdot Stab\ k^{D_{\underline{n}}}\ .$$

(Bruhat double class decomposition of $Gl(k^E)$).

2.5   *R*-subgroups of the linear group

A combinatorial simplified version of the root decomposition of the Lie algebra of $Gl(k^E)$ (cf. [5], p.185) is given in this section. We introduce a class of subgroups of $Gl(k^E)$ (The R-subgroups) giving rise to natural coordinates for $\Sigma(M, D)$. This class is a particular case of R-subgroups in the setting of reductive groups schemes (cf. [23], Exp. XXII).

Let us identify the Lie algebra $\mathscr{G}l(k^E) = Lie(Gl(k^E))$ with the Lie algebra of $E \times E$-matrices $M_E(k)$ with entries in k. Denote by $(E_{ij})_{(i,j) \in E \times E}$ the canonical basis of $M_E(k) = M_{E \times E}(k)$. The set $(E_{ij})_{(i,j) \in E \times E}$ is a basis of eigenvectors of $M_E(k)$ under the adjoint action by the subgroup $T \subset M_E(k)$ of diagonal matrices.

Definition 2.29 *Let E be a finite set. Write*

$$R(E) = E \times E - \Delta(E) .$$

A subset $\mathscr{R} \subset R(E)$ is **closed** *if:*

$$(i,j) , (j,k) \in \mathscr{R} , i \neq k \longrightarrow (i,k) \in \mathscr{R} \ (cf. [4], Ch. VI).$$

Associate with the closed subset $\mathscr{R}$ of $R(E)$ the k-subspace:

$$\mathscr{G}l_{\mathscr{R}} = Vect((E_{ij})_{(i,j) \in \Delta(E) \cup \mathscr{R}}) \subset M_E(k) .$$

Observe that $[E_{ij}, E_{kl}] = \delta_{jk} E_{il}$ if $i \neq l$. Thus clearly $\mathscr{G}l_{\mathscr{R}}$ is a Lie subalgebra of $M_E(k)$ stable under T. To this subalgebra corresponds a subgroup $Gl_{\mathscr{R}} \subset Gl(k^E)$.

We say that $Gl_{\mathscr{R}}$ **is a R-subgroup of** $Gl(k^E)$. We shall see that there is a canonical bijection between the set of combinatorial flags $Drap(E)$ and a particular class of closed subsets of $R(E)$.

Definition 2.30 *Let $D \in Drap_r(E)$ be a maximal length flag of E. Write*

$$R_D = R(E) \cap Gr(\omega_D) ,$$

where $Gr(\omega_D) \subset E \times E$ denotes the graph of the order relation ω_D. From the definition of ω_D easily results that R_D is a closed subset of $R(E)$. Given $D' \in Drap(E)$ write

$$R_{D'} = \bigcup_{D' \subset D \in Drap_r(E)} R_D .$$

If $D' = (H_1 \cdots \subset H_l \subset H_{l+1})$ ($H_{l+1} = E$), then $E = \coprod_{0 \leq \beta \leq l} (H_{\beta+1} - H_\beta)$.

It is easily seen that:

$$R_{D'} = (\coprod_{0 \leq \beta \leq l} R(H_{\beta+1} - H_\beta) \coprod (\coprod_{1 \leq \beta \leq l} (H_\beta - H_{\beta-1}) \times (E - H_\beta)).$$

Where $H_0 = \emptyset$. Observe that

$$\coprod_{1 \leq \beta \leq l} (H_\beta - H_{\beta-1}) \times (E - H_\beta) = \coprod_{1 \leq \beta \leq l} \coprod_{\beta \leq \alpha \leq l} (H_\beta - H_{\beta-1}) \times (H_{\alpha+1} - H_\alpha) =$$

$$= \coprod_{1 \leq \alpha \leq l} \coprod_{1 \leq \beta \leq \alpha} (H_\beta - H_{\beta-1}) \times (H_{\alpha+1} - H_\alpha) = \coprod_{1 \leq \alpha \leq l} H_\alpha \times (H_{\alpha+1} - H_\alpha).$$

Remark that the subgroup $Stab \ k^D \subset Gl(k^E)$ for $D \in Drap(E)$ is characterized as the set of matrices (a_{ij}) satisfying $(i,j) \notin (R_D \cup \Delta(E) \implies a_{ij} = 0$. On the other hand, *Lie* $(Stab \ k^D) = Vect((E_{ij})_{(i,j) \in (R_D \cup \Delta(E))})$ thus $[E_{ij}, E_{kl}] = \delta_{jk} E_{il} \in Vect((E_{ij})_{(i,j) \in (R(D) \cup \Delta(E))})$ for $E_{ij}, E_{kl} \in Vect((E_{ij})_{(i,j) \in R_D \cup \Delta(E)})$. It follows that R_D is a closed set of the set of roots $R(E)$, and thus that

$$Gl_{R_D} = Stab \ (k^D) .$$

Definition 2.31 *The correspondence $D \mapsto R_D$ gives rise to a mapping $Drap(E) \longrightarrow \mathscr{P}(R(E))$ between the combinatorial flags and a class of closed set of roots. The image of this mapping is called* **the class $P(R(E))$ of parabolic subsets** *of $R(E)$. Thus the parabolic subsets of $R(E)$ are closed subsets and correspond to the stabilizers of the flags adapted to the canonical basis e_E of k^E.*

Thus the set of combinatorial flags may be identified with the parabolics subsets of the set of roots $R(E)$, and a group theoretical description of $Drap(E)$ is obtained. The latter appears as a geometrical realization of the former.

The class of parabolic subsets $P(R(E)) \subset \mathscr{P}(R(E))$ is characterized by:

$$P(R(E)) = \{ \, \mathscr{R} \in \mathscr{P}(R(E)) \mid \mathscr{R} \text{ closed and } R(E) = \mathscr{R} \cup \mathscr{R}^{opp} \, \} \, ,$$

where $\mathscr{R}^{opp} = \{ \, (i,j) \in R(E) \mid (j,i) \in \mathscr{R} \, \}$. It is clear that $\mathscr{R}^{opp}$ is closed if $\mathscr{R}$ is closed. Thus $\mathscr{R}$ is parabolic if and only if $\mathscr{R}^{opp}$ is parabolic (cf. loc. cit.). The parabolic subsets of $R(E)$ are the closed subsets corresponding to the stabilizers of the flags adapted to the canonical basis e_E of k^E. They are naturally indexed by the combinatorial flags $D \in Drap(E)$.

Define two closed subsets of the parabolic set R_D by:

$$R_D^u = R_D - R_D \cap R_D^{opp} \quad (resp. \ R_D^s = R_D \cap R_D^{opp}).$$

Clearly we have:

$$R_D^s = \coprod_{0 \leq \beta \leq l} R_{(H_{\beta+1} - H_\beta)}$$

$$(resp. \ R_D^u = \coprod_{1 \leq \beta \leq l} (H_\beta - H_{\beta-1}) \times (E - H_\beta) = \coprod_{1 \leq \beta \leq l} H_\beta \times (H_{\beta+1} - H_\beta)).$$

Definition 2.32 *Let $D = (H_1 \cdots \subset H_l \subset H_{l+1}) \in Drap(E)$. Write*

$$D^{opp} = (H_l^\perp \cdots \subset H_1^\perp \subset E) \in Drap(E).$$

Proposition 2.33 *The following relations between closed subsets of $R(E)$ hold:*

$$R_D^{opp} = R_{D^{opp}} \ (resp.(R_D^s)^{opp} = R_{D^{opp}}^s \, , \text{ and } (R_D^u)^{opp} = R_{D^{opp}}^u).$$

Proof *Let us prove the last formula. The two others follow easily from this one. We know that:*

$$R_D^u = \coprod_{1 \leq \beta \leq l} (H_\beta - H_{\beta-1}) \times (E - H_\beta) = \coprod_{1 \leq \beta \leq l} H_\beta \times (H_{\beta+1} - H_\beta))$$

Thus

$$R_{D^{opp}}^u = \coprod_{1 \leq \beta \leq l} H_\beta^\perp \times (H_{\beta-1}^\perp - H_\beta^\perp) = \coprod_{1 \leq \beta \leq l} (E - H_\beta) \times (H_\beta - H_{\beta-1}) = (R_D^u)^{opp} \, .$$

Let $U(D) \subset Gl_{R_D^u}$ be the subgroup defined by

$$U(D) = \{(a_{ij}) \in Gl_{R_D^u} \mid a_{ii} = 1 \ (i \in E) \} \,,$$

and observe that:

$$(i,j) \in R_D^u \,, \ (j,k) \in R_D^s \Rightarrow (i,k) \in R_D^u$$

$$(resp. \ (i,j) \in R_D^s \,, \ (j,k) \in R_D^u \Rightarrow (i,k) \in R_D^u \ ).$$

We conclude that the Lie subalgebra $\mathscr{U}(D) = Lie(U(D)) \subset \mathscr{P}(D) = Lie(P(D))$ is a Lie algebra ideal. Thus, the subgroup $U(D)$ is an invariant subgroup of $P(D)$. Let $Gl_{R_D^s} \ltimes U(D)$ denote the semi-direct product defined by the action of $Gl_{R_D^s}$ on $U(D)$ given by $g \mapsto int(g)$. Thus we obtain **Levi's decomposition** of $P(D)$ (cf. [23], Exp. XXVI):

$$P(D) = Gl_{R_D^s} \ltimes U(D) \,.$$

As an example let us determine $U(D)$ for $D = (H \subset E)$. In this case

$$R_D^u = H \times H^\perp,$$

and:

$$(i,j), (k,l) \in H \times H^\perp \Longrightarrow E_{ij} \times E_{kl} = 0 \,.$$

It results that the Lie subalgebra $M_{H \times H^\perp}(k) \subset M_{E \times E}(k)$ is abelian. The elements ν of $M_{H \times H^\perp}(k)$ being considered as endomorphisms of k^E, by imposing $\nu(k^{H^\perp}) = 0$ on ν. Thus $U(D) = \{Id_E + \nu \mid \nu \in M_{H \times H^\perp}(k)\}$ is canonically isomorphic to the vector group $Vect(H \times H^\perp)$ whose underlying k-variety is $\mathbb{A}(M_{H \times H^\perp}(k))$.

Proposition 2.34

1. *The group $U(D^{opp})$ stabilizes the open subset $U_D \subset Grass_{|H|}(k^E)$, and its action is transitive.*

2. *The group $P(D^{opp}) = Stab \ D^{opp}$ acts transitively on U_D. Actually U_D is a $P(D^{opp})$-Schubert cell.*

Proof *By definition $Stab \ D^{opp}$ stabilizes $k^{H^\perp}$. As we have for $g \in Stab \ D^{opp}$*

$$g(S \cap k^{H^\perp}) = 0 \Longleftrightarrow S \cap k^{H^\perp} = 0 \,,$$

we deduce that $rk \ (\pi_H)_S = |H| \Longleftrightarrow rk \ (\pi_H)_{g(S)} = |H|$. A fortiori we obtain that $S \in U_D \Longrightarrow g(S) \in U_D$, for $g \in Stab \ D^{opp}$, by definition of U_D. As $U(D^{opp}) \subset Stab \ D^{opp}$ we conclude that $U(D^{opp})$ stabilizes U_D.

Let us see that the action of $U(D^{opp})$ on U_D is transitive. Given $S, S' \in U_D$, the lifted basis $\tilde{e}_H(S)$ (resp. $\tilde{e}_H(S')$) may be written $\tilde{e}_H(S) = ((e_i, \nu(e_i))_{i \in H}$ (resp. $\tilde{e}_{H'}(S) = ((e_i, \nu'(e_i))_{i \in H})$ with $\nu, \nu' \in M_{H^\perp \times H}(k)$. Then: $(Id_E + (\nu' - \nu)) \in U(D^{opp})$ and $(Id_E + (\nu' - \nu))(S) = S'$. Observe that U_D is the Schubert cell defined by $S \cap k^{H^\perp} = 0$ for $S \in Grass_{|H|}(k^E)$. This achieves the proof of the proposition.

In fact the open set U_D is parametrized by the R-subgroup $U(D^{opp}) \subset Gl(k^E)$.

Proposition 2.35 *The action of $U(D^{opp})$ on U_D is simply transitive.*

Proof *We recall that the canonical coordinates on U_D $D = (H \subset E)$ define an isomorphism $U_D \longrightarrow M_{H^\perp \times H}(k)$ (cf. 1.20) by $S \mapsto m(S)$. Denote by $A(S) \in M_{E \times H}(k)$ the H-normalized matrix associated to $m(S)$. Thus for $g \in U(D^{opp})$ the canonical coordinates of $g(S)$ are obtained by normalizing the product matrix $g \times A(S)$. Observe that if $S_0 = k^H$, the matrix $g \times A(S_0)$ is H-normalized. Thus $U(D^{opp}) \longrightarrow M_{H^\perp \times H}(k)$, defined by $g \mapsto m(g \times A(S_0))$, is an isomorphism of k-varieties. This suffices to prove that the action of $U(D^{opp})$ is simply transitive.*

Remark 2.36 *Consider $Gl(k^H)$ for $H \in Grass(E)$ as a subgroup of $Gl(k^E)$. In fact $g \in Gl(k^H)$ may be extended to an endomorphism of k^E, by defining $g|_{k^{H^\perp}} = Id_{k^{H^\perp}}$.*

Proposition 2.37 *Let $D = (H_1 \cdots \subset H_l \subset H_{l+1})$, and $\underline{n} = typ(D)$.*

1. *$P(D^{opp})$ stabilizes U_D.*

2. *The action of $U(D^{opp})$(resp. $P(D^{opp})$) on U_D is transitive, thus U_D is a $P(D^{opp})$-Schubert cell.*

Proof *Let $\mathscr{D} = (S_1 \cdots \subset S_l \subset S_{l+1}) \in Drap_{\underline{n}}(E)$, by definition 1.11:*

$$\mathscr{D} \in U_D \Longleftrightarrow rk \ (\pi_{H_\beta})_{S_\beta} = |H_\beta| \ ,$$

for all $1 \leqq \beta \leqq l$. From proposition 2.34, in view of $P(D^{opp}) \subset P((H_\beta^\perp \subset E))$, one obtains that $P(D^{opp})$ stabilizes U_D.

Observe that $U(D^{opp})$ contains $U(((H_{\beta+1} - H_\beta) \subset H_{\beta+1})) \subset Gl(k^{H_{\beta+1}})$. From proposition 2.34, by induction on the cardinal $|E|$, there exists $g \in U(D^{opp})$ with $g \cdot \mathscr{D} = k^D$. This proves the transitivity of the action of $P(D^{opp})$ on U_D.

2.5.1   The parabolic subgroups of the linear group

Let $(i, j) \in R(E)$. Denote by $G^{(i,j)} \subset Gl(k^E)$ the subgroup of matrices (a_{kl}) satisfying: $a_{kk} = 1$ (resp. $a_{kl} = 0$ for $(k, l) \notin (\Delta(E) \cup \{(i,j)\})$).
Thus $G^{(i,j)} = \{ Id_{k^E} + \lambda E_{ij} \mid \lambda \in k \}$. It may be noted that if $\mathcal{G}^{(i,j)}$ denotes the Lie algebra of $G^{(i,j)}$, then:

$$\mathcal{G}^{(i,j)} = Vect(E_{ij}),$$

and

$$\mathcal{G}l(k^E) = M_E(k) = \mathcal{T} \oplus (\underset{(i,j) \in R(E)}{\oplus} \mathcal{G}^{(i,j)}),$$

where $\mathcal{T} = Lie(T)$. This direct sum decomposition is given by the diagonalisation of the adjoint action $t \mapsto Ad(t)$ of T on the Lie algebra $\mathcal{G}l(k^E)$. The set $R(E)$ canonically indexes the non trivial characters $(\lambda_{(i,j)})_{(i,j) \in R(E)}$ of T (**The roots defined by** T) in this decomposition. The component $\mathcal{T}$ corresponds to the trivial character of T and is, precisely, the Lie algebra $Lie(T)$ of T. Thus T is a maximal torus of $Gl(k^E)$, and the above direct sum decomposition is the root decomposition of $\mathcal{G}l(k^E)$ under the adjoint action of T.

The following relations hold between the roots:

$$\lambda_{(i,j)} \cdot \lambda_{(j,k)} = \lambda_{(i,k)}$$

(resp.$\lambda_{(i,j)} \cdot \lambda_{(j,i)} = 1$). The product $\lambda_{(i,j)} \cdot \lambda_{(k,l)}$ of the two root characters $\lambda_{(i,j)}$ and $\lambda_{(k,l)}$ is a root character if and only if $j = k$ and $i \neq l$.

Definition 2.38 *A set of roots $\mathscr{S}$ is* **closed** *if:*

$$\lambda, \lambda' \in \mathscr{S} \text{ , and } \lambda \cdot \lambda' \in R(E) \Longrightarrow \lambda \cdot \lambda' \in \mathscr{S} .$$

A set of roots $\mathscr{S}$ is **parabolic** *if $\mathscr{S}$ is closed, and $\mathscr{S} \cup \mathscr{S}^{-1} =$ the set of roots defined by T. Clearly*

$$\mathscr{S} \text{ is closed (resp. parabolic)} \Longleftrightarrow \mathscr{S} = (\lambda_{(i,j)})_{(i,j) \in \mathscr{R}}$$

with $\mathscr{R} \subset R(E)$ closed (resp. parabolic).

If k is an algebraically closed field, then a maximal torus $T \subset Gl(k^E)$ is obtained as a subgroup whose elements stabilize the vectors of a base of k^E. A maximal torus of $Gl(k^E)$ is conjugated to the canonical torus defined by the basis e_E. Thus the root characters defined by the adjoint action of T may be indexed by $R(E)$, namely: $(\lambda_{ij})_{(i,j) \in R(E)}$. Let $(G^{(i,j)})_{(i,j) \in R(E)}$ be the set of root subgroups defined by T.

As a particular case of the general definition of a parabolic subgroup we have:

A subgroup $P \subset Gl(k^E)$ is **parabolic** if:

1) the group $P_{k'}$ obtained by the algebraic closure $k \longrightarrow k'$ contains a maximal torus T;

2) the subset $(G^{(i,j)})_{(i,j) \in \mathscr{R}}$ of the root subgroups contained in $P_{k'}$ is indexed by a parabolic subset $\mathscr{R} \subset R(E)$ and P is generated by the subgroups $(G^{(i,j)})_{(i,j) \in \mathscr{R}}$ and T.

Remark 2.39 *It can be seen later that there is a $Gl(k^E)$-homogeneous k-projective variety $Par(Gl(k^E))$ whose sections with values in a field extension $k \longrightarrow k'$, correspond to the parabolic subgroups of $Gl(k'^E)$). There is natural isomorphism $Drap(k^E) \longrightarrow Par(Gl(k^E))$ defined by $\mathscr{D} \mapsto Stab \mathscr{D}$. $typ(P) = typ(\mathscr{D})$ should be written if the parabolic subgroup P corresponds to the flag $\mathscr{D}$. $Par_{\underline{n}}(Gl(k^E))$ denotes the variety of parabolic subgroups of type $\underline{n}$.*

A parabolic subgroup P of $Gl(k^E)$ is its own normalizer $P = Norm_P$, as results from the following proposition.

Proposition 2.40 $Norm_{P(D)} = P(D)$ *is available.*

Proof *It may be supposed that k is algebraically closed. Let $e_{D_{\underline{n}}} = \underset{1 \leq \beta \leq l}{\otimes} \wedge$ e_{H_β} (cf. 1.12), and $L = Vect(e_{D_{\underline{n}}})$. The parabolic subgroup $P(D_{\underline{n}})$ is the stability group of L. Clearly $g(e_{D_{\underline{n}}}) = \lambda \cdot e_{D_{\underline{n}}}$ ($\lambda \in k$) implies that $g(\wedge e_{H_\beta}) = \lambda_\beta \cdot \wedge e_{H_\beta}$ ($\lambda_\beta \in k$) and thus, that $g(k^{H_\beta}) \subset k^{H_\beta}$. From $P(D_{\underline{n}}) = Stab\ D_{\underline{n}} = \underset{1 \leq \beta \leq l}{\cap} Stab\ k^{H_\beta}$ follows that $g \in P(D_{\underline{n}})$. It is concluded that $g(L) = L \Longleftrightarrow g \in P(D_{\underline{n}})$.*

Consider the representation of $V_{e_{D_{\underline{n}}}}$ of $Gl(k^E)$ generated by $e_{D_{\underline{n}}}$. It is well-known that there is only one 1-dimensional subspace L of $V_{e_{D_{\underline{n}}}}$ stabilized by $P(D_{\underline{n}})$, namely $L = Vect(e_{D_{\underline{n}}})$, i.e. the weight vectors for $P(D_{\underline{n}})$ form a 1-dimensional subspace of $V_{e_{D_{\underline{n}}}}$. Let $\Gamma \in Norm_{P(D_{\underline{n}})}$. Then

$$(\forall g \in P(D_{\underline{n}}))\ \Gamma^{-1} \cdot g \cdot \Gamma \in P(D_{\underline{n}}) \Rightarrow g(\Gamma \cdot e_{D_{\underline{n}}}) = \Gamma(g' \cdot e_{D_{\underline{n}}}) = \lambda_{g'}(\Gamma \cdot e_{D_{\underline{n}}}) \ ,$$

where $g' \in P(D_{\underline{n}})$. Thus $P(D_{\underline{n}})$ stabilizes $L' = Vect(\Gamma \cdot e_{D_{\underline{n}}})$ and it may be concluded $L' = L$, that is to say $\Gamma \cdot e_{D_{\underline{n}}} = \lambda \cdot e_{D_{\underline{n}}}$ ($\lambda \in k$). It may be deduced that $\Gamma \in P(D_{\underline{n}})$. This achieves the proof.

2.5.2 Unipotent R-subgroups

By definition a unipotent R-group G is a subgroup of the form $G = Gl_R \cap U(D)$, where $R \subset R_D$ is a closed subset, with D a combinatorial maximal length flag. Then G is the unipotent radical of a R-subgroup of a $P(D)$ with D of maximal length. Given $D' \in Drap(E)$, if $D' \subset D$ is a maximal length flag, we have $R_{D'}^u \subset R_D^u$. Thus $U(D')$ is a unipotent R-group.

A unipotent R-subgroup G generated by a family of subgroups $(G^{(i,j)})_{(i,j) \in R \subset R_D}$ with R is a closed subset. The following proposition is a consequence of the general theory of unipotent k-groups in $Gl(k^E)$ (cf. [10], §II).

Proposition 2.41 *We assume the notation above. The product in $Gl(k^E)$ induces an isomorphism of k-varieties:*

$$\prod_{(i,j) \in R} G^{(i,j)} \simeq G \subset Gl(k^E).$$

In view of

$$R_{D^{opp}}^u = \prod_{1 \leq \beta \leq l} (H_{\beta+1} - H_\beta) \times H_\beta \ ,$$

proposition 2.41 implies the product decomposition:

$$\prod_{l \geq \beta \geq 1} U(((H_{\beta+1} - H_\beta) \subset H_{\beta+1})) \simeq U(D^{opp}) \,,$$

the isomorphism being given by the product in $Gl(k^E)$.

There is an important corollary of this decomposition. One has defined the canonical coordinates in U_D (cf. definition 1.20). In fact, this parametrization corresponds to the parametrization of U_D in terms of the principal action of $U(D^{opp})$ on U_D (cf. proposition 2.37).

One may embed $M_{(H_{\beta+1}-H_\beta) \times H_\beta}$ in $M_{E \times E}$ by completing a $(H_{\beta+1}-H_\beta) \times H_\beta$-matrix m with zeros in a $E \times E$-matrix m', which one also denotes by m. Write $u(m) = 1_E + m \in U(((H_{\beta+1}-H_\beta) \subset H_{\beta+1}))$. There is an isomorphism:

$$u_D : \mathbb{A}_D = \prod_{1 \leq \beta \leq l} M_{(H_{\beta+1}-H_\beta) \times H_\beta}(k) \longrightarrow \prod_{1 \leq \beta \leq l} U(((H_{\beta+1} - H_\beta) \subset H_\beta))$$

$$\simeq U(D^{opp}) \,,$$

defined by $u_D : (m_\beta) \mapsto u_D((m_\beta)) = u(m_l) \times \cdots \times u(m_1)$.

Let $(m_D(S_\beta))$ denote the canonical coordinates of $\mathscr{D} = (S_1 \cdots \subset S_l \subset k^E) \in U_D$. The element $u_D((m_D(S_\beta))) \in U(D^{opp})$ satisfies:

$$m_D(u_D((m_D(S_\beta))) \cdot k^{H_\beta}) = m_D(S_\beta) \,.$$

This means that if one H_β-normalizes the matrix whose columns are the columns of $u_D((m_D(S_\beta)))$ indexed by H_β, one obtains a $E \times H_\beta$-matrix whose $(H_{\beta+1} - H_\beta) \times H_\beta$-submatrix is given by $m_D(S_\beta))$. Thus it follows that:

$$u_D((m_D(S_\beta))) \cdot k^D = \mathscr{D} \,.$$

The above formula is a particular case of the following one. Let $U(D^{opp}) \longrightarrow U_D$ be defined by $u \mapsto u \cdot k^D$. This morphism is, in fact, an isomorphism, which can be seen by proving that the composed morphism

$$\mathbb{A}_D \simeq U(D^{opp}) \longrightarrow U_D \simeq \mathbb{A}_D$$

reduces to the identity morphism of $\mathbb{A}_D$. Given $(m_\beta) \in \mathbb{A}_D$, write $\mathscr{D} = (S_1 \cdots \subset S_l \subset k^E) = u_D((m_\beta)) \cdot k^D$. Then

$$m_D(u_D((m_\beta)) \cdot k^{H_\beta}) = m_\beta \,,$$

i.e. $m_D(S_\beta) = m_\beta$, which proves what has been asserted.

Proposition 2.42   *1) The morphism $U(D^{opp}) \longrightarrow U_D$ defined by $u \mapsto u \cdot k^D$ is a k-isomorphism ;*

2) *the action of $U(D^{opp})$ on U_D is principal and, thus, induces an isomorphism of varieties $U(D^{opp}) \simeq U_D$ defined by $g \mapsto g \cdot k^D$ (**The $U(D^{opp})$-parametrization of U_D**).*

The reciprocal of the isomorphism $U(D^{opp}) \longrightarrow U_D$ is given by $\mathscr{D} = (S_1 \cdots \subset S_l \subset k^E) \mapsto u(m_D(S_l)) \times u(m_D(S_{l-1})) \cdots u(m_D(S_1))$.

Definition 2.43 *Let $D = (E_1 \cdots \subset E_r \subset E)$ ($|E| = r + 1$) be a maximal length flag (resp. $D' = (H_1 \cdots \subset H_l \subset E) \in Drap_{\underline{n}}(k^E)$), ω_D the order associated with D, and $\omega_{D \cap H_\beta}$ the induced order on H_β. Denote by e_D (resp. $e_{D \cap H_\beta}$) the corresponding ordered basis of k^E (resp. k^{H_β}), and by $D'_{D \cap H_\beta} \in Drap(H_\beta)$ the maximal length flag defined by $e_{D \cap H_\beta}$.*

It may be observed that with the notation of 1.4 one has $D'_{D \cap H_\beta} = \varphi(D \cap H_\beta)$, i.e. $D'_{D \cap H_\beta}$ is the flag of H_β defined by the chain $D \cap H_\beta$.

Corollary 2.44 *There is a section $(\tilde{e}_{D \cap H_\beta})_{1 \leq \beta \leq l}$ on U_D of the product fiber bundle*

$$\prod_{1 \leq \beta \leq l} Stief(\xi_{n_\beta}) \longrightarrow Drap_{\underline{n}}(k^E) ,$$

where $\xi_{\underline{n}} = (\xi_{n_1} \subset \cdots \subset \xi_{n_l} \subset \mathscr{O}^E_{Drap_{\underline{n}}(k^E)})$, defined by $(\tilde{e}_{D \cap H_\beta}) : \mathscr{D} = (S_1 \cdots \subset S_l \subset k^E) \mapsto (u_D((m_D(S_\beta))) \cdot e_{D \cap H_\beta})_{1 \leq \beta \leq l}$. The following relation holds $\tilde{e}_{D \cap H_\beta}(\mathscr{D}) = (\pi_{H_\beta})^{-1}_{S_\beta}(e_{D \cap H_\beta})$.

Denote by $\mathscr{D}_{\omega_{D \cap H_\beta}}$ the maximal length flag section of $Drap(\xi_{n_\beta})$ defined by $\tilde{e}_{\omega_{D \cap H_\beta}}$.

Remark 2.45 *The total ordering ω_D of E is coherent with D' if*

$$i \in H_\beta \implies \{i' \in E \mid i' \underset{\omega_D}{\leq} i\} \subset H_\beta .$$

This is equivalent supposing $D' \subset D$, in which case the section $(\tilde{e}_{D \cap H_\beta})$ gives a section of $Stief(\xi_{\underline{n}}) \subset \prod_{1 \leq \beta \leq l} Stief(\xi_{n_\beta})$.

2.5.3 Closed set of roots defined by a couple of flags

In this section we establish the relationship between the parametrization by normalization of a Schubert cell, with that one given by an unipotent R-subgroup.

Definition 2.46 *Let D be a maximal length flag and $D' \in Drap(E)$. With the couple (D, D') one associates the closed subset $R(D, D') \subset R^u_D$ given by*

$$R(D, D') = \{(i, j) \in R^u_D \mid (i, j) \notin R_{D'}\} .$$

Define the unipotent R-subgroup

$$U(D, D') = Gl_{R(D,D')} \cap U(D) \subset U(D) \ .$$

The set $R(D, D')$ is closed. If D' is of maximal length one has, then $R(D, D') = R_D \cap R_{D'^{opp}}$ is closed. From the definition formula $R_{D'} = \cup R_{D''}$, where D'' runs on the set of maximal length flags containing D' as a subflag, it follows that $R(D, D') = \cap R(D, D'')$ is closed in general.

Remark 2.47 *Let $D' = (H \subset E)$. From $R(E) - R(D') = \cap(R(E) - R(D''))$ we deduce $R(E) - R(D') = \{ (j, i) \in R(E) \mid j \in H^\perp, i \in H \}$. Thus, $R(D, D') = R_D \cap \{ (j, i) \in R(E) \mid j \in H^\perp, i \in H \} = \{ j \in H^\perp \mid j \underset{\omega_D}{<} i \}$. Hence, $R(D, D') = \coprod\limits_{i \in H} H^\perp \cap E_{\tau_D(i)} \times \{i\} \subset H^\perp \times H$.*

Definition 2.48 *Let $|E| = r + 1$, $D = (E_1 \cdots \subset E_r \subset E)$, $D' = (H_1 \cdots \subset H_l \subset H_{l+1}) \in Drap_{\underline{n}}(E)$ $(\underline{n} = (n_1 < \cdots < n_l < n_{l+1}))$. Suppose E ordered by ω_D. Let $\tau_D : E \simeq [\![1, |E|]\!]$ be the order preserving bijection given by the ordering ω_D of E, i.e. $\tau_D(i) = \inf\{\alpha \in [\![1, |E|]\!] \mid i \in E_\alpha \}$.*

Write:

1. *$H[\ \leqq i\ ] = \{x \in H \mid x \leqq i \}$;*

2. *$R_\beta(D, D') = \coprod\limits_{i \in H_\beta} (H_{\beta+1}[\ \leqq i\ ] - H_\beta[\ \leqq i\ ]) \times \{i\} \subset (H_{\beta+1} - H_\beta) \times H_\beta$.*
   The set $R_\beta(D, D')$ is closed.

Proposition 2.49 $R(D, D') = \coprod\limits_{1 \leqq \beta \leqq l} R_\beta(D, D') \subset R_{D'^{opp}}^u = \coprod\limits_{1 \leqq \beta \leqq l} (H_{\beta+1} - H_\beta) \times H_\beta$.

Proof *One knows that:*

$$R_{D'} = \left(\coprod\limits_{0 \leq \beta \leq l} R_{(H_{\beta+1} - H_\beta)} \right) \coprod \left(\coprod\limits_{1 \leq \beta \leq l} H_\beta \times (H_{\beta+1} - H_\beta) \right) ,$$

$R_{D'}^{opp} = R_{D'^{opp}}$ *thus,* $R(E) = R_{D'} \cup R_{D'^{opp}}$. *Then* $R(E) - R_{D'} = \coprod\limits_{1 \leqq \beta \leqq l} (H_{\beta+1} - H_\beta) \times H_\beta$, *and finally*

$$R(D, D') = R_D \cap (R(E) - R_{D'}) = \coprod\limits_{1 \leqq \beta \leqq l} \left(\coprod\limits_{i \in H_\beta} \{ (j, i) \in (H_{\beta+1} - H_\beta) \times \{i\} \mid j \underset{\omega}{<} i \} \right) =$$

$$= \coprod\limits_{1 \leqq \beta \leqq l} \left(\coprod\limits_{i \in H_\beta} E_{\tau_D(i)} \cap (H_{\beta+1} - H_\beta) \times \{i\} \right) = \coprod\limits_{1 \leqq \beta \leqq l} R_\beta(D, D') \ .$$

Now suppose that $D' = (H \subset E)$. Then one has:

1. $E[\leq i] = E_{\tau_D(i)}$ (by definition of τ_D);

2. $E[\leq i] - H[\leq i] = H^\perp \cap E_{\tau_D(i)}$;

3. $R(D, D') = \coprod_{i \in H} H^\perp \cap E_{\tau_D(i)} \times \{i\} \subset H^\perp \times H$
   (**The set $R(D, D')$ for $D' = (H \subset E)$**) ;

4. $U(D, D') \subset U((H^\perp \subset E)) = U(D'^{opp})$.

Let

$$M_{R(D,D')}(k) = \{ \, m = (a_{ij}) \in M_{H^\perp \times H}(k) \mid (i,j) \notin R(D, D') \Rightarrow a_{ij} = 0 \, \} \, .$$

One has

$$U(D, D') = \{ \, 1_E + m \in U(D'^{opp}) \mid m \in M_{R(D,D')}(k) \} \, .$$

The following proposition explains how the $U(D'^{opp})$-parametrization of $U_{D'}$ induces a $U(D, D')$-parametrization of $\Sigma(M, D') \hookrightarrow U_D$.

The relative position matrix $M = M(D, D')$ is characterized by the vector $(|E_j \cap H|)_{1 \leq j \leq r+1}$. Let $H = (i_1 < \cdots < i_n)$. Then one has that $|E_{\tau_D(i_\alpha)} \cap H| = \alpha$, and that $(\tau_D(i_\alpha))_{1 \leq \alpha \leq n}$ are the "increasing points" of the chain $D \cap H$. Thus $S \in \Sigma(M, D) \Leftrightarrow \dim k^{E_{\tau_D(i_\alpha)}} \cap S = \alpha$, and $(\tau_D(i_\alpha))_{1 \leq \alpha \leq n}$ are the increasing points of the chain $k^D \cap S$. A basis $f_S = (f_{i_\alpha})_{1 \leq \alpha \leq n}$ of S may be obtained satisfying $k^{E_{\tau_D(i_\alpha)}} \cap S = Vect((f_{i_1}, \cdots, f_{i_\alpha}))$. Let $M(f_S) = (\xi_{ji_\alpha})$ denote the $E \times H$-matrix whose column vectors are given by the components of the f_{i_α}'s.

The matrix (ξ_{ji_α}) satisfies the equations $\xi_{ji_\alpha} = 0$ for $i_\alpha < j$, and $\xi_{ji_\alpha} \neq 0$ for $j = i_\alpha$. One deduces that a H-normalized matrix may be obtained by right multiplying (ξ_{ji_α}) by a $n \times n$-lower triangular matrix B. Hence the lifted basis $\tilde{e}_H = (\pi_H)_S^{-1}(e_H) \, (e_H = (e_{i_\alpha}))$, may be uniquely written as:

$$\tilde{e}_{i_\alpha} = e_{i_\alpha} + \sum_{(j, i_\alpha) \in (H^\perp \cap E_{\tau_D(i_\alpha)}) \times \{i_\alpha\}} \xi_{ji_\alpha} e_j \ (i_\alpha \in H) \, .$$

Thus, $m_{D'}(S) = (\xi_{ji_\alpha}) \in M_{R(D,D')}(k)$ and $u(m_{D'}(S)) \in U(D, D')$. Thus it has been proved:

Proposition 2.50 *Keep the above notation. Let $M = M(D, D')$. Remark that D' is the center of $\Sigma(M, D)$, i.e. $D' = D_M$.*

1. *$\Sigma(M, D) \subset U_{D'}$;*

2. *the morphism $U(D, D') \longrightarrow \Sigma(M, D)$ defined by $g \mapsto g \cdot k^H$ is an isomorphism of k-varieties;*

3. *$\Sigma(M, D)$ is principal homogeneous under $U(D, D')$.*

(The $U(D, D')$-parametrization of a Schubert cell $\Sigma(M, D) \subset Grass(K^E)$)

Keep the notation of 1.3.2, with $D' = D$, and the notation of corollary 2.44 as well. The following proposition, which we give without proof, explains how $\Sigma(M(\beta), D)$ is decomposed in terms of the fibrations $(U_{D^\beta} \to U_{D^{\beta+1}})_{1 \leq \beta \leq l}$.

Proposition 2.51 *Let $M(\beta) = M(D, D^\beta)$. Clearly $M(1) = M(D, D')$. It may be assumed that $\Sigma(M(\beta), D) \subset U_{D^\beta}$, and that $\Sigma(M, D) = \Sigma(M(1), D)$ decomposes in a sequence of locally trivial fibrations $(\Sigma(M(\beta), D) \to \Sigma(M(\beta + 1), D))_{1 \leq \beta \leq l}$ with typical fiber $\Sigma(M_\beta, D'_{D \cap H_\beta}) \subset Grass_{n_\beta}(k^{n_{\beta+1}})$, where $M_\beta = M(D'_{D \cap H_\beta}, (H_\beta \subset H_{\beta+1}))$ and $D'_{D \cap H_\beta}$ denotes the flag given by $D \cap H_\beta$. (This decomposition being induced by $(U_{D^\beta} \to U_{D^{\beta+1}})_{1 \leq \beta \leq l}$).*

The restriction $(\xi_{n_{\beta+1}})_\Sigma$ of the $\mathcal{O}_{Drap_{n^{\beta+1}}(k^E)}$-module $\xi_{n_{\beta+1}}$ to $\Sigma = \Sigma(M(\beta + 1), D) \subset U_{\beta+1} \subset Drap_{n^{\beta+1}}(k^E)$ is endowed with a maximal length flag $\mathcal{D}_{D \cap H_{\beta+1}}$ defining a section of $Drap(\xi_{n_{\beta+1}})_\Sigma$ (cf. corollary 2.44). The sections of the fiber bundle $\Sigma(M(\beta), D) \longrightarrow \Sigma(M(\beta + 1), D)$ given by the sections $\mathcal{S}$ of $Grass_{n_\beta}(\xi_{n_{\beta+1}})_\Sigma$ are satisfying:

1) $\mathcal{S}$ intersects the submodules of $\mathcal{D}_{D \cap H_{\beta+1}}$ in a locally free direct factor;

2) $M(\mathcal{D}_{D \cap H_{\beta+1}}, (\mathcal{S} \subset \xi_{n_{\beta+1}})) = M_\beta$.

Remark 2.52 *In fact, $(\xi_{n_{\beta+1}})_\Sigma$ is endowed with the ordered basis $\tilde{e}_{D \cap H_{\beta+1}}$ defining the flag $\mathcal{D}_{D \cap H_{\beta+1}}$, and a section $\mathcal{H}_\beta$ of $Grass_{n_\beta}(\xi_{n_{\beta+1}})_\Sigma$. Thus, the fiber bundle $\Sigma(M(\beta), D) \longrightarrow \Sigma(M(\beta + 1), D)$ is isomorphic as a $\Sigma(M(\beta + 1), D)$-scheme to the **relative Schubert cell** $\Sigma(M_\beta, \mathcal{D}_{D \cap H_{\beta+1}})$, whose center is given by $(\mathcal{H}_\beta \subset \xi_{n_{\beta+1}})$, i.e. $M(\mathcal{D}_{D \cap H_{\beta+1}}, (\mathcal{H}_\beta \subset \xi_{n_{\beta+1}})) = M_\beta$ and $\mathcal{H}_\beta$ is adapted to $\tilde{e}_{D \cap H_{\beta+1}}$.*

One has:

1. $R_\beta(D, D') = R(D'_{D \cap H_{\beta+1}}, (H_\beta \subset H_{\beta+1}))$
$= \coprod_{i \in H_\beta} (E_{\tau_D(i)} \cap (H_{\beta+1} - H_\beta)) \times \{i\} \subset (H_{\beta+1} - H_\beta) \times H_\beta$.

2. $U_\beta(D, D') = U(D'_{D \cap H_{\beta+1}}, (H_\beta \subset H_{\beta+1})) \subset Gl(k^{H_{\beta+1}})$;

3. $U_\beta(D'^{opp}) = U(((H_{\beta+1} - H_\beta) \subset H_{\beta+1}))$.

To the decomposition of $R(D, D')$:

$$R(D, D') = \coprod_{1 \leq \beta \leq l} R_\beta(D, D') \subset \coprod_{1 \leq \beta \leq l} (H_{\beta+1} - H_\beta) \times H_\beta = R^u_{D'^{opp}} \ ,$$

corresponds to the product decomposition:

$$U(D, D') = \prod_{l \leq \beta \leq 1} U_\beta(D, D') \subset \prod_{l \leq \beta \leq 1} U_\beta(D'^{opp}) = U(D'^{opp}) \ .$$

Denote by $\mathbb{A}_{R(D,D')} = \mathbb{A}(\prod_{1\leq\beta\leq l} M_{R_\beta(D,D')}) \subset \mathbb{A}(\prod_{1\leq\beta\leq l} M_{(H_{\beta+1}-H_\beta)\times H_\beta}(k))$, where $M_{R_\beta(D,D')} = \{ m = (a_{ij}) \in M_{(H_{\beta+1}-H_\beta)\times H_\beta}(k) \mid (i,j) \notin R_\beta(D,D') \Rightarrow a_{ij} = 0\}$, the affine subspace being defined by $\prod_{1\leq\beta\leq l} M_{R_\beta(D,D')}$.

Proposition 2.53 *Let* $D' = (H_1 \subset \cdots H_l \subset H_{l+1}) \in Drap(E)$ *and* $M = M(D,D')$. *It is noted that* D' *is the center of* $\Sigma(M,D)$, *i.e.* $D' = D_M$. *The restriction morphism given by the embedding:*

$$\mathbb{A}(\prod_{l\geq\beta\geq 1} M_{R_\beta(D,D')}(k)) \hookrightarrow \mathbb{A}(\prod_{l\geq\beta\geq 1} M_{(H_{\beta+1}-H_\beta)\times H_\beta}(k)) \simeq U(D'^{opp}) \longrightarrow U_{D'}$$

induces an isomorphism

$$\mathbb{A}(\prod_{l\geq\beta\geq 1} M_{R_\beta(D,D')}(k)) \simeq U(D,D') \simeq \Sigma(M,D) \ (cf. \ proposition \ 2.42).$$

Thus one has:

1. $\Sigma(M,D) \subset U_{D'}$;

2. *the morphism* $U(D,D') \longrightarrow \Sigma(M,D)$ *defined by* $g \mapsto g \cdot k^H$ *is an iso-morphism of* k-*varieties;*

3. $\Sigma(M,D)$ *is principal homogeneous under* $U(D,D')$.

(The $U(D,D')$-parametrization of a Schubert cell $\Sigma(M,D) \subset Grass(k^E)$)

Proof *Let us prove the first assertion by induction on* l. *From 2.50, it results for* $l = 1$. *Let* $1 \leq \beta$. *Suppose* $\prod_{l\geq\beta'\geq\beta+1} M_{R_{\beta'}(D,D'^{\beta+1})}(k) \simeq \Sigma(M(\beta + 1), D)$ *(cf. proposition 2.51). Given a section* $(S_\beta \subset S_{\beta+1} \cdots S_l \subset S_{l+1})$ *of* $\Sigma(M(\beta), D) \to \Sigma(M(\beta + 1), D)$ *we may write* $u((m_{\beta'})_{l\geq\beta'\geq\beta+1}) \cdot k^{D'^{\beta+1}} = (S_{\beta+1} \cdots S_l \subset S_{l+1})$ *with* $(m_{\beta'})_{l\geq\beta'\geq\beta+1} \in \prod_{l\geq\beta'\geq\beta+1} M_{R_{\beta'}(D,D'^{\beta+1})}(k)$, *and, thus,* $u((m_{\beta'})_{l\geq\beta'\geq\beta+1}) \times u(m_{D'^\beta}(S_\beta)) \cdot k^{H_\beta} = S_\beta$. *As* S_β *defines a section of* $\Sigma(M_\beta, \mathscr{D}_{D\cap H_{\beta+1}})$ *one obtains* $u(m_{D'^\beta}(S_\beta)) \in M_{R_\beta(D,D'^\beta)}(k)$. *We conclude that* $u((m_{\beta'})_{l\geq\beta'\geq\beta+1}) \times u(m_{D'^\beta}(S_\beta)) \cdot k^{D'^\beta} = (S_\beta \subset S_{\beta+1} \cdots S_l \subset S_{l+1})$. *This achieves the proof of the first assertion. The other statements follow easily from the first one.*

2.6   The linear group parabolics variety

Denote by $q_{\underline{n}} : Gl(k^E) \longrightarrow Gl(k^E)/P(D_{\underline{n}})$ the quotient morphism. The morphism $p_{\underline{n}} : Gl(k^E) \longrightarrow Drap_{\underline{n}}(k^E)$ defined by $x \mapsto x \cdot k^{D_{\underline{n}}}$ factors as: $p_{\underline{n}} = \tilde{p}_{\underline{n}} \circ q_{\underline{n}}$. We prove here that

Proposition 2.54 *The induced morphism* $\tilde{p}_{\underline{n}} : Gl(k^E)/P(D_{\underline{n}}) \longrightarrow Drap_{\underline{n}}(k^E)$ *is an isomorphism of k-varieties.*

Given $g \in Gl(k^E)$ denote by $\Delta_{n_\beta}(g)$ the n_β-th diagonal minor of g. Let

$$\mathcal{U}_{\underline{n}} = \{\, g \in Gl(k^E) \mid \Delta_{n_\beta}(g) \neq 0 \ (1 \leq \beta \leq l) \,\} \, .$$

Clearly $\mathcal{U}_{\underline{n}}$ is an open subvariety of $Gl(k^E)$.

Lemma 2.55   *1. If* $g \in \mathcal{U}_{\underline{n}}$, *g may be written uniquely as* $g = u \cdot g'$, *with* $u \in U(D_{\underline{n}}^{opp})$ *and* $g' \in P(D_{\underline{n}})$;

2. $(w \cdot \mathcal{U}_{\underline{n}})_{\overline{w} \in \mathfrak{S}_E/\mathfrak{S}_{\underline{n}}}$ *is an open covering of* $Gl(k^E)$.

Proof   *The first statement is easily verified. Let us prove the second one. It is clear that for* $w \in \mathfrak{S}$, $w \cdot \mathcal{U}_{\underline{n}}$ *is an open subvariety of* $Gl(k^E)$. *From 5.21 it follows that* $(w \cdot \mathcal{U}_{\underline{n}})_{\overline{w} \in \mathfrak{S}_E/\mathfrak{S}_{\underline{n}}}$ *is an open covering of* $Gl(k^E)$.

Proof of 2.54
    Lemma 2.55 shows that the restriction $(\tilde{p}_{\underline{n}})_{q_{\underline{n}}(\mathcal{U}_{\underline{n}})}$ *induces an isomorphism* $q_{\underline{n}}(\mathcal{U}_{\underline{n}}) \simeq U(D_{\underline{n}}^{opp}) \cdot k^{D_{\underline{n}}}$, *between an open subvariety of the integral variety* $Gl(k^E)/P(D_{\underline{n}})$ *and an open subvariety of the integral variety* $Drap_{\underline{n}}(k^E)$. *We conclude that* $p_{\underline{n}}$ *is a birational morphism, and that* $\forall w \in \mathfrak{S}_E$:

$$(\tilde{p}_{\underline{n}})_{q_{\underline{n}}(w(\mathcal{U}_{\underline{n}}))} : q_{\underline{n}}(w(\mathcal{U}_{\underline{n}})) \longrightarrow U(w(D_{\underline{n}})^{opp}) \cdot k^{w(D_{\underline{n}})}$$

is an isomorphism too. This implies that $\tilde{p}_{\underline{n}}$ *is a quasi-finite morphism. Thus we may apply Zariski's main theorem (cf. [9], p.6) and deduce that* $\tilde{p}_{\underline{n}}$ *is an isomorphism of k-varieties.*

Definition 2.56 *A parabolic subgroup P of the A-group* $Gl(A^E)$ *may be defined following the pattern of definition 2.38 and thus also the functor*

$Par(Gl(k^E))(A) = $ *the parabolic subgroups of* $Gl(A^E)$ *(A a k − algebra),*

may be defined without any reference to $Drap(k^E)$. *By proposition 2.40 we know that* $P(D_{\underline{n}})$ *is equal to its own normalizer and thus that* $Gl(k^E)/P(D_{\underline{n}}) \longrightarrow Par(Gl(k^E))$ *induced by* $g \mapsto int(g)(P(D_{\underline{n}}))$ *is a functorial isomorphism. We conclude that* $Par(Gl(k^E))$ *is a representable functor isomorphic to the homogeneous space* $Gl(k^E)/P(D_{\underline{n}})$ *(cf. [23], Exp. XXVI, Corollaire 3.6).*

From proposition 2.54 it follows

Corollary 2.57 *The morphism $Drap_{\underline{n}}(k^E) \longrightarrow Par(Gl(k^E))$ defined in 2.39 is an isomorphism.*

2.7 Big cell defined by a maximal length flag

Taking into account the cell decomposition $Drap(k^E)_{\underline{n}} = \coprod_{M \in Relpos_{(\underline{m},\underline{n})}} \Sigma(M, D)$ with $D \in Drap_{\underline{m}}(E)$, which is also the $P(D)$-orbit decomposition, and the isomorphism $Gl(k^E)/P(D') \simeq Drap_{\underline{n}}(k^E)$ $(D' \in Drap_{\underline{n}}(E))$ implying that $Drap_{\underline{n}}(k^E)$ is an irreducible k-variety, one can state the following

Definition 2.58 *There is only one k^D-Schubert cell which is an open sub-variety of $Drap_{\underline{n}}(k^E)$, namely the cell containing the* **generic point** *of $Drap_{\underline{n}}(k^E)$. Denote by $M^{bc}_{(\underline{m},\underline{n})} \in Relpos_{(\underline{m},\underline{n})}$ the relative position matrix defining the open cell of the k^D-Schubert cell decomposition. One calls $\Sigma(M^{bc}_{(\underline{m},\underline{n})}, D)$* **the big (open) cell** *of $Drap_{\underline{n}}(k^E)$* **defined by** *$D$. One can say that a couple of flags (D, D') satisfying $M^{bc}_{(\underline{m},\underline{n})} = M(D, D')$ is in* **transversal position** *(cf. loc. cit., Exp. XXVI, 4.). By definition two flags D and D' are* **incident** *(cf. [50], p.2) if there exists a flag D'' satisfying: $D, D' \subset D''$.*

Proposition 2.59 *A couple of flags (D, D') is in transversal position if and only if D and $(D')^{opp}$ are incident.*

Proof *The condition $D, (D')^{opp} \subset D''$ implies that $U((D')^{opp}) \subset P(D'')$. Thus*
$$U_{D'} = U((D')^{opp}) \cdot k^{D'} \subset P(D'') \cdot k^{D'} \subset P(D) \cdot k^{D'}, \text{ i.e. } \Sigma(M(D, D'), D) = P(D) \cdot k^{D'} \text{ is an open subvariety and the couple } (D, D') \text{ is in transversal position.}$$
Suppose D is of maximal length and that $\Sigma(M(D, D'), D) \subset U_D \subset Drap_{\underline{n}}(k^E)$ is an open subvariety. On the other hand, one knows that $\Sigma(M(D, D'), D)$ is a closed subvariety of U_D thus $\Sigma(M(D, D'), D) = U_D$. It is deduced that $U((D')^{opp}) = U(D, D')$, and a fortiori that $R^u_{(D')^{opp}} = R(D, D') \subset R_D$. This implies that the two flags D and $(D')^{opp}$ are incident flags. If $\Sigma(M(D, D'), D)$ is open there exists a maximal length flag $D \subset \overline{D}$ with $\Sigma(M(\overline{D}, D'), \overline{D})$ open. One concludes from the above reasoning that $(D')^{opp}, D \subset \overline{D}$, i.e. $(D')^{opp}$ and D are incident flags.

Proposition 2.60 *Let $D \in Drap_{\underline{r}}(k^E)$ be a maximal length flag of E ($|E| = r+1$). There exists a unique $D' = (H_1 \cdots \subset H_l \subset E) \in Drap_{\underline{n}}(k^E)$ satisfying $M(D, D') = M^{bc}_{(\underline{r},\underline{n})}$, characterized by $\tau_D(H_\beta) = (r + 2 - n_\beta < \cdots < r + 1)$.*

Proof *From lemma 2.19 it follows that given a relative position matrix $M \in Relpos_{(r,\underline{n})}(E)$, and a maximal length flag D of E, there exists a unique flag $D' \in Drap_{\underline{n}}(E)$ such that $M(D, D') = M$. Apply this result to $M^{bc}_{(r,\underline{n})}$ and D, and let D' denote the unique flag satisfying $M(D, D') = M^{bc}_{(r,\underline{n})}$. Thus one obtains*

$$R(D, D') = \coprod_{1 \leq \beta \leq l} R_\beta(D, D') = \coprod_{1 \leq \beta \leq l} (\coprod_{i \in H_\beta} (E_{\tau_D(i)} \cap (H_{\beta+1} - H_\beta)) \times \{i\} =$$

$$\coprod_{1 \leq \beta \leq l} (H_{\beta+1} - H_\beta) \times H_\beta = R^u_{D'^{opp}} \, ,$$

and one deduces that: $(H_{\beta+1} - H_\beta) \subset E_{\tau_D(i)}$ for $i \in H_\beta$, and thus that $\tau_D(H_\beta) = (r + 2 - n_\beta < \cdots < r + 1)$.

2.7.1   The embedding of a Schubert cell in a big Schubert cell

The open subvariety $U_{D'} \subset Drap_{\underline{n}}(k^E)$ contains as a closed subvariety every Schubert cell $\Sigma(M, D)$, with D some maximal length flag, and center $D_M = D'$, i.e. $M = M(D, D')$. It may be observed that the set of flags of maximal length in transversal position with D' is those of flags containing $(D')^{opp}$ as a subflag. In other terms incident to $(D')^{opp}$ by 2.59. On the other hand, given a maximal length flag D and a minimal gallery $\Gamma(D, D')$ (cf. [50]), this gallery may be completed into a minimal gallery $\Gamma(D, D') \subset \Gamma(D^{tr}, D')$, where D^{tr} is in transversal position with D', and thus $(D')^{opp} \subset D^{tr}$. Thus $\Sigma(M, D) \subset \Sigma(M(D^{tr}, D'), D^{tr}) = U_{D'}$. By proposition 2.35, $U_{D'}$ is principal homogeneous under the unipotent R-subgroup $U((D')^{opp})$ defined by the closed set $R^u_{(D')^{opp}} = \coprod_{1 \leq \beta \leq l} (H_{\beta+1} - H_\beta) \times H_\beta$ and by 1.3.3 one knows that there is an isomorphism of k-varieties: $U_{D'} \simeq \mathbb{A}_{D'} = \mathbb{A}(\prod_{1 \leq \beta \leq l} M_{(H_{\beta+1} - H_\beta) \times H_\beta}(k))(\simeq U((D')^{opp}))$, giving rise to a canonical parametrization of $U_{D'}$. Denote by $(\xi_{ij})_{(i,j) \in R^u_{(D')^{opp}}}$ the canonical coordinates given by the above isomorphism. The embedding $\Sigma(M, D) \hookrightarrow U_{D'}$ is defined by the set of equations $(\xi_{ij} = 0)_{(i,j) \in (R^u_{(D')^{opp}} - R(D,D'))}$. This embedding corresponds by the isomorphism $U_{D'} \simeq \mathbb{A}_{D'}$ to an embedding $\mathbb{A}_{R(D,D')} \hookrightarrow \mathbb{A}_D \simeq \mathbb{A}_{\underline{n}}$, where $\mathbb{A}_{R(D,D')}$ is given by the coordinates $(m_\beta) = (a_{\beta;ij})$ satisfying the set of equations $a_{\beta;ij} = 0$ for $\sigma_\beta(i) < j \leq n_{\beta+1} - n_\beta$.

Proposition 2.61 *Let $\sigma_\beta(i) = |E_{\tau_D(i)} \cap (H_{\beta+1} - H_\beta)|$. The image of $\mathbb{A}_{R(D,D')}$ by $\mathbb{A}_D \simeq \mathbb{A}_{\underline{n}}$ is given by the $(m_\beta) = (a_{\beta;ij})$ satisfying the set of equations: $a_{\beta;ij} = 0$ for $\sigma_\beta(i) < j \leq n_{\beta+1} - n_\beta$.*

Chapter 3

Resolution of Singularities of a Schubert Variety

A smooth resolution of singularities for a Schubert variety in a flag variety is constructed in terms of a Configurations variety directly obtained from its Relative Position Matrix. This is a canonical smooth resolution whose construction is suggested by our indexation of Schubert varieties in Flag varieties by Relative Position Matrices. We explicit a canonical decomposition of this variety as a sequence of locally trivial fibrations with Grassmannians as typical fibers. A schematic version of this construction is also given.

We proceed first to define a class of subvarieties of products of Grassmannians in terms of the incidence relation between subspaces. These varieties give rise to examples of smooth resolutions of Schubert varieties by minimal generalized galleries. The first examples of smooth resolutions by means of these varieties appear in [12] and [13]. In [12] an application to the construction of an invariant differential form dual to a Schubert cycle in $Grass_n(k^E)$ is given.

Let us explain another motivation of the smooth resolutions introduced in this Chapter. In [49] R.Thom attempts to give a description of the singularities of a differentiable function by means of an iterative procedure involving functions into Grassmannians. These functions appear as locally classifying mappings associated with the family of tangent spaces to the graph of this function. Its generic singularities are thus described by a "Stratification" defined by the Pull-Backs of Special Schubert varieties by these classifying mappings. What R.Thom introduces loosely as "La Ventilation d'une Singularité" in loc.cit. amounts looking at a Classifying Mapping at a singular point of a Schubert variety through its Smooth Resolution defined in this chapter and in [13].

3.1   Relative position matrix associated configurations variety

Definition 3.1 *Given* $M = (m_{\alpha\beta}) \in \text{Relpos}_{(\underline{m},\underline{n})} \subset \mathbb{N}^{(\lambda+1)\times(l+1)}$ *with* $\underline{m} = (m_1 < \cdots < m_\lambda < r + 1)$, $\underline{n} = (n_1 < \cdots < n_l < r + 1)$ *define a* **weighted graph** $\Lambda(M)$ *as follows. The set of vertices* $\text{Vert } \Lambda(M)$ *of* $\Lambda(M)$ *is given by*

$$\text{Vert } \Lambda(M) = [\![1, \lambda + 1]\!] \times [\![1, l + 1]\!],$$

and the set of edges $\text{Edg } \Lambda(M)$ *by*

$$\begin{aligned}\text{Edg } \Lambda(M) \quad = \quad & \{((\alpha,\beta),(\alpha+1,\beta))|\ 1 \leqslant \alpha \leq \lambda,\ 1 \leq \beta \leq l+1\} \cup \\ & \{((\alpha,\beta),(\alpha,\beta+1))|\ 1 \leqslant \alpha \leq \lambda+1,\ 1 \leq \beta \leq l\}.\end{aligned}$$

Define a **weight function** *on the set of vertices* $\text{Vert } \Lambda(M)$ *by:*

$$p:\ \text{Vert } \Lambda(M) \to \mathbb{N},\ p(\alpha,\beta) = m_{\alpha\beta}.$$

A $\Lambda(M)$-**configuration** *of* $\text{Grass}(I_{r+1}) = \coprod \text{Grass}_n(I_{r+1})$ *is a mapping*

$$f:\ [\![1, \lambda + 1]\!] \times [\![1, l + 1]\!] \to \text{Grass}(I_{r+1})$$

satisfying:

(1) $|f(\alpha,\beta)| = \text{card } f(\alpha,\beta) = m_{\alpha\beta} = p(\alpha,\beta)$;

(2) $f(\alpha,\beta) \subset f(\alpha+1,\beta)$ *(resp.* $f(\alpha,\beta+1)$*) if* $\alpha \leqslant \lambda$ *(resp.* $\beta \leqslant l$*).*

The inclusion relation seems better suited to desoube this relation. In fact it is a usual mathematical term relation of $\text{Grass}(I_{r+1}) = \mathscr{P}(I_{r+1})$ and $J \to |J|$ naturally defines a weighted graph structure on $\text{Grass}(I_{r+1})$. A $\Lambda(M)$-configuration φ of $\text{Grass}(I_{r+1})$ is a morphism $\varphi: \Lambda(M) \to \text{Grass}(I_{r+1})$ of weighted graphs. A $\Lambda(M)$-configuration φ is given by a matrix

$$(J_{\alpha\beta}) \in \text{Grass}(I_{r+1})^{(\lambda+1)\times(l+1)}$$

satisfying $|J_{\alpha\beta}| = m_{\alpha\beta}$, and $J_{\alpha\beta} \subset J_{\alpha+1,\beta}$ (resp. $J_{\alpha\beta} \subset J_{\alpha\beta+1}$) for $1 \leqslant \alpha \leqslant \lambda$ (resp. $1 \leqslant \beta \leqslant l$).

Denote by

$$\text{Conf}^{comb}(\Lambda(M)) = \text{Conf}(\Lambda(M),\ \text{Grass}(I_{r+1}))$$

the set of $\Lambda(M)$-configurations of $\text{Grass}(I_{r+1})$ (**The Combinatorial Configurations**). Remark that $m_{\lambda+1\ l+1} = r + 1$ and $m_{1\ l+1} = m_1$, $m_{2\ l+1} = m_2, \cdots, m_{\lambda\ l+1} = m_\lambda$ (resp. $m_{\lambda+1\ 1} = n_1$, $m_{\lambda+1\ 2} = n_2, \cdots, m_{\lambda+1\ l} = n_l$). Thus there are natural mappings

$$p_1 = p_1(M)\quad :\quad \text{Conf}(\Lambda(M), \text{Grass}(I_{r+1})) \to \text{Drap}_{\underline{m}}(I_{r+1})$$

$$\left(\ \text{resp. } p_2 = p_2(M)\quad :\quad \text{Conf}(\Lambda(M), \text{Grass}(I_{r+1})) \to \text{Drap}_{\underline{n}}(I_{r+1})\ \right),$$

defined by

$$p_1((J_{\alpha\beta})) = (J_{1\ l+1} \subset \cdots J_{\lambda\ l+1} \subset I_{r+1})$$
$$\left(\text{resp. } p_2((J_{\alpha\beta})) = (J_{\lambda+1\ 1} \subset \cdots J_{\lambda+1\ l} \subset I_{r+1}) \right),$$

where $J_{\lambda+1\ l+1} = I_{r+1}$. Given $D, D' \in \mathrm{Drap}(I_{r+1})$ write

$$\mathrm{Conf}^{\mathrm{comb}}(\Lambda(M), D) = \mathrm{Conf}(\Lambda(M), \mathrm{Grass}(I_{r+1}))_D := p_1^{-1}(D)$$

$$\left(\text{resp. } \mathrm{Conf}^{\mathrm{comb}}(\Lambda(M), D, D') := p_1^{-1}(D) \cap p_2^{-1}(D') \right).$$

Definition 3.2 *Following the same pattern as above define a $\Lambda(M)$-configuration φ of the k-variety $Grass(k^{r+1})$. Denote by*

$$\mathrm{Conf}(\Lambda(M), \mathrm{Grass}(k^{r+1})),$$

the set of the $\Lambda(M)$-configurations of $Grass(k^{r+1})$. If no confusion arises one can write $Conf(\Lambda(M)) = \mathrm{Conf}(\Lambda(M), \mathrm{Grass}(k^{r+1}))$.

By definition there is a canonical inclusion:

$$\mathrm{Conf}(\Lambda(M)) \subset \prod_{(\alpha,\beta) \in [\![1,\lambda+1]\!] \times [\![1,l+1]\!]} Grass_{m_{\alpha\beta}}(k^{r+1}).$$

Proposition 3.3 *$Conf(\Lambda(M))$ is a projective variety.*

In the next section it is shown that $Conf(\Lambda(M))$ is a k-variety canonically decomposed in a sequence of fiberings, each one with a smooth base and typical fiber a Grassmannian, and thus a k-variety in Serre's sense.

Proof *There is an embedding of $\displaystyle\prod_{(\alpha,\beta) \in [\![1,\lambda+1]\!] \times [\![1,l+1]\!]} Grass_{m_{\alpha\beta}}(k^{r+1})$ in a product of projective spaces, given by the product of the corresponding Plücker embeddings.*
By the proof of 1.9 one can conclude that the image of the subset $\mathrm{Conf}(\Lambda(M))$ is characterized by the vanishing of a set of homogeneous equations in the Plücker coordinates of the factors. Thus $\mathrm{Conf}(\Lambda(M))$ is a closed subvariety of a product of grassmannians. (The equations defining the embedding correspond to the set of inclusions defining $\mathrm{Conf}(\Lambda(M))$ as a subset of this product of grassmannians.) This achieves the proof.

Definition 3.4 *To p_1 (resp.p_2) it corresponds the morphism*

$$\pi_1 = \pi_1(M) : \mathrm{Conf}(\Lambda(M)) \longrightarrow \mathrm{Drap}_{\underline{m}}(k^{r+1})$$

(resp.

$$\pi_2 = \pi_2(M) : \mathrm{Conf}(\Lambda(M)) \longrightarrow Drap_{\underline{n}}(k^{r+1}))$$

defined by $\pi_1((S_{\alpha\beta})) = (S_{1\ l+1} \subset \cdots S_{\lambda\ l+1} \subset k^{r+1})$ *(resp.* $\pi_2((S_{\alpha\beta})) = (S_{\lambda+1\ 1} \subset \cdots S_{\lambda+1\ l} \subset k^{r+1})$*;* π_1 *and* π_2 *are induced by the canonical projections of the product. Thus it is immediate that* π_1 *and* π_2 *are k-morphisms.*

Let $\mathscr{D} \in Drap_{\underline{m}}(k^{r+1})$ and $\mathscr{D}' \in Drap_{\underline{n}}(k^{r+1})$. Write

$$\mathrm{Conf}(\Lambda(M), \mathscr{D}) := \mathrm{Conf}(\Lambda(M))_{\mathscr{D}} := \pi_1^{-1}(\mathscr{D}))$$

$$\left(resp.\ \ \mathrm{Conf}(\Lambda(M), \mathscr{D}, \mathscr{D}') := \pi_1^{-1}(\mathscr{D})) \cap \pi_2^{-1}(\mathscr{D}'))\right).$$

From the above proposition it results that $\mathrm{Conf}(\Lambda(M), \mathscr{D})$ is a k-projective variety. This property also follows from the natural fibering of $\mathrm{Conf}(\Lambda(M), \mathscr{D})$ as it is shown in the next section.

Notation 3.5 *Write:*
a) $\pi(M) = (\pi_1, \pi_2)$ *and denote by* $\pi(M, \mathscr{D})$ *the restriction of* $\pi(M)$ *to* $\mathrm{Conf}(\Lambda(M))_{\mathscr{D}}$ *;*
b) $\hat{\Sigma}(M) = \mathrm{Conf}(\Lambda(M))$ *(resp.*$\hat{\Sigma}(M, \mathscr{D}) = \mathrm{Conf}(\Lambda(M), \mathscr{D})$*).*

Let $\mathscr{D} = k^D$ (resp. $\mathscr{D}' = k^{D'}$). Define the **Combinatorial Fiber** $\mathrm{Conf}^{comb}(\Lambda(M), \mathscr{D}, \mathscr{D}')$ of (π_1, π_2) as the image of the mapping

$$\mathrm{Conf}^{comb}(\Lambda(M), D, D') \longrightarrow \mathrm{Conf}^{comb}(\Lambda(M), \mathscr{D}, \mathscr{D}')\ ,$$

defined by $\varphi \mapsto (k^{\varphi(\alpha, \beta)})$. If no confusion arises denote by $\mathrm{Conf}^{comb}(\Lambda(M), D, D')$ this image.

3.1.1   The Configurations Canonical Section

Definition 3.6 *There is a natural embedding*

$$\theta(M) : \ \Sigma(M) \hookrightarrow \mathrm{Conf}(\Lambda(M))$$

of the **universal Schubert cell** $\Sigma(M)$ **of type** M *into the variety of* $\Lambda(M)$-*configurations, defined by* $(\mathscr{D}, \mathscr{D}') \mapsto \mathscr{M}(\mathscr{D}, \mathscr{D}') = (W_\alpha \cap V_\beta)$*, where* $\mathscr{D} = (W_1 \subset \cdots \subset W_\lambda \subset k^{r+1})$ *(resp.* $\mathscr{D}' = (V_1 \subset \cdots \subset V_l \subset k^{r+1})$*).*

Proposition 3.7 *The mapping* $\theta(M)$ *is induced by a k-morphism.*

Proof *Let* $\Sigma_{\alpha\beta}(M) \subset Drap_{\underline{m}}(k^{r+1}) \times Drap_{\underline{n}}(k^{r+1})$ *(resp.* $\Sigma'_{\alpha\beta}(M) \subset Grass_{m_\alpha}(k^{r+1}) \times Grass_{n_\beta}(k^{r+1}))$ *be the Schubert cell defined by* $\dim W_\alpha \cap V_\beta = m_{\alpha\beta}$ *(resp.* $\dim W \cap V = m_{\alpha\beta}$*), where* $D = (W_1 \cdots \subset W_\lambda \subset k^{r+1})$*, and* $D' = (V_1 \cdots \subset V_l \subset k^{r+1})$*. Remark that* $\Sigma(M) = \cap \Sigma_{\alpha\beta}(M)$*.*
The (α, β)-*component of* $\theta(M) = (\theta_{\alpha\beta}(M))$ *is obtained as the composition of the projection of* $\Sigma_{\alpha\beta}(M)$ *on* $\Sigma'_{\alpha\beta}(M)$ *followed by the mapping*

$\Sigma'_{\alpha\beta}(M) \longrightarrow Grass_{m_{\alpha\beta}}(k^{r+1})$, defined by $(W,V) \mapsto W \cap V$. In view of the above remark, it suffices to prove that this mapping is induced by a morphism $\theta'_{\alpha\beta}(M) : \Sigma'_{\alpha\beta}(M) \longrightarrow Grass_{m_{\alpha\beta}}(k^{r+1})$. Let it be proved that the Plücker coordinates of $W \cap V$ are given by homogeneous polynomials in the Plücker coordinates of W and V. We keep the notation of the proof of 1.9. The Plücker homogeneous coordinates of W (resp. V) are given by $(\Delta_H(e_1))_{H \in Grass_{m_\alpha}(I_{r+1})}$ (resp.$(\Delta_J(e_2))_{J \in Grass_{n_\beta}(I_{r+1})}$, where e_1 (resp.e_2) denotes a basis of W (resp. V). A system of defining linear equations is obtained for $W \cap V$ as follows. The kernel of the linear mapping

$$\Phi : k^{r+1} \longrightarrow \overset{m+1}{\bigwedge} k^{r+1} \times \overset{n+1}{\bigwedge} k^{r+1}$$

defined by:

$$v \longmapsto ((e_{11} \wedge ... \wedge e_{1m}) \wedge v, (e_{21} \wedge ... \wedge e_{2n}) \wedge v),$$

is equal to $W \cap V$. Denote by ρ the rank of Φ. Let $S \in Grass_{m_{\alpha\beta}}(k^{r+1})$ and $f = (f_k)_{1 \leq k \leq m_{\alpha\beta}}$ be a basis of S. The condition $S = Ker\ \Phi$ may be written as follows. Let

$$\Psi : k^{r+1} \longrightarrow \overset{m+1}{\bigwedge} k^{r+1} \times \overset{n+1}{\bigwedge} k^{r+1} \times \overset{m_{\alpha\beta}}{\bigwedge} k^{r+1}$$

be defined by:

$$v \longmapsto ((e_{11} \wedge ... \wedge e_{1m}) \wedge v, (e_{21} \wedge ... \wedge e_{2n}) \wedge v, (f_1 \wedge ... \wedge f_{m_{\alpha\beta}}) \wedge v).$$

Thus $S = Ker\ \Phi$ if and only if: $Ker\ \Phi = Ker\ \Psi$. This last condition may be written, as in 1.9, in terms of the vanishing of the $(\rho + 1)$-minors of the linear system obtained from Ψ. The minors of this system are polynomials in $(\Delta_H(e_1))_{H \in Grass_{m_\alpha}(I_{r+1})}$ (resp.$(\Delta_J(e_2))_{J \in Grass_{n_\beta}(I_{r+1})}$ and in the Plücker coordinates $(\Delta_K(f))_{K \in Grass_{m_{\alpha\beta}}(I_{r+1})}$ of S. More precisely: $S = Ker\ \Phi \Leftrightarrow$ the set $(M_\mu((\Delta_H(e_1)), (\Delta_J(e_2)); (\Delta_K(f)))$ of the $(\rho+1)$-minors linear in $(\Delta_K(f))$ vanishes. Consider now the linear system in the variables $(X_K)_{K \in Grass_{m_{\alpha\beta}}(I_{r+1})}$ given by $(M_\mu((\Delta_H(e_1)), (\Delta_J(e_2)); (X_K)) = 0)$; $(\Delta_K(f))$ is clearly a solution of this linear system. Let (X_K^0) also be a solution. Then the vector

$$\omega = \sum_{K \in Grass_{m_{\alpha\beta}}(I_{r+1})} X_K^0 e_{i_1(K)} \wedge ... \wedge e_{i_n(K)} \ ,$$

satisfies $\omega \wedge f_k = 0$ for all $1 \leq k \leq m_{\alpha\beta}$. Thus there exists $\lambda \in k$ with

$$\omega = \lambda \cdot (\sum_{K \in Grass_{m_{\alpha\beta}}(I_{r+1})} \Delta_K(f) e_{i_1(K)} \wedge ... \wedge e_{i_n(K)}) = \lambda \cdot f_1 \wedge \cdots \wedge f_{m_{\alpha\beta}}.$$

Thus the subspace of solutions of this linear system is one dimensional. It follows immediately that the coordinates $(\Delta_K(f))$ of S are homogeneous polynomials in the coefficients of $(M_\mu((\Delta_H(e_1)),(\Delta_J(e_2));(X_K)) = 0)$, which are themselves homogeneous polynomials in $(\Delta_H(e_1))$ and $(\Delta_J(e_2))$. This achieves the proof of the proposition.

Proposition 3.8 *The morphism $\theta(M) : \Sigma(M) \longrightarrow Conf(\Lambda(M))$ defines an open embedding, and we have:*

$$\hat{\Sigma}'(M) = \pi(M)^{-1}(\Sigma(M)) = \operatorname{Im} \Theta(M)$$

$$(\text{resp. } \hat{\Sigma}'(M,\mathscr{D}) = \pi(M,\mathscr{D})^{-1}(\Sigma(M,\mathscr{D})) = \operatorname{Im} \Theta(M,\mathscr{D})).$$

Proof *Let us prove the first assertion. The condition on*

$$(W_{\alpha\beta})_{(\alpha,\beta)\in[\![1,\lambda+1]\!]\times[\![1,l+1]\!]} \in Conf(\Lambda(M),\operatorname{Grass}(k^{r+1}))$$

to be in the image of $\theta(M)$ is given by:

$$W_{\alpha\beta} = W_{\lambda+1\beta} \cap W_{\alpha l+1}$$

$((\alpha,\beta) \in [\![1,\lambda]\!] \times [\![1,l]\!])$. Write

$$\Sigma'_{(\alpha,\beta)} = \{(W,W') \in Grass_{m_{\alpha+1\beta}}(k^{r+1}) \times Grass_{n_{\alpha\beta+1}}(k^{r+1}) \mid m_{\alpha+1\beta} <$$

$$rk(W \cap W')\},$$

and $q_{(\alpha,\beta)} : Conf(\Lambda(M)) \longrightarrow Grass_{m_{\alpha+1\beta}}(k^{r+1}) \times Grass_{m_{\alpha\beta+1}}(k^{r+1})$ for the canonical projection. Let $(e_i)_{1\leq i\leq m_{\alpha+1\beta}}$ $(resp.(e'_i)_{1\leq i\leq m_{\alpha\beta+1}})$ be a basis of W (resp. W'). Define $\Phi : k^{r+1} \longrightarrow \overset{m_{\alpha+1\beta}}{\wedge}(k^{r+1}) \times \overset{m_{\alpha\beta+1}}{\wedge}(k^{r+1})$ by

$$\Phi : v \longmapsto ((e_1 \wedge ... \wedge e_{m_{\alpha+1\beta}}) \wedge v, (e'_1 \wedge ... \wedge e'_{m_{\alpha\beta+1}}) \wedge v) .$$

As in 1.9 a linear system is obtained whose matrix coefficients are given by the Plücker coordinates of (W,W'). The condition $m_{\alpha\beta} < rk\ ker\ \Phi$ may be stated as the vanishing of a set of minors of this system. Thus $\Sigma'_{(\alpha,\beta)}$ is a Zariski closed subset of $Grass_{m_{\alpha+1\beta}}(k^{r+1}) \times Grass_{n_{\alpha\beta+1}}(k^{r+1})$. On the other hand, it is clear that

$$\operatorname{Im} \theta(M) = Conf(\Lambda(M)) - \underset{(\alpha,\beta)\in[\![1,\lambda]\!]\times[\![1,l]\!]}{\cup} q_{(\alpha,\beta)}^{-1}(\Sigma'_{(\alpha,\beta)}),$$

thus one concludes that $\operatorname{Im} \theta(M)$ is an open set in $Conf(\Lambda(M))$. The morphism (π_1,π_2) gives a left inverse of $\theta(M)$, hence $\theta(M)$ may be seen as a section of

$$(\pi_1,\pi_2) : \operatorname{Conf}(\Lambda(M)) \to \operatorname{Drap}(k^{r+1}) \times \operatorname{Drap}(k^{r+1})$$

along the subscheme $\Sigma(M) \hookrightarrow \mathrm{Drap}(k^{r+1}) \times \mathrm{Drap}(k^{r+1})$. *This proves that* $\theta(M)$ *is an open embedding.*

A simple geometrical argument proves the second assertion. Let $(\mathscr{D}, \mathscr{D}') \in \Sigma(M)$, *with* $\mathscr{D} = (W_1 \subset \cdots \subset W_\lambda \subset k^{r+1})$ *and* $\mathscr{D}' = (V_1 \subset \cdots \subset V_l \subset k^{r+1})$. *If* $\pi(M)((W_{\alpha\beta})) = (\mathscr{D}, \mathscr{D}')$, *then* $W_{\alpha\ l+1} = W_\alpha$ $(1 \leqslant \alpha \leqslant l)$ *and* $W_{\lambda+1\ \beta} = V_\beta$ $(1 \leqslant \beta \leqslant l)$.

As $(W_{\alpha\beta})$ *is a* $\Lambda(M)$-*configuration, one has*

$$\forall\ (\alpha, \beta), \qquad \dim\ W_{\alpha\beta} = m_{\alpha\beta},$$

and $W_{\alpha\beta} \subset W_{\alpha\ l+1} \cap W_{\lambda+1\ \beta} = W_\alpha \cap V_\beta$. *On the other hand,* $(\mathscr{D}, \mathscr{D}') \in \Sigma(M) \Rightarrow \dim\ W_\alpha \cap V_\beta = m_{\alpha\beta}$ *by definition of* $\Sigma(M)$. *It is concluded that*

$$\forall\ (\alpha, \beta), \qquad W_{\alpha\beta} = W_\alpha \cap V_\beta, \text{ i.e.}$$

$$(W_{\alpha\beta}) = \mathscr{M}(\mathscr{D}, \mathscr{D}').$$

This achieves the proof of the proposition

Corollary 3.9 *The restriction of the embedding* $\theta(M)$ *to* $\Sigma(M, \mathscr{D})$ *induces an open embedding*

$$\theta(M, \mathscr{D}):\ \Sigma(M, \mathscr{D}) = \Sigma(M)_\mathscr{D} \hookrightarrow \mathrm{Conf}(\Lambda(M))_\mathscr{D} = \mathrm{Conf}(\Lambda(M), \mathscr{D}),$$

defined by $\mathscr{D}' \mapsto \mathscr{M}(\mathscr{D}, \mathscr{D}')$.

3.2 Fiber decomposition of a Grassmannian Configurations Variety

Given a type of relative position $M = (m_{\alpha\beta}) \in Relpos_{\underline{m}}(I_{r+1}) \cap \mathbb{N}^{(\lambda+1)\times 2}$, i.e. a type of relative position of a flag of type $\underline{m} = (m_1 < \cdots < m_\lambda < r+1)$ and length $l(\underline{m}) = \lambda$ and a flag of length 1, and a combinatorial flag $D = (H_1 \cdots \subset H_\lambda \subset I_{r+1})$, write $Conf(\Lambda(M), D) = Conf(\Lambda(M), k^D)$. To M is associated the sequence of relative position matrices: $M_\alpha = (m_{\alpha'\beta})_{1\leq\alpha'\leq\alpha}$ $(1 \leq \alpha \leq \lambda+1)$. Let $(D_\alpha)_{1\leq\alpha\leq\lambda+1}$ be the sequence of truncated flags of D defined as in 1.3.2, and $\underline{m}_\alpha = typ(D_\alpha)$.
Define

$$\hat{\Sigma}^{(\alpha)} = \hat{\Sigma}^{(\alpha)}(M, D) := \hat{\Sigma}(M_\alpha, D_\alpha) = \mathrm{Conf}(\Lambda(M_\alpha),$$

$$\mathrm{Grass}(k^{H_\alpha}))_{D_\alpha}\ (1 \leqslant \alpha \leqslant \lambda+1)\ .$$

We have the following identification:

$$\hat{\Sigma}^{(\alpha)} = \hat{\Sigma}^{(\alpha)}(M, D) = \left(\left(\prod_{\alpha'=1}^{\alpha} \mathrm{Grass}_{m_{\alpha'1}}(k^{H_{\alpha'}}) \right) \times \mathrm{Grass}_{m_\alpha}(k^{r+1}) \right) \cap \mathrm{Chain}(k^{H_\alpha}),$$

where $\mathrm{Chain}(k^{H_\alpha})$ denotes the k-variety of chains of subspaces of k^{H_α} .

Clearly $\hat{\Sigma}^{(\lambda+1)} = \hat{\Sigma}(M, D) = \mathrm{Conf}(\Lambda(M), k^D)$. For every $1 \leqslant \alpha \leqslant \lambda$ the projection morphism $\prod\limits_{\alpha'=1}^{\alpha+1} \mathrm{Grass}_{m_{\alpha'1}}(k^{H_{\alpha'}}) \longrightarrow \prod\limits_{\alpha'=1}^{\alpha} \mathrm{Grass}_{m_{\alpha'1}}(k^{H_{\alpha'}})$ induces a morphism

$$p^{(\alpha+1)} : \ \hat{\Sigma}^{(\alpha+1)} \longrightarrow \hat{\Sigma}^{(\alpha)}.$$

We shall now show

Proposition 3.10   *1)* $p^{(\alpha+1)}$ *defines a locally trivial fibration with typical fiber a grassmannian.*

2) The $(\hat{\Sigma}^{(\alpha)})_{1 \leq \alpha \leq \lambda+1}$ *are integral and smooth k-varieties. In particular* $\hat{\Sigma}(M, \mathcal{D})$, $\mathcal{D}$ *a flag of type* $\underline{m}$ *of* k^{r+1}, *is an integral and smooth k-variety.*

Proof   *There is a natural morphism:* $q_{\alpha} : \hat{\Sigma}^{(\alpha)} \longrightarrow \mathrm{Grass}_{m_{\alpha 1}}(k^{H_{\alpha}}) \hookrightarrow$ $\mathrm{Grass}_{m_{\alpha 1}}(k^{H_{\alpha}+1})$. *On the other hand, one knows that there is a locally trivial fibration*

$$\mathrm{Grass}_{(m_{\alpha+11}-m_{\alpha 1})}(\mathscr{O}^{H_{\alpha}+1}_{\mathrm{Grass}_{m_{\alpha 1}}(k^{H_{\alpha}+1})}/\xi_{m_{\alpha 1}}) \longrightarrow \mathrm{Grass}_{m_{\alpha 1}}(k^{H_{\alpha}+1}).$$

The fiber on the section $J_{\alpha 1}$ *of* $\mathrm{Grass}_{m_{\alpha 1}}(k^{H_{\alpha}+1})$ *on some* $X = Spec(A)$, *where* A *denotes k-algebra, is given by the set of sections* $J_{\alpha+11}$ *of* $\mathrm{Grass}_{m_{\alpha+11}}(k^{H_{\alpha}+1})$ *satisfying* $J_{\alpha 1} \subset J_{\alpha+11} \subset \mathscr{O}^{H_{\alpha}+1}_{\mathrm{Grass}_{m_{\alpha 1}}(k^{H_{\alpha}+1})}$.
From this description it follows that there is a canonical isomorphism:

$$q_{\alpha}^{*}(\mathrm{Grass}_{(m_{\alpha+11}-m_{\alpha 1})}(\mathscr{O}^{H_{\alpha}+1}_{\mathrm{Grass}_{m_{\alpha 1}}(k^{H_{\alpha}+1})}/\xi_{m_{\alpha 1}})) \simeq \hat{\Sigma}^{(\alpha+1)} \ .$$

This implies that $p^{(\alpha+1)}$ *defines a locally trivial fibration with typical fiber a grassmannian.*
The total space of a locally trivial fiber bundle, with base and typical fiber smooth and integral k-varieties, is a smooth and integral k-variety. It may be observed that $\hat{\Sigma}^{(1)} = \mathrm{Grass}_{m_{11}}(k^{H_1})$. *Thus the second assertion results immediately by induction from the first one.*

Remark 3.11 *Let* $M \in \mathbb{N}^{(\lambda+1) \times (l+1)}$ *be a type of relative position matrix of a chain of type* $\underline{m}$ *and a flag of type* $\underline{n}$ *(cf. 1.4). Given a chain* $D \in Chain_{\underline{m}}(E)$, *the Schubert cell* $\Sigma(M) \subset Drap_{\underline{n}}(k^E) \times Drap_{\underline{n}}(k^E)$ *(resp.* $\Sigma(M, D) \subset Drap_{\underline{n}}(k^E)$) *may be defined following the pattern of 2.10, as well as the corresponding Schubert variety. The varieties* $Conf(\Lambda(M))$ *(resp.* $Conf(\Lambda(M), D) = Conf(\Lambda(M), k^D)$) *may also be defined. The preceding constructions of this chapter may be carried out for M and D; the above statements hold for a chain relative position type matrix M and a chain D. In this case* $k^D \in Chain(k^{r+1})$ *denotes the chain of* k^{r+1} *corresponding to D. It is easy to see that their proofs may be reduced to that of equivalent statements involving only flags, and relative position types of flags.*

3.3 Fiber decomposition of a Flags Configurations Variety

Let $D = (H_1 \subset \cdots \subset H_\lambda \subset I_{r+1})$ (resp. $D' = (J_1 \subset \cdots \subset J_l \subset I_{r+1})$), with typ $D = \underline{m}$ (resp. typ $D' = \underline{n}$). Write $\mathscr{M}(D, D') = (H_{\alpha\beta}) = (H_\alpha \cap J_\beta)$ (resp. $M(D, D') = (m_{\alpha\beta}) = (|H_{\alpha\beta}|) = (|H_\alpha \cap J_\beta|)$), and $D'^\beta = (J_\beta \subset \cdots \subset J_l \subset I_{r+1})$. M is associated with two sequences of relative positions matrices. The first of these sequences is a sequence of types of relative positions of flags, and the second one of chains.

1. $M(\beta) - M(D, D'^\beta) = (m_{\alpha\beta'})_{\beta \leqslant \beta' \leqslant l+1} \in \mathbb{N}^{(\lambda+1)\times(l+2-\beta)}$, for $(l \geq \beta \geq 1)$;

2. $M^\beta = M(D^{\bullet\beta+1}, (J_\beta \subset J_{\beta+1})) = (m_{\alpha\beta'})_{\beta \leqslant \beta' \leqslant \beta+1} \in Relpos\,(J_{\beta+1}) \cap \mathbb{N}^{(\lambda+1)\times 2}$, for $(1 \leq \beta \leq l)$,

where $D^{\bullet\beta+1} = (H_{1\beta+1} \subset \cdots \subset H_{\lambda\beta+1} \subset J_{\beta+1})$, thus M^β denotes the relative position matrix defined by the two chains $D^{\bullet\beta+1}, (J_\beta \subset J_{\beta+1}) \in Chain(J_{\beta+1})$.

Let $p_{\alpha\beta}$ be the composed morphism

$$\hat\Sigma(M, D) = \mathrm{Conf}(\Lambda(M), D) \hookrightarrow \prod_{(\alpha,\beta)\in[\![1,\lambda+1]\!]\times[\![1,l+1]\!]} Grass_{m_{\alpha\beta}}(k^{r+1}) \longrightarrow$$

$$Grass_{m_{\alpha\beta}}(k^{r+1})\,.$$

The last arrow denotes the canonical projection on the (α, β)-factor. We say that $p_{\alpha\beta}$ is induced by the corresponding canonical projection.

Definition 3.12 *Denote by $p^*_{\alpha\beta}(\xi_{m_{\alpha\beta}})$ the pullback of the tautological module $\xi_{m_{\alpha\beta}}$ associated with $Grass_{m_{\alpha\beta}}(k^{r+1})$ by $p_{\alpha\beta}$. Write*

$$(\mathscr{H}_{\alpha\beta}) = (p^*_{\alpha\beta}(\xi_{m_{\alpha\beta}}))$$

This family of submodules of $\mathscr{O}^{r+1}_{\hat\Sigma(M,D)}$ is characterized as follows. Given a section $s : X \longrightarrow \hat\Sigma(M, D)$ the fiber $(\mathscr{H}_{\alpha\beta})_s$ on s is given by the $\Lambda(M)$-configuration $(s^*(\mathscr{H}_{\alpha\beta}))$ of $\mathscr{O}_{\hat\Sigma(M,D)}$-direct factor submodules of $\mathscr{O}^{r+1}_{\hat\Sigma(M,D)}$, i.e. the $\Lambda(M)$-configuration given by s. Let us write $\hat\Sigma_\beta := \hat\Sigma(M(\beta), D(\beta)) = \mathrm{Conf}(\Lambda(M(\beta)), D(\beta)) \subset \prod_{(\alpha,\beta')\in[\![1,\lambda+1]\!]\times[\![\beta,l+1]\!]} Grass_{m_{\alpha\beta}}(k^{r+1})$. $\mathscr{D}^{\bullet\beta+1}$ denotes the chain of $\mathscr{O}_{\hat\Sigma_{\beta+1}}$-modules defined as follows:

$$\mathscr{D}^{\bullet\beta+1} = (\mathscr{H}_{1\,\beta+1} \subset \cdots \subset \mathscr{H}_{\lambda+1\beta+1}).$$

The inclusion of weighted graphs $\Lambda(M(\beta + 1)) \hookrightarrow \Lambda(M(\beta))$ defines a morphism

$$\mathrm{pr}_\beta : \ \mathrm{Conf}(\Lambda(M(\beta)), D(\beta)) \longrightarrow \mathrm{Conf}(\Lambda(M(\beta + 1)), D(\beta + 1))$$

$$(resp. \ \hat{\Sigma}_\beta \longrightarrow \hat{\Sigma}_{\beta+1}) \ .$$

$\hat{\Sigma}_\beta$ is now described as a fiber bundle over $\hat{\Sigma}_{\beta+1}$.

Let $\varphi(D^{\bullet\beta+1})$ (resp. $\varphi(\mathscr{D}^{\bullet\beta+1})$) be the flag defined by $D^{\bullet\beta+1}$ (resp. $\mathscr{D}^{\bullet\beta+1}$), and $Stief(\varphi(\mathscr{D}^{\bullet\beta+1}))$ the Stiefel variety of ordered basis of $\mathscr{H}_{\lambda+1\beta+1}$ adapted to the flag $\varphi(\mathscr{D}^{\bullet\beta+1})$, defined the following pattern of definition 1.15. It is known that $Stief(\varphi(\mathscr{D}^{\bullet\beta+1}))$ is right principal under the natural action of $Stab \ k^{\varphi(D^{\bullet\beta+1})} \subset Gl(k^{r+1})$.

$(M^\beta, D^{\bullet\beta+1})$ is associated to the sequence of Schubert cells $(\hat{\Sigma}^{(\alpha)}(M^\beta, D^{\bullet\beta+1}))_{1\leqslant\alpha\leqslant\lambda+1}$ and the sequence of morphisms $(\hat{\Sigma}^{(\alpha+1)}(M^\beta, D^{\bullet\beta+1})) \longrightarrow \hat{\Sigma}^{(\alpha)}(M^\beta, D^{\bullet\beta+1}))_{1\leqslant\alpha\leqslant\lambda}$. There is a natural left action of $Stab \ k^{\varphi(D^{\bullet\beta+1})}$ on $\hat{\Sigma}^{(\alpha)}(M^\beta, D^{\bullet\beta+1}))$ for $1 \leqslant \alpha \leqslant \lambda+1$ and the morphisms $\hat{\Sigma}^{(\alpha+1)}(M^\beta, D^{\bullet\beta+1})) \longrightarrow \hat{\Sigma}^{(\alpha)}(M^\beta, D^{\bullet\beta+1}))$ are $Stab \ k^{\varphi(D^{\bullet\beta+1})}$-equivariants.

Definition 3.13 *Write*

$$\hat{\Sigma}(M^\beta, \mathscr{D}^{\bullet\beta+1}) = Stief(\varphi(\mathscr{D}^{\bullet\beta+1})) \wedge_{Stab \ k^{\varphi(D^{\bullet\beta+1})}} \hat{\Sigma}(M^\beta, D^{\bullet\beta+1}))$$

$$(resp. \ \hat{\Sigma}_\beta^{(\alpha)} = \hat{\Sigma}^{(\alpha)}(M^\beta, \mathscr{D}^{\bullet\beta+1}) = Stief(\varphi(\mathscr{D}^{\bullet\beta+1}))\wedge_{Stab \ k^{\varphi(D^{\bullet\beta+1})}}$$

$$\hat{\Sigma}^{(\alpha)}(M^\beta, D^{\bullet\beta+1})) \ .$$

Let $pr_\beta^{(\alpha+1)} : \hat{\Sigma}^{(\alpha+1)}(M^\beta, \mathscr{D}^{\bullet\beta+1}) \longrightarrow \hat{\Sigma}^{(\alpha)}(M^\beta, \mathscr{D}^{\bullet\beta+1})$ *be defined by*

$$pr_\beta^{(\alpha+1)} := Stief(\varphi(\mathscr{D}^{\bullet\beta+1})) \wedge_{Stab \ k^{\varphi(D^{\bullet\beta+1})}} (\hat{\Sigma}^{(\alpha+1)}(M^\beta, D^{\bullet\beta+1}) \longrightarrow$$

$$\hat{\Sigma}^{(\alpha)}(M^\beta, D^{\bullet\beta+1}))$$

The following proposition results easily from the above definition.

Proposition 3.14   *1. The canonical morphism* $pr_\beta : \hat{\Sigma}(M^\beta, \mathscr{D}^{\bullet\beta+1}) \longrightarrow$ $\hat{\Sigma}_{\beta+1}$ *defines a locally trivial fibration with typical fiber* $\hat{\Sigma}(M^\beta, D^{\bullet\beta+1})$.

2. The morphism $pr_\beta^{(\alpha+1)} : \hat{\Sigma}^{(\alpha+1)}(M^\beta, \mathscr{D}^{\bullet\beta+1}) \longrightarrow \hat{\Sigma}^{(\alpha)}(M^\beta, \mathscr{D}^{\bullet\beta+1})$ *defines a locally trivial fibration with typical fiber* $Grass_{(m_{\alpha+1\beta}-m_{\alpha\beta})}(k^{H^{\alpha+1\beta+1}}/k^{H^{\alpha\beta}})$.

3. A canonical isomorphism of $\hat{\Sigma}_{\beta+1}$-*schemes exists:*

$$i_\beta : \hat{\Sigma}_\beta \longrightarrow \hat{\Sigma}(M^{\beta+1}, \mathscr{D}^{\bullet\beta+1}),$$

defined functorially by

$$i_\beta : (\mathscr{H}_{\alpha\beta'})_{l+1\geqslant\beta'\geqslant\beta} \mapsto ((\mathscr{H}_{\alpha\beta'})_{l+1\geqslant\beta'\geqslant\beta+1}; (\mathscr{H}_{\alpha\beta'})_{\beta+1\geqslant\beta'\geqslant\beta}).$$

Observe that $M(1) = M$ and that $\hat{\Sigma}_l := \text{Conf}(\Lambda(M(l)), D)$ is a smooth projective k-variety. We deduce from this and 3.10 the

Corollary 3.15 *There is a sequence of locally trivial fibrations:*

$$\hat{\Sigma}(M, D) = \hat{\Sigma}_1 \longrightarrow \hat{\Sigma}_2 \cdots \longrightarrow \hat{\Sigma}_l = \hat{\Sigma}(M(l), D) \ ,$$

each fibration with typical fiber integral and smooth. Thus $\hat{\Sigma}(M, D)$ is an integral smooth k-variety.

From the decomposition of $\hat{\Sigma}(M, D)$ in a sequence of fibrations $(\hat{\Sigma}_\beta, \text{pr}_\beta)$ a further refined sequence of fibrations $(\hat{\Sigma}_\beta^{(\alpha)}, \text{pr}_\beta^{(\alpha)})$ $(2 \leq \alpha \leq \lambda + 1, \ 1 \leq \beta \leq l)$ is obtained. Observe that $\hat{\Sigma}_\beta^{(1)} = \text{Grass}_{m_{1\beta}}(\mathscr{H}_{m_{1\beta+1}})$. Define $pr_\beta^{(1)} : \hat{\Sigma}_\beta^{(1)} \longrightarrow \hat{\Sigma}_{\beta+1}^{(\lambda+1)}$ as the canonical morphism defining the fiber bundle structure of the relative grassmannian $\text{Grass}_{m_{1\beta}}(\mathscr{H}_{m_{1\beta+1}})$ on $\hat{\Sigma}_{\beta+1}^{(\lambda+1)}$. For $1 < \beta \leq l$ there is a sequence of locally trivial fiberings each one with a typical fiber grassmannian:

$$\cdots \overset{pr_{\beta-1}^{(1)}}{\longrightarrow} \hat{\Sigma}_\beta^{(\lambda+1)} \longrightarrow \hat{\Sigma}_\beta^{(\lambda)} \cdots \longrightarrow \hat{\Sigma}_\beta^{(2)} \longrightarrow \hat{\Sigma}_\beta^{(1)} \overset{pr_\beta^{(1)}}{\longrightarrow} \hat{\Sigma}_{\beta+1}^{(\lambda+1)} \cdots \ .$$

Coherence of $(\hat{\Sigma}_\beta, \text{pr}_\beta)$ with $(\hat{\Sigma}_\beta^{(\alpha)}, \text{pr}_\beta^{(\alpha)})$.

There are:

1. $\hat{\Sigma}_\beta^{(\lambda+1)} = \hat{\Sigma}_\beta$ (as a particular case $\hat{\Sigma}_1^{(\lambda+1)} = \hat{\Sigma}_1 = \hat{\Sigma}(M, D)$);

2. $\text{pr}_\beta = \text{pr}_\beta^{(1)} \circ \text{pr}_\beta^{(2)} \circ \cdots \circ \text{pr}_\beta^{(l+1)} : \hat{\Sigma}_\beta^{(\lambda+1)} = \hat{\Sigma}_\beta \longrightarrow \hat{\Sigma}_{\beta+1}^{(\lambda+1)} = \hat{\Sigma}_{\beta+1}.$

A decomposition of $\hat{\Sigma}(M, D) = \text{Conf}(\Lambda(M), D)$ in a sequence of locally trivial smooth fibrations with grassmannians as fibers is thus obtained.

Remark 3.16 *Let $Stief(\xi_{\underline{m}}) \longrightarrow \text{Drap}_{\underline{m}}(k^{r+1})$ be as in definition 1.15. It may be written $\hat{\Sigma}(M) = Stief(\xi_{\underline{m}}) \wedge_{Stab \ k^D} \hat{\Sigma}(M, D)$ (The Universal Schubert Variety of type M as a contracted product defined by a Stiefel variety of adapted basis of a flag variety).*

Corollary 3.17 *The projection $\pi_1 : \hat{\Sigma}(M) \longrightarrow \text{Drap}_{\underline{m}}(k^{r+1})$ defines a locally trivial fibration with typical fiber $\hat{\Sigma}(M, D)$. Thus $\hat{\Sigma}(M)$ is an integral smooth projective k-variety.*

It is known that $\theta(M) : \Sigma(M) \longrightarrow \hat{\Sigma}(M)$ (resp.$\theta(M, D) : \Sigma(M, D) \longrightarrow \hat{\Sigma}(M, D)$) is an open embedding, thus it is deduced from the proposition that the Zariski closure of Im $\theta(M)$ (resp.Im $\theta(M, D)$) is equal to $\hat{\Sigma}(M)$ (resp.$\hat{\Sigma}(M, D)$).

These results may be resumed:

Proposition 3.18   *1)* $\hat{\Sigma}(M)$ *is a projective, integral, and smooth k-variety;*

2) $\theta(M) : \Sigma(M) \hookrightarrow \hat{\Sigma}(M)$ *is a dense open embedding;*

3) $\pi(M)$ *factors by* $\overline{\Sigma}(M) \hookrightarrow \mathrm{Drap}(k^{r+1}) \times \mathrm{Drap}(k^{r+1})$;

4) the induced morphism $\pi(M) : \hat{\Sigma}(M) \to \overline{\Sigma}(M)$ *, which is also denoted by* $\pi(M)$*, is a* **resolution of singularities of** $\overline{\Sigma}(M)$ *. This means that the variety* $\hat{\Sigma}(M)$ *is projective smooth and integral and a section exists, namely* $\theta(M)$*, of* $\pi(M)$ *on* $\Sigma(M)$ *with* $\mathrm{Im}\ \theta(M) \subset \hat{\Sigma}(M)$ *a dense open subvariety, and on the other hand, there is:*

$$\hat{\Sigma}'(M) = \pi(M)^{-1}(\Sigma(M)) = \mathrm{Im}\ \theta(M)$$

(resp. $\hat{\Sigma}'(M, \mathscr{D}) = \pi(M, \mathscr{D})^{-1}(\Sigma(M, \mathscr{D})) = \mathrm{Im}\ \theta(M, \mathscr{D}))$.

It follows that $\pi(M)$ *is a birational morphism.*

Remark 3.19 *In general The morphism* $\pi(M)$ *is not a resolution of singularities as it is usually understood (cf. [30]). As a matter of fact it may not induce an isomorphism on the smooth subvariety of* $\Sigma(M)$*. In the next chapter the singular locus of* $\overline{\Sigma}(M)$ *in terms of* $\hat{\Sigma}(M)$ *is discussed in detail.*

Clearly the diagonal action of $\mathrm{Gl}(k^{r+1})$ on $\mathrm{Grass}(k^{r+1})^{(\lambda+1)\times(l+1)}$ leaves $\hat{\Sigma}(M)$ stable and thus there is an induced action of $\mathrm{Gl}(k^{r+1})$ on $\hat{\Sigma}(M)$. It results from this that the morphism $\pi(M)$ is $\mathrm{Gl}(k^{r+1})$-equivariant, if $\overline{\Sigma}(M) \subset \mathit{Drap}(k^{r+1} \times \mathit{Drap}(k^{r+1})$ is endowed with the action induced by the left diagonal action of $\mathrm{Gl}(k^{r+1})$. It can be deduced that:

Theorem 3.20 *The morphism*

$$\pi(M) : \hat{\Sigma}(M) \longrightarrow \overline{\Sigma}(M)$$

is a smooth resolution. Moreover $\pi(M)$ *is* $\mathrm{Gl}(k^{r+1})$*-equivariant.*

An equivariant resolution of singularities of $\overline{\Sigma}(M, \mathscr{D})$ is now described. The restriction $\theta(M, \mathscr{D})$ of the embedding $\theta(M)$ to $\Sigma(M, \mathscr{D}) \subset \overline{\Sigma}(M, \mathscr{D})$ induces a dense open embedding

$$\theta(M, \mathscr{D}) : \ \Sigma(M, \mathscr{D}) = \Sigma(M)_{\mathscr{D}} \hookrightarrow \hat{\Sigma}(M, \mathscr{D}) = \mathrm{Conf}(\Lambda(M), \mathscr{D}),$$

defined by $\mathscr{D}' \mapsto \mathscr{M}(\mathscr{D}, \mathscr{D}')$. Thus there is

$$\hat{\Sigma}'(M, \mathscr{D}) = \pi(M, \mathscr{D})^{-1}(\Sigma(M, \mathscr{D})) = \mathrm{Im}\ \theta(M, \mathscr{D})).$$

As $\mathrm{Conf}(\Lambda(M), \mathscr{D})$ is stable under the action of $P(\mathscr{D}) = \mathrm{Stab}(\mathscr{D}) \subset \mathrm{Gl}(k^{r+1})$, one may consider the induced action of $P(\mathscr{D})$ on $\hat{\Sigma}(M, \mathscr{D}) = \mathrm{Conf}(\Lambda(M), \mathscr{D})$.

By definition of $\Sigma(M, \mathscr{D})$ it is immediate that the natural action of $P(\mathscr{D}) \subset \mathrm{Gl}(\mathrm{k}^{r+1})$ on $\mathrm{Drap}(k^{r+1})$ leaves $\Sigma(M, \mathscr{D})$ stable and, a fortiori, its closure $\overline{\Sigma}(M, \mathscr{D})$ as well. On the other hand, the embedding $\theta(M, \mathscr{D}) : \Sigma(M, \mathscr{D}) \hookrightarrow \mathrm{Conf}(\Lambda(M), \mathscr{D})$ is $P(\mathscr{D})$-equivariant as it is easily seen.

This embedding may be seen as a section of the morphism

$$(\pi_2)_{\mathscr{D}} : \mathrm{Conf}(\Lambda(M), \mathscr{D}) \to \mathrm{Drap}(k^{r+1})$$

induced by π_2. Also $(\pi_2)_{\mathscr{D}}$ is $P(\mathscr{D})$-equivariant. Thus it has been proved

Theorem 3.21 *The morphism*

$$\pi(M, \mathscr{D}) = (\pi_2)_{\mathscr{D}} : \hat{\Sigma}(M, \mathscr{D}) \longrightarrow \overline{\Sigma}(M, \mathscr{D})$$

is a smooth resolution. Moreover $\pi(M, \mathscr{D})$ is $P(\mathscr{D})$-equivariant.

3.4   The Schematic point of view

The description of the above constructions in scheme theory is summarized here. The reader is referred to [24], Chap I, section 9, Foncteurs Representables elementaires, for a detailed description of the Grassmannians (resp. Flag, Stiefel) functors and their representability. In this setting the general linear group Schubert cells (resp. varieties) correspond to the linear group scheme Schubert cells (resp. schemes) over a base scheme.

Let S be a base scheme playing the role of the field k, and $\mathscr{M}$ a locally free $\mathscr{O}_S$-module of rank $r+1$ playing the role of the k-vector space k^E. Define the grassmannian functor $\mathscr{G}rass_n(\mathscr{M})$ by:

$$\Gamma(S', \mathscr{G}rass_n(\mathscr{M})) = \{\ \mathscr{S} \subset$$
$$\mathscr{M}_{S'} \text{ a submodule} \mid \mathscr{M}_{S'}/\mathscr{S} \text{ locally free }, rank\ \mathscr{S} = n\ \},$$

where $S' \longrightarrow S$ denotes an S-scheme. The above condition on $\mathscr{S}$ is equivalent to:

"the submodule $\mathscr{S} \subset \mathscr{M}_{S'}$ is locally a direct factor"

(resp. locally there exists a basis (e_i) of $\mathscr{M}$ adapted to $\mathscr{S}$). An S-morphism $f : S' \longrightarrow S''$ is associated with the mapping

$$\Gamma(S'', \mathscr{G}rass_n(\mathscr{M})) \longrightarrow \Gamma(S', \mathscr{G}rass_n(\mathscr{M}))$$

given by the pull-back: $\mathscr{S} \mapsto f^*(\mathscr{S})$.

It is easy to see that $\mathscr{G}rass_n(\mathscr{M})$ is representable by a smooth projective S-scheme $Grass_n(\mathscr{M})$ (cf. 1.1.1).

The functor $\mathscr{D}rap_{\underline{m}}(\mathscr{M})$ is defined as the subfunctor of $\displaystyle\prod_{1 \leq i \leq \lambda+1} \mathscr{G}rass_{m_i}(\mathscr{M})$

whose sections $(\mathscr{S}_i)_{1 \leq i \leq \lambda+1}$ satisfy the inclusions $\mathscr{S}_i \subset \mathscr{S}_{i+1}$, and is thus representable by a smooth projective S-scheme $Drap_{\underline{m}}(\mathscr{M})$(cf. 1.10). Plücker coordinates $(\Delta_H(\mathscr{S}))$ of a locally free rank n submodule $\mathscr{S} \subset \mathscr{M}$ may be defined, locally in S, as in the case $S = Spec\ (k)$. One has:

Lemma 3.22 *A locally free rank n submodule $\mathscr{S} \subset \mathscr{M}$ defines a section of $Grass_n(\mathscr{M})$ if and only if the ideal $\mathscr{I}(\Delta_H(\mathscr{S}))$ generated by the Plücker coordinates of $\mathscr{S}$ is the unit ideal .*

Let $\mathscr{S} \subset \mathscr{M}$ be locally a direct factor, and $\mathscr{S}' \subset \mathscr{S}$ a submodule. Then $\mathscr{S}'$ is locally a direct factor of $\mathscr{S}$ if and only if it is a direct factor of $\mathscr{M}$. (cf. [23], Exp. XXVI, 4.5.)

Definition 3.23 *Two $\mathscr{O}_S$-modules $\mathscr{H}, \mathscr{H}' \subset \mathscr{M}$ are in **standard position** (**std position**) if the quotient module $\mathscr{M}/\mathscr{H} \cap \mathscr{H}'$ is locally free. It is said that the couple of flags $(\mathscr{D}, \mathscr{D}')$ of $\mathscr{M}$, i.e. $\mathscr{D}$ and $\mathscr{D}'$ are respectively sections of $Drap_{\underline{m}}(\mathscr{M})$, and $Drap_{\underline{m}'}(\mathscr{M})$, is in a standard position if their terms are two by two in standard position.*
Denote by $Stand(\mathscr{M}) \subset Drap(\mathscr{M}) \times Drap(\mathscr{M})$ the subfunctor whose sections are the couples $(\mathscr{D}, \mathscr{D}')$ in standard position.

There is

Proposition 3.24 *$Stand(\mathscr{M})$ is a representable subfunctor. The group scheme $Gl(\mathscr{M}) = Aut_{\mathscr{O}_S}(\mathscr{M})$ stabilises $Stand(\mathscr{M})$ under the diagonal action. Let $\mathscr{M} = \mathscr{O}_S^{r+1}$, and $Relpos(\mathscr{O}_S^{r+1}) = Relpos(I_{r+1}) \times S$. There is a canonical isomorphism*

$$Stand(\mathscr{O}_S^{r+1})/Gl(\mathscr{O}_S^{r+1}) \simeq Relpos(\mathscr{O}_S^{r+1}).$$

(cf. loc. cit. 4.5.3.)
Let some definitions be introduced before giving the proof of the proposition. To $E \in Grass(I_{r+1})$ is associated the section of $Grass(\mathscr{O}_S^{r+1})$ given by the submodule $(\mathscr{O}_S^E \subset O_S^{r+1})$, and to $D = (E_1 \subset ... \subset E_l \subset I_{r+1}) \in Drap_{\underline{n}}(I_{r+1})$, the section $\mathscr{O}_S^D$ of $Drap_{\underline{n}}(\mathscr{O}_S^{r+1})$ defined by the flag:
$\mathscr{O}_S^D = (\mathscr{O}_S^{E_1} \subset ...\mathscr{O}_S^{E_l} \subset \mathscr{O}_S^{r+1})$.
Given $M \in Relpos(I_{r+1})$ denote by the same symbol the section of $Relpos(\mathscr{O}_S^{r+1})$ that it defines.

Definition 3.25 *Let $M \in Relpos_{(\underline{m},\underline{n})}$. Denote by $\Sigma(M) \subset Stand(\mathscr{O}_S^{r+1})$ the fiber of the canonical morphism $Stand(\mathscr{O}_S^{r+1}) \longrightarrow Relpos(\mathscr{O}_S^{r+1})$ over M, and by $\overline{\Sigma}(M)$ denote its schematic closure in $Drap_{\underline{m}}(\mathscr{O}_S^{r+1}) \times Drap_{\underline{n}}(\mathscr{O}_S^{r+1})$.*
Given a section $\mathscr{D}$ of $Drap_{\underline{m}}(\mathscr{O}_S^{r+1})$ let $\Sigma(M, \mathscr{D}) \subset Drap_{\underline{n}}(\mathscr{O}_S^{r+1})$ be the fiber of $\Sigma(M) \subset Stand(\mathscr{O}_S^{r+1}) \longrightarrow Drap_{\underline{n}}(\mathscr{O}_S^{r+1})$ over $\mathscr{D}$, and $\overline{\Sigma}(M, \mathscr{D})$ its

schematic closure in $Drap_{\underline{n}}(\mathscr{O}_S^{r+1})$.
Let

$$\hat{\Sigma}(M) \subset \prod_{(\alpha,\beta)\in\Lambda(M)} Grass_{m_{\alpha\beta}}(\mathscr{O}_S^{r+1}),$$

be the repesentable subfunctor defined as follows. The sections of $\hat{\Sigma}(M)$ over S' are characterized as the set of matrices $(\mathscr{H}_{\alpha\beta})_{(\alpha,\beta)\in\Lambda(M)}$ of submodules of $\mathscr{O}_S^{r+1}$ satisfying:

1. *the quotient module $\mathscr{O}_S^{r+1}/\mathscr{H}_{\alpha\beta}$ is locally free;*

2. *rk $\mathscr{H}_{\alpha\beta} = m_{\alpha\beta}$;*

3. *there are inclusions $\mathscr{H}_{\alpha\beta} \hookrightarrow \mathscr{H}_{\alpha+1\ \beta}$ (resp. $\mathscr{H}_{\alpha\beta} \hookrightarrow \mathscr{H}_{\alpha\ \beta+1}$).*

Write $\hat{\Sigma}(M,\mathscr{D})$ for the fiber over $\mathscr{D}$ of $\pi(M) : \hat{\Sigma}(M) \longrightarrow Drap_{\underline{m}}(\mathscr{O}_S^{r+1})$ induced by the projection morphism $\prod_{(\alpha,\beta)\in\Lambda(M)} Grass_{m_{\alpha\beta}}(\mathscr{O}_S^{r+1}) \longrightarrow Drap_{\underline{m}}(\mathscr{O}_S^{r+1})$ (cf. 3.4). There is a canonical morphism $\theta(M) : \Sigma(M) \longrightarrow \hat{\Sigma}(M)$ associating with the section $(\mathscr{D},\mathscr{D}')$ of $\Sigma(M)$ the $\Lambda(M)$-configuration $(\mathscr{H}_{\alpha\beta}) = (\mathscr{H}_\alpha \cap \mathscr{J}_\beta)$, where $\mathscr{D} = (\mathscr{H}_1 \subset ...\mathscr{H}_\lambda \subset \mathscr{H}_{\lambda+1} = \mathscr{O}_S^{r+1})$ (resp. $\mathscr{D}' = (\mathscr{J}_1 \subset ...\mathscr{J}_\lambda \subset \mathscr{J}_{\lambda+1} = \mathscr{O}_S^{r+1})$). Let $\theta(M,\mathscr{D})$ be the restriction of $\theta(M)$ to $\Sigma(M,\mathscr{D}) \hookrightarrow \Sigma(M)$. The explicit expression of $\theta(M)$ may be obtained as in the proof of proposition 3.7.

From the definition of $\hat{\Sigma}(M)$ as a subfunctor of the product of grassmannians defined in terms of inclusions, it immediately follows that $\hat{\Sigma}(M)$ is a closed S-subscheme of an S-projective scheme. Thus it follows that $\hat{\Sigma}(M)$ (resp. $\hat{\Sigma}(M,\mathscr{D})$) is a projective S-scheme.

The following proposition results from the proof of 2.3 and taking into account the following remark: let $\mathscr{S} \subset \mathscr{M}$ be a direct factor locally and $\mathscr{S}' \subset \mathscr{S}$ a submodule, then $\mathscr{S}'$ is a direct factor locally of $\mathscr{S}$ if and only if it is a direct factor of $\mathscr{M}$.

Proposition 3.26 *Denote by $\hat{\Sigma}'(M) \subset \hat{\Sigma}(M)$ the image of $\theta(M)$. The subscheme $\hat{\Sigma}'(M) \subset \hat{\Sigma}(M)$ is an open subscheme. Let $(\mathscr{H}_{\alpha\beta})$ be a section in the image of $\hat{\Sigma}'(M)$, then there exists an open subscheme U of S and a basis $(f_i)_{1\leqslant i\leqslant r+1}$ of $\mathscr{O}_U^{r+1}$ and a couple $(D,D') \in Drap_{\underline{m}}(I_{r+1}) \times Drap_{\underline{m}}(I_{r+1})$ such that if it is written*

$$\mathscr{M}(D,D') = (H_{\alpha\beta}) = (H_\alpha \cap J_\beta),$$

where $D = (H_1 \subset \cdots \subset H_\lambda \subset I_{r+1})$, $D' = (J_1 \subset \cdots \subset J_l \subset I_{r+1})$, then

$$\mathscr{H}_{\alpha\beta} = \mathscr{O}_S^{H_{\alpha\beta}}.$$

where $\mathscr{O}_U^{H_{\alpha\beta}}$ is defined in terms of the indexed basis (fi).

Proof of 3.24 *The first statement may be proved as proposition 3.8. In view of the above proposition the proof of the second statement follows from the same argument as in the proof of 2.6.*

Notation 3.27 *Let*

$$(\pi_1, \pi_2) : \hat{\Sigma}(M) \longrightarrow Drap_{\underline{m}}(\mathscr{O}_S^{r+1}) \times_S Drap_{\underline{n}}(\mathscr{O}_S^{r+1})$$

be the morphism induced by the projection morphism
$$\prod_{(\alpha,\beta)\in\Lambda(M)} Grass_{m_{\alpha\beta}}(\mathscr{O}_S^{r+1}) \longrightarrow Drap_{\underline{m}}(\mathscr{O}_S^{r+1}) \times_S Drap_{\underline{n}}(\mathscr{O}_S^{r+1}).$$ *Write*
$\pi(M) = (\pi_1, \pi_2)$ and denote by $\pi(M, \mathscr{D})$ the restriction of $\pi(M)$ to $\hat{\Sigma}(M, \mathscr{D})$ (cf. 3.4).

The decomposition of $\hat{\Sigma}(M)$ in terms of a sequence of fibrations, namely $(\hat{\Sigma}_\beta, \mathrm{pr}_\beta)$ and $(\hat{\Sigma}_\beta^{(\alpha)}, \mathrm{pr}_\beta^{(\alpha)})$, may be easily transposed into the scheme theoretic frame. It may be noted that the fiber of $\mathrm{pr}_\beta^{(\alpha)}$ is a grassmannian. On the other hand, this sequence of fibrations induces a decomposition $(\hat{\Sigma}'^{(\alpha)}_\beta, \mathrm{pr}'^{(\alpha)}_\beta)$ of the open subscheme $\hat{\Sigma}'(M)$ with the fiber of $\mathrm{pr}'^{(\alpha)}_\beta$ being a big cell of the fiber of $\mathrm{pr}_\beta^{(\alpha)}$, i.e. a big cell of a grassmannian. It will be proved that a big cell in a grassmannian is relatively schematically dense in that grassmannian (cf. [23], Exp. XXII, Proposition 4.1.2, and [27], Ch. IV, 11.10.). Thus we can now obtain the following basic proposition.

Proposition 3.28 *The open subscheme $\hat{\Sigma}'(M) \subset \hat{\Sigma}(M)$ is relatively schematically dense.*

Define

$$\hat{\Sigma}'(M, \mathscr{D}) = \hat{\Sigma}'(M) \cap \hat{\Sigma}(M, \mathscr{D}).$$

Let $P(\mathscr{D}) = Stab_{Gl(\mathscr{O}_S^{r+1})}$. We have then

Proposition 3.29 *The open subscheme $\hat{\Sigma}'(M)$ (resp. $\hat{\Sigma}'(M, \mathscr{D})$ is stable under the natural action of $Gl(\mathscr{O}_S^{r+1})$ (resp. $P(\mathscr{D})$). The morphisms $\theta(M)$ and $\pi(M)$ (resp. $\theta(M, \mathscr{D})$ and $\pi(M, \mathscr{D})$) are $Gl(\mathscr{O}_S^{r+1})$-equivariant (resp. $P(\mathscr{D})$-equivariant). $\pi(M)$ (resp. $\pi(M, \mathscr{D})$) defines a left inverse of $\theta(M)$ (resp. $\theta(M, \mathscr{D})$).*

From the transitivity of schematic closures (cf. [24], 6.10.) and proposition 3.28 it follows that the morphism $\pi(M)$ (resp. $\pi(M, \mathscr{D})$) factors through the closed embedding

$$\overline{\Sigma}(M) \hookrightarrow Drap_{\underline{m}}(\mathscr{O}_S^{r+1}) \times_S Drap_{\underline{n}}(\mathscr{O}_S^{r+1})$$

$$(resp. \ \overline{\Sigma}(M, \mathscr{D}) \hookrightarrow Drap_{\underline{n}}(\mathscr{O}_S^{r+1})).$$

Denote by $\pi(M)$ (resp. $\pi(M, \mathscr{D})$) the induced morphism $\hat{\Sigma}(M) \longrightarrow \overline{\Sigma}(M)$ (resp. $\hat{\Sigma}(M, \mathscr{D}) \longrightarrow \overline{\Sigma}(M, \mathscr{D})$. From proposition 3.4 it follows that this morphism is $Gl(\mathscr{O}^{r+1})$-equivariant (resp. $P(\mathscr{D})$-equivariant).
One has

$$\pi(M)^{-1}(\Sigma(M)) = Im \ \theta(M)$$

(resp.

$$\pi(M, \mathscr{D})^{-1}(\Sigma)(M)) = Im \ \theta(M, \mathscr{D})),$$

and a fortiori that the morphism $\theta(M)$ (resp. $\theta(M, \mathscr{D})$) induces an isomorphism

$$\pi(M)^{-1}(\Sigma(M)) = \hat{\Sigma}'(M)$$

(resp.

$$\pi(M, \mathscr{D})^{-1}(\Sigma(M)) = \hat{\Sigma}'(M, \mathscr{D})).$$

Theorem 3.30 *The morphism*

$$\pi(M, \mathscr{D}) = (\pi_2)_{\mathscr{D}} : \hat{\Sigma}(M, \mathscr{D}) \longrightarrow \overline{\Sigma}(M, \mathscr{D})$$

is a smooth resolution of singularities. Moreover $\pi(M, \mathscr{D})$ is $P(\mathscr{D})$-equivariant.

Theorem 3.31 *The morphism $\theta(M) : \Sigma(M) \longrightarrow \hat{\Sigma}(M)$ (resp. $\theta(M, \mathscr{D}) : \Sigma(M, \mathscr{D}) \longrightarrow \hat{\Sigma}(M, \mathscr{D})$) defines a section of the projective morphism $\pi(M)$ (resp. $\pi(M, \mathscr{D})$) over $\Sigma(M) \subset \overline{\Sigma}(M)$ (resp. $\Sigma(M, \mathscr{D}) \subset \overline{\Sigma}(M, \mathscr{D})$) and its image $Im \ \theta(M)$ (resp. $Im \ \theta(M, \mathscr{D})$) is an open set schematically dense in $\hat{\Sigma}(M)$ (resp. $\hat{\Sigma}(M, \mathscr{D})$). It is said the morphism*

$$\pi(M) : \hat{\Sigma}(M) \longrightarrow \overline{\Sigma}(M)$$

(resp.

$$\pi(M, \mathscr{D}) : \hat{\Sigma}(M, \mathscr{D}) \longrightarrow \overline{\Sigma}(M, \mathscr{D})) \ .$$

with $\hat{\Sigma}(M)$ (resp. $\hat{\Sigma}(M, \mathscr{D})$) a smooth S-scheme, is a smooth resolution of singularities of $\overline{\Sigma}(M)$ (resp. $\overline{\Sigma}(M, \mathscr{D})$).
Moreover $\pi(M)$ is $Gl(\mathscr{O}_S^{r+1})$-equivariant (resp. $\pi(M, \mathscr{D})$ is $P(\mathscr{D})$-equivariant).

Chapter 4

The Singular Locus of a Schubert Variety

By the following developments a detailed description of the infinitesimal structure of a $\hat{\Sigma}(M)$ is obtained, with an application to the determination of the singular locus of $\overline{\Sigma}(M)$. The calculations carried out in this chapter are quite involved and some proofs are outlined or simply omitted. The reader is referred to [15] and [16] for details. The main result amounts determine the Zariski's tangent space (resp. Nash tangent space) at a point of a Flag Schubert variety by a combinatorial procedure thus allowing the **characterization of the Singular cells** contained in a Schubert variety. This determination depends on the combinatorial structure of the smooth resolutions constructed in the preceding chapter, more precisely on the combinatorial fibers, $\mathrm{Conf}^{\mathrm{comb}}(\Lambda(M), D, D')$.

It is known that the smooth resolution associated with a Schubert variety in the former chapter contains an open subvariety isomorphic to the corresponding Schubert cell and that this cell is isomorphic to its pull-back. Hence, this pull-back is equal to this open subvariety. Anyway it may happen that the restriction of the resolving morphism to the localization along a "critical" Schubert cell, contained in the smooth open set of the Schubert variety, might not define an isomorphism, i.e. this smooth resolution might not be a strict resolution of singularities in Hironaka's sense (cf. [34]). The critical Schubert cells contained in the smooth open set of a Schubert variety are determined by the Zariski's tangent space at a point, combinatorially calculated, and this without any characteristic restrictions on the base field k.

The notation of the preceding chapter are retained.

4.1 The Schubert cells contained in a Schubert variety

Definition 4.1 *There is an* **order relation** *on* $Relpos(I_{r+1})$. *Given the relative position matrices* $M = (m_{\alpha\beta}), M' = (m'_{\alpha\beta}) \in Relpos_{(m,n)}(I_{r+1}) \subset \mathbb{N}^{(\lambda+1)\times(l+1)}$, *write* $M \leq M'$ *if* $m_{\alpha\beta} \leq m'_{\alpha\beta}$, *i.e. the order induced by the product order of* $\mathbb{N}^{(\lambda+1)\times(l+1)}$.

Proposition 4.2 *Let* x *be the point of* $Drap(k^{r+1})$ *defined by* $\mathscr{D}'$. *Then:*

$$x \in \overline{\Sigma}(M, D) \Longleftrightarrow M \leq M' = M(D, D'),$$

If $x \in \overline{\Sigma}(M, D)$ *then* $\Sigma(M', D) \subset \overline{\Sigma}(M, D)$.

Proof *It is recalled that the schematic image of* $Im\ \pi_2$ *is equal to* $\overline{\Sigma}(M, D)$. *Thus as* $k^{D'} \in \overline{\Sigma}(M, D)$ *there exists* $(\mathscr{H}_{\alpha\beta}) \in Conf(\Lambda(M); \mathscr{D}, \mathscr{D}')$ *with* $\pi_2((\mathscr{H}_{\alpha\beta})) = k^{D'}$. *This implies that* $M \leq M(\mathscr{D}, \mathscr{D}') = M(D, D')$.
To see that $M = (m_{\alpha\beta}) \leq M' = (m_{\alpha'\beta'}) \Longrightarrow x \in \overline{\Sigma}(M, D)$ *it suffices to prove that* $Conf^{comb}(\Lambda(M), D, D') \neq \emptyset$. *In other there is a* $\Lambda(M)$-*configuration* φ *subordinated to the* $\Lambda(M')$-*configuration* $(H_\alpha \cap J_\beta)$, *i.e. satisfying* $\varphi(\alpha, \beta) \subset H_\alpha \cap J_\beta$. *The graph* $\Lambda(M)$ *may be well-ordered by the lexicographical order obtained with* β *as the first variable. By considering the inequalities* $m_{\alpha+1\beta} + m_{\alpha\beta+1} - m_{\alpha\beta} \leq m_{\alpha+1\beta+1} \leq m'_{\alpha+1\beta+1}$ *and proceeding by induction, it is easy to see that the choice of* $H_{11} \subset H_1 \cap J_1$ *with* $|H_{11}| = m_{11}$ *may be completed in a* $\Lambda(M)$-*configuration* φ *subordinated to* $(H_\alpha \cap J_\beta)$. *The last assertion follows from the* $Stab\ k^D$-*stability of* $\overline{\Sigma}(M, D)$.

Corollary 4.3 *The closure of* $\Sigma(M, D)$ *is given by* $\overline{\Sigma}(M, D) = \cup \Sigma(M(D, D'), D)$ *where* D' *runs on the set* $\{D' \mid M \leq M(D, D')\}$.

4.2 A smoothness criterium for a Schubert variety

Notation 4.4 *Given a smooth* k-*variety* X *and a point* $x \in X$ *it is denoted by* $T(X)_x$ *the tangent space to* X *in* x, *i.e.* $T(X)_x = Hom_{k-vect}((\Omega^1_{X/k})_x, k)$, *where* $\Omega^1_{X/k}$ *denotes the* $\mathscr{O}_X$-**module of differentials** *of* X.
To a morphism $f : X \longrightarrow Y$ *the differential mapping* $T(f)_x : T(X)_x \longrightarrow T(Y)_y$ $(y = f(x))$ *is associated, defined as follows. Let* $f^*(\Omega^1_{Y/k}) \longrightarrow \Omega^1_{X/k}$ *be the induced morphism of* $\mathscr{O}_X$-*modules, then* $T(f)_x : Hom_{k-vect}((\Omega^1_{X/k})_x, k) \longrightarrow Hom_{k-vect}(f^*(\Omega^1_{Y/k})_y, k)$ *is the dual* k-*vector space mapping of* $f^*(\Omega^1_{Y/k})_x \longrightarrow (\Omega^1_{X/k})_x$.
The tangent space $T(X)_x$ *may be calculated in terms of* **dual numbers**

$$Hom_k^x \left(Spec(k[\epsilon]/(\epsilon^2)), X \right) \simeq T(X)_x.$$

The left term denotes the set of k-morphisms $f : Spec(k[\epsilon]/(\epsilon^2)) \longrightarrow X$ giving
x by composition with $Spec(k) \longrightarrow Spec(k[\epsilon]/(\epsilon^2))$.
The tangent $\mathcal{O}_X$-module may be written as

$$\mathcal{T}^1_{X/k} = Hom_{\mathcal{O}_X}(\Omega^1_{X/k}, \mathcal{O}_X).$$

Thus $T(X)_x = \mathcal{T}^1_{X/k} \otimes_{\mathcal{O}_X}(\mathcal{O}_X/m_{X,x})$, where $m_{X,x} \subset \mathcal{O}_X$ denotes the maximal
ideal of the local ring $\mathcal{O}_{X,x}$.

The reader is referred to the SGA III, vol I (cf. [22]), for the above procedure
for calculating the tangent space to a k-scheme X by means of the functor
that it defines.

Definition 4.5

   Let X be a smooth N-dimensional k-variety, $Y \subset X$ an integral d-
dimensional k-subvariety , smooth at the generic point ξ_Y, and $Y' \subset Y$ an
open smooth subvariety. Denote by $\tilde{Y} \longrightarrow Y$ the Y-variety obtained as the
Zariski closure of $Grass_d(\mathcal{T}^1_{Y'/k}) \subset Grass_d(\mathcal{T}^1_{X/k})_Y$. $\tilde{Y}$ is called **the Nash
transform** of Y. Given $y \in Y$ the fiber $\tilde{Y}_y$ satisfies: $\tilde{Y}_y \subset Grass_d(\mathcal{T}^1_{X/k})_y = Grass_d(T(X)_y)$. Let $T^{nash}(Y)_y \subset T(X)_y$ denote the minimal k-subspace
$S \subset T(X)_y$ satisfying $\tilde{Y}_y \subset Grass_d(S)$ **(the Nash tangent space** of Y at
y).

Observe that the existence of the subspace S results from the fact that
given two subspaces S', S'' of $T(X)_y$ one has $Grass_d(S') \cap Grass_d(S'') = Grass_d(S \cap S'')$. Loosely speaking $T^{nash}(Y)_y$ is the subspace of $T(X)_y$ gener-
erated by the limiting subspaces of Y' at y. The following inequality holds:

$$dim\ Y \leq dim_k T^{nash}(Y)_y \ ,$$

where $dim\ Y$ denotes the dimension of the variety Y. It may be recalled that
$dim\ Y$ is also given by the transcendance degree of the function field $k(Y)$ of
Y over k. From the above definition the following result is obtained:

Proposition 4.6

   1) Let Y be a subvariety of the smooth variety X, as the preceding def-
      inition, and $Z \subset X$ be a smooth k-subvariety containing Y. Then
      $T^{nash}(Y)_y \subset T(Z)_y$.

   2) Let $(Z_i)_{1 \leq i \leq n}$ be a family of smooth subvarieties X containing Y. Then

$$T^{nash}(Y)_y \subset \cap T(Z_i)_y.$$

The following result is a corollary of the jacobian criterium.

Proposition 4.7

The notation of the preceding proposition is retained. Let $y \in Y$. The follow-ing statements are equivalent:

1) *There exists a family of smooth subvarieties $(Z_i)_{1 \leq i \leq n}$ of X containing Y with $T^{nash}(Y)_y = \cap T(Z_i)_y$.*

2) *There exists a smooth subvariety $Y \subset Z$ with $T^{nash}(Y)_y = T(Z)_y$*

Under the above hypothesis $dim\, T^{nash}(Y)_y = dim\, Y$ implies $dim\, Y = dim\, Z$. Hence we have the **smoothness criterium** of Y in y:

Proposition 4.8

Let it be supposed that $T^{nash}(Y)_y$ satisfies one of the two equivalent conditions above.

Then the variety Y is smooth in $y \iff dim_k\, T^{nash}(Y)_y = dim\, Y$.

Proof It suffices to prove that "$dim_k\, T^{nash}(Y)_y = dim\, Y \implies Y$ is smooth in y". It may be supposed that $Z = Spec(A)$ (resp. that $Y = Spec(B)$), where A (resp. B) is an integral finite type k-algebra, and that $dim(A) = dim(B) = dim_k\, T^{nash}(Y)_y$. Here $dim(A)$ (resp. $dim(B)$) denotes the ring dimension of A (resp. B). Let $f : A \longrightarrow B$ be the k-algebra morphism corresponding to the embedding $Y \hookrightarrow Z$. From the formula $dim(A) = dim(A_{\mathfrak{p}}) + dim(A/\mathfrak{p})$ (cf. [48], Ch 3, Proposition 15), it is deduced that $dim(A_{\mathfrak{p}}) = 0$ and thus $\mathfrak{p} = Ker\,(f) \subset A$ is the zero prime ideal and f is an isomorphism. This clearly proves that Y is smooth in y.

Notation 4.9 1) *For the sake of briefness, put $\overline{\Sigma} = \overline{\Sigma}(M, D)$,*

$$\left(resp.\ \Sigma = \Sigma(M, D), Conf = Conf(\Lambda(M), D),\ Conf_{D'}^{comb}\right.$$

$$= Conf^{comb}(\Lambda(M), D, D'), Conf_{D'} = Conf(\Lambda(M); D, D'))$$

2) *Given a subvariety $Z \subset Y$ denote by Z_y **the germ of** Z in $y \in Y$ (resp. **the local scheme of** Z at y (cf. [24])).*

The following proposition exhibits the connection between the tangent space at a point x, $T(Conf(\Lambda(M), D))_x$ of the smooth resolution $Conf(\Lambda(M), D)$, and the Nash tangent space $T^{nash}(\overline{\Sigma}(M, D))_{\pi_2(x)}$.

Proposition 4.10

Given by a point x of $Conf$, one has

$$Im\, T(\pi_2)_x \subset T^{nash}(\overline{\Sigma})_{\pi_2(x)}\ .$$

Proof *Let V be a valuation ring and $f : Spec(V) \longrightarrow Conf$ a morphism sending the generic point ξ to the generic point of $Conf$ (resp. the special point ξ_0 to x). Thus the image $T(\pi_2)_\xi(T(Conf)_\xi)$ defines a section $\sigma_K :$ $Spec(K) \longrightarrow Grass_d(\mathcal{T}^1_{Drap_{\underline{n}}(k^{r+1})/k})$ over $Spec(K)$, where K denotes the quotient field of V. By the valuation criterion of property there is a section σ_V on $Spec(V)$ extending σ_K. Observe that $\pi_2(\xi)$ is the generic point of Σ. Thus $\sigma_V(\xi_0)$ defines a subspace S of $T(Drap_{\underline{n}}(k^{r+1}))_{\pi_2(x)}$ necessarily contained in $T^{nash}(\overline{\Sigma})_{\pi_2(x)}$. Given $v \in T(Conf)_x$ let it be assumed that $T(\pi_2)_x(L_v) \neq 0$ ($L_v = Vect_k(v)$). We prove now that:*

$$T(\pi_2)_x(L_v) \subset T^{nash}(\overline{\Sigma})_{\pi_2(x)} \ .$$

Lift L_v to a section $\tilde{L}_v \subset (\mathcal{T}^1_{Conf/k})_V$. Thus $T(\pi_2)_\xi((\tilde{L}_v)_\xi) \subset T(\pi_2)_\xi(T(Conf)_\xi)$. This clearly gives: $T(\pi_2)_x(L_v) \subset S \subset T^{nash}(\overline{\Sigma})_{\pi_2(x)}$. Thus $Im(T(\pi_2)_x \subset T^{nash}(\overline{\Sigma})_{\pi_2(x)}$.

Definition 4.11

1) *Let $\mathscr{D}' \in \overline{\Sigma}(M, D)$. Define*

$$\tilde{T}(\overline{\Sigma}(M, D))_{\mathscr{D}'} = Vect_k\left(\bigcup_{\varphi \in \pi_2^{-1}(\mathscr{D}')} Im(T(\pi_2)_x) \right) \subset T(Drap_{\underline{n}}(k^{r+1}))_{\mathscr{D}'} \ ,$$

*as the subspace of $T(Drap_{\underline{n}}(k^{r+1}))_{\mathscr{D}'}$ generated by the images of the differentials $T(\pi_2)_x$ of π_2 at the points x of the fiber $\pi_2^{-1}(\mathscr{D}') = Conf(\Lambda(M), D)_{\mathscr{D}'}$. $\tilde{T}(\overline{\Sigma}(M, D))_{\mathscr{D}'}$ is called the $Conf$-**tangent space** of $\overline{\Sigma}(M, D))$ at $\mathscr{D}'$.*

2) *Assume $\mathscr{D}' = k^{D'}$ ($D' \in Drap_{\underline{n}}(I_{r+1})$). Define*

$$\tilde{T}^{comb}(\overline{\Sigma}(M, D))_{\mathscr{D}'} = Vect_k\left(\bigcup_{\varphi \in Conf(\Lambda(M), D, D')} Im(T(\pi_2)_{k^\varphi}) \right) \ ,$$

*the subspace of $\tilde{T}(\overline{\Sigma}(M, D))_{\mathscr{D}'}$ generated by the images of the differentials $T(\pi_2)_{k^\varphi}$ of π_2 at the points φ of the combinatorial fiber $Conf(\Lambda(M), D, D')$ over D'. $\tilde{T}^{comb}(\overline{\Sigma}(M, D))_{\mathscr{D}'}$ is called the **combinatorial $Conf$-tangent space** of $\overline{\Sigma}(M, D))$ at $\mathscr{D}'$.*

From the above proposition it follows that:

$$\tilde{T}^{comb}(\overline{\Sigma}(M, D))_{\mathscr{D}'} \subset \tilde{T}(\overline{\Sigma}(M, D))_{\mathscr{D}'} \subset T^{nash}(\overline{\Sigma}(M, D))_{\mathscr{D}'} \ .$$

Remark 4.12   1) *The Nash tangent space $T^{nash}(\overline{\Sigma}(M, D))_y$ at a point $y \in \overline{\Sigma}(M, D)$ satisfies the following equivariance property relative to $Stab \ k^D \subset Gl(k^{r+1})$:*

$$T(g)_y(T^{nash}(\overline{\Sigma}(M, D))_y) = T^{nash}(\overline{\Sigma}(M, D))_{g(y)} \ \ (g \in Gl(k^{r+1})) \ .$$

2) *The subspace $\tilde{T}(\overline{\Sigma}(M,D))_y$ satisfies the same property of equivariance, and $\tilde{T}^{comb}(\overline{\Sigma}(M,D))_y = \tilde{T}(\overline{\Sigma}(M,D))_y$ for $y = k^{D'}$, as it results from the main theorem proved below.*

By combining the above three propositions the **combinatorial smoothness criterium** of $\overline{\Sigma}(M,D)_{\mathscr{D}'}$ at $\mathscr{D}'$ is obtained:

Proposition 4.13 *Let $y \in \overline{\Sigma}(M,D)$ given by $\mathscr{D}'$, and $(Z_i)_{1\leq i\leq n}$ be a family of smooth subvarieties of $Drap_{\underline{n}}(k^{r+1})$ satisfying:*

1) $\overline{\Sigma}_y \subset (Z_i)_y$;

2) $\displaystyle\bigcap_{1\leq i\leq n} T(Z_i)_y = \tilde{T}^{comb}(\overline{\Sigma})_y.$

Then one has;

a) $\tilde{T}^{comb}(\overline{\Sigma}(M,D))_{\mathscr{D}'} = \tilde{T}(\overline{\Sigma}(M,D))_{\mathscr{D}'} = T^{nash}(\overline{\Sigma}(M,D))_{\mathscr{D}'}$;

b) $\overline{\Sigma}$ is smooth at $y \Longleftrightarrow dim_k \tilde{T}^{comb}(\overline{\Sigma})_y = dim_k \overline{\Sigma}$.

In the following sections $\tilde{T}^{comb}(\overline{\Sigma}(M,D))_{\mathscr{D}'}$ is calculated. We show a family of smooth subvarieties of $Drap_{\underline{n}}(k^{r+1})$ satisfying the conditions of the **smoothness criterium** at the point y corresponding to $\mathscr{D}'$. Given $\varphi \in Conf(\Lambda(M), D, D')$ a canonical basis $\tilde{\mathcal{B}}_{(\varphi,D)}$ of the tangent space $T(Conf(\Lambda(M), D))_{k^\varphi}$ is constructed, and its image by the differential $T(\pi_2)_{k^\varphi}$ is calculated in terms of the combinatorics of $Conf(\Lambda(M), D, D')$. The set

$$\tilde{\mathcal{B}}_{(D,D')} = \bigcup_{\varphi\in Conf(\Lambda(M),D,D')} T(\pi_2)_{k^\varphi}(\tilde{\mathcal{B}}_{(\varphi,D)})$$

is a basis of $\tilde{T}^{comb}(\overline{\Sigma}(M,D))_{\mathscr{D}'}$. The next step is to express $\tilde{T}^{comb}(\overline{\Sigma}(M,D))_{\mathscr{D}'}$ as an intersection of tangent spaces to smooth subvarieties containing the point y given by $\mathscr{D}'$ using the expression of $\tilde{\mathcal{B}}_{(D,D')}$, and thus to prove that $\tilde{T}^{comb}(\overline{\Sigma}(M,D))_{\mathscr{D}'} = T^{nash}(\overline{\Sigma}(M,D))_{\mathscr{D}'}$. This result is also obtained by means of the combinatorics of $Conf(\Lambda(M), D, D')$.

4.3 Calculation of the tangent spaces of some configurations varieties

The following canonical isomorphisms are obtained from remark 4.4. The reader is referred to the SGA III, vol I (cf. [22]), for the procedure for calculating the tangent space to a k-scheme X by means of the functor that it defines and to [15] and [16] for details.

1) The tangent space $T(Drap_{\underline{n}}(k^{r+1}))_{\mathscr{D}}$ $(\mathscr{D} = (\mathscr{H}_1 \cdots \subset \mathscr{H}_l \subset k^{r+1})$ is given by:

$$T(Drap_{\underline{n}}(k^{r+1}))_{\mathscr{D}} = Ker\left(\prod_{1 \leq \beta \leq l} Hom_{k-vect}(\mathscr{H}_\beta, k^{r+1}/\mathscr{H}_\beta) \xrightarrow{\longrightarrow}\right.$$

$$\left.\prod_{1 \leq \beta,\beta' \leq l} Hom_{k-vect}(\mathscr{H}_{inf(\beta',\beta)}, k^{r+1}/\mathscr{H}_{sup(\beta',\beta)})\right). \quad \text{In fact there is an}$$

isomorphism:

$$Ker\left(\prod_{1 \leq \beta \leq l} Hom_{k-vect}(\mathscr{H}_\beta, k^{r+1}/\mathscr{H}_\beta) \xrightarrow{\longrightarrow}\right.$$

$$\left.\prod_{1 \leq \beta \leq l} Hom_{k-vect}(\mathscr{H}_\beta, k^{r+1}/\mathscr{H}_{\beta+1})\right)$$

$$\simeq Ker\left(\prod_{1 \leq \beta \leq l} Hom_{k-vect}(\mathscr{H}_\beta, k^{r+1}/\mathscr{H}_\beta) \xrightarrow{\longrightarrow}\right.$$

$$\left.\prod_{1 \leq \beta \leq l} Hom_{k-vect}(\mathscr{H}_{inf(\beta',\beta)}, k^{r+1}/\mathscr{H}_{sup(\beta',\beta)})\right).$$

2) To the embedding

$$i_{\mathrm{Conf}(\Lambda(M))} : \mathrm{Conf}(\Lambda(M)) \hookrightarrow \prod_{(\alpha,\beta)} Grass_{m_{\alpha\beta}}(k^{r+1}),$$

it corresponds the associated differential at the point $(\mathscr{H}_{\alpha\beta}) \in \mathrm{Conf}(\Lambda(M))$:

$$T(i_{\mathrm{Conf}(\Lambda(M))})_{(\mathscr{H}_{\alpha\beta})} : \mathrm{T}(\mathrm{Conf}(\Lambda(M)))_{(\mathscr{H}_{\alpha\beta})} \longrightarrow$$

$$T(\prod_{(\alpha,\beta)} Grass_{m_{\alpha\beta}}(k^{r+1}))_{(\mathscr{H}_{\alpha\beta})}$$

where

$$T(\prod_{(\alpha,\beta)} Grass_{m_{\alpha\beta}}(k^{r+1}))_{(\mathscr{H}_{\alpha\beta})} \simeq \prod_{(\alpha,\beta)\in\Lambda(M)} Hom_{k-vect}(\mathscr{H}_{\alpha\beta}, k^{r+1}/\mathscr{H}_{\alpha\beta}).$$

The tangent space $T(Conf(\Lambda(M))_{(\mathscr{H}_{\alpha\beta})}$ is given by:

$$T(Conf(\Lambda(M))_{(\mathscr{H}_{\alpha\beta})} = Ker\left(\prod_{(\alpha,\beta)\in\Lambda(M)} Hom_{k-vect}(\mathscr{H}_{\alpha\beta}, k^{r+1}/\mathscr{H}_{\alpha\beta}) \xrightarrow{\longrightarrow}\right.$$

$$\prod_{((\alpha,\beta),(\alpha',\beta'))\in Edg\ \Lambda(M)} Hom_{k-vect}(\mathscr{H}_{\alpha\beta} \cap \mathscr{H}_{\alpha'\beta'}, k^{r+1}/\mathscr{H}_{(\alpha,\beta)} +$$

$$\left.\mathscr{H}_{(\alpha',\beta')})\right).$$

3) The tangent space $T(\Sigma(M, \mathscr{D}))_{\mathscr{J}} \subset T(Grass_n(k^{r+1}))_{\mathscr{J}}$ of $\Sigma(M, \mathscr{D}) \subset Grass_n(k^{r+1})$ at $\mathscr{J}$ where $\mathscr{D} = (\mathscr{H}_1 \cdots \subset \mathscr{H}_\lambda \subset k^{r+1})$ is obtained as the following subspace:

$$Ker \left(Hom_{k-vect}(\mathscr{J}, k^{r+1}/\mathscr{J}) \longrightarrow \right.$$

$$\left. \prod_{1 \leq \alpha \leq \lambda+1} Hom_{k-vect}(\mathscr{H}_\alpha \cap \mathscr{J}, k^{r+1}/(\mathscr{H}_\alpha + \mathscr{J})) \right) .$$

More generally the tangent space $T(\Sigma(M, \mathscr{D}))_{\mathscr{D}'}$ $(\mathscr{D}' = (\mathscr{J}_1 \cdots \subset \mathscr{J}_l \subset k^{r+1}))$ of a Schubert cell $\Sigma(M, \mathscr{D}) \subset Drap(k^{r+1})$ is given by the following subspace of $T(Drap_{\underline{n}}(k^{r+1}))_{\mathscr{D}'}$:

$$\bigcap_{1 \leq \beta \leq l} Ker \left(T(Drap_{\underline{n}}(k^{r+1}))_{\mathscr{D}'} \longrightarrow \right.$$

$$\left. \prod_{1 \leq \alpha \leq \lambda+1} Hom_{k-vect}(\mathscr{H}_\alpha \cap \mathscr{J}_\beta, k^{r+1}/(\mathscr{H}_\alpha \cap \mathscr{J}_{\beta+1} + \mathscr{J}_\beta)) \right).$$

4.4 The Young indexation of the tangent basis to a Schubert cell at a combinatorial point

The aim of this section and the two following ones is the determination of a canonical basis of the tangent space to a configuration variety at a "combinatorial point", i.e. at a point given by a $\Lambda(M)-$ configuration "adapted" to the canonical basis of k^{r+1}. The canonical isomorphism:

$$T(Drap_{\underline{n}}(k^{r+1}))_{k^{D'}} = Ker \left(\prod_{1 \leq \beta \leq l} Hom_{k-vect}(k^{J_\beta}, k^{r+1}/k^{J_\beta}) \overset{\longrightarrow}{\longrightarrow} \right.$$

$$\left. \prod_{1 \leq \beta \leq l} Hom_{k-vect}(k^{J_\beta}, k^{r+1}/k^{J_{\beta+1}}) \right),$$

gives rise to:

Proposition 4.14 *There is an isomorphism:*

$$T(Drap_{\underline{n}}(k^{r+1}))_{k^{D'}} \simeq \prod_{1 \leq \beta \leq l} Hom_{k-vect}(k^{J_\beta}, k^{J_{\beta+1}-J_\beta})$$

induced by the linear mapping defined as follows. To an element

$$(\nu_\beta)_{1 \leq \beta \leq l} \in Ker \left(\prod_{1 \leq \beta \leq l} Hom_{k-vect}(k^{J_\beta}, k^{r+1}/k^{J_\beta}) \overset{\longrightarrow}{\longrightarrow} \right.$$

$$\prod_{1 \leq \beta \leq l} Hom_{k-vect}(k^{J_\beta}, k^{r+1}/k^{J_{\beta+1}}) \Big)$$

is associated the element:

$$(\nu'_\beta)_{1 \leq \beta \leq l} \in \prod_{1 \leq \beta \leq l} Hom_{k-vect}(k^{J_\beta}, k^{J_{\beta+1}-J_\beta}), \text{ given by}$$

$$\nu'_\beta = \nu_\beta - \pi_{J^\perp_{\beta+1}} \circ \nu_\beta .$$

The set $\coprod_{1 \leq \beta \leq l} J_\beta \times (J_{\beta+1} - J_\beta)$ *indexes a basis of* $T(Drap_{\underline{n}}(k^{r+1}))_{kD'}$. *More precisely*

$$\mathcal{B}_{D'} := \coprod_{1 \leq \beta \leq l} (E^\beta_{ij})_{(i,j) \in J_\beta \times (J_{\beta+1}-J_\beta)}$$

is a basis of $T(Drap_{\underline{n}}(k^{r+1}))_{kD'}$ **(The combinatorial basis of** $T(Drap_{\underline{n}}(k^{r+1}))_{kD'}$**)** .

As a particular case of this proposition it results that $\mathcal{B}_J : (E_{ij})_{(i,j) \in J \times J^\perp}$ is the **combinatorial basis of** $T(Grass_n(k^{r+1}))_{k^J}$.

Notation 4.15 *Let* $\varphi \in Conf(\Lambda(M), D)$ *with* $p_2(M, D) : \varphi \mapsto D'$. *If no confusion arises denote by* $T(\pi_2(M, D))_{k\varphi}$ *the composition of* $T(\pi_2(M, D))_{k\varphi}$ *with the identification* $T(Drap_{\underline{n}}(k^{r+1}))_{kD'} \simeq \prod_{1 \leq \beta \leq l} Hom_{k-vect}(k^{J_\beta}, k^{J_{\beta+1}-J_\beta})$.

Definition 4.16 *A* D'**-Young data** *is a couple of sequences of chains* $(\underline{\mathcal{D}}, \underline{\mathcal{D}}^*) = ((\mathcal{D}_\beta), (\mathcal{D}^*_\beta))$, *where* $(\mathcal{D}_\beta) \in \prod_{1 \leq \beta \leq l} Chain(J_\beta)$, *with* $\mathcal{D}_\beta = (K_{1\beta} \cdots \subset K_{\lambda_\beta \beta} \subset J_\beta)$ *(resp.* $(\mathcal{D}^*_\beta) \in \prod_{1 \leq \beta \leq l} Chain(J_{\beta+1})$, *with* $\mathcal{D}^*_\beta = (K^*_{1\beta} \cdots \subset K^*_{\lambda_\beta \beta} \subset J_{\beta+1})$ *satisfying* $J_\beta \subset K^*_{1\beta})$. *To a* D'*-Young data* $(\underline{\mathcal{D}}, \underline{\mathcal{D}}^*)$ *are associated:*

1) $Y^+_\beta(\underline{\mathcal{D}}, \underline{\mathcal{D}}^*) = \bigcup_{1 \leq i \leq \lambda_\beta} K_{i\beta} \times (J_{\beta+1} - K^*_{i\beta}) \subset J_\beta \times (J_{\beta+1} - J_\beta);$

2) $Y^+(\underline{\mathcal{D}}, \underline{\mathcal{D}}^*) = (Y^+_\beta(\underline{\mathcal{D}}, \underline{\mathcal{D}}^*))_{1 \leq \beta \leq l}$, *with* $\coprod Y^+_\beta \subset \coprod J_\beta \times (J_{\beta+1} - J_\beta);$

3) $Y^-_\beta(\underline{\mathcal{D}}, \underline{\mathcal{D}}^*) = J_\beta \times (J_{\beta+1}/J_\beta) - Y^+_\beta(\underline{\mathcal{D}}, \underline{\mathcal{D}}^*);$

4) $Y^-(\underline{\mathcal{D}}, \underline{\mathcal{D}}^*) = (Y^-_\beta(\underline{\mathcal{D}}, \underline{\mathcal{D}}^*))_{1 \leq \beta \leq l}.$

Remark 4.17 *The subset* $Y^+(\underline{\mathcal{D}}, \underline{\mathcal{D}}^*)$ *(resp.* $Y^-(\underline{\mathcal{D}}, \underline{\mathcal{D}}^*)$*) of the set of indices* $\coprod J_\beta \times (J_{\beta+1} - J_\beta)$ *defines a subfamily of the basis* $\mathcal{B}_{D'}$. *A basis* $\subset \mathcal{B}_{D'}$ *of the tangent space to a Schubert cell* $\Sigma(M, D)$ *at* $k^{D'} \in \Sigma(M, D)$ *is defined by means of Young data.*

4.5 The tangent basis to a Configuration variety at a combinatorial configuration

Remark that $T(Conf(\Lambda(M))_{k^\varphi} \subset \bigoplus_{(\alpha,\beta)} T(Grass_{m_{\alpha\beta}}(k^{r+1}))_{k^{\varphi(\alpha,\beta)}}$, where (α,β) runs on the vertices of $\Lambda(M)$. The above description of $T(Conf(\Lambda(M))_{k^\varphi}$, where $\varphi \in Conf^{comb}(\Lambda(M))$, may be applied in order to determine a combinatorial basis of this tangent space. With $\varphi \in Conf^{comb}(\Lambda(M))$ is associated $(k^{\varphi(\alpha,\beta)}) \in Conf(\Lambda(M))$. On the other hand, there is a canonical basis $\coprod_{(\alpha,\beta)} \mathcal{B}_{\varphi(\alpha,\beta)}$ of the tangent space $\bigoplus_{(\alpha,\beta)} T(Grass_{m_{\alpha\beta}}(k^{r+1}))_{k^{\varphi(\alpha,\beta)}}$ of the product $\prod_{(\alpha,\beta)} Grass_{m_{\alpha\beta}}(k^{r+1}))_{k^\varphi}$ at $(k^{\varphi(\alpha,\beta)})$, where

$$\mathcal{B}_{\varphi(\alpha,\beta)} = (E^{(\alpha,\beta)}_{(i,j)})_{(i,j)\in\varphi(\alpha,\beta)\times\varphi(\alpha,\beta)^\perp}$$

is the canonical basis of $T(Grass_{m_{\alpha\beta}}(k^{r+1}))_{k^{\varphi(\alpha,\beta)}}$ identified with the set of elementary matrices indexed by $\varphi(\alpha,\beta) \times \varphi(\alpha,\beta)^\perp$. Write $R^{(\alpha,\beta)}_\varphi = \varphi(\alpha,\beta) \times \varphi(\alpha,\beta)^\perp$, and let

$$\mathbb{E}(\varphi) = \coprod_{(\alpha,\beta)} \{(\alpha,\beta)\} \times R^{(\alpha,\beta)}_\varphi .$$

Thus $\mathbb{E}(\varphi)$ indexes the basis

$$(E^{(\alpha,\beta)}_{(i,j)})_{((\alpha,\beta),(i,j))\in\mathbb{E}(\varphi)}$$

of

$$T(\prod_{(\alpha,\beta)} Grass_{m_{\alpha\beta}}(k^{r+1}))_{k^\varphi} \simeq \bigoplus_{(\alpha,\beta)} T(Grass_{m_{\alpha\beta}}(k^{r+1}))_{k^{\varphi(\alpha,\beta)}} \simeq \bigoplus_{(\alpha,\beta)} k^{R^{(\alpha,\beta)}_\varphi} .$$

The first isomorphism given by $\bigoplus_{(\alpha,\beta)} T(p_{\alpha\beta})_{k^\varphi}$, where $p_{\alpha\beta}$: $\prod_{(\alpha,\beta)} Grass_{m_{\alpha\beta}}(k^{r+1}) \longrightarrow Grass_{m_{\alpha\beta}}(k^{r+1})$, denotes the canonical projection.

Definition 4.18 *An* **equivalence relation** *is defined on the set* $\mathbb{E}(\varphi)$ *as follows. Write* $((\alpha,\beta),(i,j)) \sim ((\alpha',\beta'),(i',j'))$, *if there is a sequence of vertices of* $\Lambda(M)$:

$$((\alpha_\rho,\beta_\rho))_{0\leq\rho\leq r} : (\alpha,\beta) = (\alpha_0,\beta_0),(\alpha_1,\beta_1),\cdots,(\alpha_r,\beta_r) = (\alpha',\beta')$$

satisfying:

(a) $((\alpha_\rho,\beta_\rho),(\alpha_{\rho+1},\beta_{\rho+1}))$ $(0 \leq \rho \leq r-1)$ *is an edge of* $\Lambda(M)$, *i.e.* $((\alpha_\rho,\beta_\rho))_{0\leq\rho\leq r}$ *is* **a path of the graph** $\Lambda(M)$ **with origin** (α_0,β_0) *and* **extremity** (α_r,β_r).

(b) $(i,j) = (i',j')$ *and* $(i,j) \in \bigcap_{0 \leqq \rho \leqq r} R_\varphi^{(\alpha_\rho, \beta_\rho)}$.

Denote by $\mathbb{B}(\varphi) = \mathbb{E}(\varphi)/ \sim$ **the set of equivalence classes defined by** **"$\sim$".**

Observe that to an equivalence class is associated a couple $(i,j)_\mathscr{C}$, characterized by: $((\alpha, \beta), (i,j)) \in \mathscr{C} \Rightarrow (i,j) = (i,j)_\mathscr{C}$.

   An equivalence class $\mathscr{C} \in \mathbb{B}(\varphi)$ defines an element

$$E_\mathscr{C} = \underset{((\alpha,\beta),(i,j)) \in \mathscr{C}}{\Sigma} E_{(i,j)}^{(\alpha,\beta)}.$$

of $\underset{(\alpha,\beta)}{\bigoplus} T(Grass_{m_{\alpha\beta}}(k^{r+1}))_{k^{\varphi(\alpha,\beta)}}$. Let $D = (H_1 \subset \ldots \subset H_{\lambda+1})$. Write

$$\mathbb{B}(\varphi, D) = \{ \ \mathscr{C} \in \mathbb{B}(\varphi) \mid \mathscr{C} \cap (\underset{1 \leqq \alpha \leqq \lambda+1}{\coprod} \{(\alpha, l+1)\} \times R_\varphi^{(\alpha, l+1)}) = \emptyset \ \}$$

$\Big($resp.

$$\mathbb{B}^c(\varphi, D) = \{ \ \mathscr{C} \in \mathbb{B}(\varphi) \mid \mathscr{C} \cap (\underset{1 \leqq \alpha \leqq \lambda+1}{\coprod} \{(\alpha, l+1)\} \times R_\varphi^{(\alpha, l+1)}) \neq \emptyset \ \} \Big),$$

and

$$\mathcal{B}_{(\varphi,D)} = (E_\mathscr{C})_{\mathscr{C} \in \mathbb{B}(\varphi,D)} \ (resp. \ \mathcal{B}^c_{(\varphi,D)} = (E_\mathscr{C})_{\mathscr{C} \in \mathbb{B}^c(\varphi,D)} \ , \mathcal{B}_\varphi = (E_\mathscr{C})_{\mathscr{C} \in \mathbb{B}(\varphi)}) \ .$$

The following relation results immediatly from the definitions above

$$\mathcal{B}_\varphi = \mathcal{B}_{(\varphi,D)} \coprod \mathcal{B}^c_{(\varphi,D)} \ .$$

   The following is deduced easily:

Proposition 4.19 *It is also denoted by* $p_{\alpha\beta} : Conf(\Lambda(M)) \longrightarrow$ $Grass_{m_{\alpha\beta}}(k^{r+1})$ *the morphism induced by the canonical projection.*

   1) Let $\varphi \in Conf^{comb}(\Lambda(M))$. $\mathcal{B}_\varphi$ *is a basis of the tangent space*

$$T(Conf(\Lambda(M)))_{k^\varphi} \subset \underset{(\alpha,\beta)}{\bigoplus} T(Grass_{m_{\alpha\beta}}(k^{r+1}))_{k^{\varphi(\alpha,\beta)}}$$

   (The φ**-tangent basis of** $Conf(\Lambda(M))$ **at** k^φ**).**

   2) Let $\varphi \in Conf^{comb}(\Lambda(M), D)$. *Then*

$$Vect_k(\mathcal{B}_{(\varphi,D)}) = T(Conf(\Lambda(M), D))_{k^\varphi} \subset \underset{(\alpha,\beta)}{\bigoplus} T(Grass_{m_{\alpha\beta}}(k^{r+1}))_{k^{\varphi(\alpha,\beta)}} \ .$$

$\mathcal{B}_{(\varphi,D)} \subset \mathcal{B}_\varphi$ *is called the* (φ, D)**-tangent basis of** $Conf(\Lambda(M), D)$ **at** k^φ.

3) *The matrix* $\mathbb{M}(p_{\alpha\beta})_{k^\varphi}$ *of the differential* $T(p_{\alpha\beta})_{k^\varphi}$:
$T(Conf(\Lambda(M)))_{k^\varphi} \longrightarrow T(Grass_{m_{\alpha\beta}}(k^{r+1}))_{k^{\varphi(\alpha,\beta)}}$ *calculated with respect to the basis* $\mathcal{B}_\varphi$ *and* $\mathcal{B}_{\varphi(\alpha,\beta)}$ *admits the following simple description:*

$$T(p_{\alpha\beta})_{k^\varphi}(E_\mathscr{C}) = E_{(i,j)}^{(\alpha,\beta)} \ \ if \ \ \mathscr{C} \cap \{(\alpha,\beta)\} \times R_\varphi^{(\alpha,\beta)} \neq \emptyset,$$

thus necessarily $\mathscr{C} \cap \{(\alpha,\beta)\} \times R_\varphi^{(\alpha,\beta)} = \{((\alpha,\beta),(i,j)_\mathscr{C})\}$, *otherwise* $T(p_{\alpha\beta})_{k^\varphi}(E_\mathscr{C}) = 0$.

Let $\varphi \in Conf^{comb}(\Lambda(M), D)$. Denote by $\pi_2(M, D)$ the restriction of $\pi_2 :$ $Conf(\Lambda(M)) \longrightarrow Grass_n(k^{r+1})$ to $Conf(\Lambda(M)), D)$, and by $T(\pi_2(M,D))_{k^\varphi}$ its differential at k^φ.

Definition 4.20   *1) Write:*

$$\tilde{\mathcal{B}}_{(\varphi,D)} = T(\pi_2(M,D))_{k^\varphi}(\mathcal{B}_{(\varphi,D)}) - \{0\},$$

$$\left(resp. \ \tilde{\mathcal{B}}_{(\varphi,D)}^c - T(\pi_2(M,D))_{k^\varphi}(\mathcal{B}_{(\varphi,D)}^c) - \{0\}\right).$$

2) $\tilde{\mathcal{B}}_{(D,D')} = \bigcup\limits_{\varphi \in Conf^{comb}(\Lambda(M),D,D')} \tilde{\mathcal{B}}_{(\varphi,D)}$

$\left(resp. \ \tilde{\mathcal{B}}_{(D,D')}^c = \bigcap\limits_{\varphi \in Conf^{comb}(\Lambda(M),D,D')} \tilde{\mathcal{B}}_{(\varphi,D)}^c \right).$

Observe that $\tilde{\mathcal{B}}_{(D,D')}$ *(resp.* $\tilde{\mathcal{B}}_{(D,D')}^c$*) is a linearly independant set of vectors. It is recalled that by definition* $\tilde{\mathcal{B}}_{(D,D')}$ *generates the combinatorial* $Conf(\Lambda(M), D)$*-tangent space* $\tilde{T}^{comb}(\overline{\Sigma}(M,D))_{k^{D'}}$ *(cf. definition 4.11).* $\tilde{\mathcal{B}}_{(D,D')}$ *is called the* **canonical tangent basis** *of* $\tilde{T}^{comb}(\overline{\Sigma}(M,D))_{k^{D'}}$.

3) One calls

$$N(Conf(\Lambda(M), D))_{k^\varphi} = Vect_k(\mathcal{B}_{(\varphi,D)}^c)$$

the (φ, D)-**normal space** *to* $Conf(\Lambda(M), D))$ *at* k^φ, *and* $\mathcal{B}_{(\varphi,D)}^c \subset \mathcal{B}_\varphi$ *the* (φ, D)-**normal basis** *at* k^φ .

4) Write $\tilde{\mathcal{B}}_{(\varphi,D)}^{(\alpha,\beta)} = T(p_{\alpha\beta})_{k^\varphi}(\mathcal{B}_{(\varphi,D)}) - \{0\}$ *(resp.* $\tilde{\mathcal{B}}_{(\varphi,D)}^{c(\alpha,\beta)} = T(p_{\alpha\beta})_{k^\varphi}(\mathcal{B}_{(\varphi,D)}^c) - \{0\}$*).*

Remark 4.21 *Let* $\varphi \in Conf^{comb}(\Lambda(M), D)$.

1) From $\mathcal{B}_\varphi = \mathcal{B}_{(\varphi,D)} \coprod \mathcal{B}_{(\varphi,D)}^c$ *it follows the following direct sum decomposition:*

$$T(Conf(\Lambda(M)))_{k^\varphi} = T(Conf(\Lambda(M), D))_{k^\varphi} \bigoplus N(Conf(\Lambda(M), D))_{k^\varphi} .$$

2) *The fiber of $\pi_1(M) : Conf(\Lambda(M)) \longrightarrow Drap_{\underline{m}}(k^{r+1})$ on k^D identifies with $Conf(\Lambda(M), D)$ by definition, and the differential mapping:*

$$T(\pi_1(M))_{k^\varphi} : T(Conf(\Lambda(M)))_{k^\varphi} \longrightarrow T(Drap_{\underline{m}}(k^{r+1}))_{k^D},$$

induces an isomorphism:

$$T(\pi_1(M))_{k^\varphi}|_N : N(Conf(\Lambda(M), D))_{k^\varphi} \simeq T(Drap_{\underline{m}}(k^{r+1}))_{k^D} .$$

In fact $\pi_1(M)$ defines a fiber bundle structure.

4.6   The canonical basis of a Combinatorial Tangent Space

The tangent space $T(Drap_{\underline{n}}(k^{r+1}))_{\mathscr{D}}$ $(\mathscr{D} = (\mathscr{H}_1 \cdots \subset \mathscr{H}_l \subset k^{r+1}))$ is given by:

$$T(Drap_{\underline{n}}(k^{r+1}))_{\mathscr{D}} \simeq Ker\Big(\prod_{1 \leq \beta \leq l} Hom_{k-vect}(\mathscr{H}_\beta, k^{r+1}/\mathscr{H}_\beta) \overset{\longrightarrow}{\longrightarrow}$$

$$\prod_{1 \leq \beta, \beta' \leq l} Hom_{k-vect}(\mathscr{H}_{inf(\beta',\beta)}, k^{r+1}/\mathscr{H}_{sup(\beta',\beta)})\Big). \quad \text{We have seen that for}$$

$\mathscr{D} = k^{D'}$ $(D' = (J_1 \cdots \subset J_l \subset k^{r+1}))$ this isomorphism gives rise to the isomorphism:

$$T(Drap_{\underline{n}}(k^{r+1}))_{k^{D'}} \simeq Ker\Big(\prod_{1 \leq \beta \leq l} Hom_{k-vect}(k^{J_\beta}, k^{r+1}/k^{J_\beta}) \overset{\longrightarrow}{\longrightarrow}$$

$$\prod_{1 \leq \beta \leq l} Hom_{k-vect}(k^{J_\beta}, k^{r+1}/k^{J_{\beta+1}})\Big) \text{ and finally to an isomorphism}$$

$$T(Drap_{\underline{n}}(k^{r+1}))_{k^{D'}} \simeq \prod_{1 \leq \beta \leq l} Hom_{k-vect}(k^{J_\beta}, k^{J_{\beta+1}-J_\beta}) ,$$

induced by the canonical projections:

$$\prod_{1 \leq \beta \leq l} Hom_{k-vect}(k^{J_\beta}, k^{r+1}/k^{J_\beta}) \longrightarrow Hom_{k-vect}(k^{J_\beta}, k^{J_{\beta+1}-J_\beta}) .$$

On the other hand, it is known that the canonical basis of this space is given by

$$\mathcal{B}_{D'} := \coprod_{1 \leq \beta \leq l} (E_{ij}^\beta)_{(i,j) \in J_\beta \times (J_{\beta+1}-J_\beta)}.$$

The matrix $\mathbb{M}(\pi_2(M))_{k^\varphi}$ of the differential

$$T(\pi_2(M))_{k^\varphi} : T(Conf(\Lambda(M)))_{k^\varphi} \longrightarrow$$

$$T(Drap_{\underline{n}}(k^{r+1}))_{kD'} \quad (\pi_2(M) : k^\varphi \mapsto k^{D'})$$

calculated with respect to the basis $\mathcal{B}_\varphi$ and $\mathcal{B}_{D'}$, admits the following simple description:

$$T(\pi_2(M))_{k^\varphi}(E_\mathscr{C}) = E_{ij}^\beta \; if \; \mathscr{C} \cap \{(\lambda+1,\beta)\} \times (J^\beta \times J^{\beta+1}/J^\beta) \neq \emptyset,$$

thus necessarily $\mathscr{C} \cap \{(\lambda+1,\beta)\} \times (J^\beta \times J^{\beta+1}/J^\beta) = \{((\lambda+1,\beta),(i,j))\}$, where (i,j) is the element $(i,j)_\mathscr{C}$ associated with the class $\mathscr{C}$, otherwise $T(\pi_2(M))_{k^\varphi}(E_\mathscr{C}) = 0$. The subset

$$\bigcup_{\mathscr{C} \in \mathbb{B}^c(\varphi,D)} \mathscr{C} \subset \mathbb{E}(\varphi)$$

is the set of elements equivalent to an element of $\coprod_{1 \leq \alpha \leq \lambda+1} \{(\alpha,l+1)\} \times R_\varphi^{(\alpha,l+1)}$. The image of the tangent vectors $\mathcal{B}_{(\varphi,D)}^c = (E_\mathscr{C})_{\mathscr{C} \in \mathbb{B}^c(\varphi,D)} \subset T(Conf(\Lambda(M))_{k^\varphi}$ by $T(\pi_1)_{k^\varphi}$ is a linearly independant set of vectors. The subset $\mathbb{B}_{(\varphi,D)}$ indexes a basis $\mathcal{B}_{(\varphi,D)} = (E_\mathscr{C})_{\mathscr{C} \in \mathbb{B}(\varphi,D)}$ of the tangent space $T(Conf(\Lambda(M),D)_{k^\varphi}$. There is a natural isomorphism $T(Conf(\Lambda(M))_{k^\varphi}/T(Conf(\Lambda(M),D)_{k^\varphi} \simeq Vect_k((E_\mathscr{C})_{\mathscr{C} \in \mathbb{B}^c(\varphi,D)})$.
Observe that the image of $\mathcal{B}_{(\varphi,D)}$ by $T(\pi_2(M,D))_{k^\varphi}$ is the same as its image by $T(\pi_2)_{k^\varphi}$ where $\pi_2 = \pi_2(M)$. One has

$$\mathcal{B}_{D'} = T(\pi_2)_{k^\varphi}(\mathcal{B}_{(\varphi,D)})/\{0\} \coprod T(\pi_2)_{k^\varphi}(\mathcal{B}_{(\varphi,D)}^c)/\{0\} = \tilde{\mathcal{B}}_{(\varphi,D)} \coprod \tilde{\mathcal{B}}_{(\varphi,D)}^c.$$

Proceding with $p_{\alpha\beta} : Conf(\Lambda(M)) \longrightarrow Grass_{m_{\alpha\beta}}(k^{r+1})$ instead of π_2 one obtains

1) $\tilde{\mathcal{B}}_{(\varphi,D)}^{(\alpha,\beta)} \bigcup \tilde{\mathcal{B}}_{(\varphi,D)}^{c(\alpha,\beta)} = \mathcal{B}_{\varphi(\alpha,\beta)}$;

2) $\tilde{\mathcal{B}}_{(\varphi,D)}^{(\alpha,\beta)} \bigcap \tilde{\mathcal{B}}_{(\varphi,D)}^{(\alpha,\beta)} = \emptyset$.

Where $\mathcal{B}_{\varphi(\alpha,\beta)}$ denotes the canonical basis of $T(Grass_{m_{\alpha\beta}}(k^{r+1}))_{k^{\varphi(\alpha,\beta)}}$ indexed by $R_\varphi^{(\alpha,\beta)}$. This set decomposes into the disjointed union of the set of indices given by $\tilde{\mathcal{B}}_{(\varphi,D)}^{(\alpha,\beta)}$ and that given by $\tilde{\mathcal{B}}_{(\varphi,D)}^{c(\alpha,\beta)}$.

Definition 4.22 *Let $\varphi \in Conf^{comb}(\Lambda(M),D,D')$. Define $N_\varphi^{(\alpha,\beta)} \subset R_\varphi^{(\alpha,\beta)}$ by*

$$\{(\alpha,\beta)\} \times N_\varphi^{(\alpha,\beta)} = (\bigcup_{\mathscr{C} \in \mathbb{B}^c(\varphi,D)} \mathscr{C}) \cap (\{(\alpha,\beta)\} \times R_\varphi^{(\alpha,\beta)})$$

$$(resp. \; T_\varphi^{(\alpha,\beta)} = R_\varphi^{(\alpha,\beta)} - N_\varphi^{(\alpha,\beta)}) \; .$$

One has

$$\tilde{\mathcal{B}}^{(\alpha,\beta)}_{(\varphi,D)} = (E^{(\alpha,\beta)}_{(i,j)})_{(i,j)\in T^{(\alpha,\beta)}_{\varphi}} \quad \left(resp.\ \tilde{\mathcal{B}}^{c(\alpha,\beta)}_{(\varphi,D)} = (E^{(\alpha,\beta)}_{(i,j)})_{(i,j)\in N^{(\alpha,\beta)}_{\varphi}}\right).$$

As a particular case one obtains

$$J_\beta \times J_\beta^\perp = N^{(\lambda+1,\beta)}_{\varphi} \coprod T^{(\lambda+1,\beta)}_{\varphi},$$

and thus

$$J_\beta \times (J_{\beta+1} - J_\beta) = N^{(\lambda+1,\beta)}_{\varphi} \cap (J_\beta \times (J_{\beta+1} - J_\beta)) \coprod T^{(\lambda+1,\beta)}_{\varphi} \cap (J_\beta \times (J_{\beta+1} - J_\beta)).$$

From these equalities it follows that:

Proposition 4.23

$$\tilde{\mathcal{B}}^c_{(\varphi,D)} = \coprod_{1\leq\beta\leq l} (E^\beta_{ij})_{(i,j)\in N^{(\lambda+1,\beta)}_{\varphi}\cap(J_\beta\times(J_{\beta+1}-J_\beta))}$$

$\left(resp.\right.$

$$\tilde{\mathcal{B}}_{(\varphi,D)} = \coprod_{1\leq\beta\leq l} (E^\beta_{ij})_{(i,j)\in T^{(\lambda+1,\beta)}_{\varphi}\cap(J_\beta\times(J_{\beta+1}-J_\beta))}\Big),$$

thus $\mathcal{B}_{D'} = \tilde{\mathcal{B}}_{(\varphi,D)} \coprod \tilde{\mathcal{B}}^c_{(\varphi,D)}$.

Proof *It suffices to prove the first statement. The composition of*
$T(\pi_2(M,D))_{k^\varphi}$ *with* $T(Drap_{\underline{n}}(k^{r+1}))_{kD'} \longrightarrow Hom_{k-vect}(k^{J_\beta}, k^{J_\beta^\perp})$ *is pre-*
cisely $T(\pi_{(\lambda+1,\beta)})_{k^\varphi}$, *thus the image of* $\tilde{\mathcal{B}}^c_{(\varphi,D)}$ *minus* $\{0\}$ *by the above pro-*
jection gives $\tilde{\mathcal{B}}^{c(\lambda+1,\beta)}_{(\varphi,D)} = (E^{(\alpha,\beta)}_{(i,j)})_{(i,j)\in N^{(\lambda+1,\beta)}_{\varphi}}$. *It is concluded that the image*
of $\tilde{\mathcal{B}}^c_{(\varphi,D)}$ *by* $T(Drap_{\underline{n}}(k^{r+1}))_{kD'} \longrightarrow Hom_{k-vect}(k^{J_\beta}, k^{J_{\beta+1}-J_\beta})$ *minus* $\{0\}$ *is*
given by $(E^{(\alpha,\beta)}_{(i,j)})_{(i,j)\in N^{(\lambda+1,\beta)}_{\varphi}\cap(J_\beta\times(J_{\beta+1}-J_\beta))}$, *and finally that*

$$T'(\pi_2)_{k^\varphi}(\mathcal{B}^c_{(\varphi,D)}) = \coprod_{1\leq\beta\leq l} (E^\beta_{ij})_{(i,j)\in N^{(\lambda+1,\beta)}_{\varphi}\cap(J_\beta\times(J_{\beta+1}-J_\beta))}.$$

From the definition of $\tilde{\mathcal{B}}_{(D,D')}$ $\left(resp.\ \tilde{\mathcal{B}}^c_{(D,D')}\right)$ and the above proposition
we deduce:

Proposition 4.24

$$\tilde{\mathcal{B}}^c_{(D,D')} = \coprod_{1\leq\beta\leq l} (E^\beta_{ij})_{(i,j)\in \bigcap_{\varphi\in Confcomb(\Lambda(M),D,D')} N^{(\lambda+1,\beta)}_{\varphi}\cap(J_\beta\times(J_{\beta+1}-J_\beta))}$$

$\left(resp.\right.$

$$\tilde{\mathcal{B}}_{(D,D')} = \coprod_{1\leq\beta\leq l} (E^\beta_{ij})_{(i,j)\in \bigcup_{\varphi\in Confcomb(\Lambda(M),D,D')} T^{(\lambda+1,\beta)}_{\varphi}\cap(J_\beta\times(J_{\beta+1}-J_\beta))}\Big),$$

and $\mathcal{B}_{D'} = \tilde{\mathcal{B}}_{(D,D')} \coprod \tilde{\mathcal{B}}^c_{(D,D')}$.

In next sections it is aimed at giving another expression of the basis $\tilde{\mathcal{B}}_{(D,D')} \subset \mathcal{B}_{D'}$ of $\tilde{T}^{comb}(\overline{\Sigma}(M,D))$.

The following equality results immediately from definition 4.18:

$$N_\varphi^{(\alpha,\beta)} = \bigcup_{\gamma \in Path(D,(\alpha,\beta))} \mathcal{P}(\varphi,\gamma),$$

where $\mathcal{P}(\varphi,\gamma) = \bigcap_{0\leqq\rho\leqq r} R_\varphi^{(\alpha_\rho,\beta_\rho)}$ $(\gamma = (\alpha_\rho,\beta_\rho)_{0\leqq\rho\leqq r})$, and $Path(D,(\alpha,\beta))$ denotes **the set of paths** of $\Lambda(M)$ **issued from** $\{(1, l+1), \cdots, (\lambda, l+1), (\lambda+1, l+1)\}$ **and extremity** (α,β).

For the proof of the following proposition the reader is referred to [15] (see also [16]).

Proposition 4.25 *The following formula holds:*

$$N_\varphi^{(\alpha,\beta)} = R_\varphi^{(\alpha,\beta)} \bigcap \left(\bigcap_{\beta\leqq\beta'\leqq l} \left(\bigcup_{1\leqq\alpha'\leqq\alpha} \varphi(\alpha',\beta') \times \varphi(\alpha',\beta'+1)^\perp \right) \right) =$$

$$= R_\varphi^{(\alpha,\beta)} \bigcap \left(\bigcap_{\beta\leqq\beta'\leqq l} \left(\bigcup_{1\leqq\alpha'\leqq\alpha} R_\varphi^{(\alpha',\beta')} \right) \right).$$

Notation 4.26 *Write:*

$$\mathcal{N}_\varphi^{(\alpha,\beta)} = \bigcap_{\beta\leqq\beta'\leqq l} \left(\bigcup_{1\leqq\alpha'\leqq\alpha} \varphi(\alpha',\beta') \times \varphi(\alpha',\beta'+1)^\perp \right) = \bigcap_{\beta\leqq\beta'\leqq l} \left(\bigcup_{1\leqq\alpha'\leqq\alpha} R_\varphi^{(\alpha',\beta')} \right),$$

thus: $N_\varphi^{(\alpha,\beta)} = R_\varphi^{(\alpha,\beta)} \bigcap \mathcal{N}_\varphi^{(\alpha,\beta)}$.

4.7 A family of smooth varieties associated with a point of a Schubert variety

Let y be a point in the closure $\overline{\Sigma}(M,D)$, and $\mathscr{D}'$ the flag to which it corresponds. Assume that $y \in \Sigma(M',D)$ with $M \leq M'$, and let D' be a combinatorial flag such that $M' = M(D,D')$. A family of smooth subvarieties of $Drap_{\underline{n}}(k^{r+1})$ is introduced whose germs at a point $y \in \overline{\Sigma}(M,D)$ contains $\overline{\Sigma}(M,D)_y$ and such that the intersection of the corresponding family of tangent spaces is given by $Vect(\tilde{\mathcal{B}}_{(D,D')})$. Recall that the set of relative position matrices $M' \in \mathbb{N}^{(\lambda+1)\times(l+1)}$ satisfying $M \leq M'$ indexes the set of cells $\Sigma(M',D)$ contained in $\overline{\Sigma}(M,D)$. Given $M' = (m'_{\alpha\beta}) \in Relpos(I_{r+1}) \cap \mathbb{N}^{(\lambda+1)\times(l+1)}$ with $M \leq M'$ write

$$GI(M,M') = \{(\alpha,\beta) \in [\![1,\lambda]\!] \times [\![1,l]\!] \mid m'_{\alpha\beta} = m_{\alpha\beta}\}$$

(The set of generic indices of the relative position matrix
$M \leq M'$**).** To $(\alpha, \beta) \in GI(M, M')$ is associated the relative po-
sition matrix $M_{\alpha\beta} = \begin{pmatrix} m_\alpha & r+1 \\ m_{\alpha\beta} & n_\beta \end{pmatrix}$, where it is written $m_\alpha = m_{\alpha l+1}$ (*resp.* $n_\beta = n_{\lambda+1\beta}$), between a couple of subspaces in $\mathrm{Grass}_{m_\alpha}(k^{r+1}) \times \mathrm{Grass}_{n_\beta}(k^{r+1})$. The set $(M_{\alpha\beta})_{(\alpha,\beta) \in GI(M,M')}$ gives rise to a family of subvarieties $(\Sigma^\natural(M_{\alpha\beta}, D_\alpha)_{(\alpha,\beta) \in GI(M,M')}$ of $\mathrm{Drap}_{\underline{n}}(k^{r+1})$ associated to a couple
$(D, D') \in \mathrm{Drap}_{\underline{m}}(I_{r+1}) \times \mathrm{Drap}_{\underline{n}}(I_{r+1})$ defined as follows. Let $D = (H_1 \subset \cdots \subset H_\lambda \subset I_{r+1})$ (*resp.* $D' = (J_1 \subset \cdots \subset J_l \subset I_{r+1})$, and $M' = M(D, D')$.
Thus one has $M_{\alpha\beta} = \begin{pmatrix} |H_\alpha| & |I_{r+1}| \\ |H_\alpha \cap J_\beta| & |J_\beta| \end{pmatrix}$.

Definition 4.27

Let $D_\alpha = (H_\alpha \subset I_{r+1}) \in \mathrm{Grass}_{|H_\alpha|}(I_{r+1})$,

$$\Sigma(M_{\alpha\beta}, D_\alpha) \subset \mathrm{Grass}_{|J_\beta|}(k^{r+1}),$$

the Schubert cell defined by $(M_{\alpha\beta}, D_\alpha)$, *and denote by* p_β :
$\mathrm{Drap}_{\underline{n}}(k^{r+1}) \longrightarrow \mathrm{Grass}_{|J_\beta|}(k^{r+1})$ $(\underline{n} = (|J_1| < \ldots < |J_l| < r+1))$, *the*
canonical morphism induced by the projection:

$$\left(\mathrm{Drap}_{\underline{n}}(k^{r+1}) \subset\right) \prod_{\beta'=1}^{l} \mathrm{Grass}_{|J_{\beta'}|}(k^{r+1}) \longrightarrow \mathrm{Grass}_{|J_\beta|}(k^{r+1}).$$

Define:

$$\Sigma^\natural(M_{\alpha\beta}, D) = (p_\beta)^{-1}\left(\Sigma(M_{\alpha\beta}, D_\alpha)\right) = \Sigma(M_{\alpha\beta}, D_\alpha) \times_{\mathrm{Grass}_{|J_\beta|}(k^{r+1})} \mathrm{Drap}_{\underline{n}}(k^{r+1}).$$

Clearly $\Sigma^\natural(M_{\alpha\beta}, D)$ is a k-smooth locally closed subvariety of
$\mathrm{Drap}_{\underline{n}}(k^{r+1})$, as the pull-back of a k-smooth locally closed subvariety of
$\mathrm{Grass}_{|J_\beta|}(k^{r+1})$ by the morphism p_β.

Remark 4.28 *The family of subvarieties* $(\Sigma^\natural(M_{\alpha\beta}, D))_{(\alpha,\beta) \in GI(M,M')}$ *satis-*
fies:

1) $(\forall (\alpha, \beta) \in GI(M, M'))$ $\Sigma(M', D) \subset \Sigma^\natural(M_{\alpha\beta}, D)$.

2) $\Sigma^\natural(M_{\alpha\beta}, D)$ *is stable under the action of* $\mathrm{Stab}\, k^D$.

3) $(\forall (\alpha, \beta) \in GI(M, M'))$ $\Sigma(M', D) \subset \Sigma^\natural(M_{\alpha\beta}, D)$.

4) *The tangent space* $T(\Sigma^\natural(M_{\alpha\beta}, D))_{\mathscr{D}'}$ $(\mathscr{D}' = (\mathscr{J}_1 \cdots \subset \mathscr{J}_l \subset k^{r+1}))$ *is*
   given by:
   $$T(\Sigma^\natural(M_{\alpha\beta}, D))_{\mathscr{D}'} \simeq$$
   $$\simeq \mathrm{Ker}\left(T(\mathrm{Ker}(\mathrm{Drap}_{\underline{n}}(k^{r+1}))_{\mathscr{D}'} \longrightarrow\right.$$

$$\prod_{\beta \leq \beta'} Hom_{k-vect}\left(\mathcal{H}_\alpha \cap \mathcal{J}_\beta, k^{r+1}/\mathcal{J}_{\beta'} + \mathcal{H}_\alpha \cap \mathcal{J}_{\beta'+1}\right),$$

where $k^D = (\mathcal{H}_1 \subset \cdots \subset \mathcal{H}_\lambda \subset k^{r+1})$.

Definition 4.29 *Given a flag $\mathcal{D}'$ satisfying $M \leq M(k^D, \mathcal{D}') = M'$, i.e. the point x corresponding to $\mathcal{D}'$ belongs to $\overline{\Sigma}(M, D)$ and determines the Schubert cell $\Sigma(M', D) \subset \overline{\Sigma}(M, D)$. Define:*

$$\overline{T}(\Sigma(M, D))_{\mathcal{D}'} = \bigcap_{(\alpha,\beta) \in GI(M,M')} T(\Sigma^\natural(M_{\alpha\beta}, D))_{\mathcal{D}'}$$

(The (M, M')-tangent space to $\overline{\Sigma}(M, D)$ at $\mathcal{D}' \in \Sigma(M', D)$).

Remark 4.30 *If $M = M(k^D, \mathcal{D}') = M'$ one has $GI(M, M') = \Lambda(M)$ and thus $\Sigma(M, D) = \bigcap_{(\alpha,\beta) \in GI(M,M')} \Sigma^\natural(M_{\alpha\beta}, D)$. It follows that $T(\Sigma(M, D))_{\mathcal{D}'} = \bigcap_{(\alpha,\beta) \in GI(M,M')} T(\Sigma^\natural(M_{\alpha\beta}, D))_{\mathcal{D}'}$, i.e. the (M, M')-tangent space to $\overline{\Sigma}(M, D)$ at $\mathcal{D}' \in \Sigma(M, D)$ is the tangent space $T(\Sigma(M, D))_{\mathcal{D}'}$.*

From 2) of the 4.28 remark it results:

$$(\forall g \in Stab \; k^D) \quad T(g)_{\mathcal{D}'}\left(\overline{T}(\overline{\Sigma}(M, D))_{\mathcal{D}'}\right) = \overline{T}(\overline{\Sigma}(M, D))_{g(\mathcal{D}')},$$

where $T(g)_{\mathcal{D}'}$ denotes the differential of the translation morphism defined by g. Thus $dim_k \overline{T}(\overline{\Sigma}(M, D))_{\mathcal{D}'}$ is independent of $\mathcal{D}' \in \Sigma(M', D)$.
The next task is to prove the equality

$$\tilde{T}^{comb}(\overline{\Sigma}(M, D))_{\mathcal{D}'} = \overline{T}(\overline{\Sigma}(M, D))_{\mathcal{D}'},$$

between the combinatorial $Conf(\Lambda(M), D)$-tangent space and the (M, M')-tangent space to $\overline{\Sigma}(M, D)$ at $\mathcal{D}' \in \Sigma(M', D)$ for $\mathcal{D}' = k^{D'}$. According to the smoothness criterion this implies that

$$T^{nash}(\overline{\Sigma}(M, D))_{\mathcal{D}'} = \tilde{T}^{comb}(\overline{\Sigma}(M, D))_{\mathcal{D}'} = \overline{T}(\overline{\Sigma}(M, D))_{\mathcal{D}'}$$

and thus allows to decide if $\Sigma(M', D)$ is singular in $\overline{\Sigma}(M, D))$, without hypothesis on k.

4.8 A combinatorial basis of a (M, M')-tangent space

From the formula 4.28, 4), a basis $\mathcal{B}^{\natural(\alpha,\beta)}_{(D,D')}$ of $T(\Sigma^\natural(M_{\alpha\beta}, D))_{\mathcal{D}'}$ $((\alpha,\beta) \in GI(M, M'))$ is obtained which is in fact an indexed subset of $\mathcal{B}_{D'}$. It results that the intersection $\mathcal{B}_{(D,D')} = \cap \mathcal{B}^{\natural(\alpha,\beta)}_{(D,D')}$ is a basis of $\overline{T}(\overline{\Sigma}(M, D))_{\mathcal{D}'}$. One has

$$\mathcal{B}^{\natural(\alpha,\beta)}_{(D,D')}, \mathcal{B}_{(D,D')} \subset \mathcal{B}_{D'} = \coprod_{1 \leq \beta \leq l} (E^\beta_{ij})_{(i,j) \in J_\beta \times (J_{\beta+1} - J_\beta)},$$

thus each one of these bases is determined by its index set which is given in terms of the following D'-Young-Data. For the sake of simplifying notation $J_{\beta+1}/J_\beta = J_{\beta+1} - J_\beta$ is written.

Definition 4.31 *The Young D'-data* $(\underline{d}(\alpha,\beta),\underline{d}^*(\alpha,\beta))_{(\alpha,\beta)\in GI(M,M')}$

given by $\underline{d}(\alpha,\beta) = (d_{\beta'}(\alpha,\beta))_{1\leq\beta'\leq l}$ *(resp.* $\underline{d}^*(\alpha,\beta) = (d^*_{\beta'}(\alpha,\beta)_{1\leq\beta'\leq l})$ *is introduced defined by:*

$$d_{\beta'}(\alpha,\beta) = \begin{cases} (H_\alpha \cap J_\beta \subset J_{\beta'}) & \text{for} \quad \beta \leq \beta' \leq l \\ (\emptyset \subset J_{\beta'}) & \text{for} \quad 1 \leq \beta' < \beta \end{cases}$$

(resp.

$$d^*_{\beta'}(\alpha,\beta) = \begin{cases} (J_{\beta'} \subset J_{\beta'} \cup J_{\beta'+1} \cap H_\alpha) & \text{for} \quad \beta \leq \beta' \leq l \\ (J_{\beta'} \subset J_{\beta'+1}) & \text{for} \quad 1 \leq \beta' < \beta \end{cases}$$ *).*

Let
$$Y^+_{\beta'}(\alpha,\beta) = Y^+_{\beta'}((M,D,D'),(\alpha,\beta)) = Y^+_{\beta'}(\underline{d}(\alpha,\beta),\underline{d}^*(\alpha,\beta))$$

(resp.

$$Y^-_{\beta'}(\alpha,\beta) = Y_{\beta'}((M,D,D'),(\alpha,\beta)) = Y_{\beta'}(\underline{d}(\alpha,\beta),\underline{d}^*(\alpha,\beta)))\ .$$

For $\beta \leq \beta' \leq l$ *one has:*

$$Y^+_{\beta'}(\alpha,\beta) = H_\alpha \cap J_\beta \times (J_{\beta'+1}/(J_{\beta'} \cup H_\alpha \cap J_{\beta'+1})) \subset J_{\beta'} \times (J_{\beta'+1}/J_{\beta'})$$

(resp.

$$Y^-_{\beta'}(\alpha,\beta) = J_{\beta'}\times(J_{\beta'+1}/J_{\beta'})-Y^+_{\beta'}(\alpha,\beta) = H_\alpha\cap J_\beta\times((H_\alpha\cap J_{\beta'+1})/(H_\alpha)\cap J_{\beta'}))$$

$$\coprod (J_\beta/H_\alpha \cap J_\beta) \times (J_{\beta'+1}/J_{\beta'}))\ ,$$

$$Y^+(\alpha,\beta) = \coprod_{1\leq\beta'\leq l} Y^+_{\beta'}(\alpha,\beta) = \coprod_{\beta\leq\beta'\leq l} Y^+_{\beta'}(\alpha,\beta) =$$

$$\coprod_{\beta\leq\beta'\leq l} H_\alpha \cap J_\beta \times (J_{\beta'+1}/(J_{\beta'} \cup H_\alpha \cap J_{\beta'+1})) =$$

$$= H_\alpha \cap J_\beta \times (I_{r+1}/(J_\beta \cup H_\alpha))$$

Lemma 4.32 *Let* $(\alpha,\beta) \in GI(M,M')$. *The indexed subset*

$$B^{\natural(\alpha,\beta)}_{(D,D')} := \coprod_{1\leq\beta'\leq l} (E^{\beta'}_{ij})_{(i,j)\in Y^-_{\beta'}(\alpha,\beta)} \subset \mathcal{B}_{D'} \ (cf.4.4)$$

is a basis of $T(\Sigma^\natural(M_{\alpha\beta},D))_{kD'}$ **(The combinatorial basis of** $T(\Sigma^\natural(M_{\alpha\beta},D))_{kD'}$**).**

The (M, M')-tangent space:
$$\overline{T}(\overline{\Sigma}(M, D))_{k^{D'}} = \bigcap_{(\alpha, \beta) \in GI(M, M')} T(\Sigma^{\natural}(M_{\alpha\beta}, D))_{k^{D'}} \quad (M' = M(D, D')) \text{ may}$$
be endowed with a canonical basis indexed by means of Young D-data.

Definition 4.33 *1) Define:*

$$\overline{H}_{\alpha\beta} = \bigcup_{\{ (\alpha', \beta') \in GI(M, M') \mid (\alpha', \beta') \leq (\alpha, \beta) \}} H_{\alpha'\beta'}(= H_{\alpha'} \cap J_{\beta'}),$$

and $\overline{\mathcal{M}}(D, D') = (\overline{H}_{\alpha\beta})$ *(resp.* $\overline{M}(D, D') = (\overline{m}_{\alpha\beta}) = (|\overline{H}_{\alpha\beta}|)$ *.*

2) With (M, D, D') *is associated the* D'-Young data $\underline{\mathcal{D}}(D, D') = \underline{\mathcal{D}}(M, D, D') = (\mathcal{D}_\beta(D, D'))$ *(resp.* $\underline{\mathcal{D}}^*(D, D') = \underline{\mathcal{D}}^*(M, D, D') = (\mathcal{D}_\beta^*(D, D'))$ *defined by:*

$$\mathcal{D}_\beta(D, D') = (\overline{H}_{1\beta} \cdots \subset \overline{H}_{\lambda\beta} \subset \overline{H}_{\lambda+1\beta}).$$

(resp.

$$\mathcal{D}_\beta^*(D, D') = (\cdots \subset J_\beta \cup J_{\beta+1} \cap H_\alpha \subset \cdots) \ (1 \leq \alpha \leq \lambda + 1)) \ .$$

Write:

$$Y_\beta^+(D, D') = Y_\beta^+(M, D, D') = Y_\beta^+(\underline{\mathcal{D}}(D, D'), \underline{\mathcal{D}}^*(D, D')) =$$

$$\bigcup_{1 \leq \alpha \leq \lambda} \overline{H}_{\alpha\beta} \times J_{\beta+1}/(J_\beta \cup J_{\beta+1} \cap H_\alpha),$$

and $Y_\beta^-(D, D') = Y_\beta^-(M, D, D') = (J_\beta \times (J_{\beta+1}/J_\beta) - Y_\beta^+(D, D')$ *(resp.*

$$Y^+(D, D') = Y^+(M, D, D') = \coprod_{1 \leq \beta \leq l} Y_\beta^+(D, D') \ , \ Y^-(D, D')$$

$$= Y^-(M, D, D') =$$

$$\coprod_{1 \leq \beta \leq l} Y_\beta^-(D, D') = \coprod_{1 \leq \beta \leq l} (J_\beta \times (J_{\beta+1}/J_\beta) - Y_\beta^+(D, D')) \ .$$

From the following easy to check lemma it results that a combinatorial basis of the (M, M')-tangent space to $\overline{\Sigma}(M, D))$ at $k^{D'}$ indexed by the above Young data may be obtained.

Lemma 4.34   *1)* $Y_{\beta'}^-(M, D, D') = \bigcap\limits_{(\alpha,\beta)\in GI(M,M')} Y_{\beta'}^-((M, D, D'), (\alpha, \beta))$ *(resp.*

$Y^-(M, D, D') = \bigcap\limits_{(\alpha,\beta)\in GI(M,M')} Y^-((M, D, D'), (\alpha, \beta))$ *).*

2) The indexed subset

$$\overline{\mathcal{B}}_{(D,D')} := \coprod_{1\leq\beta'\leq l} (E_{ij}^\beta)_{(i,j)\in Y_\beta^-(M,D,D')} \subset \coprod_{1\leq\beta'\leq l} (E_{ij}^\beta)_{(i,j)\in J_\beta\times(J_{\beta+}-J_\beta)}$$

is a basis of $\overline{T}^{comb}(\overline{\Sigma}(M, D))_{kD'}$ **(The combinatorial basis of** $\overline{T}(\overline{\Sigma}(M, D))_{kD'}$**).**

It may thus be written:

$$\overline{\mathcal{B}}_{(D,D')} = \bigcap_{(\alpha,\beta)\in GI(M,M')} \mathcal{B}_{(D,D')}^{\natural(\alpha,\beta)} .$$

4.9   Schubert variety Nash tangent spaces and its singular locus

In order to obtain a characterization of the singular locus $LS\overline{\Sigma}(M, D)$ of $\overline{\Sigma}(M, D)$ applying the **combinatorial smoothness criterium (cf. proposition 4.13)**, it is proved that the (M, M')-tangent space $\overline{T}(\overline{\Sigma}(M, D))_{kD'}$ to $\overline{\Sigma}(M, D)$ is equal to $\tilde{T}^{comb}(\overline{\Sigma}(M, D))_{kD'}$. The following proposition reduces the proof of the equality to a combinatorial verification.

Proposition 4.35 *The following assertions are equivalent:*

1) $\bigcap\limits_{(\alpha,\beta)\in GI(M,M')} T(\Sigma^\natural(M_{\alpha\beta}, D))_{kD'} = \tilde{T}^{comb}(\overline{\Sigma}(M, D))_{kD'}$;

2) $\overline{\mathcal{B}}_{(D,D')} = \tilde{\mathcal{B}}_{(D,D')}$;

3) $\overline{T}(\overline{\Sigma}(M, D))_{kD'} = \tilde{T}^{comb}(\overline{\Sigma}(M, D))_{kD'}$;

4) $Y_\beta^+(M, D, D') = \bigcap\limits_{\varphi\in Conf^{comb}(\Lambda(M),D,D')} \mathcal{N}_\varphi^{(\lambda+1,\beta)}\cap(J_\beta\times(J_{\beta+1}/J_\beta))$ $(1 \leq \beta \leq l)$;

5) $Y_\beta^-(M, D, D') = \bigcup\limits_{\varphi\in Conf^{comb}(\Lambda(M),D,D')} T_\varphi^{(\lambda+1,\beta)}\cap(J_\beta\times(J_{\beta+1}/J_\beta))$ $(1 \leq \beta \leq l)$;

6) $Y^+(M, D, D') = \coprod\limits_{1\leq\beta\leq l} \bigcap\limits_{\varphi\in Conf^{comb}(\Lambda(M),D,D')} \mathcal{N}_\varphi^{(\lambda+1,\beta)} \cap (J_\beta \times (J_{\beta+1}/J_\beta))$ $(1 \leq \beta \leq l)$.

The last assertion is equivalent to the equality of the index set of $\overline{\mathcal{B}}_{(D,D')}$ with the index set of $\tilde{\mathcal{B}}_{(D,D')}$. The reader is referred to [15] for the proof of the following proposition which is the main verification to be accomplished in the proof of the fundamental equality:

Proposition 4.36 *1)* $Y_\beta^+(M, D, D') = \bigcap\limits_{\varphi \in Conf^{comb}(\Lambda(M), D, D')} \mathcal{N}_\varphi^{(\lambda+1, \beta)} \cap$

$(J_\beta \times (J_{\beta+1}/J_\beta))$.

2) $Y_\beta^-(M, D, D') = (J_\beta \times (J_{\beta+1}/J_\beta)) - \bigcap\limits_{\varphi \in Conf^{comb}(\Lambda(M), D, D')} \mathcal{N}_\varphi^{(\lambda+1, \beta)} \cap$

$(J_\beta \times (J_{\beta+1}/J_\beta))$.
Where $Y_\beta^+(M, D, D') = \bigcup\limits_{1 \leq \alpha \leq \lambda} \overline{H}_{\alpha\beta} \times J_{\beta+1}/(J_\beta \cup J_{\beta+1} \cap H_\alpha)$.

Remark 4.37 *The proof of the above equalities results from the exactness of the properties of φ when a $\Lambda(M)$-configuration φ is interpreted as a functor from the category defined by $\Lambda(M)$ to the category $\mathscr{P}(I_{r+1})$.*

From propositions 4.35, and 4.36 it results that the family of subvarieties $(\Sigma^\natural(M_{\alpha\beta}, D))_{(\alpha,\beta) \in GI(M, M')}$ satisfies $\bigcap\limits_{(\alpha,\beta) \in GI(M, M')} T(\Sigma^\natural(M_{\alpha\beta}, D))_{k^{D'}} = \tilde{T}^{comb}(\overline{\Sigma}(M, D))_{k^{D'}}$ and thus the combinatorial smoothness criterium (cf. proposition 4.13) applies and it is concluded that:

$\overline{\Sigma}(M, D)$ *is smooth at* $k^{D'} \iff dim_k \overline{T}(\overline{\Sigma}(M, D))_{k^{D'}} = dim_k \Sigma(M, D)$.

It is convenient to re-state this assertion in the following form.

Theorem 4.38

$\overline{\Sigma}(M, D)$ *is smooth at* $k^{D'} \iff |Y^-(M, D, D')| = dim_k \Sigma(M, D)$.

Remark 4.39 *1) The dimension $dim_k \Sigma(M, D)$ may be calculated easily from M, and the condition $|Y^-(M, D, D')| = dim_k \Sigma(M, D)$ may be written in terms of $\mathfrak{S}_{r+1}$.*

2) The set $GI(M, M')$ is not arbitrary as it follows from the proof of proposition 4.36.

Chapter 5

The Flag Complex

It is shown in this chapter and the following ones that the construction of the smooth varieties $Conf(\Lambda(M), D)$ is in fact a particular case of a general one in the setting of building theory. For this aim it is introduced here the building theory terminology.

The **Flag Complex** is the Tits Building of the general linear group $GL(k^{r+1})$ so named by Mumford in [42]. It also plays an important role in the contruction of a natural compactification of a reductive group [43], and in that of smooth compactifications of symmetric spaces [44]. The term Flag Complex refers to its underlying abstract simplicial complex, i.e. its combinatorial structure. The aim of both this chapter and the next one is to introduce the building setting in the case of the linear group, and to construct a family of configurations varieties, which give rise to smooth resolutions of Schubert varieties, in terms of the building geometry introduced by Tits in [50]. This chapter may be seen as a guide suggesting the following general constructions of the next chapters. The set of flags of k^{r+1} adapted to the canonical basis, i.e. the combinatorial flags, is endowed with the structure of a simplicial complex, namely the first barycentric subdivision of the combinatorial $(r+1)$-simplex. On the other hand, there is a simplicial complex naturally associated to the symmetric group $\mathfrak{S}_{r+1}$ (**The Coxeter complex**). The natural action of $\mathfrak{S}_{r+1}$ on the combinatorial flags induces an isomorphism between these two complexes. There is a geometrical interpretation of these complexes as a subdivision of an euclidean space by simplicial cones, given by a finite set of hyperplanes, whose equations are given by the roots of $Gl(r+1)$, as defined in the preceding chapter. A geometrical realization of the **Cayley Diagram** of the symmetric group is obtained from the geometrical interpretation of the Coxeter Complex. This construction establishes a correspondence between the paths issued from a point of the former with the galleries issued from a chamber of the latter.

The set of flags $Drap(k^{r+1})$ of k^{r+1} endowed with a simplicial complex structure forms the **Flag complex** that contains the complex of adapted flags as a subcomplex (**The Canonical Apartment**). Both of them are **Buildings** (cf. loc. cit.), so the general definitions of abstract buildings apply. Generalized galleries in these complexes can then be defined. **Minimal generalized galleries** are defined in terms of the combinatorial geometry of the Apartment. In the next chapter it is shown how their associated typical graphs give rise to a family of smooth resolutions of Schubert varieties.

5.1 Buildings and galleries

We define an abstract building as follows. The reader is referred to [4] and [50] for details. Let A be a set and $\mathrm{Ch}(A) \subset \mathcal{P}(A)$ a subset of the class of subsets of A. The elements C of $\mathrm{Ch}(A)$ are called chambers of A. A subset $F \subset C$ of a chamber C is called a facet, and the cardinal of $C \setminus F$ is the **codimension of F in C**.

A **gallery** $\Gamma = (C = C_0, C_1, \ldots, C_n = C')$ **of length** n between two chambers C and C' of A is a sequence of $n + 1$ chambers such that C_i and C_{i+1} have a common facet F of codimension 1. Then, either $C_i = C_{i+1}$, or $C_i \cap C_{i+1} = F$. The chamber C is called the origin (resp. left extremity) of the gallery Γ and C' is called the end (resp. right extremity) of Γ. The set $\{C, C'\}$ is the set of extremities of the gallery Γ. Then, Γ is a gallery between C and C' of length n. A **minimal gallery** Γ **between** C **and** C' **of length** n is a gallery such that there is no gallery between C and C' of length $< n$.

The pair $(A, \mathrm{Ch}(A))$ is a **building** if:

- $A = \cup_{C \in \mathrm{Ch}(A)} C$;

- There is at least one gallery between two chambers C and C'.

Observe that a building is naturally endowed with a structure of a simplicial complex. In a building $\mathcal{A} = (A, \mathrm{Ch}(A))$, the codimension of a facet F is independent of the chamber C containing F.

A **sub-building** $\mathcal{A}'$ of $\mathcal{A} = (A, \mathrm{Ch}(A))$ is a couple $(D, \mathrm{Ch}(A) \cap \mathcal{P}(D))$, where D is a subset of A and $\mathcal{P}(D)$ is the class of subsets of D, so that $(D, \mathrm{Ch}(A) \cap \mathcal{P}(D))$ satisfies the two properties of buildings above. With a chamber C is associated the sub-building $\Delta(C)$ formed by the set of facets of C.

A **morphism of buildings** $f : (A, \mathrm{Ch}(A)) \to (B, \mathrm{Ch}(B))$ is a mapping $f : A \to B$ inducing for every chamber C an isomorphism between $\Delta(C)$ and $\Delta(f(C))$.

An **apartment** is a building A so that every facet of codimension 1 is contained in exactly two chambers.

The set of facets of a building $\mathcal{A}$ is naturally ordered by inclusion. Two facets are **incident** if $F \cup F'$ is a facet, or equivalently if F and F' are contained in some chamber C of the building $\mathcal{A}$. A set Φ of facets of the building $\mathcal{A}$ is a **sub-complex** of $\mathcal{A}$ if $F \subset F'$ and $F \in \Phi$ implies $F' \in \Phi$.

Given two chambers C and C' of $\mathcal{A}$ the **distance** $d(C, C')$ between C and C' is the length of a minimal gallery between C and C'. This length is independent of the minimal gallery.

If the galleries $\Gamma^1 = (C_0^1, \ldots, C_m^1)$ and $\Gamma^2 = (C_0^2, \ldots, C_n^2)$ satisfy $C_m^1 = C_0^2$, then the sequence $\Gamma^1 \circ \Gamma^2 = (C_0^1, \ldots, C_m^1 = C_0^2, \ldots, C_n^2)$ is a gallery called the **composed gallery of Γ^1 and Γ^2**.

Definition 5.1 *A sequence of facets $\gamma = (F_n, \ldots, F_0)$ of a building $\mathcal{A}$ is a* **generalized gallery (gg)** *of $\mathcal{A}$ if it satifies one of the following conditions:*

(i) *For $n > i \geqslant 1$, $i \equiv 1 \pmod{2}$, $F_{i+1} \supset F_i \subset F_{i-1}$, γ is called a closed gallery;*

(ii) *For $n > i \geqslant 2$, $i \equiv 0 \pmod{2}$, with $F_{n+1} = 0$, $F_i \subset F_{i-1} \supset F_{i-2}$, γ is called a right open gallery;*

(iii) *For $n > i \geqslant 1$, $i \equiv 1 \pmod{2}$, with $F_{n+1} = 0$, $F_{i+1} \supset F_i \subset F_{i-1}$, γ is called a left open gallery;*

(iv) *For $n > i \geqslant 1$, $i \equiv 1 \pmod{2}$, $F_{i+1} \subset F_i \supset F_{i-1}$, γ is called an open gallery.*

If γ is a gallery satisfying the conditions (i) (resp. (ii), (iii), (iv)), we denote it respectively as follows

(i)' $\gamma = (F_r \supset F_{r-1} \subset F_{r-2} \ldots F_1 \subset F_0)$;

(ii)' $\gamma = (F_r \supset F_{r-1} \subset F_{r-2} \ldots F_0 \supset F_0)$;

(iii)' $\gamma = (F_r \subset F_{r-1} \supset F_{r-1} \subset F_{r-2} \ldots F_1 \subset F_0)$;

(iv)' $\gamma = (F_r \subset F_{r-1} \supset F_{r-1} \subset F_{r-2} \ldots F_1 \supset F_0)$.

Remark 5.2 *Another indexation of the facets composing a* **generalized gallery** *γ is possible. In fact this is the usual notation we employ in this work. Denote by:*

i)" $\gamma = (F_r \supset F_r' \subset F_{r-1} \ldots F_1 \supset F_1' \subset F_0)$ *a closed gallery;*

ii)" $\gamma = (F_r \supset F_r' \subset F_{r-1} \supset F_{r-1}' \ldots F_1 \supset F_1')$ *a right open gallery;*

iii)" $\gamma = (F_{r+1}' \subset F_r \supset F_r' \subset F_{r-1} \ldots F_1' \subset F_0)$ *a left open gallery;*

vi)" $\gamma = (F_{r+1}' \subset F_r \supset F_r' \subset F_{r-1} \ldots F_0 \supset F_0')$ *an open gallery.*

Observe that given an ordered set $(E, \prec)$, the above definition may be reformulated in terms of the order $\prec$, i.e. by replacing $\subset$ *or* $\supset$ respectively by $\prec$ *or* $\succ$, and the facets by elements of E. We thus obtain the definition of a generalized gallery in E.

A generalized gallery is **non-stammering** if all the facets inclusions are strict. From now on all the generalized galleries we consider are implicitly supposed to be non-stammering. On the other hand, it is immediate that a generalized gallery may be reduced to a non-stammering gallery defined by the same set of facets.

If no confusion arises, with the standard terminology of buildings, we say simply gallery for generalized gallery. With the definition of 5.1, by definition a gallery γ in $\mathcal{A}$ is issued from F if $F = F_n$, and F_n and F_0, or (F_n, F_0), are the extremities of γ. Write $E_1(\gamma) = F_n$ (resp. $E_2(\gamma) = F_0$, $E(\gamma) = (E_1(\gamma), E_2(\gamma))$), for the left extremity(origin) (resp. right extremity(end), the extremities) of γ.

Notation 5.3 *Given a building I (resp. a building I, and $F, F' \in I$) Gall_I (resp. $\mathrm{Gall}_I(F)$, $\mathrm{Gall}_I(F, F')$) is defined as the set of generalized galleries of I (resp. of generalized galleries of I issued from F, generalized galleries with extremities (F, F')).*

5.2   The simplex barycentric subdivision

An important example of building may be given. Denote by $\mathcal{P}(I_{r+1})$ (resp. $\mathcal{P}^*(I_{r+1})$) the set of subsets of I_{r+1} (resp. the set of non-void subsets of I_{r+1}).

Definition 5.4 *Write $\Delta^{(r)} = \Delta(I_{r+1}) = \mathcal{P}^*(I_{r+1})$. Let $\Delta^{(r)}$ be endowed with the symmetrization of the inclusion relation.* (**The combinatorial** r**-simplex**)

Let $\Delta^{(r)'} = \Delta'(I_{r+1})$ be the set of combinatorial flags $D = (J_1 \subset J_2 \subset \cdots \subset J_l \subset I_{r+1})$ of I_{r+1}. Write (I_{r+1}) for $(\emptyset \subset I_{r+1})$. (I_{r+1}) is included in $\Delta'(I_{r+1})$. Let

$$Vert(\Delta^{(r)'}) = \{\, (J \subset I_{r+1}) \mid J \in \mathcal{P}^*(I_{r+1}) - \{I_{r+1}\} \,\}$$

(**The set of vertices of** $\Delta^{(r)'}$). *It is said that $Vert(D) = \{J_1, ..., J_l, I_{r+1}\}$ is the set of vertices of D. The inclusion relation*

$$Vert(D) \subset Vert(D')$$

defines an order on $\Delta^{(r)'}$. In this case we write $D \subset D'$. By definition two flags D and D' are **incident** *if there is a maximal length flag D'' with $D \subset D''$*

and $D' \subset D''$ (cf. [50]). We endow $\Delta^{(r)'}$ with the inclusion order relation.
(The first barycentrical subdivision of $\Delta^{(r)}$)
   Define

$$typ : \Delta^{(r)'} \longrightarrow typ(I_{r+1})$$

*as follows. Let $D = (J_1 \subset J_2 \subset \cdots J_l \subset I_{r+1}) \in \Delta^{(r)'}$ with card $J_1 = n_1$, card $J_2 = n_2, \cdots$, card $J_l = n_l$. Write typ $D = \underline{n} = (n_1 < n_2 < \cdots < n_l < r+1)$ (resp. typ $((I_{r+1})) = (r+1))$ (**The type of a flag D**).*

The set of **combinatorial flags** is naturally endowed with a building structure. Observe that $Drap(I_{r+1}) = \Delta^{(r)'}$. Write $\Delta_{\underline{n}}^{(r)'} = Drap_{\underline{n}}(I_{r+1})$. Let it be proved that $\Delta^{(r)'}$ is a building with the set of maximal length flags $Drap_{\underline{r}}(I_{r+1})$ $(\underline{r} = (1 < \cdots < r < r+1)$ as the set of chambers $Ch\ \Delta^{(r)'}$. This results from the bijection: $\Delta^{(r)'} \simeq \Delta^{(r)'}(p_1, \cdots, p_{r+1})$, where $\Delta^{(r)}(p_1, \cdots, p_{r+1})$ denotes the convex **envelope** (also called the convex **hull**) of a set of affinely independent points $p_1, \cdots p_{r+1}$ in some affine space, and $\Delta^{(r)'}(p_1, \cdots, p_{r+1})$, the set of simplices of the first barycentrical subdivision of $\Delta^{(r)}(p_1, \cdots, p_{r+1})$ having the barycenter $(p_1 + \cdots + p_{r+1})/(r+1)$ as a vertex. The bijection being induced by the natural bijection $i \mapsto p_i$. By this correspondence the chambers of $\Delta^{(r)'}$ correspond to the r-dimensional simplices of $\Delta^{(r)'}(p_1, \cdots, p_{r+1})$. Given two r-dimensional simplices in $\Delta^{(r)'}(p_1, \cdots, p_{r+1})$ it is clear that there exists a sequence of r-dimensional simplices such that two succesive simplices have a common $(r-1)$-face. This shows that the second axiom defining a building holds for $\Delta^{(r)'}$, the first one is trivially satisfied by $\Delta^{(r)'}$.

Remark 5.5   *1) The set of vertices $Vert(\Delta^{(r)'})$ is given by $Vert(\Delta^{(r)'}) = Grass(I_{r+1})$.*

   *2) In fact $\Delta^{(r)'}$ is a **flag complex**. This means that $Vert(\Delta^{(r)'}) = Grass(I_{r+1})$ endowed with the incidence relation " **generates**" $\Delta^{(r)'}$. More precisely the flags (resp. facets) D of $\Delta^{(r)'}$ are given by the subsets of $Grass(I_{r+1})$ whose elements are two by two incidents (cf. loc. cit.).*

5.2.1   Simplex barycentric subdivision automorphisms group

The symmetric group $\mathfrak{S}_{r+1}$ acts naturally on $\Delta^{(r)'}$. This group may be characterized as the group of building automorphisms of $\Delta^{(r)'}$ which preserve the type of the facets. Let

$$typ_{\Delta^{(r)'}} = \Delta^{(r)'}/\mathfrak{S}_{r+1},$$

and

$$Relpos_{\Delta^{(r)'}} = (\Delta^{(r)'} \times \Delta^{(r)'})/\mathfrak{S}_{r+1}.$$

One has the following identifications:

$$Relpos(I_{r+1}) = Relpos_{\Delta^{(r)'}}$$

and

$$typ_{\Delta^{(r)'}} = typ(I_{r+1}) = \{\underline{n} = (n_1 < n_2 < \cdots < n_l < r+1)$$

$$| \; n_1, n_2, \cdots, n_l \in \mathbb{N}\} \cup \{r+1\}$$

(The typical simplex of $\Delta^{(r)'}$). In fact $typ(I_{r+1})$ is endowed with a canonical building structure with only one chamber C given by $C = \{1, \cdots, r+1\}$, and typ is a building morphism. One has $\Delta(C) \simeq typ(I_{r+1})$.

5.3 Combinatorial roots, and hyperplanes

Let $D = (J_1 \subset \cdots \subset J_l \subset I_{r+1})$. Write: $(\forall \, i \in I_{r+1}) \, \alpha_D(i) := \min\{\alpha | \, i \in J_\alpha\}$. We associate to D the set

$$R_D(I_{r+1}) := \{(i,j) \in I_{r+1} \times I_{r+1}/\Delta | \; \alpha_D(i) \leqslant \alpha_D(j)\}.$$

The closed subset of roots $R_D \subset E \times E/\Delta(E)$ associated with a flag $D \in Drap(E)$ has yet been defined. It is easy to see that $R_D(I_{r+1})$ is equal to the parabolic set R_D associated to $D \in Drap(I_{r+1})$. Recall that $Gl_{R_D} = Stab\,(k^D)$ thus $Gl_{R_D \cap R_{D'}} = Stab\,(k^D) \cap Stab\,(k^{D'})$. If D is a maximal length flag, i.e. $D \in Ch\,\Delta^{(r)'}$ then by the bijection $Ch\,\Delta^{(r)'} \simeq Ord(I_{r+1})$ it corresponds to D an order $\underset{D}{\leqslant}$ of I_{r+1} which is precisely ω_D and whose graph is given by $R_D(I_{r+1}) \cup \Delta(I_{r+1})$.

With $(i,j) \in I_{r+1} \times I_{r+1}/\Delta(I_{r+1})$ is associated a subcomplex of $\Delta^{(r)'}$ defined by

$$\mathscr{A}_{(i,j)} := \{D \in \Delta^{(r)'} | \; (i,j) \in R_D(E)\} \subset \Delta^{(r)'}.$$

The subcomplex $\mathscr{A}_{(i,j)}$ may be characterized as the set of flags $D \in \Delta^{(r)'}$ incident to some maximal length flag D' such that $i \underset{D'}{\leqslant} j$. One has that $\Delta^{(r)'} = \mathscr{A}_{(i,j)} \cup \mathscr{A}_{(j,i)}$. The transposition $(ij) \in \mathfrak{S}_{r+1}$ sends $\mathscr{A}_{(i,j)}$ to $\mathscr{A}_{(j,i)}$. $\mathscr{A}_{(i,j)}$ is called a **combinatorial root of $\Delta^{(r)'}$**. From the definition of a combinatorial root it results that $\{D\} = \underset{(i,j) \in R_D(I_{r+1})}{\bigcap} \mathscr{A}_{(i,j)}$.

The intersection subcomplex $H_{ij} = \mathscr{A}_{(i,j)} \cap \mathscr{A}_{(j,i)}$ is given by the set of flags invariant by the transposition (ij). H_{ij} is called the **combinatorial hyperplane of $\Delta^{(r)'}$** (resp. **wall of $\Delta^{(r)'}$**) defined by the combinatorial root $\mathscr{A}_{(i,j)}$. Denote by $\mathscr{H}$ the set of combinatorial hyperplanes $(H_{ij})_{(i,j) \in R(I_{r+1})}$. Write:

$$\partial \mathscr{A}_{(i,j)} = H_{ij} \quad (\text{resp. } \partial \mathscr{A}_{(j,i)} = H_{ji}),$$

i.e. H_{ij} is the hyperplane determined by $\mathscr{A}_{(i,j)}$. A hyperplane is the wall of exactly two roots (cf. [4]).

Definition 5.6 *Let* $D \in Ch \ \Delta^{(r)'} = \Delta^{(r)'}_{\underline{r}}$. $H \in \mathcal{H}$ *is a* **bounding hyperplane** *of* D *if* $H = H_{i_\alpha i_{\alpha+1}}$ *where* $\omega_D = (i_1 < \cdots i_\alpha < i_{\alpha+1} \cdots i_{r+1})$ *denotes the total order of* I_{r+1} *defined by* D *(cf. Definition 2.22).*

Let $\mathcal{H}_D = \{ H \in \mathcal{H} | \ D \in H \}$

$(resp. \ \mathcal{H}_D(D') = \{ H \in \mathcal{H} | \ D \in H, \ D' \notin H \}, \ \mathcal{H}(D') = \{ H \in \mathcal{H} | \ D' \notin H \})$.

Let $D, D' \in \Delta^{(r)'}$. The hyperplane H_{ij} **separates** D and D' if $H_{ij} \notin \mathcal{H}_D \cup \mathcal{H}_{D'}$ and if one of the following statements holds:

1. $i \underset{D}{\leqslant} j$ and $j \underset{D'}{\leqslant} i$,

2. $j \underset{D}{\leqslant} i$ and $i \underset{D'}{\leqslant} j$.

(i.e. either $(i,j) \in R_D$ and $(i,j) \notin R_{D'}$, or $(i,j) \notin R_D$ and $(i,j) \in R_{D'}$) In that case we have that D belongs to one of the combinatorial roots determined by H_{ij} and D' to the other. Denote by $\mathcal{H}(D, D')$ the set of hyperplanes which separates D and D'.

If $D \in Ch \ \Delta^{(r)'} = \Delta^{(r)'}_{\underline{r}}$ ($\underline{r} = (1 < 2 < \cdots < r+1)$), and $D' \in \Delta^{(r)'}$ write

$$d(D, D') := |\mathcal{H}(D, D')| = |R(D, D')|$$

(The combinatorial distance between D and D'). It is recalled that $R(D, D')$ is the set of couples $(i, j) \in R_D$ such that $(i, j) \notin R_{D'}$. Otherwise stated the set of couples (i, j) in R_D such that the wall $H_{(i,j)}$ separates D and D'. Remark that $(i, j) \in R_D$ if and only if $D \in \mathcal{A}_{(i,j)}$. This definition of the distance between a maximal length flag D, i.e. a chamber of $\Delta^{(r)'}$, and a flag D' is equivalent to the definition given in 5.1 if D' is also a maximal length flag.

Retain the notation of 2.5.3. The following proposition, establishing a relation between the set of couples $R(D, D')$ and the Coxeter group $(\mathfrak{S}_{r+1}, S_D)$ $(S_D = \{(12), \cdots, (rr + 1)\})$ (cf. [4]), results from general considerations about Coxeter complexes exposed in the next chapters, or may be checked directly for the flag complex $Drap(I_{r+1})$.

Proposition 5.7
Retain the above notation.

1. *Let* (D, D') *be a couple of maximal length flags then* $l_{S_D}(w(D, D')) = |R(D, D')|$.

2. *For* (D, D') *a couple of flags with* D *of maximal length we have:* $R(D, D') = R(D, proj_{D'} \ D)$.

3. *For* (D, D') *as in 2– we have* $l_{S_D}(w(D, proj_{D'} \ D)) = |R(D, D')|$.

5.4 The Star Complex defined by a flag in the Simplex barycentric subdivision

The general terminology of simplicial complexes may be adapted to a building complex $\Delta^{(r)'}$. Let $D \in \Delta^{(r)'}$. As a particular case of a general definition in the theory of buildings (cf. [50], p.1) the set is introduced

$$St_D := \{D' \in \Delta^{(r)'} \mid D \subset D'\}$$

(the star of D in $\Delta^{(r)'}$). This set endowed with the relation of inclusion between flags is a building, whose set of chambers $Ch\ St_D$ is given by the maximal length flags incident to D. Recall that the flags D' (resp. simplices) of $\Delta^{(r)'}$ are classified according to their type.

Definition 5.8 *Given a flag $D \in Drap(I_{r+1})$ of type t and a type $s \in$ $typ(I_{r+1})$ define $\Sigma_D(s) \subset Drap(I_{r+1})$ (resp. $\Sigma_t(s) \subset Drap(I_{r+1}) \times Drap(I_{r+1})$) as the set of combinatorial flags given by:*

$$\Sigma_D(s) \quad := \quad \{\ D' \in Drap_s(I_{r+1}) \mid D \subset D'\}$$
$$(resp.\ \Sigma_t(s) \quad := \quad \{\ (D, D') \in Drap_t(I_{r+1}) \times Drap_s(I_{r+1}) \mid D \subset D'\}).$$

One has that $\Sigma_t(s) \subset Drap_t(k^{r+1}) \times Drap_s(k^{r+1})$ is the graph of the order relation $D \subset D'$. Write:
$$St_D = \coprod_{t \subset s} \Sigma_D(s)\ .$$

The group of type preserving automorphisms of St_D is given by the stabilizer of D: $\mathfrak{S}_D \subset \mathfrak{S}_{r+1}$. There is a building isomorphism

$$St_D \simeq \prod \Delta^{(r_\alpha)'}$$

of St_D with a product of buildings given by barycentrical subdivisions of combinatorial simplices. It would be seen that in fact $\Delta^{(r_\alpha)'}$ (resp. St_D) may be obtained in terms of the system of Coxeter $(\mathfrak{S}_{r+1}, S)$ (resp. $(\mathfrak{S}_D, S_D)$), where S (resp. S_D) denotes the canonical set of generating transpositions.

5.5 The Simplex barycentric subdivision geometric realization

Let $\mathbb{A}^r \subset \mathbb{R}^{r+1}$ be the r-dimensional affine subspace of the euclidian space $\mathbb{R}^{r+1}$, defined by the equation: $x_1 + \cdots x_{r+1} = 1$. Write $p_1 = e_1, \cdots, p_{r+1} = e_{r+1}$, where $(e_1, \cdots, e_{r+1})$ denotes the canonical basis of $\mathbb{R}^{r+1}$. Clearly there is $p_1, \cdots, p_{r+1} \in \mathbb{A}^r$. Denote by $\Delta^{(r)}(p_1, \cdots, p_{r+1}) \subset \mathbb{A}^r$ the r-dimensional regular simplex with vertices $p_1, \cdots p_{r+1}$, i.e. the **convex hull** of the points $p_1, \cdots p_{r+1}$. Given $(i, j) \in I_{r+1} \times I_{r+1}/\Delta(I_{r+1})$ let $\overline{H}_{ij} \subset \mathbb{R}^{r+1}$ be the hyperplane defined by the equation $x_j - x_i = 0$, and $H_{ij} \subset \mathbb{A}^r$ the affine hyperplane obtained as the intersection $H_{ij} = \overline{H}_{ij} \cap \mathbb{A}^r$. Clearly one has $\overline{H}_{ij} = \overline{H}_{ji}$ (resp. $H_{ij} = H_{ji}$). Write

$$\mathscr{H} := \{H_{ij}| \ (i,j) \in I_{r+1} \times I_{r+1}/\Delta\}.$$

The family $\mathscr{H}$ is indexed by the set of pairs $\{\{i,j\}| \ i,j \in I_{r+1}, \ (i \neq j)\}$. This family is the family of symmetry hyperplanes of $\Delta^{(r)}(p_1, \cdots, p_{r+1})$. The pair $\{i,j\}$ corresponds to the hyperplane $H_{ij} = H_{ji}$ which may be characterized as the symmetry hyperplane of $\Delta^{(r)}$ defined by the r affinely independant points

$$(p_i + p_j)/2, p_1, \cdots, \hat{p}_i, \cdots, \hat{p}_j, \cdots, p_{r+1}.$$

The orthogonal reflexion $\overline{s} = \overline{s}_{ij} = \overline{s}_{ji}$ defined by the hyperplane $\overline{H}_{ij} = \overline{H}_{ji}$ is given by

$$\overline{s} : (x_1, \cdots, x_i, \cdots, x_j, \cdots, x_{r+1}) \mapsto (x_1, \cdots, x_j, \cdots, x_i, \cdots, x_{r+1}),$$

and leaves the simplex $\Delta^{(r)}(p_1, \cdots, p_{r+1})$: $x_1 + \cdots x_{r+1} = 1, \ 0 \leqslant x_1, \cdots, x_{r+1}$ invariant. The restriction $s = s_{ij} = s_{ji}$ of $\overline{s}$ to $\mathbb{A}^r$ is clearly the orthogonal reflexion of $\mathbb{A}^r$ defined by $H_{ij} = H_{ji}$. Remark that

$$(\forall \ H \in \mathscr{H}), \quad (p_1 + \cdots + p_{r+1})/(r+1) \in H.$$

Let $\mathscr{C}(\mathscr{H})$ be the set of **simplicial cones** obtained from the decomposition of $\mathbb{A}^r$ in terms of the equivalence relation defined by $\mathscr{H}$ (cf. [4], Ch.V, §1). By definition **the carrier** *supp* F of a cone $F \in \mathscr{C}(\mathscr{H})$ is the affine subspace generated by F. Write *dim* $F = $ *dim supp* F. A chamber C of $\mathscr{C}(\mathscr{H})$ is by definition a cone F with *dim* $F = r$. Moreover the **set of chambers** Ch $\mathscr{C}(\mathscr{H}) \subset \mathscr{C}(\mathscr{H})$ is equal to the set of connected components of $\mathbb{A}^r - \bigcup H_{ij}$. An order relation on $\mathscr{C}(\mathscr{H})$ is defined by

$$F' \prec F \quad \text{if} \quad F' \subset \overline{F}.$$

Two cones F and F' are **incident** if there exists $C \in$ Ch $\mathscr{C}(\mathscr{H})$ with $\overline{C} \supset F, F'$. Two chambers C, C' are **adjacent** if their closures contain a common cone of dimension $r-1$.

There is a bijection

$$\Delta^{(r)'} \xrightarrow{\sim} \mathscr{C}(\mathscr{H})$$

compatible with the order relations of $\Delta^{(r)'}$ and $\mathscr{C}(\mathscr{H})$ defined as follows. Given $J \subset I_{r+1}$ let Env$\{p_i| \ i \in J\}$ be the facet of $\Delta^{(r)}(p_1, \cdots, p_{r+1})$ with vertices $\{p_i| \ i \in J\}$, i.e. Env$\{p_i| \ i \in J\} = $ convex hull in $\mathbb{A}^r$ of the set $\{p_i| \ i \in J\}$. Write

$$\sigma(J) := \text{Env}\{p_i| \ i \in J\}.$$

Given a facet $\sigma(J)$ of $\Delta^{(r)}(p_1, \cdots, p_{r+1})$ let bar $\sigma(J)$ denote the barycenter of $\sigma(J)$. Denote by $\mathscr{C}_{\sigma(J)}$ the open ray issued from bar $\sigma(I_{r+1}) = (p_1 + \cdots + p_{r+1})/(r+1)$ and determined by bar $\sigma(J)$. It is easy to see that $\mathscr{C}_{\sigma(J)} \in \mathscr{C}(\mathscr{H})$, i.e. that if $\mathscr{L}$ is the line determined by $\mathscr{C}_{\sigma(J)}$ then: $\mathscr{L} = \bigcap_{\substack{H \in \mathscr{H} \\ \mathscr{C}_{\sigma(J)} \subset H}} H$, and $\mathscr{C}_{\sigma(J)} = D_H \cap \mathscr{L}$ where D_H denotes a half space defined by some $H \in \mathscr{H}$ containing $\mathscr{C}_{\sigma(J)}$. The vertices of $\Delta^{(r)'}$ correspond to the faces of $\Delta^{(r)}$. It is said that two faces J, J' of $\Delta^{(r)}$ are **incident** if $(J \subset I_{r+1}), (J' \subset I_{r+1})$ are incident in $\Delta^{(r)'}$, i.e. if $J \subset J'$ or $J' \subset J$. One has

$$J, \ J' \in \Delta^{(r)'} \text{ incident} \quad \Longleftrightarrow \quad \mathscr{C}_{\sigma(J)}, \ \mathscr{C}_{\sigma(J')} \in \mathscr{C}(\mathscr{H}) \text{ incident.}$$

Given $D = (J_1 \subset \cdots \subset J_l \subset I_{r+1}) \in \Delta^{(r)'}$ the elements of $\text{Vert}(D) - \{J_1, \cdots, J_l, I_{r+1}\} \subset \Delta^r$ are two by two incident. Consequently the elements of $\{\mathscr{C}_{\sigma(J_1)}, \cdots, \mathscr{C}_{\sigma(J_l)}\}$ are two by two incident in $\mathscr{C}(\mathscr{H})$. Let then $\mathscr{C}_{\sigma(D)} \in \mathscr{C}(\mathscr{H})$ be the unique facet such that $\mathscr{C}_{\sigma(D)} \subset \text{Env}(\mathscr{C}_{\sigma(J_1)} \cup \cdots \cup \mathscr{C}_{\sigma(J_l)})$ and $\mathscr{C}_{\sigma(J_1)}, \cdots, \mathscr{C}_{\sigma(J_l)} \subset \overline{\mathscr{C}}_{\sigma(D)}$, i.e. $\mathscr{C}_{\sigma(D)}$ is incident to the $\mathscr{C}_{\sigma(J_i)}$'s. The bijection $\Delta^{(r)'} \xrightarrow{\sim} \mathscr{C}(\mathscr{H})$ is defined by

$$D \mapsto \mathscr{C}_{\sigma(D)} \ .$$

Let $\mathfrak{S}'_{r+1} \subset O(\mathbb{A}^r)$ be the subgroup of the orthogonal group of $\mathbb{A}^r$, generated by the set of orthogonal reflexions $(s_{ij}) = (s_H)_{H \in \mathscr{H}}$ defined by the symmetry hyperplanes of the simplex $\Delta^{(r)'}(p_1, \cdots, p_{r+1})$. Thus $\mathfrak{S}'_{r+1}$ fixes its barycenter $(p_1 + \cdots + p_{r+1})/(r + 1)$.

Proposition 5.9 *There is a natural isomorphism:* $\mathfrak{S}'_{r+1} \simeq Aut(I_{r+1}) = \mathfrak{S}_{r+1}$.

Proof *It is clear that $\mathfrak{S}'_{r+1}$ stabilises the set of vertices $\{p_1, \cdots, p_{r+1}\}$ and that the action of $\mathfrak{S}'_{r+1}$ on this set characterizes the action of $\mathfrak{S}'_{r+1}$ on $\mathbb{A}^r$. Thus there is a monomorphism*

$$\mathfrak{S}'_{r+1} \longrightarrow Aut(I_{r+1}) = \mathfrak{S}_{r+1}.$$

The image of this monomorphism contains the set of transpositions of $\mathfrak{S}_{r+1}$. It is concluded that it is an epimorphism and thus an isomorphism.

From the fact that $\mathfrak{S}'_{r+1} \subset O(\mathbb{A}^r)$ is generated by the set of orthogonal reflexions $(s_H)_{H \in \mathscr{H}}$ defined by $\mathscr{H}$, it is deduced that $\mathfrak{S}_{r+1}$ acts naturally on $\mathscr{C}(\mathscr{H})$ through the isomorphism $\mathfrak{S}_{r+1} \xrightarrow{\sim} \mathfrak{S}'_{r+1}$. One has $(\forall \ w \in \mathfrak{S}_{r+1})$ $w(\mathscr{C}_{\sigma(D)}) = \mathscr{C}_{\sigma(w(D))}$. One finally gets

Proposition 5.10 *The correspondence $D \to \mathscr{C}_{\sigma(D)}$ defines a $\mathfrak{S}_{r+1}(\simeq \mathfrak{S}'_{r+1})$-equivariant order preserving bijection*

$$\Delta^{(r)'} \simeq \mathscr{C}(\mathscr{H}).$$

A building structure is associated to the set of cones $\mathscr{C}(\mathscr{H})$.

Definition 5.11 *Denote by* $Vert(\mathscr{C}(\mathscr{H})) \subset \mathscr{C}(\mathscr{H})$ *the set of* 1*-dimensional cones, i.e. rays, and given* $F \in \mathscr{C}(\mathscr{H})$, *by* $Vert(F) \subset Vert(\mathscr{C}(\mathscr{H}))$ *the set of cones* $F' \in \mathscr{C}(\mathscr{H})$ *contained in* $\overline{F}$. *It results from* 5.10 *that the couple* $(Vert(\mathscr{C}(\mathscr{H})), (Vert(C))_{C \in Ch\mathscr{C}(\mathscr{H})})$ *defines a building structure on* $Vert(\mathscr{C}(\mathscr{H}))$. *There is a natural bijection between* $\mathscr{C}(\mathscr{H})$ *and the set of facets of* $(Vert(\mathscr{C}(\mathscr{H})), (Vert(C))_{C \in Ch\mathscr{C}(\mathscr{H})})$. *If no confusion arises this building is denoted by* $\mathscr{C}(\mathscr{H})$.

5.6   The Cayley Diagram of the Symmetric group

The reader is referred to [18], 6.2 (see also [35] and [19]) for the definition and properties of the Cayley Diagram of a group, given in terms of a set of generators and a set of relations. The symmetric group $\mathfrak{S}_{r+1}$ is, according to loc. cit., Ch. 6, §2 defined by the set of elementary transpositions $S = \{s_{12}, \cdots, s_{rr+1}\}$ and the following set of relations:

- $s_{ii+1}^2 = 1$ $(1 \leq i \leq r)$;

- $(s_{ii+1}s_{i+1i+2})^3 = 1$ $(1 \leq i \leq r - 1)$;

- $(s_{ii+1}s_{kk+1})^2 = 1$ $(i \leq k - 2)$.

The Cayley Diagram of $\mathfrak{S}_{r+1}$ consists of the vertices and edges of a "uniform polytope" whose two-dimensional faces are hexagons and squares representing the set of defining relations of $\mathfrak{S}_{r+1}$, the hexagons corresponding to the second type and the squares to the third type of relations. The geometric realization $\Delta^{(r)'}(p_1, \cdots, p_{r+1})$ of the Coxeter complex gives rise to the following one of the Cayley Complex. Let $\mathbb{S}^{r-1} = \mathbb{S}^{r-1}(p_1, \cdots, p_{r+1})$ be the sphere centered in the barycenter $(p_1 + \cdots + p_{r+1})/(r + 1)$ of $\Delta^{(r)}(p_1, \cdots, p_{r+1})$, and containing $\{p_1, \cdots, p_{r+1}\}$. To a chamber $\mathscr{C}_{\sigma(D)}$ (resp. maximal length flag $D \in \Delta^{(r)'}$) it corresponds the spherical simplex $\mathscr{C}_{\sigma(D)} \cap \mathbb{S}^{r-1}$. Denote by $\beta_D \in \mathscr{C}_{\sigma(D)} \cap \mathbb{S}^{r-1}$ the intersection point of the "barycentric ray of the chamber" $\mathscr{C}_{\sigma(D)}$ with $\mathbb{S}^{r-1}$.

Thus the Cayley Diagram of $(\mathfrak{S}_{r+1}, S)$ is the graph whose set of **vertices** is $(\beta_D)_{D \in \Delta_r^{(r)'}}$, and the set of **edges** is the set of segments $([\beta_D, \beta_{D'}])_{(D,D') \in adj(\Delta_r^{(r)'} \times \Delta_r^{(r)'})}$, where $adj(\Delta_r^{(r)'} \times \Delta_r^{(r)'})$ denotes the graph of the adjacency relation. This graph is contained in a "uniform polytope" whose faces are regular hexagons and regular squares. Observe that the set of the generating reflexions S of $\mathfrak{S}_{r+1}'$ is given by the the set of reflexions defined by the bounding hyperplanes of the chamber $\mathscr{C}_{D_0} : x_1 < x_2 \cdots < x_{r+1}$. It follows that the paths of the Cayley Diagram issued from β_{D_0} correspond to the galleries issued from $\mathscr{C}_{D_0}$, and thus to the words in S.

5.7 The geometric realization of the Chambers, Roots, and Hyperplanes

Denote by $Im_{\mathscr{C}(\mathscr{H})}(R(\Delta^{(r)'}))$ the image in $\mathscr{C}(\mathscr{H})$ of the set of the combinatorial roots

$$R(\Delta^{(r)'}) = (\mathscr{A}_{(i,j)})_{(i,j)\in I_{r+1}\times I_{r+1}/\Delta},$$

by the building isomorphism $\Delta^{(r)'} \simeq \mathscr{C}(\mathscr{H})$. Let

$$\overline{D}_{(i,j)} := \{(x_1,\cdots,x_{r+1}) \in \mathbb{A}^r \mid x_j - x_i \geqslant 0\}.$$

(resp.

$$D_{(i,j)} := \{(x_1,\cdots,x_{r+1}) \in \mathbb{A}^r \mid x_j - x_i > 0\})$$

be the closed half space (resp. open half space) of $\mathbb{A}^r$ defined by $x_j - x_i \geqslant 0$ (resp. $x_j - x_i > 0$). From the the definition of the isomorphism $\Delta^{(r)'} \simeq \mathscr{C}(\mathscr{H})$ it results that the image of the subcomplex $\mathscr{A}_{(i,j)}$ is given by

$$Im_{\mathscr{C}(\mathscr{H})}(\mathscr{A}_{(i,j)}) = \overline{D}_{(i,j)} \cap \mathscr{C}(\mathscr{H}) = \{F \in \mathscr{C}(\mathscr{H}) \mid F \subset \overline{D}_{(i,j)}\}.$$

Finally the result is

$$Im_{\mathscr{C}(\mathscr{H})}(R(\Delta^{(r)'}) = (\overline{D}_{(i,j)} \cap \mathscr{C}(\mathscr{H})).$$

It may be recalled that $Drap_r(I_{r+1}) = \mathrm{Ch}\,\Delta^{(r)'}$. Let the image of the set of chambers $Im_{\mathscr{C}(\mathscr{H})}(\mathrm{Ch}\,\Delta^{(r)'}) \subset \mathscr{C}(\mathscr{H})$ of $\Delta^{(r)'}$ be determined. It begins by calculating the image $Im_{\mathscr{C}(\mathscr{H})}(D_r)$ of $D_r = (I_1 \subset \cdots \subset I_{r+1})$. Write $\mathrm{Ch}\,(\mathscr{A}_{(i,j)}) := \mathscr{A}_{(i,j)} \cap \mathrm{Ch}\,\Delta^{(r)'}$. Observe that the canonical flag D_r defined by the total order of I_{r+1} is determined in terms of a set of combinatorial roots

$$\{D_r\} = \bigcap_{(i,j)\in R(D_r)} \mathrm{Ch}\,(\mathscr{A}_{(i,j)}) = \mathrm{Ch}\,(\mathscr{A}_{(1,2)}) \cap \cdots \cap \mathrm{Ch}\,(\mathscr{A}_{(r,r+1)}).$$

It follows that

$$Im_{\mathscr{C}(\mathscr{H})}(D_r) = D_{(1,2)} \cap \cdots \cap D_{(r,r+1)} =$$

$$\{(x_1,\cdots,x_{r+1}) \in \mathbb{A}^r \mid x_2 - x_1 > 0,\cdots,x_{r+1} - x_r > 0\}.$$

It may be recalled that the set of maximal length flags $Drap_r(I_{r+1})$ corresponds to the set of total orderings $Ord(I_{r+1})$ of I_{r+1}. Given $D \in \Delta_r^{(r)'} = Drap_r(I_{r+1})$, let $(i_1 < i_2 < \cdots < i_r < i_{r+1})$ be the order $\underset{D}{\leqslant}$ of I_{r+1} defined by D, i.e. ω_D, and $w \in \mathfrak{S}_{r+1}$ defined by $w(\alpha) = i_\alpha$ $(1 \leqslant \alpha \leqslant r + 1)$. Then $w(D_r) = D$. From the characterization of the image $Im_{\mathscr{C}(\mathscr{H})}(D_r)$ it is deduced

$$\mathrm{Im}_{\mathscr{C}(\mathscr{H})}(D) \;\; = \;\; \mathrm{Im}_{\mathscr{C}(\mathscr{H})}(w(D_{\underline{r}})) = \{(x_1,\cdots,x_{r+1})$$

$$\mid x_{i_2} - x_{i_1} > 0,\cdots,x_{i_{r+1}} - x_{i_r}0\}$$

$$= D_{(w(1),w(2))} \cap D_{(w(2),w(3))} \cap \cdots \cap D_{(w(r),w(r+1))}.$$

Finally the image of the set of maximal length flags $\Delta_{\underline{r}}^{(r)'}$ by the geometrical realisation is given by

$$\mathrm{Im}_{\mathscr{C}(\mathscr{H})}(\Delta_{\underline{r}}^{(r)'}) = \Big(\bigcap_{1\leqslant\alpha\leqslant l} D_{(w(\alpha),w(\alpha+1))} \Big)_{w\in\mathfrak{S}_{r+1}}.$$

As the set $\Delta_{\underline{r}}^{(r)'}$ is $\mathfrak{S}_{r+1}$-principal homogeneous, for every chamber $\mathscr{C}$ of $\mathscr{C}(\mathscr{H})$ there is an order $(i_1 < \cdots < i_{r+1})$ of I_{r+1} such that $\mathscr{C}$ is characterized by the inequalities

$$\mathscr{C} : \; x_{i_1} < x_{i_2} < \cdots < x_{i_r} < x_{i_{r+1}}.$$

Thus the set $S_{\mathscr{C}}$ of reflexions defined by the walls of $\mathscr{C}$ (resp. the set $\mathscr{H}_{\mathscr{C}}$) is given by

$$S_{\mathscr{C}} = (s_{i_\alpha i_{\alpha+1}}) \quad \Big(\text{resp. } \mathscr{H}_{\mathscr{C}} = (H_{i_\alpha i_{\alpha+1}}) \Big) \; (1 \leqslant \alpha \leqslant r).$$

By definition a **wall of the chamber** $\mathscr{C}$ is a hyperplane $H \in \mathscr{C}(\mathscr{H})$ given by the carrier of a cone $F \subset \overline{\mathscr{C}}$ of codimension 1. It is clear that the set of equations defining the walls of $\mathscr{C}$ is given by $(x_{i_\alpha+1} - x_{i_\alpha} = 0)$. If $\mathscr{C} = \mathscr{C}_D$ then the walls of $\mathscr{C}$ correspond to the bounding hyperplanes of D.

5.8   The Flag complex

Definition 5.12 *There is a natural order on $Drap(k^{r+1})$ and a mapping $typ : Drap(k^{r+1}) \longrightarrow typ(I_{r+1})$ defined following the pattern of 5.4. This order and the set of chambers given by the maximal length flags $Drap_{\underline{r}}(k^{r+1})$ define a building structure on $Drap(k^{r+1})$. This results from the following facts:*

1) *Given two flags $\mathscr{D}$ and $\mathscr{D}'$ of k^{r+1} there exists a **basis** e_E **adapted** to both (cf. 2.3).*

2) *To a basis e_E of k^{r+1} is associated a mapping $\psi_E : Drap(E) \longrightarrow Drap(k^{r+1})$ defined by $\psi_E : D \mapsto k^D$, identifying $Drap(E)$ to a subcomplex of $Drap(k^{r+1})$ (Apartment defined by ψ_E).*

It results that two maximal length flags in $Drap(I_{r+1})$ may be joined by a sequence of chambers with two succesive having a common $(r-1)$-facet. Denote this building by $I(Drap(k^{r+1}))$. It may be noted that here $Vert(I(Drap(k^{r+1})) = Grass(k^{r+1})$.

From 1) it follows that the two flags $\mathscr{D}$ and $\mathscr{D}'$ are contained in some $\psi_E(Drap(E))$, i.e. in some apartment. On the other hand, it is known that $Drap(E)$ is a building and thus it results that the two flags may be joined by a gallery.

The mapping typ gives rise to a building morphism $typ : I = I(Drap(k^{r+1})) \longrightarrow typ(I_{r+1})$ inducing a mapping

$$Gall_I \longrightarrow Gall_{typ(I_{r+1})}$$

defined by $\gamma \mapsto typ\,\gamma$, where $typ\,\gamma$ denotes the gallery of $typ(I_{r+1})$ defined by the types of the facets of γ. No confusion arises if this mapping is also denoted by "typ".

Write $gall_I = Gall_{typ(I_{r+1})}$, and given $g \in gall_I$ let

$$Gall_I(g) := typ^{-1}(g)$$

(resp.

$$\text{Gall}_I(g, \mathscr{D}) := \{\gamma \in Gall_I(g) \mid E_1(\gamma) = \mathscr{D} \}$$

if $\mathscr{D} \in Drap(k^{r+1})$).

Definition 5.13 *Define e_1 (resp. e_2) $: gall_I \longrightarrow typ(I_{r+1})$ as the **left** (**resp. right**) **extremity mapping** of $gall_I = Gall_{typ(I_{r+1})}$, and write $e = (e_1, e_2)$ for the **extremities mapping**.*

Definition 5.14 *An **apartment** A of $I(Drap(k^{r+1}))$ is by definition the subcomplex given by the image of a building morphism $\psi_E : Drap(E) \longrightarrow Drap(k^{r+1})$ (cf. definition 2.5). The set of apartments $Ap(I(Drap(k^{r+1})))$ is in bijection with the class of sets $\{L_1, \ldots, L_{r+1}\}$ of $r+1$ independant one dimensional subspaces of k^{r+1}.*

The apartment A defined by $\{L_1, \ldots, L_{r+1}\}$ is given by the set of flags adapted to the direct sum decomposition $k^{r+1} = L_1 \oplus \cdots \oplus L_{r+1}$.

Remark 5.15   *1) The condition 1- of definition 5.12 may be translated as follows:*

   "A couple of facets (D, D') is always contained in an apartment".

   *2) The class of sets $\{L_1, \ldots, L_{r+1}\}$ of $r+1$ independant one dimensional subspaces of k^{r+1} corresponds to the set of direct sum decompositions $k^{r+1} = L_1 \oplus \cdots L_r \oplus L_{r+1}$ with $\dim_k L_i = 1$. This set is the set of k-points of the k-**variety of Stiefel decompositions** of k^{r+1} (cf. [24], 9.10).*

Definition 5.16 *A generalized gallery of $Chain(I_{r+1})$ (resp. $Chain(k^{r+1})$) is defined in terms of the "$\subset$" relation between chains following the same pattern as in the definition of a generalized gallery in an ordered set (cf. Definition 1.23).*

To a gallery γ^{ch} of chains a generalized gallery may be associated by considering the reduction of the image $\varphi(\gamma^{ch})$ to a non-stammering gallery.

5.8.1   Flag complex automorphisms group

The action of $Gl(k^{r+1})$ on $Drap(k^{r+1})$ factors through the action of the projective group $PGl(k^{r+1}) = Gl(k^{r+1})/Center$ (The adjoint group of $Gl(k^{r+1})$). More precisely we have that the group of type preserving automorphisms of $I(Drap(k^{r+1}))$ is given by $PGl(k^{r+1})$ (The **Fundamental Theorem of projective Geometry**) (cf. [11] and [50]). Let $\mathscr{A}$ be an apartment given by the $(r+1)$-subspaces $\{L_1, \cdots, L_{r+1}\}$. Write $N(L_1, \cdots, L_{r+1})$ for the stabilizer $Stab\ \{L_1, \cdots, L_{r+1}\}$. There is an invariant subgroup $T(L_1, \cdots, L_{r+1})$ of $N(L_1, \cdots, L_{r+1})$ defined by the automorphisms fixing each subspace $L_1, \cdots, L_{r+1}$. The action of $N(L_1, \cdots, L_{r+1})$ on $\mathscr{A}$ factors through the quotient $N(L_1, \cdots, L_{r+1})/T(L_1, \cdots, L_{r+1})$ which is easily seen to be isomorphic with $\mathfrak{S}_{r+1}$. This group is isomorphic to the group of type preserving automorphisms of $\Delta^{(r)'} \simeq \mathscr{A}$. This last isomorphism makes correspond to $H \in Grass(I_{r+1})$ the subspace $Vect((L_i)_{i \in H})$ of k^{r+1}. Thus the action of $\mathfrak{S}_{r+1}$ on $\mathscr{A}$ is characterized as follows. The image of $Vect((L_i)_{i \in H})$ by $\sigma \in \mathfrak{S}_{r+1}$ is $Vect((L_i)_{i \in \sigma(H)})$.

5.8.2   The Star complex defined by a flag in the Flag Complex

The flag complex is a particular case of a simplicial complex. A simplicial complex is by definition a set K endowed with a class of finite subsets, called the simplices of K, and such that every non empty subset of a simplex is also a simplex. The general definition of the star of a simplex applies to the Flag Complex.

Definition 5.17 *Given a flag $\mathscr{D} \in Drap = Drap(k^{r+1})$ of type t and a type $s \in typ(I_{r+1})$ define:*

1. *$\Sigma_{\mathscr{D}}(s) \subset Drap(k^{r+1})$ (resp. $\Sigma_t(s) \subset Drap(k^{r+1}) \times Drap(k^{r+1})$) as the subvariety whose set of k-points is given by*

$$(\Sigma_{\mathscr{D}}(s))(k) \quad := \quad \{\mathscr{D}' \in Drap_s(k^{r+1})|\ \mathscr{D} \subset \mathscr{D}'\}$$
$$(\text{resp. } (\Sigma_t(s))(k) \quad := \quad \{(\mathscr{D}, \mathscr{D}') \in Drap_t(k^{r+1}) \times Drap_s(k^{r+1})|\ \mathscr{D} \subset \mathscr{D}'\}).$$

2.

$$St_{\mathscr{D}} = \{\ \mathscr{D}' \in Drap(k^{r+1})|\mathscr{D} \subset \mathscr{D}'\ \}$$

(**The Star of $\mathscr{D}$**).

3. If $\mathscr{D} = k^D$ it is written $\Sigma_{\mathscr{D}}(s) = \Sigma_D(s)$ if no confusion arises.

The star complexes are the keystone in the construction of generalized galleries.

1. Observe that there is the following decomposition:

$$\mathrm{St}_{\mathscr{D}} = \coprod_{t \subset s} (\Sigma_{\mathscr{D}}(s))(k).$$

2. We have $\Sigma_t(s) \subset Drap_t(k^{r+1}) \times Drap_s(k^{r+1})$ is the graph of the order relation $\mathscr{D} \subset \mathscr{D}'$.

The variety $\Sigma_{\mathscr{D}}(s)$ may be seen as a Schubert variety. Let typ $\mathscr{D} = t = (m_1 < \cdots < m_\lambda < m_{\lambda+1} = r + 1)$, $s = (n_1 < \cdots < n_l < n_{l+1} = r + 1)$, and $M := (\inf(m_\alpha, n_\beta)) \in \mathbb{N}^{(\lambda+1) \times (l+1)}$, then

$$\Sigma_{\mathscr{D}}(s) = \Sigma(M, \mathscr{D}) = \overline{\Sigma}(M, \mathscr{D}).$$

Let $\pi_1 = \pi_1(t, s)$ (resp. $\pi_2 = \pi_2(t, s)$) : $\Sigma_t(s) \longrightarrow Drap(k^{r+1})$ be the morphism induced by the first (resp. second) projection $Drap(k^{r+1}) \times Drap(k^{r+1}) \longrightarrow Drap(k^{r+1})$. Let $t' \subset s$, $\pi_1(t, s \supset t')$ is defined thus: $\Sigma_t(s) \longrightarrow Drap_{t'}(k^{r+1})$ as the composition of $\pi_1 = \pi_1(t, s)$ followed by the natural morphism $Drap_s(k^{r+1}) \longrightarrow Drap_{t'}(k^{r+1})$. It is observed that $\Sigma_t(s) = \Sigma(M)$.
From 1.14 it is deduced

$$\pi_1(t, s) : \Sigma_t(s) \longrightarrow Drap_t(k^{r+1})$$

is a locally trivial fiber bundle with typical fiber $\Sigma_{\mathscr{D}_t}(s)$ where $\mathscr{D}_t = (k^{m_1} \subset \cdots k^{m_\lambda} \subset k^{r+1})$.

Proposition 5.18 *The Schubert variety $\Sigma_{\mathscr{D}}(s)$ is isomorphic to a product of flag varieties.*

Proof *Let $t = \underline{m}$ (resp.$s = \underline{n}$), and $(\mathscr{D}, \mathscr{D}') \in \Sigma_{\mathscr{D}}(s)$, $\mathscr{D} = (V_1 \subset \cdots V_\lambda \subset k^{r+1})$ (resp. $\mathscr{D}' = (W_1 \subset \cdots, W_l \subset k^{r+1}))$. As $\mathscr{D} \subset \mathscr{D}'$ there exists an increasing sequence $(i_\alpha)_{1 \leq \alpha \leq \lambda}$ with $W_{i_\alpha} = V_\alpha$. Write $V_0 = \{0\}$, $\mu_\alpha = \dim V_{\alpha+1} - \dim V_\alpha$, and $\mathscr{D}_\alpha = (W_{i_\alpha+1}/V_\alpha \subset \cdots W_{i_{\alpha+1}-1}/V_\alpha \subset V_{\alpha+1}/V_\alpha)$, for $0 \leq \alpha \leq \lambda$.
Then there is*

$$\Sigma_{\mathscr{D}}(s) \simeq \prod_{0 \leq \alpha \leq \lambda} \mathrm{Drap}_{\underline{\nu}_\alpha}(k^{\mu_\alpha}),$$

where $\underline{\nu}_\alpha = typ \, \mathscr{D}_\alpha \in typ(I^{\mu_\alpha})$.

Corollary 5.19 *If* $\mathrm{St}_{\mathscr{D}}$ *is endowed with the induced ordering between flags there is an order preserving isomorphism:*

$$\mathrm{St}_{\mathscr{D}} \simeq \prod Drap(k^{r_\alpha+1}).$$

The right hand term is a product of buildings and thus endowed with a canonical building structure, giving rise to a building structure on $\mathrm{St}_{\mathscr{D}}$. *The group of type preserving isomorphisms of* $\mathrm{St}_{\mathscr{D}}$ *is given by* $\prod PGl(k^{r_\alpha+1})$.

5.8.3 The centered Bruhat decomposition

An improvement of the Bruhat decomposition is obtained as a corollary of proposition 2.53 (cf. 2.28) by relating it to the building $\Delta^{(r)'}$.

Proposition 5.20 *Let* $D \in Drap_{\underline{r}}(E)$ $(E \simeq I_{r+1})$ *be a maximal length flag.*

1) The Bruhat decomposition may be re-written as:

$$Gl(k^E) = \coprod_{\overline{w} \in \mathfrak{S}_E/\mathfrak{S}_{\underline{n}}} U(D, \overline{w} \cdot D_{\underline{n}}) \cdot w(D, \overline{w} \cdot D_{\underline{n}}) \cdot P(D_{\underline{n}}) \ .$$

Write $\overline{w} \cdot D_{\underline{n}} = w \cdot D_{\underline{n}}$.

2)

$$Drap_{\underline{n}}(k^E) = \coprod_{M \in Relpos_{(\underline{r},\underline{n})}(E)} U(D, D_M) \cdot w(D, D_M) \cdot k^{D_{\underline{n}}} \ ,$$

where $\Sigma(M, D) = U(D, D_M) \cdot w(D, D_M) \cdot k^{D_{\underline{n}}} = U(D, D_M) \cdot k^{D_M}$ **(The centered Bruhat decomposition of** $Gl(k^E)$ *(resp.* $Drap_{\underline{n}}(k^E)$*)).*

3) There is an isomorphism of k-varieties $U(D, D_M) \simeq \Sigma(M, D)$ *given by:* $u \mapsto u \cdot k^{D_M}$.

From the inclusion $U(D, \overline{w} \cdot D_{\underline{n}}) \subset U((\overline{w} \cdot D_{\underline{n}})^{opp})$ one obtains:

Corollary 5.21 *With the above notation there is an open covering:*

$$Gl(k^E) = \bigcup_{\overline{w} \in \mathfrak{S}_E/\mathfrak{S}_{\underline{n}}} U((\overline{w} \cdot D_{\underline{n}})^{opp}) \cdot P(\overline{w} \cdot D_{\underline{n}})$$

(The big cell open covering of $Gl(k^E)$**).**

Remark 5.22   *1) As* $dim_k \, U(D, D_M) = |R(D, D_M)|$ *it results from the above two propositions and the definition of* $R(D, D_M)$ *that:*

$dim_k \, \Sigma(M, D) = $ *number of hyperplanes in* $\Delta^{(r)'}$ *separating D from*

D_M, *and that* $dim_k \, \Sigma(M, D) = l_{S_D}(w(D, proj_{D_M} D)$.

2) Let $M^{bc} = M^{bc}_{(\underline{r},\underline{n})}$ the relative position matrix defining the big cell $\Sigma(M^{bc}, D)$ of the decomposition $Drap_{\underline{n}}(k^E) = \coprod_{M \in Relpos_{(\underline{r},\underline{n})}(E)} U(D, D_M) \cdot k^{D_M}$. On the other hand, it follows from the proof of proposition 2.60 that $R(D, D_{M^{bc}}) = R^u_{D_{M^{bc}}}$. Thus:

$$dim_k \, Drap_{\underline{n}}(k^{r+1}) = |R(D, D_{M^{bc}})| = |\mathcal{H}(D_{M^{bc}})| \, .$$

The following proposition is deduced from the above remark.

PROPOSITION - DEFINITION 5.23 A minimal gallery $\Gamma(D, D')((D, D') \in Drap_{\underline{r}}(I_{r+1}) \times Drap_{\underline{n}}(I_{r+1}))$ satisfies length $(\Gamma(D, D')) \leq |\mathcal{H}(D')|$. If the equality holds then the distance $d(D, D')$ is maximal and it is said that the couple (D, D') is in **transversal position**. In this case a minimal gallery $\Gamma(D, D')$ crosses all the hyperplanes of $\Delta^{(r)'}$ not containing D'.

It follows from proposition 5.18 and the above proposition the

Proposition 5.24 Let $D \in Drap_s(I_{r+1}) \cap St_{D'}$. Then:

$$dim_k \, \Sigma_{D'}(D) = |\mathcal{H}_{D'}(D)| \, .$$

5.9 The Retraction of the Flag Complex on an Apartment

The building Complex as defined in [42] admits a retraction on the finite subcomplex given by an Apartment. This retraction is also a combinatorial Building morphism which transforms galleries into galleries, and gives another interpretation of the Bruhat decomposition. It results from definition 2.21 that a maximal lenght flag $D \in Drap_{\underline{r}}(I_{r+1})$ defines a natural bijection:

$$\coprod_{\underline{n} \in typ(I_{r+1})} Relpos_{(\underline{r},\underline{n})} \simeq \Delta^{(r)'} \, ,$$

given by $M \mapsto D_M$, whose reciprocal mapping is given by $D' \mapsto M(D, D')$. On the other hand, there is a mapping $I(Drap(k^{r+1})) \longrightarrow \coprod_{\underline{n} \in typ(I_{r+1})} Relpos_{(\underline{r},\underline{n})}$, associating to a flag $\mathscr{D} \in I(Drap(k^{r+1}))$ the relative position matrix $M(k^D, \mathscr{D})$.

Definition 5.25 *Define :*

$$\rho_D : I(Drap(k^{r+1})) \longrightarrow \coprod_{\underline{n} \in typ(I_{r+1})} Relpos_{(\underline{r},\underline{n})} \simeq \Delta^{(r)'}$$

(**The retraction of** $I(Drap(k^{r+1}))$ **on the apartment** $\Delta^{(r)'}$ **with center in** D).

The following proposition results immediatly from Bruhat decomposition.

Proposition 5.26

Retain the above notation. Let $M = M(k^D, \mathscr{D})$. There exists a unique $\alpha(\mathscr{D}) \in U(D, D_M)$ so that $\alpha(\mathscr{D}) \cdot \mathscr{D} = k^{D_M}$ with $D_M = \rho_D(\mathscr{D})$.

Proposition 5.27

Keep the above notation. The mapping $\rho_D : I(Drap(k^{r+1}) \longrightarrow \Delta^{(r)'}$ defines a building morphism.

Proof *It suffices to prove that:*

$$\mathscr{D}' \subset \mathscr{D} \Longrightarrow k^{D_{M'}} = \rho_D(\mathscr{D}') \subset \rho_D(\mathscr{D}) = k^{D_M} .$$

This results immediatly from

$$\rho_D(\mathscr{D}') = \alpha(\mathscr{D}) \cdot \mathscr{D}' \subset \alpha(\mathscr{D}) \cdot \mathscr{D} = \rho_D(\mathscr{D}) .$$

5.9.1 The Flag Complex Retraction on an Apartment and the Bruhat decomposition

The Bruhat decomposition of $Drap(k^{r+1})$ defined by k^D may be written:

$$Drap(k^{r+1}) = \coprod_{\underline{n} \in typ(I_{r+1})} \coprod_{M \in Relpos_{(\underline{r},\underline{n})}} \Sigma(M, D)$$

$$= \coprod_{\underline{n} \in typ(I_{r+1})} \coprod_{M \in Relpos_{(\underline{r},\underline{n})}} U(D, D_M) \cdot k^{D_M} .$$

It follows from the definition of ρ_D that the restriction $\rho_D|_{\Sigma(M,D)}$ is equal to the *constant mapping defined by* D_M. Thus:

$$\rho_D^{-1}(D') = \Sigma(M(D, D'), D) ,$$

and one has that this fiber is principal under the action of $U(D, D_M)$.

5.10 The Flag Complex and the Parabolic subgroups of the Linear Group

The canonical isomorphism $Drap(k^{r+1}) \longleftrightarrow Par(Gl(k^{r+1}))$ given by $\mathscr{D} \mapsto P_{\mathscr{D}} = Stab\ \mathscr{D}$ allows another interpretation of $I(Drap(k^{r+1}))$ in terms of parabolic subgroups of $Gl(k^{r+1})$. Denote by $I(Gl(k^{r+1}))$ the set of parabolic subgroups endowed with the relation "$\prec$" defined as the opposite of the inclusion relation between parabolics. The set of chambers of $Drap(k^{r+1})$ correspond to the set of minimal parabolic subgroups of $Gl(k^{r+1})$, and it results that two subgroups P *and* Q are incident if $P \cap Q$ is a parabolic subgroup. The set of apartments corresponds to the set of conjugates of the subgroup of diagonal matrices. Let T be a subgroup conjugate to the subgroup of diagonal matrices. It is clear that to T it corresponds a Stiefel decomposition $k^{r+1} = L_1 \oplus \cdots \oplus L_{r+1}$. The appartment $\mathscr{A}_T$ corresponding to T is thus formed by the image of the mapping $\Delta^{(r)'} \longleftrightarrow \mathscr{A}_T$ defined by $H \mapsto P_{\mathscr{H}}$ where $\mathscr{H} = \bigoplus_{i \in H} L_i$.

Proposition 5.28 *The set of conjugates by $Gl(k^{r+1})$ of the subgroup of diagonal matrices contained in $P_{\mathscr{D}}$ is homogeneous under conjugation by $P_{\mathscr{D}}$.*

Proof *Given two basis $\tilde{e} = (\tilde{e}_i)_{1 \le i \le r+1}$ and $\tilde{e}' = (\tilde{e}'_j)_{1 \le j \le r+1}$ adapted to $\mathscr{D} = (\mathscr{H}_1 \subset \cdots \mathscr{H}_\lambda \subset k^{r+1})$, i.e. satisfying $\mathscr{H}_\alpha = \bigoplus_{L_i \subset \mathscr{H}_\alpha} L_i$ ($L_i = k\tilde{e}_i$) $\left(\text{resp. } \mathscr{H}_\alpha = \bigoplus_{L'_j \subset \mathscr{H}_\alpha} L'_j \ (L'_j = k\tilde{e}'_j)\right)$ there exists a renumbering of $\tilde{e}'$, namely $\tilde{e}'' = (\tilde{e}'_{j_\beta})_{1 \le \beta \le r+1}$ with :*

"the automorphism $\sigma(\tilde{e}, \tilde{e}'') : \tilde{e}_i \mapsto \tilde{e}'_{j_1}, \cdots, \tilde{e}_{r+1} \mapsto \tilde{e}'_{j_{r+1}}$ belongs to $P_{\mathscr{D}}$ ".

It is immediate to see that this suffices to prove the proposition.

Proposition 5.29 *Let $\mathscr{D}, \mathscr{D}' \in Drap(k^{r+1})$. The set of conjugates T of the subgroup of diagonal matrices satisfying $T \subset P_{\mathscr{D}} \cap P_{\mathscr{D}'}$ is homogeneous under conjugation by $P_{\mathscr{D}} \cap P_{\mathscr{D}'}$. In other terms the set of Stiefel decompositions of k^{r+1} "adapted to both $\mathscr{D}$ and $\mathscr{D}'$ " is homogeneous under $P_{\mathscr{D}} \cap P_{\mathscr{D}'}$.*

Proof *Following the pattern of the proof of Corollary 2.4 it is obtained that given two basis $\tilde{e} = (\tilde{e}_i)_{1 \le i \le r+1}$ and $\tilde{e}' = (\tilde{e}'_j)_{1 \le j \le r+1}$ adapted to both $\mathscr{D} = (\mathscr{H}_1 \subset \cdots \mathscr{H}_\lambda \subset k^{r+1})$ and $\mathscr{D}' = (\mathscr{J}_1 \subset \cdots \mathscr{J}_l \subset k^{r+1})$ there exists disjoint decompositions:*

$$\tilde{e} = \coprod_{(\alpha,\beta) \in [\![1,\lambda+1]\!] \times [\![1,l+1]\!]} \tilde{e}_{\alpha\beta} \left(\text{resp. } \tilde{e}' = \coprod_{(\alpha,\beta) \in [\![1,\lambda+1]\!] \times [\![1,l+1]\!]} \tilde{e}'_{\alpha\beta}\right),$$

satisfying:

1) *$k^{r+1} = \bigoplus W_{\alpha\beta}$ ($W_{\alpha\beta} = Vect(\tilde{e}_{\alpha\beta})$) $\left(\text{resp. } k^{r+1} = \bigoplus W'_{\alpha\beta} \ (W'_{\alpha\beta} = Vect(\tilde{e}'_{\alpha\beta}))\right)$;*

2) *$|\tilde{e}_{\alpha\beta}| = |\tilde{e}'_{\alpha\beta}|$;*

3) *$\mathscr{H}_{\alpha'} = \bigoplus_{W_{\alpha\beta} \subset \mathscr{H}_{\alpha'}} W_{\alpha\beta} = \bigoplus_{W'_{\alpha\beta} \subset \mathscr{H}_{\alpha'}} W'_{\alpha\beta}$ $\left(\text{resp. } \mathscr{J}_{\beta'} = \bigoplus_{W_{\alpha\beta} \subset \mathscr{J}_{\beta'}} W_{\alpha\beta} = \bigoplus_{W'_{\alpha\beta} \subset \mathscr{J}_{\beta'}} W'_{\alpha\beta}\right)$;*

4) $\{ (\alpha\beta) \in [\![1, \lambda + 1]\!] \times [\![1, l + 1]\!] \mid W_{\alpha\beta} \subset \mathscr{H}_{\alpha'} \} = \{ (\alpha\beta) \in [\![1, \lambda + 1]\!] \times [\![1, l + 1]\!] \mid W'_{\alpha\beta} \subset \mathscr{H}_{\alpha'} \};$

5) $\{ (\alpha\beta) \in [\![1, \lambda + 1]\!] \times [\![1, l + 1]\!] \mid W_{\alpha\beta} \subset \mathscr{J}_{\beta'} \} = \{ (\alpha\beta) \in [\![1, \lambda + 1]\!] \times [\![1, l + 1]\!] \mid W'_{\alpha\beta} \subset \mathscr{J}_{\beta'} \}.$

After reordering the basis $\tilde{e}'$ there exists $f : k^{r+1} \longrightarrow k^{r+1}$ satisfying $f(\tilde{e}_{\alpha\beta}) = \tilde{e}'_{\alpha\beta}$ and thus $f(W_{\alpha\beta}) = W'_{\alpha\beta}$. This implies that $f(\mathscr{H}_{\alpha'}) = \mathscr{H}_{\alpha'}$ (resp. $f(\mathscr{J}_{\beta'}) = \mathscr{J}_{\beta'}$), i.e. $f \in P_{\mathscr{D}} \cap P_{\mathscr{D}'}$.

Given T and T', basis $\tilde{e}$ and $\tilde{e}'$ are chosen defining respectively the Stiefel decompositions corresponding to T and T'. If the above construction is applied to $\tilde{e}$ and $\tilde{e}'$, there results $f : k^{r+1} \longrightarrow k^{r+1}$ satisfying $f(T) = T'$ and $f \in P_{\mathscr{D}} \cap P_{\mathscr{D}'}$.

5.10.1   Invariance of the Convex hull of two flags with respect to the Apartment containing them

Definition 5.30 *The convex hull (resp. envelope) $Env(D, D')$ of two flags $D, D' \in I(Drap(I_{r+1})) = \Delta^{(r)'}$ is defined as the subcomplex of $\Delta^{(r)'}$ given by :*

$$Env(D, D') = \bigcap_{\{(i,j) \in R(I_{r+1}) \mid D, D' \in \mathscr{A}_{(i,j)}\}} \mathscr{A}_{(i,j)} .$$

Given $\mathscr{D}, \mathscr{D}' \in I(Drap(k^{r+1}))$ their convex hull $Env^I(\mathscr{D}, \mathscr{D}')$ in $I(Drap(k^{r+1}))$ is defined by: if $\mathscr{D}, \mathscr{D}'$ are adapted to a Stiefel decomposition corresponding to the subgroup T and

$$\text{if } \mathscr{D} = k^D, \mathscr{D}' = k^{D'} \text{ then } Env^I(\mathscr{D}, \mathscr{D}') = Env^{\mathscr{A}_T}(\mathscr{D}, \mathscr{D}') ,$$

where

$$Env^{\mathscr{A}_T}(\mathscr{D}, \mathscr{D}') = \text{image of } Env(D, D') \text{ by } \Delta^{(r)'} \longrightarrow \mathscr{A}_T ,$$

and $\mathscr{A}_T$ denotes the apartment of flags adapted to the Stiefel decomposition corresponding to T. Observe that $\{(i, j) \in R(I_{r+1}) \mid D, D' \in \mathscr{A}_{(i,j)}\} = R_D \cap R_{D'}$, and it is recalled that given two flags $\mathscr{D}$ and $\mathscr{D}'$ there always exists a Stiefel decomposition adapted to both flags.

From the following proposition it results that this definition is well posed.

Proposition 5.31 *Let $T, T' \subset P_{\mathscr{D}} \cap P_{\mathscr{D}'}$ then $Env^{\mathscr{A}_T}(\mathscr{D}, \mathscr{D}') = Env^{\mathscr{A}_{T'}}(\mathscr{D}, \mathscr{D}')$.*

Proof *From the equality $Gl_{R_D \cap R_{D'}} = Stab\ (k^D) \cap Stab\ (k^{D'})$ (cf. [23], Exp XXII, 5.4.5.), the fact that $P_{\mathscr{D}}$ (resp. $P_{\mathscr{D}'}$) is its own normalizer, and that its Lie algebra is generated by $Lie(T)$ (resp. $Lie(T')$) and $(E_{(i,j)})_{(i,j) \in R_D}$ (resp. $(E'_{(i,j)})_{(i,j) \in R_{D'}}$),* *where* $(E_{(i,j)})_{(i,j) \in R(E)}$ *(resp.* $(E'_{(i,j)})_{(i,j) \in R(E)}$ *) denotes a basis of* $Lie(Gl(k^{r+1})$ *formed of eigenvectors for* T *(resp.* T' *), corresponding to the eigenvalues indexed by* $R(E) = E \times E - \Delta$ *, it is deduced that:*

$$Env^{\mathscr{A}_T}(\mathscr{D}, \mathscr{D}') = Env^{\mathscr{A}_{T'}}(\mathscr{D}, \mathscr{D}') = \{ \mathscr{D}'' \in I(Drap(k^{r+1})) \mid P_{\mathscr{D}} \cap P_{\mathscr{D}'} \subset P_{\mathscr{D}''} \}.$$

According to the general definition given in 5.1 a gallery in a building $\Gamma(C, C') = (C = C_0, C_1, \ldots, C_n = C')$ of length n between two chambers C and C' is minimal if it is of minimal length. This general definition is equivalent for $\Delta^{(r)'} = I(Drap(I_{r+1}))$ to the following one.

Definition 5.32 *A gallery:*

$$\Gamma(D, D') = (D = D_0, D_1, \ldots, D_n = D') \subset \Delta_r^{(r+1)'}$$

where D_i, D_{i+1} *have precisely a common* $(r-1)$*-length flag (resp. a flag of codimension 1)* $D_{ii+1} \subset D_i, D_{i+1}$*, is minimal if the set of hyperplanes it crosses is equal to* $\mathscr{H}(D, D')$*. We say that* $\Gamma(D, D')$ **crosses the hyperplane** H *if there exists* $0 \le i \le n$ *satisfying* $D_{ii+1} \in H$*, i.e. if* $\omega_{D_i} = (i_1 < \cdots i_\alpha < i_{\alpha+1} \cdots < i_{r+1})$ *(resp.* $\omega_{D_{i+1}} = (j_1 < \cdots j_\alpha < j_{\alpha+1} \cdots < j_{r+1})$*) then* D_{ii+1} *is invariant under the transposition* $(i_\alpha, i_{\alpha+1}) = (j_\alpha, j_{\alpha+1})$*. Thus* H *is a bounding hyperplane of both* D_i *and* D_{i+1}*.*

Proposition 5.33 *Keep the above notation. A gallery* $\Gamma(D, D') \subset \Delta^{(r)'}$ *is minimal according to Definition 5.32* $\Longleftrightarrow$ $\Gamma(D, D')$ *is of minimal length.*

The proof of this proposition is a particular case of a general assertion about Coxeter complexes and will be given later. The main point is that galleries in $\Delta^{(r)'}$ issued from D correspond to words in the set of generators defined by D, $S_D \subset \mathfrak{S}_{r+1}$, and minimal galleries to minimal length words. More precisely the following proposition resumes several equivalent definitions whose equivalence will be proved in all generality in the next chapters.

Proposition 5.34 *Keep the above notation. Let* $\Gamma(D, D')$ *be a gallery* $\Gamma(D, D')$ *in* $\Delta^{(r)'}$*, satisfying* $D_i \ne D_{i+1}$*, i.e.* $\Gamma(D, D')$ *is an injective gallery, then the following assertions are equivalent.*

1) $\Gamma(D, D')$ *is minimal according to Definition 5.32.*

2) The set of the hyperplanes crossed by $\Gamma(D, D')$ *is equal to* $\mathscr{H}(D, D')$*.*

3) length $\Gamma(D, D') = |\mathscr{H}(D, D')|$*.*

4) $\Gamma(D, D')$ *is of minimal length.*

5) $l_{S_D}(w(D, proj_{D'} D)) =$ *the number of the hyperplanes crossed by* $\Gamma(D, D')$*.*

The following corollary justifies the introduction of the convex hull of two flags.

Corollary 5.35 *A minimal gallery $\Gamma(D, D')$ is contained in the convex hull of its extremities, i.e. $\Gamma(D, D') \subset Env(D, D')$.*

Proof   *Consider the geometrical realization of $\Gamma(D, D')$:*

$$\Gamma(D, D') = (D_0 \cdots D_n, D_{n+1}) \longrightarrow \Gamma^{geom}(D, D') = (\mathscr{C}_{\sigma(D_0)} \cdots \mathscr{C}_{\sigma(D_n)}, \mathscr{C}_{\sigma(D_{n+1})}) \ .$$

It is recalled that the geometrical realization $Env^{geom}(D, D')$ of the convex hull $Env(D, D') \subset \Delta^{(r)'}$ is given by:

$$Env^{geom}(D, D') = \bigcap_{\{(i,j) \in R(I_{r+1})| \ D,D' \in \mathscr{A}_{(i,j)}\}} \overline{D}_{(i,j)} \ .$$

If $\Gamma^{geom}(D, D') \nsubseteq Env^{geom}(D, D')$ then there exists some half space $D_{(i,j)}$ satisfying:

1) *$\mathscr{C}_{\sigma(D)} \cup \mathscr{C}_{\sigma(D')} \subset D_{(i,j)}$, i.e. $D, D' \in \mathscr{A}_{(i,j)}$;*

2) *$D_{(i,j)} \cap \Gamma^{geom}(D, D') \neq \emptyset$ (resp.$(-D_{(i,j)}) \cap \Gamma^{geom}(D, D') \neq \emptyset$). Thus it is "geometrically" clear that $\Gamma^{geom}(D, D')$ crosses the hyperplane $H_{ij} = D_{(i,j)} \cap (-D_{(i,j)}) \notin \mathscr{H}(D, D')$.*

This last assertion contradicts the minimality of $\Gamma(D, D')$.

Chapter 6

Configurations and Galleries varieties

The Configuration varieties defined by typical graphs are introduced. The galleries of types, or more generally the linear typical graphs, define a class of k-smooth and integral Configurations varieties particularly important in this work. The **minimal** galleries of types are characterized as the galleries of types whose associated Configurations variety defines a smooth resolution of a Schubert variety. A minimal generalized gallery of the Flag Complex is a general point of the Configurations variety given by its gallery of types. They are characterized by a combinatorial property. It is shown that a minimal generalized gallery in the Flag Complex is contained in the convex hull of its extremities.

Definition 6.1

1) Let J be a finite set, $t : J \to typ\ \Delta_{\underline{r}}^{(r)'}$ a mapping, $K \subset J \times J$ a subset, satisfying:
$$(i, j) \in K \Rightarrow t(i) \subset t(j).$$
We call $\Lambda = (J, K, t)$ a **typical graph**.

Let $E \subset \mathscr{P}(J)$ be a class of sets with two elements, and $t : J \to typ\ \Delta_{\underline{r}}^{(r)'}$ a mapping satisfying:
$$\{i, j\} \in E \Rightarrow t(i) \subset t(j) \quad or \quad t(j) \subset t(i).$$
We call $M = (J, E, t)$ a **symmetric typical graph**. M is a **linear typical graph** if M is linear as a graph. Denote by Λ^{sym}, the symmetric typical graph defined by Λ. By definition Λ is linear if and only if Λ^{sym} is linear.

111

2) A Λ**-configuration** *of flags of* $Drap(k^{r+1})$ *(resp.* $I(Drap(k^{r+1}))$
(resp.$Drap(I_{r+1}))$ *is a point* $(\mathcal{D}_j)_{j\in J}$ *(resp.*$(D_j)_{j\in J})$ *of the product*
$\prod\limits_{j\in J} Drap_{t(j)}(k^{r+1})$ *(resp.* $\prod\limits_{j\in J} Drap_{t(j)}(I_{r+1}))$ *satisfying:*

$$(i,j) \in K \Rightarrow (\mathcal{D}_i, \mathcal{D}_j) \ and \ \mathcal{D}_i \subset \mathcal{D}_j$$

(resp.

$$(i,j) \in K \Rightarrow (D_i, D_j) \ and \ D_i \subset D_j) \ .$$

A M**-configuration** *of flags of* $Drap(k^{r+1})$ *(resp.* $I(Drap(k^{r+1}))$
(resp.$Drap(I_{r+1}))$ $(\mathcal{D}_j)_{j\in J}$ *(resp.*$(D_j)_{j\in J})$ *is a point of the product*
$\prod\limits_{j\in J} Drap_{t(j)}(k^{r+1})$ *(resp.* $\prod\limits_{j\in J} Drap_{t(j)}(I_{r+1}))$ *satisfying*

$$\{i,j\} \in E \Rightarrow \mathcal{D}_i \subset \mathcal{D}_j \quad or \quad \mathcal{D}_j \subset \mathcal{D}_i$$

(resp.

$$\{i,j\} \in E \Rightarrow D_i \subset D_j \quad or \quad D_j \subset D_i) \ .$$

Write $Conf(\Lambda) = Conf(\Lambda, I(Drap(k^{r+1})))$ *(resp.* $Conf^{comb}(\Lambda) = Conf(\Lambda, \Delta^{(r)'}))$ *the set of* Λ*-configurations of* $I(Drap(k^{r+1}))$ *(resp.*
$\Delta^{(r)'})$. *The set* $Conf^{comb}(\Lambda)$ *is considered as a subset of* $Conf(\Lambda)$ *by*
means of the injective mapping $\varphi \mapsto k^\varphi$.

Proposition 6.2 *The set*

$$Conf(\Lambda) \subset \prod_{j\in J} Drap_{t(j)}(k^{r+1})$$

of Λ*-configurations of* $Drap(k^{r+1})$ *is the set of* k *points of a projective* k-
variety (**The** Λ**-Configurations variety**).

(Here by a projective k-variety we understand a finite type projective scheme
over k)

Proof *It results easily from the proof of 1.9 (resp.3.3) that given a section*
$(\mathcal{D}, \mathcal{D}')$ *of* $Drap_{\underline{m}}(k^{r+1}) \times Drap_{\underline{n}}(k^{r+1})$ *the condition* $\mathcal{D} \subset \mathcal{D}'$ *may be defined*
by a set of equations in the Plücker coordinates of $(\mathcal{D}, \mathcal{D}')$. *Thus we deduce*
that $Conf(\Lambda)$ *is the underlying set of an algebraic* k*-variety (see Remark 1.10).*

Two typical graphs Λ and Λ' are **equivalent** if there is an isomorphism

$$Conf(\Lambda) \simeq Conf(\Lambda'),$$

i.e. if the variety of Λ-configurations $Conf(\Lambda)$ is isomorphic to the variety of
Λ'-configurations $Conf(\Lambda')$.

The typical graph associated to a generalized gallery of types $g \in gall_I$.

Recall that to a gallery of types $g \in gall_I$ is associated a typical graph $\Lambda(g)$ whose set of vertices $Vert(\Lambda(g))$ is given by the facets of g, and the set of edges $Edg(\Lambda(g))$ by the inclusions of its facets. For instance, the gallery

$$g: \ t_{r+1} \subset s_r \supset \cdots \subset s_0 \supset t_0$$

defines a typical graph $\Lambda(g)$ whose set of vertices is given by $Vert(\Lambda(g)) = \{t_{r+1}, s_r \cdots s_0, t_0\}$, its set of edges by $Edg(\Lambda(g)) = \{(t_{r+1} \subset s_r), \cdots, (s_0 \subset t_0)\}$, and the typical weight mapping $t : \Lambda(g) \longrightarrow typ(I_{r+1})$ by the inclusion $Vert(\Lambda(g)) \subset typ \, \Delta^{(r)'}$. Write $Conf(g) = Conf(\Lambda(g))$. There are morphisms

$$\mathscr{E}_1 : Conf(g) \longrightarrow Drap(k^{r+1})$$

(resp.

$$\mathscr{E}_2 : Conf(g) \longrightarrow Drap(k^{r+1}) \ ) \ ,$$

associating to a $\Lambda(g)$-configuration γ its left extremity $\mathscr{E}_1(\gamma)$ (resp. right extremity $\mathscr{E}_2(\gamma)$).

Definition 6.3 *A* **variety of galleries** *is a variety of the form* $Conf(\Lambda(g))$. *The underlying set of this variety is the set of galleries* $Gall_I(g)$ *in the Flag Complex* $I(Drap(k^{r+1}))$ *(resp. building* $I(Gl(k^{r+1}))$ *of* $Gl(k^{r+1})$*).*

We have

Proposition 6.4 *Given a linear typical graph* Λ *there exists a gallery* $g \in gall_I$ *whose associated graph* $\Lambda(g)$ *is equivalent to* Λ. *The gallery* g *may be chosen non-stammering, i.e. with strict inclusions between facets, and in this case it is unique.*

Let us give other examples of typical graphs.

1) **(The weighted graph $\Lambda(M)$ of definition 3.1)** It is recalled that the set of vertices $Vert \, \Lambda(M)$ of $\Lambda(M)$ is given by

$$Vert \, \Lambda(M) = [\![1, \lambda + 1]\!] \times [\![1, l + 1]\!],$$

and the set of edges $Edg \, \Lambda(M)$ by

$$Edg \, \Lambda(M) = \{((\alpha, \beta), (\alpha + 1, \beta))| \ 1 \leqslant \alpha \leqslant \lambda, \ 1 \leq \beta \leq l + 1\} \cup \\ \{((\alpha, \beta), (\alpha, \beta + 1))| \ 1 \leqslant \alpha \leqslant \lambda + 1, \ 1 \leq \beta \leq l\}.$$

The weight function being defined by:

$$p: \ Vert \, \Lambda(M) \to \mathbb{N}, \ p(\alpha, \beta) = m_{\alpha\beta}.$$

may be seen as a mapping with values in $typ(Grass(I_{r+1})) \subset typ(I_{r+1})$.

2) **(The Nash typical graph)** Given a relative position matrix $M = M(D, d) = (m_{\alpha\beta})$ defined by $D = (H_1 \subset \cdots \subset H_\lambda \subset I_{r+1})$ and $d = (J \subset I_{r+1})$. M is associated with a weighted graph $I(M)$. Let $I(M)$ be the weighted graph defined as follows. The set of vertices Vert $I(M)$ is given by $[\![1, \lambda + 1]\!] \times [\![0, \lambda + 1]\!]$, and the set of oriented edges Edg $I(M)$ by the couples $((\alpha, \beta), (\alpha', \beta'))$ with $(\alpha' - \alpha, \beta' - \beta) \in \{e_1, e_2\}$, where $e_1 = (1, 0)$, $e_2 = (0, 1)$.

Define the weight function $p : I(M) \longrightarrow \mathbb{N}$ by

$$p(\alpha, \beta) = |H_\alpha \cap J \cup H_\beta| \quad (\text{where } H_0 = \emptyset)$$

and assume $I(M)$ endowed with the product order.

The configurations variety associated to this graph is a smooth resolution of singularities $\tilde{\Sigma}(M, D) \longrightarrow \overline{\Sigma}(M, D)$ such that the pull-back of the tangent submodule $\mathcal{T}^1_{\Sigma(M,D)/k} \subset \mathcal{T}^1_{\mathrm{Grass}/k}$ admits an extension as a locally trivial submodule $\tilde{\mathcal{T}}^1_{\Sigma(M,D)/k} \subset (\mathcal{T}^1_{\mathrm{Grass}/k})_{\tilde{\Sigma}(M,D)}$ (cf. [13]).

3) **(The weighted graph $\Lambda'(M)$ equivalent to $\Lambda(M)$)** Let $M \in \mathbb{N}^{(\lambda+1) \times (l+1)}$ be a relative position matrix with $\Lambda(M)$ is associated a typical (resp. weighted) $\Lambda'(M)$ graph as follows. Define the set

$$\mathrm{In}(\underline{m}_\beta) := \{1 \leqslant \alpha \leqslant \lambda + 1 | \ m_{\alpha-1 \ \beta} < m_{\alpha\beta} \quad (m_{0\beta} = 0)\}$$

the set of increasing points of $\underline{m}_\beta = (m_{1\beta} \leqslant \cdots \leqslant m_{\lambda\beta} = n \leqslant r + 1)$, where $\underline{m}_\beta$ denotes the β-row of M. Remark that

$$\mathrm{In}\ \underline{m}_\beta \subset \mathrm{In}\ \underline{m}_{\beta+1} \quad \text{for}\ \ l \geqslant \beta \geqslant 1.$$

Define $\Lambda'(M)$ by giving the set of vertices:

1. Vert $\Lambda'(M) = \coprod_{l+1 \geqslant \beta \geqslant 1} \mathrm{In}\ \underline{m}_\beta \times \{\beta\} \subset \mathrm{Vert}\Lambda(M) = [\![1, \lambda + 1]\!] \times [\![1, l + 1]\!]$; and the set of edges:

2. $\mathrm{Edg}^{(1)}\ \Lambda'(M) = (((\alpha, \beta), (\alpha, \beta + 1)))_{\substack{l \geqslant \beta \geqslant 1, \\ \alpha \in \mathrm{In}\ \underline{m}_\beta}}$;

   $\mathrm{Edg}^{(2)}\ \Lambda'(M) = \{((\alpha, \beta), (\alpha', \beta)) \in \mathrm{Vert}\ \Lambda'(M) \times \mathrm{Vert}\ \Lambda'(M)|\ \alpha \neq \lambda + 1,\ \alpha' = \inf_{\alpha < \alpha'' \in \mathrm{In}(\underline{m}_\beta)} \alpha''\}$;

   $\mathrm{Edg}\ \Lambda'(\mathrm{M}) = \mathrm{Edg}^{(1)}\ \Lambda'(M) \cup \mathrm{Edg}^{(2)}\ \Lambda'(M).$

The weight function $p' : \Lambda'(M) \longrightarrow \mathbb{N}$ is given by the restriction of $p : \Lambda(M) \longrightarrow \mathbb{N}$ to $\Lambda'(M) \subset \Lambda(M)$.

The following proposition results from the definition of $\Lambda'(M)$.

Proposition 6.5 *The variety of $\Lambda'(M)$-configurations*

$$\text{Conf}(\Lambda'(M)) \subset \prod_{\substack{(\alpha,\beta)\in\Lambda'(M), \\ m_{\alpha\beta}=p'(\alpha,\beta)}} \text{Grass}_{m_{\alpha\beta}}(k^{r+1}),$$

is canonically isomorphic to $\text{Conf}(\Lambda(M))$. *More precisely the inclusion* $\Lambda'(M) \subset \Lambda(M)$ *gives rise to a morphism of varieties* $\Phi : \text{Conf}(\Lambda(M)) \longrightarrow \text{Conf}(\Lambda'(M))$ *which is in fact an isomorphism. Thus* $\Lambda(M)$ *and* $\Lambda'(M)$ *are equivalent typical graphs.*

Proof *Let* $(\mathscr{H}_{\alpha\beta})$ *be a* $\Lambda(M)$-*configuration. For* $1 \leq \beta \leq l+1$ *we denote by* $\mathscr{D}_\beta$ *the chain defined by* $(\mathscr{H}_{\alpha\beta})_{1\leq\alpha\leq\lambda+1}$. *Thus* $\varphi(\mathscr{D}_\beta)$ *is given by* $(\mathscr{H}_{\alpha\beta})_{\alpha\in In\ \underline{m}_\beta}$ *and* $(\varphi(\mathscr{D}_\beta))$ *defines a* $\Lambda'(M)$-*configuration. The isomorphism* Φ *makes correspond* $(\varphi(\mathscr{D}_\beta))$ *to* $(\mathscr{D}_\beta)$.

Denote by $\pi'_1 = \pi'_1(M)$ (resp. $\pi'_2 = \pi'_2(M)$) the restriction of π_1 (resp. π_2) to $\text{Conf}(\Lambda'(M))$.

In the next chapter it will be proved that there is a generalized gallery $g(M)$ associated to M defining a typical graph equivalent to $\Lambda'(M)$ and thus to $\Lambda(M)$ too.

6.1 The Associated Fiber Product to a gallery of types

The variety $Conf(\Lambda(g))$ admits a description as a fiber product which is in fact a decomposition of $Conf(\Lambda(g))$ in a sequence of locally trivial fiber bundles with smooth bases and typical fiber of the form $\Sigma_{\mathscr{D}}(s)$ (resp. product of flag varieties).

Definition 6.6 *[Fiber product along a gallery of types]*
   Given a gallery of types $g : t_{r+1} \subset s_r \supset \cdots \subset s_0$, *i.e.* $g \in gall_{I(Drap(k^{r+1}))}$, *we associate to* g *the following fiber product*

$$\hat{\Sigma}(g) = \Sigma_{t_{r+1}}(s_r) \times_{Drap(k^{r+1})} \cdots \times_{Drap(k^{r+1})} \Sigma_{t_1}(s_0) \subset \prod_{1\leq i\leq r} \Sigma_{t_{i+1}}(s_i)$$

defined by the sequence $(\pi_2(t_{i+1}, s_i \supset t_i), \pi_1(t_i, s_{i-1}))_{1\leq i\leq r}$ *(cf. §5.8.2 and what follows).*
   Denote by

$$g^{(\alpha)} : t_{r+1} \subset s_r \supset \cdots \subset t_{r-\alpha+2} \subset s_{r-\alpha+1}$$

the α-*th truncated gallery of* g.

   For every $1 \leqslant \alpha \leqslant r+1$ *there is*

$$\hat{\Sigma}(g^{(\alpha)}) := \hat{\Sigma}(g^{(\alpha-1)}) \times_{Drap(k^{r+1})} \Sigma_{t_{r-\alpha+2}}(s_{r-\alpha+1})$$

the fiber product over $\mathrm{Drap}(k^{r+1})$ *of*

$$(\hat{\Sigma}(g^{(\alpha-1)}),\ \pi'_2(g^{(\alpha-1)}))\ \text{ and }\ (\hat{\Sigma}_{t_{r-\alpha+2}}(s_{r-\alpha+1}),\ \pi_1(t_{r-\alpha+2}, s_{r-\alpha+1})),$$

where $\pi'_2(g^{(\alpha-1)})$ *is induced by* $\pi_2(t_{r-\alpha+3}, s_{r-\alpha+2})$ *composed with the natural morphism*

$$\mathrm{Drap}_{s_{r-\alpha+2}}(k^{r+1}) \longrightarrow \mathrm{Drap}_{t_{r-\alpha+2}}(k^{r+1}).$$

$\hat{\Sigma}(g)\ =\ \hat{\Sigma}(g^{(r+1)})$ *is called* **the fiber product along** g, *and* $(\hat{\Sigma}(g^{(\alpha)}),\ \pi_2(g^{(\alpha)}))$, *where we write* $\pi_2(g^{(\alpha)}) = \pi_2(t_{r-\alpha+2}, s_{r-\alpha+1})$, **the fibering of** $\hat{\Sigma}(g)$. *There are two morphisms* $\pi_1(g)$, $\pi_2(g) : \hat{\Sigma}(g) \overset{\longrightarrow}{\longrightarrow} \mathrm{Drap}(k^{r+1})$ *induced respectively by* $\pi_1(t_{r+1}, s_r)$, *and* $\pi_2(t_1, s_0)$.

Remark 6.7 *Clearly*

$$(\forall\ \gamma \in \hat{\Sigma}(g))\ \ \pi(t_{r+1}, s_r)(\gamma) = E_1(\gamma)\ \ (resp.\ \pi(t_1, s_0)(\gamma) = E_1(\gamma)\ .$$

To these mappings correspond the k-morphisms:

$$\mathscr{E}_1(g) : \hat{\Sigma}(g) \longrightarrow Drap_{t_{r+1}}(k^{r+1})$$

(resp.

$$\mathscr{E}_2(g) : \hat{\Sigma}(g) \longrightarrow Drap_{s_0}(k^{r+1})\ )\ .$$

The following proposition results from the definition of $\hat{\Sigma}(g)$ as a fiber product.

Proposition 6.8 *There is a canonical isomorphism* $\hat{\Sigma(g)} \simeq Conf(\Lambda(g))$. *The underlying set of* $\hat{\Sigma}(g)$ *(resp.* $Conf(\Lambda(g))$ *) is given by* $Gall_I(g)$.

Remark 6.9 *If* $g' \in gall_I$ *is as in 5.2, (ii)″ (resp. (ii)″, (iv)″), then* $\hat{\Sigma}(g)$ *is defined as in definition 6.6 with* g' *satisfying* $t_{r+1}(g') = s_r(g') = t_r(g')$ *(resp.* $t_1(g') = s_0(g')$, $t_{r+1}(g') = s_r(g') = t_r(g')$ *and* $t_1(g') = s_0(g')$ *).*

6.1.1 Galleries issued from a flag

Given $\mathscr{D} \in Drap_{e_1(g)}(k^{r+1})$ we write:

$$Conf(\Lambda(g), \mathscr{D}) := \mathscr{E}^{-1}(\mathscr{D})\ \text{ and }\ \hat{\Sigma}(g, \mathscr{D}) = Conf(\Lambda(g), \mathscr{D})\ .$$

There is a canonical isomorphism:

$$\hat{\Sigma}(g, \mathscr{D}) = \Sigma_{\mathscr{D}}(s_r) \times_{Drap(k^{r+1})} \cdots \times_{Drap(k^{r+1})} \Sigma_{t_1}(s_0) \subset \prod_{1 \le i \le r} \Sigma_{t_{i+1}}(s_i)\ .$$

The fibering $(\hat{\Sigma}(g^{(\alpha)}),\ \pi_2(g^{(\alpha)}))$ of $\hat{\Sigma}(g)$ induces a fibering $(\hat{\Sigma}(g^{(\alpha)}, \mathscr{D}),\ \pi_2(g^{(\alpha)}, \mathscr{D}))$ of $\hat{\Sigma}(g, \mathscr{D})$ **(The fibering of** $\hat{\Sigma}(g, \mathscr{D})$**).** As

$\hat{\Sigma}(g^{(1)}, \mathscr{D}) = \Sigma_{\mathscr{D}}(s_r)$ it is deduced from proposition 5.18 that $\hat{\Sigma}(g, \mathscr{D})$ may be decomposed in a sequence of locally trivial fibrations with typical fiber a product of flag varieties. Thus:

Proposition 6.10 $\hat{\Sigma}(g, \mathscr{D})$ *is an integral projective smooth k-variety.*

It is remarked that the morphism

$$\mathscr{E}_2(g, \mathscr{D}) : \hat{\Sigma}(g, \mathscr{D}) \longrightarrow Drap_{t_0}(k^{r+1}) ,$$

induced by $\mathscr{E}_2(g)$ is a $Stab_{\mathscr{D}}$-equivariant proper morphism.

6.1.2 Relative position matrix associated to a Gallery of types

Given

$$g = g : t_{r+1} \subset s_r \supset \cdots \subset s_0 \supset t_0 \in gall_I \ (\ I = I(Drap(k^{r+1})) ,$$

and $\mathscr{D} \in Drap_{e_1(g)}(k^{r+1})$ it is deduced from the $Stab_{\mathscr{D}}$-equivariance of $\pi = \mathscr{E}_2(g, \mathscr{D}) : \hat{\Sigma}(g, \mathscr{D}) \longrightarrow Drap(k^{r+1})$ that:

$$Im\ \pi = \coprod \Sigma(M', \mathscr{D}) ,$$

where M' runs on the set of relative position matrices with $\Sigma(M', \mathscr{D}) \subset Im\ \pi$. On the other hand, $\hat{\Sigma}(g, \mathscr{D})$ being irreducible, it results that there exists one and only one $M_g \in Relpos(I_{r+1})$ satisfying $\overline{\Sigma}(M_g, \mathscr{D}) = Im\ \pi$. The relative position matrix M_g is independant of the choice of $\mathscr{D}$.

Definition 6.11 *There is a natural mapping:*

$$gall_{I(Drap(k^{r+1}))} \longrightarrow Relpos(I_{r+1})$$

defined by $g \mapsto M_g$ (**Relative position matrix associated to** g).

6.2 Minimal generalized galleries in the Flag complex

The minimal generalized galleries in the apartment $\Delta^{(r)'} \subset I(Drap(k^{r+1}))$, given by the flags adapted to the canonical basis of k^{r+1}, are defined by a birational property. The following developements are based on material that would be introduced in Chapter 9 and may be skipped in a first lecture.

Definition 6.12 *Let* $\mathscr{D}, d \in \Delta^{(r)'}$. *Write* $M = M(\mathscr{D}, d)$. *Let* $g \subset typ(\Delta^{(r)'})$ *be a gallery of types, and* $\gamma_g(\mathscr{D}, d) \subset \Delta^{(r)'}$ *a generalized gallery of type g with extremities* $(\mathscr{D}, d)$. *It is said that* $\gamma_g(\mathscr{D}, d)$ *is* **minimal** *if:*

$$dim_k\ \hat{\Sigma}(g, \mathscr{D}) = dim_k\ \Sigma(M, \mathscr{D}) .$$

A generalized gallery of types $g \subset typ(\Delta^{(r)'})$ *is* **minimal** *if g is the gallery of types of a minimal generalized gallery* $\gamma \subset \Delta^{(r)'}$.

Denote by $gall^m_{\Delta^{(r)'}} \subset gall_{\Delta^{(r)'}}$ the set of minimal galleries of types. It is observed that as $dim_k \hat{\Sigma}(g, \mathscr{D})$ increases indefinitely with the length of g, the set $gall^m_{I(Drap(k^{r+1}))}$ must be finite.

Remark 6.13 *It would be seen that this first definition of a minimal generalized gallery (**mgg**) admits a combinatorial equivalent version, i.e. solely in the terminology of building theory (cf. 9.16).*

Let $\gamma \subset \Delta^{(r)'}$ be gallery of type g. Denote by $E_1(\gamma)$ the left extremity of γ. The notation of remark 5.2 is assumed and one writes $D_i = F_i$ (*resp.* $D'_i = F'_i$). There is then:

1) $dim_k \hat{\Sigma}(g, E_1(\gamma)) = \sum_{r > i \geq 0} dim_k \Sigma_{D'_{i+1}}(D_i)$ if g is closed;

2) $dim_k \hat{\Sigma}(g, E_1(\gamma)) = \sum_{r > i \geq 1} dim_k \Sigma_{D'_{i+1}}(D_i)$ if g is right open;

3) $dim_k \hat{\Sigma}(g, E_1(\gamma)) = \sum_{r \geq i \geq 0} dim_k \Sigma_{D'_{i+1}}(D_i)$ if g is left open;

4) $dim_k \hat{\Sigma}(g, E_1(\gamma)) = \sum_{r \geq i \geq 1} dim_k \Sigma_{D'_{i+1}}(D_i)$ if g is open.

It may be easily verified that the generalized gallery $\gamma_g(\mathscr{D}, d)$ is minimal in the sens of 9.16. Consider:

- a maximal length flag $\mathscr{D} \supset \overline{\mathscr{D}}$ at maximal distance from d, thus $d(\overline{\mathscr{D}}, d) = dim_k \Sigma(M, \mathscr{D})$;

- the generalized gallery $\overline{\gamma}_g(\overline{\mathscr{D}}, d)$ obtained by composing $\mathscr{D} \supset \overline{\mathscr{D}}$ and $\gamma_g(\mathscr{D}, d)$.

The generalized gallery $\overline{\gamma}_g(\overline{\mathscr{D}}, d)$ may be completed into an adapted gallery $\Gamma(\overline{\mathscr{D}}, d)$ so that:

$$length \ \Gamma(\overline{\mathscr{D}}, d) \leqq dim_k \ \hat{\Sigma}(g, \mathscr{D}) = dim_k \ \Sigma(M, \mathscr{D}) \ ,$$

as it follows from the above equalities. Thus one has necessarily that $length \ \Gamma(\overline{\mathscr{D}}, d) = dim_k \ \Sigma(M, \mathscr{D})$, and that $\Gamma(\overline{\mathscr{D}}, d)$ is a minimal gallery. It is concluded that for all maximal length flag $\mathscr{D} \subset \overline{\mathscr{D}}$, at maximal distance from d, the generalized gallery $\overline{\gamma}_g(\overline{\mathscr{D}}, d)$ is minimal. This proves that $\gamma_g(\mathscr{D}, d)$ is minimal generalized. More precisely stated.

Proposition 6.14 *Retain the above notation and hypothesis.*

- *There exists a composed minimal gallery $\Gamma(\overline{\mathscr{D}}, d) = \Gamma_r \circ \cdots \Gamma_1$, if $\gamma_g(\mathscr{D}, d)$ is closed (resp. $\Gamma_r \circ \cdots \Gamma_2$, if $\gamma_g(\mathscr{D}, d)$ is right open, $\Gamma_{r+1} \circ \cdots \Gamma_1$ if γ is left open, $\Gamma_{r+1} \circ \cdots \Gamma_2$ if γ is open). Where $\Gamma_i \subset St_{D'_i}$. It is said that $\Gamma(\overline{\mathscr{D}}, d)$ is a gallery adapted to $\gamma_g(\mathscr{D}, d)$.*

- *One has*

$$d(\overline{\mathscr{D}}, d) = \max_{D' \in \Delta_{\Gamma}^{(r)'} \ incident \ to \ E_1(\gamma)} d(D', E_2(\gamma)) = \sum |\mathscr{H}_{D'_{i+1}}(D_i)| \ .$$

- *The generalized gallery* $\gamma_g(\mathscr{D}, d)$ *is minimal in the sens of 9.16.*

The proof of this proposition is given in the chapter about **mgg**'s in a Coxeter complex.

From proposition 5.24 and the equalities of the above proposition:

$$dim_k \ \hat{\Sigma}(g, \overline{\mathscr{D}}) = \sum dim_k \ \Sigma_{D'_{i+1}}(D_i) = \sum |\mathscr{H}_{D'_{i+1}}(D_i)|$$

it is deduced that the galleries $\Gamma_i \subset St_{D'_i}$ are minimal and of maximal length in $St_{D'_i}$ and finally that for every $\mathscr{D}$ as in proposition 6.14 the following statements hold:

1) $\mathscr{H}_{D'_{i+1}}(D_i) \cap \mathscr{H}_{D'_{i'+1}}(D_{i'}) = \emptyset$ if $i \neq i'$;

2) $\mathscr{H}(\overline{\mathscr{D}}, d) = \coprod \mathscr{H}_{D'_{i+1}}(D_i)$.

The " i's " run on a set of indices which depend on the type of the gallery γ.

Minimal generalized galleries of types are defined by the following birational property.

Proposition 6.15 *Retain the above notation. A generalized gallery* $\gamma(\mathscr{D}, d)$ *is* **minimal** *if and only if the natural morphism:*

$$\pi : \hat{\Sigma}(g, \mathscr{D}) \longrightarrow \overline{\Sigma}(M_g, \mathscr{D})$$

is **birational**.

The second statement results from the fact that $\pi : \hat{\Sigma}(g, \mathscr{D}) \longrightarrow \overline{\Sigma}(M_g, \mathscr{D})$ $(g = typ \ \gamma)$ is a birational morphism if g is a minimal generalized gallery as it results from the second part of this work.

6.3 Birational characterization of minimal generalized galleries of types

Proposition 6.16 *Let* $g \in gall_I$ *(resp.* $\mathscr{D} \in \Delta^{(r)'}$ *with* $typ \ \mathscr{D} = e_1(g)$*). The following statements are equivalent.*

1) *The morphism* $\pi : \hat{\Sigma}(g, \mathscr{D}) \longrightarrow \overline{\Sigma}(M_g, \mathscr{D})$ *is a birational morphism.*

2) *There is a minimal generalized gallery* $\gamma(\mathscr{D}, d) \subset \Delta^{(r)'}$ *whose type is* g.

Proof   *From 6.15 it follows that* 2) $\Longrightarrow$ 1). *It suffices to prove that* 1) $\Longrightarrow$ 2). *Let* $d \in \Delta^{(r)'}$ *so that* $M_g = M(\mathscr{D}, d)$. *As* π *is a surjective morphism there exists a gallery* $\gamma(\mathscr{D}, d)$ *of type* g *issued from* $\mathscr{D}$ *so that* $E_2(\gamma(\mathscr{D}, d)) = d$. *Let a maximal length flag* $\overline{\mathscr{D}} \subset \mathscr{D}$ *be chosen in* $\Delta^{(r)'}$ *at maximal distance from* d, *i.e.* $d(\overline{\mathscr{D}}, d) = dim_k \; \Sigma(M_g, \mathscr{D})$. *From* $dim_k \; \hat{\Sigma}(g, \mathscr{D}) = dim_k \; \overline{\Sigma}(M_g, \mathscr{D})$ *it results that the composed gallery* $\overline{\gamma}(\overline{\mathscr{D}}, d)$ *of* $\overline{\mathscr{D}} \subset \mathscr{D}$ *with* $\gamma(\mathscr{D}, d)$ *may be completed in a gallery (in the usual sens)* $\Gamma(\overline{\mathscr{D}}, d)$ *in* $I(Drap(k^{r+1}))$, *with length* $\Gamma(\overline{\mathscr{D}}, d) \leq d(\overline{\mathscr{D}}, d)$. *Thus* $\Gamma(\overline{\mathscr{D}}, d)$ *is a minimal gallery and thus contained in* $\Delta^{(r)'}$. *This gives* $\gamma(\mathscr{D}, d) \subset \Delta^{(r)'}$. *The result follows from 6.14.*

The above proposition is referred as the (**The birational criterium of minimality**).

Corollary 6.17   *The generalized gallery* $\gamma(\mathscr{D}, d) \subset I(Drap(k^{r+1})$ *whose type is* g *is a mgg if and only if*

1) $\pi : \hat{\Sigma}(g, \mathscr{D}) \longrightarrow \overline{\Sigma}(M_g, \mathscr{D})$ *is a birational morphism;*

2) $M(\mathscr{D}, d) = M_g$.

6.4   The convex hull of a minimal generalized gallery

The reader is referred to [4] and [50] for the proof of the following result.

Proposition 6.18   *Let* $\Gamma(\mathscr{D}, \mathscr{D}') = (\mathscr{D} = \mathscr{D}_0, \mathscr{D}_1, \ldots, \mathscr{D}_n = \mathscr{D}') \subset I(Drap(k^{r+1})$ *be a minimal length gallery, i.e. a minimal gallery (with the usual meaning) in* $I(Drap(k^{r+1}))$. *Then* $\Gamma(\mathscr{D}, \mathscr{D}') \subset Env^I(\mathscr{D}, \mathscr{D}')$.

It follows from the proposition that a minimal gallery $\Gamma(\mathscr{D}, \mathscr{D}')$ may be completed in a minimal generalized gallery $\gamma(\mathscr{D}, \mathscr{D}') = (\mathscr{D} = \mathscr{D}_0, \mathscr{D}_0 \cap \mathscr{D}_1, \mathscr{D}_1, \ldots, \mathscr{D}_n, \mathscr{D}_n \cap \mathscr{D}_{n+1}, \mathscr{D}_{n+1} = \mathscr{D}') \subset I(Drap(k^{r+1}))$. It is recalled that if $\mathscr{D}, \mathscr{D}' \in \Delta^{(r)'}$ one has that $Env^I(\mathscr{D}, \mathscr{D}') = Env^{\Delta^{(r)'}}(\mathscr{D}, \mathscr{D}')$, thus $\Gamma(\mathscr{D}, \mathscr{D}') \subset \Delta^{(r)'}$. The maximal length flags $\mathscr{D}_0, \mathscr{D}_1, \ldots, \mathscr{D}_n \subset \Delta^{(r)'}$ may be seen as chambers in $\Delta^{(r)'}$. Define $\mathscr{D}_0 \cap \mathscr{D}_1, \ldots, \mathscr{D}_{n-1} \cap \mathscr{D}_n$ as the corresponding sequence of codimension 1 common facets of two succesive chambers. Reciprocally to a minimal generalized gallery of the form

$$\gamma(\mathscr{D}, \mathscr{D}') = (\mathscr{D} = \mathscr{D}_0, \mathscr{D}_0 \cap \mathscr{D}_1, \mathscr{D}_1, \ldots, \mathscr{D}_n, \mathscr{D}_n \cap \mathscr{D}_{n+1}, \mathscr{D}_{n+1} = \mathscr{D}'),$$

is associated a gallery $\Gamma(\mathscr{D}, \mathscr{D}')$ in an obvious way. One has that $\gamma(\mathscr{D}, \mathscr{D}')$ is minimal if and only if $\Gamma(\mathscr{D}, \mathscr{D}')$ is minimal.

It would be seen in chapter 9, as a corollary of the above proposition, that:

- a minimal generalized gallery $\gamma(\mathscr{D}, d)$ of type g is the unique gallery of type g in $I(Drap(k^{r+1}))$ connecting $\mathscr{D}$ and d and is contained in the convex hull $Env^I(\mathscr{D}, d)$.

- If g is a minimal generalized gallery of types and $M(\mathscr{D}, d) = M_g$, there is a minimal gallery $\gamma(\mathscr{D}, d)$ of type g and $\gamma(\mathscr{D}, d)$ is the unique gallery of type g in $I(Drap(k^{r+1}))$ connecting $\mathscr{D}$ and d .

From this it follows, as it would be seen, the existence of a canonical section:

$$\sigma_{(g, \mathscr{D})} : \Sigma(M_g, \mathscr{D}) \longrightarrow \hat{\Sigma}(g, \mathscr{D})$$

of the natural morphism $\pi : \hat{\Sigma}(g, \mathscr{D}) \longrightarrow \overline{\Sigma}(M_g, \mathscr{D})$ on $\Sigma(M_g, \mathscr{D})$ associating to a couple $(\mathscr{D}, \mathscr{D}')$ satisfying $M(\mathscr{D}, \mathscr{D}') = M_g$ the unique mgg $\gamma_g(\mathscr{D}, \mathscr{D}')$ of type g contained in $Env^I(\mathscr{D}, \mathscr{D}')$, which establishes an isomorphism $\Sigma(M_g, \mathscr{D}) \simeq (\pi)^{-1}(\Sigma(M_g, \mathscr{D}))$.

Chapter 7

Configuration Varieties as Gallery Varieties

In this chapter it is proved that the configurations variety giving a smooth resolution of singularities for a Schubert variety is, in fact, isomorphic to a gallery variety given by a minimal generalized gallery of types. More precisely, with its relative position matrix $M \in \text{Relpos} \subset (\coprod \mathbb{N}^{(\lambda+1) \times (l+1)})$ is associated a gallery of types $g(M) \in gall_I$ and an isomorphism

$$\gamma(M) : Conf(\Lambda(M)) \simeq Gall(g(M)),$$

where $Gall(g(M)) = Gall(\Lambda(g(M)))$. Given a general $\Lambda(M)$-configuration of adapted flags to the canonical basis, chains of adapted subspaces are of use in constructing a minimal generalized gallery of type $g(M)$ in terms of this configuration. The gallery of types $g(M)$ depends solely on M. A generalized gallery of chains of adapted subspaces is first obtained from this configuration, giving rise, after reduction, to that minimal generalized gallery. Thus we obtain the above isomorphism by following the pattern of this correspondence. It is also given another example of a configurations variety (The Nash variety associated with a Schubert variety in a Grassmannian) and a corresponding isomorphic gallery variety. These two examples show, for the linear group, the role played by the geometry of the Tits building in unifying the constructions of configuration varieties.

7.1 The minimal generalized gallery associated with a Relative position matrix

Let $D = (H_1 \subset \cdots \subset H_\lambda \subset I_{r+1})$ (resp. $D' = (J \subset I_{r+1})$), with typ $D = \underline{m} = (m_1 < \cdots < m_\lambda < r+1)$ (resp. typ $D' = \underline{n} = (n < r+1)$, the type of a subspace in k^{r+1}). Write $M(D, D') = (m_{\alpha\beta}) \in \mathbb{N}^{(\lambda+1)\times 2)}$, thus

$$(m_{\alpha 2} = m_\alpha)_{1 \leqslant \alpha \leqslant \lambda} \quad (\text{resp. } (m_{\alpha 1} = |H_\alpha \cap J|)_{1 \leqslant \alpha \leqslant \lambda}, \ m_{\lambda+1 1} = n = |J|,$$

$$m_{\lambda+1 2} = r+1) .$$

The procedure is to first construct $g(M)$ assuming that $M = M(D, D') = (m_{\alpha\beta}) \in \mathbb{N}^{(\lambda+1)\times 2}$ satisfies:

1. $0 < m_{11}, m_{\alpha 1} < m_{\alpha+1\ 1}$ for $1 \leqslant \alpha \leqslant \lambda$

2. $m_{\alpha 1} < m_{\alpha 2}$ for $1 \leqslant \alpha \leqslant \lambda + 1$.

Write $J_\alpha = H_\alpha \cap J$ for $1 \leqslant \alpha \leqslant \lambda$, (resp. $J_{\lambda+1} = J$). The chain $D_1 = D \cap J = (J_1 \subset \cdots \subset J_\lambda \subset J)$ satisfies $J_\alpha \subset H_\alpha$ $(J_\alpha \neq H_\alpha)$ and is in fact a flag of J. Remark that the second column of the matrix $\mathscr{M}(D, D') = (H_{\alpha\beta})$ is given by $D_2 = D$, and the first column by $D_1 = D \cap J$. A generalized gallery is defined between D and D' in $\Delta^{(r)'} = \mathrm{Drap}(I_{r+1})$:

$$\overline{\gamma}(D, J): \ d_{\lambda+1} \subset D_\lambda \supset d_\lambda \subset \cdots \subset D_1 \supset d_1 \subset D_0 \supset d_0,$$

by

(table 1)
$$\begin{cases}
d_{\lambda+1} = D = (H_1 \subset \cdots \subset H_\lambda \subset I_{r+1}), \\[2mm]
D_\alpha = (J_1 \subset \cdots \subset J_{\lambda-\alpha+1} \subset H_{\lambda-\alpha+1} \subset \cdots \subset H_\lambda \subset I_{r+1}), \\[2mm]
d_\alpha = (J_1 \subset \cdots \subset J_{\lambda-\alpha+1} \subset H_{\lambda-\alpha+2} \subset \cdots \subset H_\lambda \subset I_{r+1}), \\[2mm]
\text{for } 2 \leqslant \alpha \leq \lambda, \text{ and} \\[2mm]
D_1 = (J_1 \subset \cdots J_\lambda \subset H_\lambda \subset I_{r+1}), \\[2mm]
d_1 = (J_1 \subset \cdots \subset J_\lambda \subset I_{r+1}), \\[2mm]
D_0 = (J_1 \subset \cdots \subset J_\lambda \subset J \subset I_{r+1}), \\[2mm]
d_0 = (J \subset I_{r+1}).
\end{cases}$$

The assumption about M makes one sure that all the chains of (Table 1) are in fact flags of I_{r+1}. The functions f defining these chains are all increasing strictly. Write typ $D_\alpha = s_\alpha$ (resp. typ $d_\alpha = t_\alpha$) for $1 \leqslant \alpha \leq \lambda$, and $t_{\lambda+1} = $ typ $d_{\lambda+1}$, and denote by $\gamma(D, J)$ the sub-gallery of $\overline{\gamma}(D, J)$ with extremities $E(\gamma(D, J)) = (d_{\lambda+1}, D_0)$.

Definition 7.1 *Let* $\overline{g}(M) := typ\ \overline{\gamma}(D, J)$ *(resp.$g(M) := typ\ \gamma(D, J)$)($M = M(D, D')$). There is*

$$\overline{g}(M):\ t_{\lambda+1} \subset s_\lambda \supset t_\lambda \subset \cdots \subset s_1 \supset t_1 \subset s_0 \supset t_0$$

(resp.

$$g(M):\ t_{\lambda+1} \subset s_\lambda \supset t_\lambda \subset \cdots \subset s_1 \supset t_1 \subset s_0).$$

Remark 7.2 *If* $\sigma((D, J)) = (D_0, J_0)$ *for* $\sigma \in \mathfrak{S}_{r+1}$ *then* $\sigma(\gamma(D, J)) = \gamma(D_0, J_0)$, *as one has* $\sigma(\mathcal{M}(D, D')) = \mathcal{M}(D_0, D_0')$. *It follows that the gallery of types* typ $\overline{\gamma}(D, J)$ *depends only on the relative position matrix* $M = M(D, D')$. *This justifies the notation* $\overline{g}(M)$ *for* typ $\overline{\gamma}(D, J)$.

Definition 7.3 (The isomorphism $\gamma(M)$**)**
    it is assumed that M satisfies conditions 1-, and 2- above, and define

$$\overline{\gamma}(M):\ \mathrm{Conf}(\Lambda(M)) \longrightarrow \mathrm{Gall}(\overline{g}(M))$$

by $\overline{\gamma(M)}:\ (\mathcal{H}_{\alpha\beta}) \mapsto \overline{\gamma}((\mathcal{H}_{\alpha\beta})) \in \mathrm{Gall}(\overline{g}(M))$, *where the gallery*

$$\overline{\gamma}((\mathcal{H}_{\alpha\beta})):\ d_{\lambda+1} \subset \mathcal{D}_\lambda \supset d_\lambda \subset \cdots \subset \mathcal{D}_0 \supset d_0$$

is obtained as follows. Let it be written

1. $\mathcal{H}_\alpha = \mathcal{H}_{\alpha 2}$ *for* $1 \leqslant \alpha \leqslant \lambda$ *(resp.* $\mathcal{H}_{\lambda+1\ 1} = k^{r+1}$*),* $\mathcal{D} = (\mathcal{H}_1 \subset \cdots \mathcal{H}_\lambda \subset k^{r+1})$ *with* typ $\mathcal{D} = \underline{m}$;

2. $\mathcal{H}_{\alpha 1} = \mathcal{J}_\alpha$ *for* $1 \leqslant \alpha \leqslant \lambda$ *(resp.* $\mathcal{J}_{\lambda+1\ 1} = \mathcal{J}$*),* $\mathcal{D}' = (\mathcal{J}_1 \subset \cdots \subset \mathcal{J}_l \subset \mathcal{J} \subset k^{r+1})$ *with* typ $\mathcal{D}' = \underline{n}$;

then:

3.

(table 2)
$$\begin{cases} d_{\lambda+1} = (\mathcal{H}_1 \subset \cdots \subset \mathcal{H}_\lambda \subset k^{r+1}) \\[2mm] \mathcal{D}_\alpha = (\mathcal{J}_1 \subset \cdots \subset \mathcal{J}_{\lambda-\alpha+1} \subset \mathcal{H}_{\lambda-\alpha+1} \subset \cdots \subset \mathcal{H}_\lambda \subset k^{r+1}), \\[2mm] d_\alpha = (\mathcal{J}_1 \subset \cdots \subset \mathcal{J}_{\lambda-\alpha+1} \subset \mathcal{H}_{\lambda-\alpha+2} \subset \cdots \subset \mathcal{H}_\lambda \subset k^{r+1}) \\[2mm] \text{for } 2 \leqslant \alpha \leqslant \lambda, \text{ and} \\[2mm] \mathcal{D}_1 = (\mathcal{J}_1 \subset \cdots \subset \mathcal{J}_\lambda \subset \mathcal{H}_\lambda \subset k^{r+1}); \\[2mm] d_1 = (\mathcal{J}_1 \subset \cdots \subset \mathcal{J}_\lambda \subset k^{r+1}); \\[2mm] \mathcal{D}_0 = (\mathcal{J}_1 \subset \cdots \subset \mathcal{J}_\lambda \subset \mathcal{J} \subset k^{r+1}); \\[2mm] d_0 = (\mathcal{J} \subset k^{r+1}). \end{cases}$$

Thus $\overline{\gamma}((\mathscr{H}_{\alpha\beta}))$ *is a generalized gallery of type* $\overline{g}(M)$ *between* $\mathscr{D} = (\mathscr{H}_1 \subset \cdots \subset \mathscr{H}_\lambda \subset k^{r+1})$ *and* $\mathscr{D}' = (\mathscr{J} \subset k^{r+1})$.

The mapping $\mathrm{Conf}(\Lambda(M)) \longrightarrow \mathrm{Gall}(\overline{g}(M))$, *defined by* $(\mathscr{H}_{\alpha\beta}) \mapsto \overline{\gamma}((\mathscr{H}_{\alpha\beta}))$, *is induced by a* k-*isomorphism, which is also denoted by*

$$\overline{\gamma} = \overline{\gamma}(M) : \ \mathrm{Conf}(\Lambda(M)) \xrightarrow{\sim} \mathrm{Gall}(\overline{g}(M)).$$

A section of $\mathrm{Gall}(\overline{g}(M))$ *comes from a unique* $\Lambda(M)$-*configuration* $(\mathscr{H}_{\alpha\beta})$. *It is clear that it is a functorial definition of the* k-*isomorphism* $\overline{\gamma}(M)$.

From the fact that a gallery of type $g(M)$ *determines a unique gallery of type* $\overline{g}(M)$ *and reciprocally, an isomorphism* $\mathrm{Conf}(\Lambda(M)) \simeq \mathrm{Gall}(g(M))$ *is deduced which is denoted by* $\gamma(M)$.

Given a relative position matrix $M \in \mathbb{N}^{(\lambda+1)\times 2}$ (without any restrictions) a gallery $g(M)$ defining a graph $\Lambda(g(M))$ equivalent to $\Lambda(M)$ may be constructed. This construction is inspired by the above one.

The preceding definitions are restated in a slighty more general frame. Given a finite set I, a flag $D \in Drap(I)$, and a subset $J \in \mathscr{P}^*(I)$, with $|J| < |I|$, let

$$I(J, D)$$

be the subflag of D whose vertices are those of D indexed by the set of increasing points $In\ \underline{m}$, where $\underline{m} \in \overline{typ}(I)$ is the type of the chain $J \cap D \in Chain(I)$. Write $D' = I(J, D)$ and $D'' = (J \subset I)$. The relative position matrix $M(D', D'')$ satisfies the set of conditions 1, but not necessarily the conditions 2.

Let it be explained how the construction giving Table 1, can be modified to obtain a gallery $\gamma(D', D'')$ between D' and D'' if $M(D', D'')$ satisfies only condition 1, i.e. between two flags whose relative position matrix satisfies the set of conditions 1 but not necessarily conditions 2. If the constructions of Table 1 is carried out, a "gallery of chains" is obtained:

$$D = d_{\lambda+1} \subset D_\lambda \supset \cdots \subset D_\alpha \supset d_\alpha \subset D_{\alpha-1} \supset d_{\alpha-1} \subset \cdots \subset D_0 \supset d_0.$$

It is remarked that if M satisfies only the inequalities of 1, without satisfying inequalities 2 there is $\lambda \geqslant \alpha \geqslant 1$ so that: $J \supset H_{\lambda-\alpha+1} \cdots \supset H_1$ and $J \not\supset H_{\lambda-\alpha+2}$, thus:

$$\lambda \geqslant \alpha' \geqslant \alpha \ \Rightarrow \ m_{\lambda-\alpha'+1\ 1} = m_{\lambda-\alpha'+1\ 2}$$
$$(\text{resp. } \alpha > \alpha'' \geqslant 1 \ \Rightarrow \ m_{\lambda-\alpha''+1\ 1} < m_{\lambda-\alpha''+1\ 2}).$$

and

$$D = \varphi(d_{\lambda+1}) = \varphi(D_\lambda) = \cdots = \varphi(D_\alpha) = \varphi(d_\alpha) \text{ is obtained,}$$

and that

$$\gamma(D', D'') : \ d_\alpha \subset D_{\alpha-1} \supset \cdots \subset D_0 \supset d_0,$$

where $d_\alpha = I(J, D)$ and $d_0 = (J \subset I)$. Thus $\gamma(D', D'')$ is a non-stammering gallery of flags with strict inclusions, i.e. a reduced gallery. Recall that the image by φ of a chain denotes its associated flag.

Definition 7.4 *If $D \neq I(J, D)$, define*

$$\overline{\gamma}(D, J) = \gamma(D, D'') : \ D \supset d_\alpha \subset D_{\alpha-1} \supset \cdots \subset D_0 \supset d_0,$$

otherwise put

$$\overline{\gamma}(D, J) = \gamma(D', D'') = \gamma(d_\alpha, D'').$$

In the latter case there is $D = I(J, D)$. Denote by $\gamma(D, J)$ the sub-gallery with extremities $E(\gamma(D, J)) = (D, \varphi(J \cap D))$. and write

$$\overline{g}(M) = typ(\overline{\gamma}(D, J))$$

(resp.

$$g(M) = typ(\gamma(D, J)) \ .$$

Following the pattern of Table 2 the above constructions allows one to define a functorial morphism

$$\overline{\gamma}(M) : Conf(\Lambda(M)) \simeq Conf(\Lambda'(M)) \longrightarrow Conf(\overline{g}(M))$$

(resp.

$$\gamma(M) : Conf(\Lambda(M)) \longrightarrow Conf(g(M)) \ .$$

A $\Lambda(g(M))$-configuration (resp. gallery of type $g(M)$) γ determines by construction a unique $\Lambda(M)$-configuration $(\mathcal{H}_{\alpha\beta})$ with

$$\gamma(M)((\mathcal{H}_{\alpha\beta})) = \gamma.$$

Thus this morphism is clearly an isomorphism. The isomorphism $\gamma(M) : Conf(\Lambda(M)) \longrightarrow Conf(g(M))$ is obtained following the same pattern as in the construction of Table 2.

Observe that $g(M)$ is a minimal generalized gallery of types, as it results from the Birational Criterium of Minimality and the fact that all the terms of its flags are obtained as the intersection of couples of terms from (D, D').

Given $M \in \mathbb{N}^{(\lambda+1) \times (l+1)}$ let a gallery of types $g(M) \in gall_I$ be defined and an isomorphism $\gamma(M) : Conf(\Lambda(M)) \simeq Conf(g(M))$, based on the preceding construction of $g(M)$ for $M \in \mathbb{N}^{(\lambda+1) \times 2}$.

Let $D = (H_1 \subset \cdots \subset H_\lambda \subset I_{r+1})$ (resp. $D' = (J_1 \subset \cdots \subset J_l \subset I_{r+1})$), and $M = M(D, D')$. Denote by (D_β), with $D_\beta = D \cap J_\beta$, the column of the matrix $\mathcal{M}(D, D') = (H_\alpha \cap J_\beta)$ indexed by β, and let

$$Conf^{comb}(\Lambda(M)) \longrightarrow \prod Chain_{\underline{m}_\beta}(J_\beta).$$

be the natural injective mapping. Recall that by definition of the graph $\Lambda'(M)$ equivalent to $\Lambda(M)$ there is a natural isomorphism $\Phi : Conf(\Lambda(M)) \simeq Conf(\Lambda'(M))$. To the configuration $(H_\alpha \cap J_\beta)$ corresponds the $\Lambda'(M)$-configuration $(\varphi(D_\beta)) = (\varphi(D \cap J_\beta))$.

Remark that to the preceding natural injective mapping corresponds the following one

$$Conf^{comb}(\Lambda'(M)) \longrightarrow \prod Drap_{m'_\beta}(J_\beta).$$

Define

$$M_\beta = M(\varphi(D \cap J_{\beta+1}), (J_\beta \subset J_{\beta+1})) \in Relpos_{\Delta'(J_{\beta+1})}.$$

Following 7.4 a gallery of types

$$g(M_\beta) = typ(\gamma(\varphi(D \cap J_{\beta+1}), J_\beta)) \in gall_{\Delta'(J_{\beta+1})},$$

and an isomorphism

$$\gamma(M_\beta) : Conf(\Lambda(M_\beta)) \simeq Conf(g(M_\beta)),$$

are associated with M_β.

Definition 7.5

Let

$$D'^\beta = (J_\beta \subset \cdots J_l \subset I_{r+1})$$

(The β-th upper truncation of D'). *Given $d \in Drap(J_\beta)$ denote by $d^\natural \in Drap(I_{r+1})$ the flag defined by*

$$Vert(d^\natural) = Vert(d) \cup \{ J_\beta, \cdots J_l, I_{r+1}\},$$

and by

$$\delta_\beta : Drap(J_\beta) \longrightarrow St_{D'^\beta},$$

the mapping which makes $d^\natural$ correspond to d; it is clear that δ_β preserves the incidence relation, thus the image $\gamma^\natural = \delta_\beta(\gamma)$ of a gallery $\gamma \subset Drap(J_\beta)$ is a gallery of $Drap(I_{r+1})$.

Let

$$g^\natural(M_\beta) = typ(\gamma^\natural(\varphi(D \cap J_{\beta+1}), J_\beta)) \in gall_I.$$

It follows from

$$E_2(\gamma(\varphi(D \cap J_{\beta+1}), J_\beta)) = \varphi(D \cap J_\beta),$$

where $\gamma(\varphi(D \cap J_{\beta+1}), J_\beta)$ is seen as a gallery in $Drap(J_{\beta+1})$, and from

$$E_1(\gamma(\varphi(D \cap J_\beta), J_{\beta-1})) = \varphi(D \cap J_\beta),$$

where $\gamma(\varphi(D \cap J_\beta), J_{\beta-1})$ is seen as a gallery in $Drap(J_\beta)$ and $\varphi(D \cap J_\beta)$ as a flag in $Drap(J_\beta)$, that

$$E_2(\gamma^\natural(\varphi(D \cap J_{\beta+1}), J_\beta)) = E_1(\gamma^\natural(\varphi(D \cap J_\beta), J_{\beta-1})) ,$$

and the sequence of galleries$(\gamma^\natural(\varphi(D \cap J_{\beta+1}), J_\beta))$ may be composed in an obvious way.

Write

$$\overline{\gamma}(D, D') = (\underset{l \geq \beta \geq 2}{\bigstar} \gamma^\natural(\varphi(D \cap J_{\beta+1}), J_\beta))) * \overline{\gamma}^\natural(\varphi(D \cap J_2), J_1)$$

(resp.

$$\gamma(D, D') = (\underset{l \geq \beta \geq 2}{\bigstar} \gamma^\natural(\varphi(D \cap J_{\beta+1}), J_\beta))) * \gamma^\natural(\varphi(D \cap J_2), J_1)).$$

Remark 7.6 *By* $\gamma * \gamma'$ *the composition of a couple of galleries* (γ, γ') *is denoted satisfying* $E_2(\gamma) = E_1(\gamma')$. *The product indexed by* $l \geq \beta \geq 2$ *denotes the composition of the sequence of galleries* $(\gamma^\natural(\varphi(D \cap J_{\beta+1}), J_\beta))$ *in decreasing order from* l *to* 2.

Let

$$\overline{g}(M) = typ(\overline{\gamma}(D, D'))$$

(resp.

$$g(M) = typ(\gamma(D, D'))).$$

Thus by adapting in an obvious way the above notation one may write

$$\overline{g}(M) = (\underset{l \geq \beta \geq 2}{\bigstar} g^\natural(M_\beta)) * \overline{g}^\natural(M_1)$$

(resp.

$$g(M) = (\underset{l \geq \beta \geq 2}{\bigstar} g^\natural(M_\beta)) * g^\natural(M_1)).$$

Given a $\Lambda(M)$-configuration $(\mathcal{H}_{\alpha\beta})$ and $1 \leq \beta_0 \leq l$ associate to it the flags

$$\mathscr{D}_{\beta_0+1} = \varphi((\mathcal{H}_{\alpha\beta_0+1})) \text{ in } Drap(\mathcal{H}_{\lambda+1\beta_0+1})$$

(resp.

$$\mathscr{D}_{\beta_0} = \varphi((\mathcal{H}_{\alpha\beta_0})) \text{ in } Drap(\mathcal{H}_{\lambda+1\beta_0})$$

defined by $In \underline{m}_{\beta_0+1}$ (resp. $In \underline{m}_{\beta_0}$), i.e. the (β_0+1)-column (resp. β_0-column) of the corresponding $\Lambda'(M)$-configuration $\Phi((\mathcal{H}_{\alpha\beta}))$. Let

$$\Gamma_{\beta_0} : Conf(\Lambda(M)) \longrightarrow Conf(g(M_{\beta_0}))$$

(resp.

$$\overline{\Gamma}_{\beta_0} : Conf(\Lambda(M)) \longrightarrow Conf(\overline{g}(M_{\beta_0}))),$$

be defined by

$$\Gamma_{\beta_0} : (\mathscr{H}_{\alpha\beta}) \mapsto \gamma(\mathscr{D}_{\beta_0+1}, \mathscr{D}'_{\beta_0})$$

(resp.

$$\overline{\Gamma}_{\beta_0} : (\mathscr{H}_{\alpha\beta}) \mapsto \overline{\gamma}(\mathscr{D}_{\beta_0+1}, \mathscr{D}'_{\beta_0}));$$

where $\gamma(\mathscr{D}_{\beta_0+1}, \mathscr{D}'_{\beta_0})$ (resp. $\overline{\gamma}(\mathscr{D}_{\beta_0+1}, \mathscr{D}'_{\beta_0})$) denotes the $g(M_{\beta_0})$–configuration (resp. $\overline{g}(M_{\beta_0})$–configuration) obtained from $\gamma(D_{\beta_0+1}, J_{\beta_0})$ (resp. $\overline{\gamma}(D_{\beta_0+1}, J_{\beta_0})$) by replacing each $H_{\alpha\beta}$ by the corresponding $\mathscr{H}_{\alpha\beta}$ in each flag of $\gamma(D_{\beta_0+1}, J_{\beta_0})$ (resp. $\overline{\gamma}(D_{\beta_0+1}, J_{\beta_0})$).

Following definition 7.5 define

$$\Gamma^{\natural}_{\beta_0} : Conf(\Lambda(M)) \longrightarrow Conf(g^{\natural}(M_{\beta_0}))$$

(resp.

$$\overline{\Gamma}^{\natural}_{\beta_0} : Conf(\Lambda(M)) \longrightarrow Conf(\overline{g}^{\natural}(M_{\beta_0}))$$

by

$$\Gamma^{\natural}_{\beta_0} : (\mathscr{H}_{\alpha\beta}) \mapsto \gamma^{\natural}(\mathscr{D}_{\beta_0+1}, \mathscr{D}'_{\beta_0}).$$

(resp.

$$\overline{\Gamma}^{\natural}_{\beta_0} : (\mathscr{H}_{\alpha\beta}) \mapsto \overline{\gamma}^{\natural}(\mathscr{D}_{\beta_0+1}, \mathscr{D}'_{\beta_0})).$$

Where $\gamma^{\natural}(\mathscr{D}_{\beta_0+1}, \mathscr{D}'_{\beta_0})$ (resp. $\overline{\gamma}^{\natural}(\mathscr{D}_{\beta_0+1}, \mathscr{D}'_{\beta_0})$) denotes the $g^{\natural}(M_{\beta_0})$–configuration (resp. $\overline{g}^{\natural}(M_{\beta_0})$–configuration) obtained from $\gamma^{\natural}(D_{\beta_0+1}, J_{\beta_0})$ (resp. $\overline{\gamma}^{\natural}(D_{\beta_0+1}, J_{\beta_0})$) by replacing each $H_{\alpha\beta}$ by the corresponding $\mathscr{H}_{\alpha\beta}$ in each flag of $\gamma^{\natural}(D_{\beta_0+1}, J_{\beta_0})$ (resp. $\overline{\gamma}^{\natural}(D_{\beta_0+1}, J_{\beta_0})$). Define

$$\overline{\gamma}(M) = (\underset{l \geq \beta \geq 2}{\bigstar} \Gamma^{\natural}(M_{\beta})) * \overline{\Gamma}^{\natural}(M_1) : Conf(\Lambda(M))$$

$$\longrightarrow (\underset{l \geq \beta \geq 2}{\bigstar} Conf(g^{\natural}(M_{\beta}))) * Conf(\overline{g}^{\natural}(M_1)) = Conf(\overline{g}(M))$$

(resp.

$$\gamma(M) = \underset{l \geq \beta \geq 1}{\bigstar} \Gamma^{\natural}_{\beta} : Conf(\Lambda(M)) \longrightarrow \underset{l \geq \beta \geq 1}{\bigstar} Conf(g^{\natural}(M_{\beta})) = Conf(g(M))).$$

Remark 7.7   *1. The product* $\underset{l \geq \beta \geq 1}{\bigstar} \Gamma^{\natural}_{\beta}$ *denotes the composition of the galleries* $(\Gamma^{\natural}_{\beta})$ *in decreasing order from* $\beta = l$ *to* $\beta = 2$. *This product is well defined since the sequence of galleries of types* $(g^{\natural}(M_{\beta}))$ *may be composed.*

2. *The sections of* $\underset{l \geq \beta \geq m}{\bigstar} Conf(g^{\natural}(M_{\beta}))$ *are given by the composite configurations* $\gamma_1 * \cdots * \gamma_m$ *defined by the sections* (γ_{β}) *of* $\underset{l \geq \beta \geq m}{\prod} Conf(g^{\natural}(M_{\beta}))$ *satisfying* $\mathscr{E}_2(\gamma_{\beta+1}) = \mathscr{E}_1(\gamma_{\beta})$.

3. *The morphism* $\overline{\gamma}(M) : Conf(\Lambda(M)) \longrightarrow Conf(\overline{g}(M))$ *given by:*

$$(\mathscr{H}_{\alpha\beta}) \mapsto \overline{\gamma}(\mathscr{D}, \mathscr{D}'),$$

where $\mathscr{D} = \pi_1((\mathscr{H}_{\alpha\beta}))$ *(resp.* $\mathscr{D}' = \pi_2((\mathscr{H}_{\alpha\beta}))$, *is obtained from the combinatorial* $\Lambda(M)$-*configuration* $\overline{\gamma}(D, D')$, *by replacing the combinatorial* $\Lambda(M)$-*configuration* $(H_{\alpha\beta}) \in Conf^{comb}(\gamma(M))$ *by the section* $(\mathscr{H}_{\alpha\beta})$ *of* $Conf(\Lambda(M))$. *That is we consider the* $H_{\alpha\beta}$ *'s as variables and we obtain a* $\overline{g}(M)$-*configuration* $\overline{\gamma}(\mathscr{D}, \mathscr{D}')$ *by specializing these variables* $(H_{\alpha\beta}) \mapsto (\mathscr{H}_{\alpha\beta})$.

From this we deduce that once $\overline{\gamma}(\mathscr{D}, \mathscr{D}')$ *is known we retrieve the* $\Lambda(M)$-*configuration* $(\mathscr{H}_{\alpha\beta})$, *and finally that* $\overline{\gamma}(M)$ *is a functorial isomorphism.*

Notation 7.8 (The fiber of $(Conf(\overline{g}(M)), \mathscr{E}(M))$ **over** $\mathscr{D} \in Drap(k^{r+1})$**)**
Write

$$(Conf(\overline{g}(M)), \mathscr{E}(M))_{\mathscr{D}} = (Conf(\overline{g}(M))_{\mathscr{D}}, \mathscr{E}(M)_{\mathscr{D}})$$

$$= (Conf(\overline{g}(M), \mathscr{D}), \mathscr{E}(M, \mathscr{D}))$$

where

$$Conf(\overline{g}(M), \mathscr{D}) := Conf(\overline{g}(M))_{\mathscr{D}} = (\mathscr{E}_1(M))^{-1}(\mathscr{D})$$

(resp.

$$\mathscr{E}(M, \mathscr{D}) := \mathscr{E}(M)_{\mathscr{D}} : Conf(\overline{g}(M), \mathscr{D}) \longrightarrow \overline{\Sigma}(M, \mathscr{D})).$$

Let

$$\overline{\gamma}(M, \mathscr{D}) := \overline{\gamma}(M)_{\mathscr{D}} : Conf(\Lambda(M), \mathscr{D}) \simeq Conf(\overline{g}(M), \mathscr{D})$$

be the induced isomorphism.

It results by construction of $\overline{\gamma}(M)$, theorem 3.30, the Birational Criterium of minimality, and the fact that $\overline{\gamma}(M)$ is contained in $Env(D, D')$ the following:

Theorem 7.9 *Keep the notation of theorem 3.30 then:*

1. The following commutativity relation holds:

$$\pi(M) = \mathscr{E}(M) \circ \overline{\gamma}(M),$$

(resp.

$$\pi(M, \mathscr{D}) = \mathscr{E}(M, \mathscr{D}) \circ \overline{\gamma}(M, \mathscr{D})),$$

where

$$\mathscr{E}(M) = (\mathscr{E}_1(M), \mathscr{E}_2(M)) : Conf(\overline{g}(M)) \longrightarrow Drap(k^{r+1}) \times Drap(k^{r+1}),$$

associates with a $\overline{g}(M)$-*configuration its left and right extremities. Thus* $\mathscr{E}(M)$ *(resp.* $\mathscr{E}(M, \mathscr{D})$*) factors through* $\overline{\Sigma}(M) \hookrightarrow Drap(k^{r+1}) \times Drap(k^{r+1})$ *(resp.* $\overline{\Sigma}(M, \mathscr{D}) \hookrightarrow Drap(k^{r+1})$*) and the induced morphism*

$$Conf(\overline{g}(M)) \longrightarrow \overline{\Sigma}(M) \ (resp.Conf(\overline{g}(M), \mathscr{D}) \longrightarrow \overline{\Sigma}(M, \mathscr{D}))$$

is a smooth resolution of singularities, i.e. a birational morphism which is denoted by $\mathscr{E}(M) = (\mathscr{E}_1(M), \mathscr{E}_2(M))$ *(resp.* $\mathscr{E}(M, \mathscr{D}) = (\mathscr{E}_1(M, \mathscr{D}), \mathscr{E}_2(M, \mathscr{D}))$ *too if no confusion arises.*

2. *The morphism*

$$\theta(\overline{g}(M)) = \overline{\gamma}(M) \circ \theta(M)$$

(resp.

$$\theta(\overline{g}(M), \mathscr{D})) = \overline{\gamma}(M, \mathscr{D}) \circ \theta(M, \mathscr{D})).$$

defines a section of $\mathscr{E}(M)$ *(resp.* $\mathscr{E}(M, \mathscr{D})$ *) over* $\Sigma(M)$ *(resp.* $\Sigma(M, \mathscr{D})$ *) and an open embedding:*

$$\theta(\overline{g}(M)) : \Sigma(M) \hookrightarrow Conf(\overline{g}(M)),$$

(resp.

$$\theta(\overline{g}(M, \mathscr{D})) : \Sigma(M, \mathscr{D}) \hookrightarrow Conf(\overline{g}(M), \mathscr{D})),$$

whose image $Im \ \theta(\overline{g}(M))$ *(resp.* $Im \ \theta(\overline{g}(M, \mathscr{D})))$ *is schematically dense. The underlying set of points of* $Im \ \theta(\overline{g}(M))$ *(resp.* $Im \ \theta(\overline{g}(M, \mathscr{D})))$ *is given by the set of galleries* $\gamma \in Gall_I(\overline{g}(M))$ *(resp.* $\gamma \in Gall_I(\overline{g}(M), \mathscr{D})$ *) with* $M(\mathscr{E}_1(\gamma), \mathscr{E}_2(\gamma)) = M$ *(resp.* $M(\mathscr{D}, \mathscr{E}_2(\gamma)) = M$ *).*

7.2 Nash smooth resolutions

The Nash typical graph gives rise to a particular smooth resolution of a Schubert variety in $Grass(k^{r+1})$ which majorates the resolution associated with $\Lambda(M)$. The main application of this construction is to furnish a universal smooth resolution of generic Boardman-Thom singularities (cf. [50]). To a relative position matrix $M = M(D, d) = (m_{\alpha\beta}) \in \mathbb{N}^{(\lambda+1)\times 2}$ defined by $D = (H_1 \subset \cdots \subset H_\lambda \subset I_{r+1}) \in Drap(I_{r+1}))$ and $d = (J \subset I_{r+1}) \in Grass(I_{r+1}))$, defining a Schubert cell $\Sigma = \Sigma(M, D) \subset X = Grass_{|J|}(k^{r+1})$, is associated a weighted graph $I(M)$. Let $Conf(I(M), D)$ be the variety of configurations defined by $I(M)$ (cf. Ch. 6). The reader is referred to [13] for details.

Proposition 7.10 *1) There is a birational morphism*

$$\pi(I(M), D) : \ Conf(I(M), D) \longrightarrow \overline{\Sigma}(M, D),$$

defined by $\pi(I(M), D) : (\mathscr{H}_{\alpha\beta})_{(\alpha,\beta)\in I(M)} \mapsto \mathscr{H}_{\lambda+10}$, *and a section* $\sigma(I(M), D)_\Sigma : \mathscr{J} \mapsto \mathscr{M}_{nash}(D, \mathscr{J}) = (k^{H_\alpha} \cap \mathscr{J} \cup k^{H_\beta})$ *inducing an isomorphism:*

$$\sigma(I(M), D)_\Sigma : \Sigma(M, D) \longrightarrow \pi(I(M), D)^{-1}(\Sigma(I(M), D)) \ .$$

2) Let $\mathcal{T}_\Sigma^1 = \mathcal{T}_{\Sigma/k}^1 \subset (\mathcal{T}_X^1)_\Sigma = (\mathcal{T}_{X/k}^1)_\Sigma$ be the tangent module to Σ. Then the pull-back module

$$\pi(I(M), D)_\Sigma^*(\mathcal{T}_\Sigma^1) \subset \pi(I(M), D)_\Sigma^*(\mathcal{T}_X^1)$$

is obtained as the restriction $(\tilde{\mathcal{T}}^1{}_\Sigma)_{\pi(I(M),D)^{-1}(\Sigma)}$ of a locally free submodule $\tilde{\mathcal{T}}^1{}_\Sigma \subset \pi(I(M), D)^*(\mathcal{T}_X)$, where $\pi(I(M), D)_\Sigma$ denotes the fiber of $\pi(I(M), D)$ over $\Sigma(M, D) \subset \overline{\Sigma}(M, D)$.

3) In dual terms there is a locally free submodule:

$$\tilde{i}_{Conf} : \tilde{\eta}_{Conf}^1 \hookrightarrow (\pi(I(M, D)))^*(\Omega_{Grass_{|J|}(k^{r+1})/k}^1) ,$$

so that:

$$(\tilde{i}_{Conf})|_{\pi(I(M,D))^{-1}(\Sigma)} = (\pi(I(M, D)))^*(i_{\Sigma/Grass_{|J|}(k^{r+1})}),$$

where

$$i_{\Sigma/Grass_{|J|}(k^{r+1})} : \eta_{\Sigma/Grass_{|J|}(k^{r+1})}^1 \hookrightarrow \Omega_{Grass_{|J|}(k^{r+1})/k}^1 \otimes_{\mathscr{O}_{Grass_{|J|}(k^{r+1})}} \mathscr{O}_\Sigma$$

denotes the canonical embedding of the co-normal submodule $\eta_{\Sigma/Grass_{|J|}(k^{r+1})}^1$ to Σ, in

$$\Omega_{Grass(k^{r+1})/k}^1 \otimes_{\mathscr{O}_{Grass(k^{r+1})}} \mathscr{O}_\Sigma.$$

The construction of two minimal generalized galleries $\gamma_{nash}(D, d)$ (resp. $\gamma'_{nash}(D, d)) \subset \Delta^{(r)'}$ and two isomorphisms are given

$$Conf(I(M), D) \simeq \hat{\Sigma}(g_{nash}(M), D) \ \left(resp. \ Conf(I(M), D) \simeq \hat{\Sigma}(g'_{nash}(M), D)\right),$$

where $g_{nash}(M) = typ \ \gamma_{nash}(D, d)$ $\left(resp. \ g'_{nash}(M) = typ \ \gamma'_{nash}(D, d)\right)$. Denote by $\Phi_{nash}(D, J) : I(M) \longrightarrow Grass(I_{r+1})$ the monotone mapping whose graph is given by:

$$\mathscr{M}_{nash}(D, J) = (H_\alpha \cap J \cup H_\beta) .$$

Recall that $I(M)$ is ordered by the product order, and that $Grass(I_{r+1})$ is ordered by the inclusion of subsets of I_{r+1}. A **linearly ordered subset** ψ **of** $I(M)$ is a strictly increasing map

$$\psi : \ [\![1, N]\!] \longrightarrow I(M).$$

With any linearly ordered subset ψ of $I(M)$ one may thus associate a chain of some subset of I_{r+1} by simply composing ψ with $\Phi_{nash}(D, d)$, i.e. $\Phi_{nash}(D, d) \circ \psi$ defines a chain of some subset of I_{r+1}.

7.2.1 The Nash minimal gallery associated with a Relative position matrix

Denote by:

1) $\overleftarrow{\square}_{(\alpha,\beta)}$ the cycle of the ordered graph $I(M)$ defined by the square whose vertices are $\{(\alpha,\beta),(\alpha+1,\beta),(\alpha,\beta+1),(\alpha+1,\beta+1)\}$, and counter clock-wise oriented, where $(\alpha,\beta) \in [\![1,\lambda]\!] \times [\![0,\lambda]\!]$;

2) $\overleftarrow{Tr}_{(\alpha,\beta)}$ the cycle defined by the triangle with the set of vertices

$$Vert\ \overleftarrow{Tr}_{(\alpha,\beta)} = \{(\alpha,\beta),(\alpha,\beta+1),(\alpha+1,\beta+1)\}$$

and counter clock-wise oriented.

3) $\overrightarrow{P}_{(1,1)}$ the complete oriented subgraph of $I(M)$ defined by the set of vertices:

$$Vert\ \overrightarrow{P}_{(1,1)} = \{\ (\alpha,\beta) \in I(M) \mid \alpha - \beta = 0\ or\ \alpha - \beta = 1\ \} \cup \{(1,0)\}\ .$$

4) $\overrightarrow{p}_{(1,1)}$ the oriented graph defined by:

$$\overrightarrow{p}_{(1,1)} = \overrightarrow{P}_{(1,1)} + \overleftarrow{Tr}_{(1,0)}$$

$\overrightarrow{P}_{(1,1)}$ may be seen as an oriented path of $I(M)$. The interval $[\![0,\lambda]\!]$ is ordered by the order opposite to its canonical order and $[\![1,\lambda]\!] \times [\![0,\lambda]\!]$ by the lexicographical order. The subset of the set of cycles $(\overleftarrow{\square}_{(\alpha,\beta)})((\alpha,\beta) \in [\![1,\lambda]\!] \times [\![0,\lambda]\!])$ indexed by the (α,β) satisfying $\alpha - \beta > 0$, is ordered totally by the order induced by the lexicographical order of $[\![1,\lambda]\!] \times [\![0,\lambda]\!]$. Given $(\alpha,\beta) \in [\![1,\lambda]\!] \times [\![0,\lambda]\!]$ satisfying $\alpha - \beta > 0$, it is written $(\alpha,\beta)^+$ for the element in this subset following (α,β).
Define recursively two sequences of oriented paths

$(\overrightarrow{P}_{(\alpha,\beta)})\{(\alpha,\beta)\in[\![1,\lambda]\!]\times[\![0,\lambda]\!] \mid \alpha-\beta>0\}$ and
$(\overrightarrow{p}_{(\alpha,\beta)})\{(\alpha,\beta)\in[\![1,\lambda]\!]\times[\![0,\lambda]\!] \mid \alpha-\beta>0\}$ as follows:

1) $\overrightarrow{P}_{(1,0)} = \overrightarrow{P}_{(1,1)} + \overleftarrow{\square}_{(1,0)}$;

2) $\overrightarrow{P}_{(\alpha,\beta)^+} = \overrightarrow{P}_{(\alpha,\beta)} + \overleftarrow{\square}_{(\alpha,\beta)^+}.$

and

1) $\overrightarrow{p}_{(2,1)} = \overrightarrow{P}_{(1,1)} + \overleftarrow{Tr}_{(2,1)}$;

2) $\overrightarrow{p}_{(\alpha,\beta)^+} = \overrightarrow{P}_{(\alpha,\beta)} + \overleftarrow{Tr}_{(\alpha,\beta)^+}.$

Clearly each of the two families of oriented graphs

$(\overrightarrow{P}_{(\alpha,\beta)})\{(\alpha,\beta)\in[\![1,\lambda]\!]\times[\![0,\lambda]\!] \mid \alpha-\beta>0\}$ and

$(\overrightarrow{p}_{(\alpha,\beta)})\{(\alpha,\beta)\in[\![1,\lambda]\!]\times[\![0,\lambda]\!] \mid \alpha-\beta>0\}$, gives rise to a family of chains of I_{r+1}, which are denoted respectively by:

$(D_{(\alpha,\beta)})\{(\alpha,\beta)\in[\![1,\lambda]\!]\times[\![0,\lambda]\!] \mid \alpha-\beta>0\}$ *and* $(d_{(\alpha,\beta)})\{(\alpha,\beta)\in[\![1,\lambda]\!]\times[\![,\lambda]\!] \mid \alpha-\beta>0\}$.

Definition 7.11 *Define the gallery of chains:*

$$\gamma_{nash}^{ch}(D,d) : D \subset D_{(1,1)} \supset d_{(1,1)} \subset D_{(1,0)} \supset d_{(1,0)} \cdots d_{(\alpha,\beta)} \subset$$

$$D_{(\alpha,\beta)+} \supset d_{(\alpha,\beta)+} \cdots d_{(\lambda,1)} \supset D_{(\lambda,0)} \supset d .$$

The gallery $\gamma_{nash}(D,d)$ *is obtained by reducing the image* $\varphi(\gamma_{nash}^{ch}(D,d))$ *to a generalized gallery with strict inclusions, i.e. a non-stammering gallery.*

Write

$$\Phi_{nash}((\mathcal{H}_{\alpha\beta})_{(\alpha,\beta)\in I(M)}) : I(M) \longrightarrow Grass(k^{r+1})$$

for the monotone mapping whose graph is given by $(\mathcal{H}_{\alpha\beta})_{(\alpha,\beta)\in I(M)}$. Clearly $\Phi_{nash}(\mathcal{M}_{nash}(D,J)) = \Phi_{nash}(D,J)$.
Let

$$Conf(I(M), D) \longrightarrow \hat{\Sigma}(g_{nash}(M), D)$$

be given by associating with $(\mathcal{H}_{\alpha\beta})_{(\alpha,\beta)\in I(M)}$ the gallery $\gamma_{nash}((\mathcal{H}_{\alpha\beta}))$ defined by the sequences of flags

$$(\mathcal{D}_{(\alpha,\beta)})\{(\alpha,\beta)\in[\![1,\lambda]\!]\times[\![0,\lambda]\!] \mid \alpha-\beta>0\} \text{ and } (d_{(\alpha,\beta)})\{(\alpha,\beta)\in[\![1,\lambda]\!]\times[\![,\lambda]\!] \mid \alpha-\beta>0\} ,$$

where

$$\mathcal{D}_{(\alpha,\beta)} = \Phi_{nash}((\mathcal{H}_{\alpha\beta})) \circ \overrightarrow{P}_{(\alpha,\beta)} \left(resp. \ d_{(\alpha,\beta)} = \Phi_{nash}((\mathcal{H}_{\alpha\beta})) \circ \overrightarrow{p}_{(\alpha,\beta)}\right) ,$$

following the same pattern as in the definition of $\gamma_{nash}(D,d)$ according to definition 7.11. It is easy to see that this morphism is, in fact, an isomorphism, and thus that $\hat{\Sigma}(g_{nash}(M), D) \longrightarrow \Sigma(M_g, D)$ is a birational morphism. From the **birational criterium of minimality** , and that the flags in γ_{nash}^{ch} belong to $Env(D,d)$, immediatly it results that the generalized gallery $\gamma_{nash}(D,d)$ is minimal.

Chapter 8

The Coxeter complex

A simplicial complex $C(W, S)$ (**The Coxeter Complex**) is associated with a Coxeter system (W, S), given solely in terms of (W, S), which is naturally endowed with a building structure. This complex admits a **canonical geometrical realization** in an euclidean space, as a decomposition of this space in simplicial cones, by means of a finite family of hyperplanes. The complex $C(W, S)$ given by the Weyl group W of a system of roots R of a complex semi-simple Lie Algebra endowed with the set of generating reflexions S, given by a system of simple roots, is a typical example of a Coxeter System. Its geometrical realization is obtained in the dual space of the real vector space generated by the set of simple roots, by means of the hyperplanes defined by the roots. It is explained how a **combinatorial realization** is provided by the set $A(R)$ of parabolic subsets of R. As an example remark that the combinatorial realization of the Coxeter Complex $C(\mathfrak{S}_{r+1}, S)$ of the general linear group $Gl(k^{r+1})$ is given by $\Delta^{(r)'} = Drap(I_{r+1})$. It is recalled that there is a correspondence $D \mapsto R_D$ associating to a combinatorial flag D a parabolic set of roots in $R(I_{r+1}) = I_{r+1} \times I_{r+1}/\Delta(I_{r+1})$. The interest of the combinatorial realization is that it is directly connected with the Tits geometry associated with the building, and appears as a Galois geometry of the characteristic equation of a generic element of the Lie algebra acting by the adjoint action (cf. [4], Note historique).

8.1 Apartment associated to a Coxeter system

Definition 8.1 *A finite Coxeter system (W, S) is the pair of a finite group W, a set of generators S of W, and a symmetric matrix $(m(s, s'))_{s.s' \in S}$ with integral coefficients $m(s, s')$, such that $m(s, s) = 1$, $m(s, s') \geqslant 2$ if $s \neq s'$, such that the group W is generated by the set S with the relations $(ss')^{m(s,s')} = 1$.*

Denote by $W_X \subset W$ the subgroup generated by X. For $X, Y \subset S$, one has the important properties $W_{X \cap Y} = W_X \cap W_Y$ and $W_X \subset W_Y$ (resp. $W_X = W_Y$) if and only if $X \subset Y$ (resp. $X = Y$). To the Coxeter system is associated a building $C(W, S)$ **(The Coxeter Complex)** defined as follows. Write

$$C(W, S) = \coprod_{X \subset S} W/W_X \; ;$$

The facets of $C(W, S)$ are given by the W_X-classes of W and ordered by the opposite ordering to the inclusion between classes. $C(W, S)$ is naturally endowed with a left action of W respecting the order structure. The set A of **vertices** of $C(W, S)$ is given by $A = \coprod_{s \in S} W/W^{(s)}$ and the set of chambers $Ch\, C(W, S)$ is defined, by $Ch\, C(W, S) := (C_w)_{w \in W}$ where

$$C_w = \{wW^{(s)} \mid s \in S\} \; (w \in W)$$

and $W^{(s)} := W_{S-\{s\}}$. The facets of $C(W, S)$ can both be described as $W_X (X \subset S)$-classes and as sets of vertices, $F_{X,w} = \{wW^{(s)} \mid s \in X\}$. Thus a chamber of $C(W, S)$ is a facet of the form $F_{S,w}$. There is a mapping associating with the set of vertices of a a facet its corresponding class:

$$j : F_{X,w} = \{ wW^{(s)} \mid s \in X\} \mapsto \bigcap_{s \in X} wW^{(s)} \; .$$

It is observed that $\bigcap_{s \in X} wW^{(s)} = wW_{S-X}$, i.e. $j(F_{X,w}) = j'(F_{X,w'})$ $\iff w^{-1}w' \in W_{S-X}$.
Denote by

$$j : facets \; of \; C(W, S) \longrightarrow \coprod_{X \subset S} W/W_X$$

the induced bijection. As a particular case one has $C_w = C'_w$ if and only if $w = w'$. Thus the set of chambers $Ch\, C(W, S) = W/W_\emptyset = W$ is **principal homogeneous under** W.

Notation 8.2 *Given two chambers C and C' let $w(C, C') \in W$ be the unique $w \in W$ defined by $w(C) = C'$.*

The complex $C(W, S)$ is in fact an **Apartment**. Remark that two differents chambers C_w and $C_{w'}$ contain a codimension one common facet F if there exists $s \in S$ such that $w' = ws$. More precisely it may be supposed that $C = C_{1_w} = \{W^{(s_1)}, \cdots, W^{(s_d)}\}$ and $F = \{W^{(s_1)}, \cdots, W^{(s_{d-1})}\}$. Write $X = \{s_d\}$. Given $C' = C_w = \{wW^{(s_1)}, \cdots, wW^{(s_d)}\}$ such that $F \subset C \cap C'$ it follows that $W_X = wW_X$, i.e. $w \in W^{(s_1)} \cap \cdots \cap W^{(s_{d-1})} = W_X$, thus $w = s_d$, and that the codimension 1 facet F is contained in exactly two chambers.

By the above bijection the inclusion of facets corresponds to the opposite of the inclusion relation. Denote a facet F by $F_{\overline{w}}$ if $j(F) = \overline{w}$. Two facets $F_{\overline{w}}$ and $F_{\overline{w}'}$ are **incident**, i.e. contained in a chamber, if $\overline{w} \cap \overline{w}' \neq \emptyset$. Thus $\overline{w} \cap \overline{w}'$ is a class.

The quotient set $C(W,S)/W$ of the set of facets of $C(W,S)$ by the action of W can be identified with the combinatorial simplex $\mathcal{P}(S)$, which can be viewed as a building in an obvious way. The quotient mapping $C(W,S) \longrightarrow C(W,S)/W$ is a building morphism.

A set of vertices $\{wW^{(s_1)}, \ldots, wW^{(s_d)}\} \subset A = \coprod_{s \in S} W/W^{(s)}$ pairwise incident has an upper bound and defines a facet F of $C(W,S)$. Thus the set of vertices A endowed with the incidence relation "generates" the apartment $C(W,S)$. This is proved using the following relation between subgroups of W:

$$W_X.(W_Y \cap W_Z) = W_X.W_Y \cap W_X.W_Z.$$

One calls $C(W,S)$ a **Flag Complex** (see [50], 1.2.1., for details).

8.1.1 Combinatorial hyperplanes

Let $T \subset W$ the set of conjugates of S in W. This set is precisely the set of order 2 elements of W (cf. [4], Ch V, 3.1) and it plays an important role in what follows.

Definition 8.3 *For any $t \in T$, the* **wall (resp. combinatorial hyperplane)**
$$L_t \subset C(W,S) = \coprod_{X \subset S} W/W_X$$
is defined as the subcomplex formed by set of classes invariant by t. Denote by $\mathcal{H}$ the set of walls of $C(W,S)$. This set is naturally indexed by the set of order two elements

A codimension one facet F of $C(W,S)$ is invariant by an order two element t if and only if $j(F) = wW_{\{s\}}$ with $t = wsw^{-1}$. It follows that, for any codimension one facet F of $C(W,S)$, there is only one wall L_t which contains F. L_t is called the **carrier** of F.

More generally the **carrier of a facet** F is defined as the intersection of subcomplexes:
$$\cap_{F \subset L_t} L_t.$$

8.1.2 Type of a facet

Definition 8.4 *The type quotient mapping* $typ : C(W,S) \longrightarrow C(W,S)/W$ *is given in terms of* **vertices** *by:*
$$typ : F = \{\, wW^{(s)} \mid s \in Y \,\} \mapsto Y \,.$$

Thus in terms of **classes***:*
$$typ : C(W,S) = \coprod_{X \in \mathcal{P}(S)} W/W_X \longrightarrow C(W,S)/W = \mathcal{P}(S) \,,$$

is defined by $\overline{w} \in W/W_X \mapsto S - X$. $typ(F)$ is called the **type of** F.

Preceding definitions concerning buildings apply to the simplex $\mathcal{P}(S)$. Thus one can speak of generalized galleries of $typ(C(W,S)) = \mathcal{P}(S)$. The image $typ(\gamma)$ of a generalized gallery γ of $C(W,S)$ is a generalized gallery of types, i.e. $typ(\gamma) \in gall_{C(W,S)} = Gall_{\mathcal{P}(S)}$.

If the facets F and F' are incident to the same chamber C and $typ(F) = typ(F')$, one has $F = F'$. Denote by $(F_Y)_{Y \in \mathcal{P}(S)}$ the set of facets incidents to $C_{1_w} = \{ W^{(s)} \mid s \in S \}$. Given $F \in C(W,S)$, $W_F = Stab\ F$ (**Stabilizer of the facet F in W**) is written. There is then $W_{F_Y} = Stab\ F_Y = W_{S-Y}$.

Notation 8.5
Write $C(W,S)$ in terms of facets:

$$facets\ of\ C(W,S) = \coprod_{t \in typ(C(W,S))} \mathfrak{F}_t \simeq \coprod W/W_X (X \subset S)\ ,$$

where $\mathfrak{F}_t = typ^{-1}(t)$, and $typ(C(W,S)) = \mathcal{P}(S)$. The bijection is given by j. The inverse map of j is given by $\bar{w} \mapsto F_{\bar{w}}$. If F is a facet whose type is given by $Y \subset S$ then $F = w(F_Y)$ where w is a representative of the class $j(F) \in W/W_{S-Y}$.

8.2   Words and galleries

The correspondence between galleries in $C(W,S)$ and words in S is established as follows. A **length n word** in S is by definition a product expression $f(1)f(2) \ldots f(n)$ in W given by a function $f : [\![1,n]\!] \to S$.

Definition 8.6 *A* **chamber generalized gallery** *is a sequence $\Gamma : C_0 \supset F_1 \subset C_1 \ldots F_{n-1} \subset C_{n-1} \supset F_n \subset C_n$, where $C_0, C_1, \ldots, C_n$ are chambers and $F_1, F_2, \ldots, F_n$ are facets of $C(W,S)$. An* **injective gallery** *verifies by definition $C_i \neq C_{i+1} (0 \leqslant i \leqslant n)$ (cf. [4]).*

Thus an injective gallery $\Gamma : C_0, C_1, \ldots, C_n$ gives rise to a chamber generalized gallery

$$\bar{\Gamma} : C_0 \supset F_1 \subset C_1 \ldots F_n \subset C_n\ ,$$

where $F_{i+1} = C_i \cap C_{i+1}$ is the common codimension 1 facet to C_i and C_{i+1}. Let $\Psi^*(\Gamma) = (L_1, \ldots, L_n)$ (**The walls crossed by Γ**) *be the sequence of walls given by the sequence of carriers of the of the codimension 1 facets $(F_1, \ldots, F_n)$ of the chamber generalized gallery $\bar{\Gamma}$. Let t_i be the reflection defined by the carrier L_i of F_i. Write $\Psi(\Gamma) = (t_1, \ldots, t_n) \in \prod^{n} W$ (*The sequence of* **reflections** associated *to Γ*)).

Given a gallery $\Gamma : C_0, C_1, \ldots, C_n$, not necessarily injective, define a sequence $t'_1, \ldots, t'_n$ of elements of $T \cup \{1\} \subset W$ as follows. If $C_i = C_{i+1}$ then $t'_{i+1} = 1$, otherwise $t'_{i+1} = t_{i+1}$. It is clear that $w(C_0, C_n) = t'_n \ldots t'_1$. Now from $t'_1, \ldots, t'_n$ a sequence $s'_1, \ldots, s'_n \in S \cup \{1\}$ is recursively defined. Write $s'_1 = t'_1$ and

$$s'_{j+1} = (s'_1 \ldots s'_j)^{-1} \, t'_{j+1}(s'_1 \ldots s'_j)(1 \leqslant j \leqslant n-1).$$

One then has $t'_j t'_{j-1} \ldots t'_1 = s'_1 s'_2 \ldots s'_j (1 \leqslant j \leqslant n)$, and as a special case $w(C_0, C_n) = s'_1 s'_2 \ldots s'_n$. Finally one can associate with the chamber generalized gallery Γ the word $s'_1 s'_2 \ldots s'_n$ in $S \cup \{1\}$. If Γ is injective this is a word in S, i.e. $s'_1, s'_2, \ldots s'_n \in S$.

Conversely, to any word $s_1 s_2 \ldots s_n$ in S, a sequence $(t_1, \ldots, t_n)$ of T may be asociated as follows:

$$t_1 = s_1, \quad t_{j+1} = (s_1 \ldots s_j) s_{j+1} (s_1 \ldots s_j)^{-1} (1 \leqslant j \leqslant n-1).$$

One then has an injective gallery Γ of $C(W, S)$ associated to the word $s_1 s_2 \ldots s_n$ defined by $C_0 = C_e = \{W^{(s)} \mid s \in S\}$, $\Gamma : C_1 = t_1(C_0), C_2 = t_2 t_1(C_0), \ldots, C_n = t_n \ldots t_2 t_1(C_0)$ (resp. $C_1 = t_1(C_0), C_2 = t_2(C_1), \ldots C_n = t_n(C_{n-1})$).

If the construction explained before is applied to Γ the word $s_1 s_2 \ldots s_n$ is obtained. The gallery Γ may be seen as a path in the **Cayley complex** (cf. [18], 3.3) associated to (W, S). The same construction may be carried out applied to any word $s'_1 \ldots s'_n$ in $S \cup \{1\}$. In this case a gallery which is not injective Γ' may be obtained.

Notation 8.7
If $w = s_1 s_2 \ldots s_n$ is a word in S one writes $\Phi(s_1, s_2 \ldots, s_n) = (t_1, \ldots, t_n)$ with $t_1, \ldots, t_n$ defined as above.

Given a gallery $\Gamma : C_0, C_1, \cdots, C_n$, and $w \in W$ denote by $w\Gamma$ the gallery given by $w(C_0)), w(C_1), \cdots, w(C_n)$. The group W operates naturally by $\Gamma \mapsto w\Gamma$ on the set of galleries Γ of $C(W, S)$. One has then $\Psi(w\Gamma) = (wt_1 w^{-1}, \ldots, wt_n w^{-1})$.

8.3 Combinatorial roots

The root subcomplexes of $C(W, S)$ are defined following [4], Ch. VI, 1.6.

Definition 8.8
Let $t \in T$, and L_t the hyperplane defined by t. To t is associated an equivalence relation on the set of chambers Ch $C(W, S)$. Given an injective gallery $\Gamma :$ $C' = C_0, \ldots, C_n = C''$ between C' and C'' we define the integer $n(\Gamma, t)$ as the number of times that $t \in T$ appears in $\Psi(\Gamma)$. The parity of this number depends only on C' and C'' and not on Γ. Write $\eta(C', C'', t) = (-1)^{n(\Gamma, t)}$ and define the relation $\eta(C', C'', t) = 1$, between the couple (C', C'').
This is an equivalence relation with two equivalence classes, namely $\mathrm{Ch}^+(t)$ and $\mathrm{Ch}^-(t)$, where $\mathrm{Ch}^+(t)$ is the class defined by $C_e = \{W^{(s)} \mid s \in S\}$.

*Remark that $\eta(C', C'', t) = 1$ if no minimal gallery Γ between C' and C''
crosses L_t, i.e. $t \notin \Psi(\Gamma)$.*

*Write $A^+(t) = \cup C(C \in \text{Ch}^+(t))$ (resp. $A^-(t) = \cup C(C \in \text{Ch}^-(t))$. More
explicitly $A^+(t)$ (resp. $A^-(t))$ denotes the subcomplex formed by the set of
facets incidents to some chamber $C \in \text{Ch}^+(t)$ (resp. $C \in \text{Ch}^-(t))$. It is
immediate that $A^+(t)$ (resp. $A^-(t))$ is a sub-building of $C(W, S)$. Each of
the subsets $A^+(t), A^-(t) \subset C(W, S)$ is called a **half** of $C(W, S)$ (resp. **half
apartment**) or a **combinatorial root** of $C(W, S)$. One has $C(W, S) =
A^+(t) \cup A^-(t)$, L_t is said to be the hyperplane given by $A^+(t)$ (resp. $A^-(t))$.
One has $L_t = A^+(t) \cap A^-(t)$ and $t(A^+(t)) = A^-(t)$ (cf. [4], IV, Ex 16), and
one writes $\partial A^+(t) = \partial A^-(t) = L_t$. Denote by Φ a subcomplex of the form
$A^+(t)$ (resp. $A^-(t))$, i.e. a root of $C(W, S)$, and write $\partial \phi = L_t$.*

The group W acts naturally on the set of roots of $C(W, S)$. To make this
action explicit the following bijection is used ι : set of roots of $C(W, S) \to
\{+1, -1\} \times T$ sending $A^+(t)$ (resp. $A^-(t))$ to the couple $(+1, t)$ (resp. $(-1, t))$.

Let Φ be a root of $C(W, S)$ with $\iota(\Phi) = (\varepsilon, t)$. For $w \in W$ one writes
$\eta(w, t) = \eta(C, w(C), t)$. One then has $\iota(w\Phi) = (\varepsilon \cdot \eta(w^{-1}, t), wtw^{-1})$ (cf. [4],
VI, §1, 1.6).

Remark 8.9 *A wall L_t (resp. root Φ) of $C(W, S)$ is a subset of the set of
vertices $\coprod_{s \in S} W/W^{(s)}$ endowed with a subcomplex structure. A facet F of L_t
(resp. root Φ) is thus a subset of L_t (resp. root Φ). Sometimes one interprets
L_t (resp. root Φ) as the set of facets (resp. classes) of $C(W, S)$ contained
in L_t (resp. root Φ). In this way L_t (resp. root Φ) becomes a subset of
$\coprod_{X \in \mathcal{P}(S)} W/W_X$.*

Definition 8.10 *It is said that the wall L_t separates the two facets F and F'
if $F \notin L_t$ and $F' \notin L_t$ and if $F \in \Phi_1$ and $F' \in \Phi_2$, where Φ_1 and Φ_2 are the
two combinatorial roots defined by the wall L_t. Denote by $\mathcal{H}(F, F')$ the **set
of walls separating** F and F'.*

8.4   Convex subcomplexes

Definition 8.11 *A subcomplex K of $C(W, S)$ is **convex** if it is an intersection of combinatorial roots,*

$$\text{i.e. } K = \bigcap_{i \in I} \Phi_i \text{ where } (\Phi_i)_{i \in I} \text{ is a family of roots of } C(W, S)$$

(cf. [50], 2.19).

Remark 8.12 *(i) As the image $f(\Phi)$ of a root Φ by an automorphism f of
$C(W, S)$ is a root of $C(W, S)$ it may be concluded that the image $f(K)$ of a
convex sub-complex K of $C(W, S)$ by f is also a convex sub-complex.*

(ii) A convex sub-complex K containing a chamber C of $C(W,S)$ is a sub-building of $C(W,S)$, i.e. K is the set of all facets F which are incident to the same chamber $C \in K$, and given two chambers C, $C' \in K$ there exists a minimal gallery Γ between C and C' contained in K.

Definition 8.13 *The **convex envelope** of a subset L of facets of $C(W,S)$ is the smallest convex sub-complex Env L containing L. If F and F' are two facets it is written $Env(F, F') = Env\{F, F'\}$. One has $Env(F, F') = \cap\Phi$ (Φ root of $C(W,S)$, $F', F \in \Phi$).*

*The **carrier** $L = L_{(F,F')}$ of $Env(F, F')$ is by definition the intersection sub-complex given by the intersection of walls H containing $Env(F, F)$ (resp. F and F'), i.e. $L = \cap H(H \in \mathcal{H}, Env(F, F') \subset H$ (resp. $F, F' \subset H$)). Remark that the set $\mathcal{H}'$ of walls defined by the roots Φ containing $Env(F, F')$ verifies $\mathcal{H} = \mathcal{H}' \coprod \mathcal{H}(F, F')$.*

Given two chambers C' and C'', and a minimal gallery between C' and C'': $\Gamma : C' = C_0 \ldots C_n = C''$, the set of walls given by $\psi(\Gamma) = (t_1, \ldots, t_n)$, i.e. $L_{t_1}, \ldots L_{t_n}$ may be characterized as the set of walls which separate C' and C'', i.e. as the set $\mathcal{H}(C', C'')$. On the other hand, the **distance** $d(C', C'')$ between C' and C'' in $C(W,S)$ is defined as the length of any minimal gallery between C' and C''. Thus

$$d(C', C'') = \text{cardinal of } \mathcal{H}(C', C'').$$

The following proposition gives a characterization of the set of chambers C belonging to $Env(C', C'')$.

Proposition 8.14 *The following three properties of a chamber C are equivalent: (i) $C \in Env(C', C'')$; (ii) every root Φ containing C' and C'' contains also C; (iii) $d(C', C'') = d(C', C) + d(C, C'')$; (iv) there exists a minimal gallery $\Gamma : C' = C_0, \ldots, C_n = C''$ and there exists $0 \leqslant i \leqslant n$ with $C_i = C$.*

Given two facets F and F', $\text{proj}_{F'} F$ is defined by means of the following

PROPOSITION - DEFINITION 8.15 *There exists $F'' \in Env(F, F')$ verifying $F'' \supset F'$ so that for every $\bar{F}' \supset F'$ with $\bar{F}' \in Env(F, F')$ there is $F'' \supset \bar{F}'$. If $F = C \in Ch\, C(W,S)$ then F'' is given by a chamber contained in $Env(F, F')$. For any minimal gallery $\Gamma : C = C_0 \ldots C_n \supset F'$ between C and F there is $C_n = F''$. Write $F'' = \text{proj}_{F'} F$ (**the projection of F on F'**) (cf. [50], 3.19).*

As a particular case of this definition for $C(\mathfrak{S}_{r+1}, S) \simeq Drap(I_{r+1})$ see definition 2.23.

Example 8.16 *Let $C = \{W^{(s)} \mid s \in S\}, F' = F_{\bar{w}}(\bar{w} \in W/W_X)$, and $w' = w(C, \text{proj}_F C)$, then $\ell(w')$ (length of w' relatively to S) $= d(C, \text{proj}_F C) =$*

$d(C, F')$ *(number of walls separating C_e and F') = cardinal of $\mathcal{H}(C, F')$.*
Every representative w of $\bar{w}$ may be uniquely written as $w = w'w''$, where
$w'' \in W_X$.

8.4.1 Star complex of a facet in the Coxeter Complex

Definition 8.17 *Given a facet F by St_F (The star of F in $C(W, S)$) is de-*
noted the complex formed by the set of all facets $F \subset F'$. The set of chambers
$Ch\ St_F \subset Ch\ C(W, S)$ *is given by the chambers incident to F.*

The following proposition shows that St_F is isomorphic to a Coxeter Complex.

Proposition 8.18 *Let $Y \subset S$.*

1) *$(W_{S-Y}, S - Y)$ is a Coxeter system.*

2) *The set of facets $F_Y \subset F$ corresponds to the set of facets F_Z with $Y \subset Z$*
   (resp. $S - Z \subset S - Y$), and thus $W_{S-Z} \subset W_{S-Y}$. An equivariant
   W_{S-Y}-isomorphism may be thus defined

$$C(W_{S-Y}, S - Y) = \coprod_{Y \subset Z} W_{S-Y}/W_{S-Z} \simeq St_{F_Y}$$

   given by $\overline{w} \mapsto w \cdot F_Z$ for W_{S-Y}/W_{S-Z}.

3) *One has $C(W_F, w(S-Y)w^{-1}) \simeq St_F\ \left(w(F_Y) = F, W_F = W_{w(S-Y)w^{-1}}\right)$.*
   It follows from this that St_F is endowed with a building structure iso-
   morphic to a Coxeter complex, independent of the choice of $w \in \bar{w}(\in$
   $W/W_{S-X}))$.

4) *The set of chambers $Ch\ St_F$ is principal homogeneous under W_F, and*
   has the property that given a minimal gallery Γ whose extremities are in
   $Ch\ St_F$ then $\Gamma \subset Ch\ St_F$. Thus the distance of two chambers of St_F is
   the same, whether measured in St_F or in $C(W, S)$.

5) *From the natural decomposition $C(W, S) = \coprod \mathfrak{F}_t$ (facets of type t) it*
   results $St_F = \coprod \mathfrak{F}_t \cap St_F$. The set of types of St_F runs on $\mathcal{P}(S - Y)$,
   as it follows from 2).

Let the set of walls containing the facet F be denoted by $\mathcal{H}_F \subset \mathcal{H}$. There
is a bijection:

$$\mathcal{H}_F \simeq \text{walls of the Star complex } St_F.$$

It has only to be proved that each $H \supset F$ defines a reflection s_H conjugate in
W_F to some reflection $s_{H'}$ defined by the boundary wall of a fixed chamber
$C^0 \supset F$.

It is first recalled that the set of chambers Ch St_F is equal to the set of chambers $C \supset F$, and that Ch St_F is principal homogeneous under $W_F =$ Stab F. Given $H \in \mathcal{H}_F$ there is a codimension 1 facet $F' \subset H$ such that $F \subset F'$. Then there is a chamber C containing F' and having H as a boundary wall. There exists $w \in W_F$ so that $w(C) = C^0$, then $w(H)$ is a boundary wall of C^0 and $s_H = w^{-1} \circ s_{w(H)} \circ w$. It is remarked that $w(H) \supset F$ because $w \in W_F$.

If $H \in \mathcal{H} - \mathcal{H}_F$ then $F \not\subset H$ (here the wall H is seen as a subcomplex of $C(W, S)$), and there exists a unique root defined by H containing F which is denoted by $\Phi_{II}(F)$. It is easily seen that Ch $St_F = \mathrm{Ch}(\cap \Phi_{II}(F)(H \in \mathcal{H} - \mathcal{H}_F))$. It is said that Ch St_F is a **convex set of chambers**. By definition that means that if $\Gamma : C_0, \ldots, C_n$ is a minimal gallery in $C(W, S)$ with $C_0, C_n \in$ Ch St_F then $\Gamma \subset St_F$.

If F_1 and F_2 are two facets of St_F denote by $\mathcal{H}_F(F_1, F_2) = \{H \in \mathcal{H}(F_1, F_2) \mid F \subset H\}$ the set of walls **separating** F_1 and F_2 **in** St_F. Then there is:

Lemma 8.19
$$\mathcal{H}_F(F_1, F_2) = \mathcal{H}(F_1, F_2) .$$

Proof *Let $H \in \mathcal{H} - \mathcal{H}_F$. If F' and F'' are two facets of St_F there exist two chambers $C', C'' \in$ Ch St_F so that $C' \supset F'$ and $C'' \supset F''$, as in a building every facet is incident to a chamber. As $C', C'' \in$ Ch St_F it is $C', C'' \in \Phi_H(F)$ and as a consequence H separates neither C' and C'' nor F' and F''.*

8.5 The set of relative position types of a Coxeter Complex

Definition 8.20
Let Relpos $C(W, S)$ be the quotient set $(C(W, S) \times C(W, S))/W$ of $C(W, S) \times C(W, S)$, by the diagonal action of W.

- *The natural decomposition $C(W, S) = \coprod \mathfrak{F}_t \simeq \coprod_{X \subset S} W/W_X$ gives rise to the decomposition*

$$
\begin{aligned}
Relpos = Relpos\, C(W, S) &= \coprod (\mathfrak{F}_t \times \mathfrak{F}_s)/W \ ((t, s) \in typ \times typ) \simeq \\
&\simeq \coprod (W/W_X \times W/W_Y)/W((X, Y) \\
&\in \mathcal{P}(S) \times \mathcal{P}(S)).
\end{aligned}
$$

- *By $\tau : C(W, S) \times C(W, S) \to (C(W, S) \times C(W, S))/W = Relpos\, C(W, S)$ the quotient map is denoted. It is said that the image $\tau(F, F')$ is the **type** of **relative position** of the facets F and F'.*
  Write

$$Relpos_{(t,s)} = Relpos_{(t,s)} C(W, S) = (\mathfrak{F}_t \times \mathfrak{F}_s)/W ,$$

thus there is the decomposition

$$\text{Relpos } C(W, S) = \coprod_{(t,s)} \text{Relpos}_{(t,s)} C(W, S) \ .$$

- Let $F_t \subset C_e$ denote the unique facet of type t contained in in the chamber C_e defined by the identity, and $W_t = \text{Stab } F_t$ for $t \in \text{typ}$. A (W_t, W_s)-**double class** *of W is by definition a W_t-orbit in the set W/W_s. Denote by $W_t \backslash W/W_s$ the set of these double classes.*

There is a bijection $(\mathfrak{F}_t \times \mathfrak{F}_s)/W \simeq \{F_t\} \times (\mathfrak{F}_s/W_t)$ defined by $O \mapsto O'$ where $O' = \{F_t\} \times \{F \in \mathfrak{F}_s \mid (F_t, F) \in O\}$. The reciprocal bijection is given by $O' \mapsto O = \underset{w \in W}{\cup} w(O')$. On the other hand there is a bijection $\{F_t\} \times (\mathfrak{F}_s/W_t) \simeq W_t \backslash W/W_s$ associating with the class of (F_t, F) the W_t-class of $\overline{w} \in W/W_s$, where $\overline{w}$ is given by $j(F) = \overline{w}$, with $(F_t, F) \in O'$. By composition a bijection is obtained

$$\text{Relpos}_{(t,s)} C(W, S) = \mathfrak{F}_t \times \mathfrak{F}_s/W \simeq W_t \backslash W/W_s$$

(cf. [23], *Exp XXVI*, 4.5.3.).

8.6 Geometrical realization of the Coxeter complex

8.6.1 Representation of a Coxeter system as a group of reflections in euclidean space

From the **geometrical representation** of the Coxeter system (W, S) (cf. [4], *Ch V*) one deduces a **geometrical realization** of the Coxeter complex $C(W, S)$, given by the simplicial cones of a suitable decomposition of some real affine euclidean space $\mathbb{A}$, and generalizing that of $C(\mathfrak{S}_{r+1}, S) \simeq \Delta^{(r)'}$. This decomposition is defined by means of a set of hyperplanes $\mathcal{H}$ of $\mathbb{A}$.

The geometrical representation of a finite Coxeter system (W, S) is described (cf. loc. cit., 4.). Let $(m(s, s'))_{s,s' \in S}$ be the Coxeter matrix (a symmetric matrix of integers verifying $m(s, s) = 1$ and $m(s, s') \geqslant 2$ if $s \neq s'$) giving W as a group with S as the set of generators and $(ss')^{m(s,s')} = 1$ as the set of defining relations. Let $\mathbb{R}^{(S)}$ be the $\mathbb{R}$-vector space with $(e_s)_{s \in S}$ as the canonical basis. Let $B(x, y)$ be the **symmetric bilinear form** defined by $B(e_s, e_s) = 1, B(e_s, e_{s'}) = -\cos(\pi/m(s, s'))$. This bilinear form is a scalar product of $\mathbb{R}^{(S)}$ which is denoted by $(x|y)$. Let $O(\mathbb{R}^{(S)})$ be the orthogonal group defined by $(x|y)$.

Denote by H_s the hyperplane orthogonal to e_s and by σ_s the orthogonal reflection defined by H_s.

Proposition 8.21 *There exists an injective homomorphism $\sigma : W \to O(\mathbb{R}^{(S)})$ defined by $\sigma(s) = \sigma_s$.*

Identify W with a subgroup of $O(\mathbb{R}^{(S)})$. Thus W is endowed with the canonical set of generators $(\sigma_s)_{s \in S}$. The couple

$$(W, (\sigma_s)_{s \in S}) = (\sigma(W), (\sigma_s)_{s \in S})$$

is called the **geometrical representation of** (W, S).

Let $\mathcal{H} = (w(H_s))_{s \in S, w \in W}$ be the set of hyperplanes of $\mathbb{R}^{(S)}$ obtained as the images of the set of hyperplanes $(H_s)_{s \in S}$ by the elements of W. The set $\mathcal{H}$ may be characterized as the set of hyperplanes H of $\mathbb{R}^{(S)}$ whose associated orthogonal reflection s_H belongs to W (**reflection hyperplanes**). This set corresponds to the set of hyperplanes (L_t) of $C(W, S)$. It is observed that the same notation $\mathcal{H}$ is used for the set of walls of $C(W, S)$ and for the set of reflection hyperplanes of $W \subset O(\mathbb{R}^{(S)})$ as there is a natural bijection between these two sets.

8.6.2 Coxeter system representation space simplicial decomposition

Definition 8.22 *Let $\mathbb{A}$ be a real affine space, $a \in \mathbb{A}$, $e_1, \ldots, e_d$ a basis of the translation vector space of $\mathbb{A}$, and $I \coprod J = \{1, \ldots, d\}$ a decomposition of the basis indexes. Any point $x \in \mathbb{A}$ may be written uniquely as $x = a + \alpha_1 e_1 + \cdots + \alpha_d e_d$ with $\alpha_1, \ldots, \alpha_d \in \mathbb{R}$. The set of $x \in \mathbb{A}$ defined by $\alpha_i = 0$ $(i \in I)$, and $\alpha_j > 0$ $(j \in J)$ is by definition a **simplicial cone**.*

Definition 8.23 *The set of reflection hyperplanes $\mathcal{H}$ endows $\mathbb{R}^{(S)}$ with an equivalence relation defined by:*

"$x \sim y$ if no hyperplane $H \in \mathcal{H}$ separates x and y".

*(It is said that H separates x and y if x and y belong to different connected components of $\mathbb{R}^{(S)} - H$.) The equivalence classes defined by this equivalence relation are simplicial cones of $\mathbb{R}^{(S)}$ which are called the **facets** defined by $\mathcal{H}$. The open facets in $\mathbb{R}^{(S)}$ are called the **chambers defined by** $\mathcal{H}$. Denote by $\mathcal{C}(\mathcal{H})$ the set of facets and by $\mathrm{Ch}\,\mathcal{C}(\mathcal{H})$ the set of chambers.*

Via the formula $s_{w(H)} = w \circ s_H \circ w^{-1}$ $(w \in W)$ the group W acts naturally on the set $\mathcal{H}$. It follows then that W operates on the set of facets $\mathcal{C}(\mathcal{H})$. If $\mathcal{F} \in \mathcal{C}(\mathcal{H})$ it holds that $w(\mathcal{F}) = \mathcal{F}$ if and only if w fixes any point in $\mathcal{F}$. One has:

Proposition 8.24 *The set $\mathrm{Ch}\,\mathcal{C}(\mathcal{H})$ is equal to the connected components of $\mathbb{R}^{(S)} - \underset{H \in \mathcal{H}}{U}\, H$, and is principal homogenous under W.*

Remark 8.25 *The definition of $\mathcal{C}(\mathcal{H})$ (resp. $\mathrm{Ch}\,\mathcal{C}(\mathcal{H})$ may be stated with any finite (resp. locally finite) family $\mathcal{H}$ of hyperplanes in any real affine space. The facets are simplicial cones. This fact will be used later.*

The **carrier** subspace $L_{\mathcal{F}} \subset \mathbb{R}^{(S)}$ of a facet $\mathcal{F}$ is defined by $L_{\mathcal{F}} = \bigcap_{H \in \mathcal{H}, \mathcal{F} \subset H} H$. The facet $\mathcal{F}$ is an open subset in the subspace $L_{\mathcal{F}}$, and then $L_{\mathcal{F}}$ may be characterized as the smallest subspace L of $\mathbb{R}^{(S)}$ containing $\mathcal{F}$. By definition **the codimension of** $\mathcal{F} \in \mathcal{C}(\mathcal{H})$ is equal to the codimension of its carrier $L_{\mathcal{F}}$ in $\mathbb{R}^{(S)}$. The closure $\bar{\mathcal{F}}$ of a facet $\mathcal{F}$ is equal to the union $\bigcup \mathcal{F}'$ of all facets $\mathcal{F}'$ contained in $\bar{\mathcal{F}}$.

The closure $\bar{C}$ of a chamber contains exactly $\ell = |S|$ codimension 1 facets. A **bounding hyperplane** H of a chamber C is by definition the carrier of a codimension 1 facet $\mathcal{F} \subset \bar{C}$. Let $C^0 = \{x \in \mathbb{R}^{(S)} \mid (e_s|x) > 0\}$; C^0 is a chamber defined by $\mathcal{H}$. The set of bounding hyperplanes of C^0 is given by $(H_s)_{s \in S}$, and as a corollary one has that W is generated by the set of orthogonal reflections $(s_{H_s})_{s \in S}$ $(s_{H_s} = \sigma_s)$.

From the fact that the set of chambers $Ch\ \mathcal{C}(\mathcal{H})$ is principal homogenous under W it results that, if C is a chamber and $S_C = (s_H)$, where H runs on the bounding hyperplanes of C, then (W, S_C) is a Coxeter system canonically isomorphic to $(W, (\sigma_s)_{s \in S})$.

8.6.3 Facets viewed as simplicial cones

The following proposition shows that there is a bijection between the set of simplicial cones $\mathcal{C}(\mathcal{H})$ and the set of facets of the Coxeter complex $C(W, S) = \coprod_{X \in \mathcal{P}(S)} W/W_X$, seen as classes in W.

Proposition 8.26

1) *For each facet $\mathcal{F} \in \mathcal{C}(\mathcal{H})$ there exists a unique facet $\mathcal{F}' \subset \bar{C}^0$ and $w \in W$, not necessarily unique, satisfying $w(\mathcal{F}) = \mathcal{F}'$, i.e. the closure $\bar{C}^0$ of C^0 is a **fundamental domain** for the action of W on $\mathbb{R}^{(S)}$.*

2) *The set of facets $\mathcal{F} \subset \bar{C}^0$ is indexed by $\mathcal{P}(S)$ and $\mathcal{F}_X$ is written for the facet given by $X \subset S$. More precisely with each subset $X \subset S$ the facet $\mathcal{F}_X \subset \bar{C}^0$ defined by $\overline{\mathcal{F}}_X = (\bigcap_{s \in X} H_s) \cap \overline{C}$ is associated. (Remark that $\bar{\mathcal{F}} = \bar{\mathcal{F}}'$ if and only if $\mathcal{F} = \mathcal{F}'$.) One has $\mathrm{Stab}\ \mathcal{F}_X = W_{S-X}$.*

3) *There is a bijection*

$$C(W, S) \xrightarrow{\sim} \mathcal{C}(\mathcal{H}) \text{ defined by } \bar{w} \mapsto C_{\mathcal{F}} = w(\mathcal{F}_X) \text{ if } \bar{w} \in W/W_{S-X}.$$

Thus $F = \{wW^{(s)} | s \in X\} \mapsto w(\mathcal{F}_X)$. By this bijection the chamber C_e corresponds to C^0. This is the unique type preserving equivariant bijection between $C(W, S)$ and $\mathcal{C}(\mathcal{H})$.

4) *By this bijection the relation of inclusion between facets $F \subset F'$ in $C(W, S)$ becomes $\mathcal{F} \subset \bar{\mathcal{F}}'$, and thus it introduces a natural structure of building on the set $\mathcal{C}(\mathcal{H})$.*

5) *The set $\mathcal{H}$ of walls of $C(W,S)$ corresponds to the set of hyperplanes H of $\mathbb{R}^{(S)}$ with $s_H \in W$ which we also denote by $\mathcal{H}$.*

Following the general definition a **subcomplex** $\mathcal{C}'$ of $\mathcal{C}(\mathcal{H})$ is by definition a set of facets of $\mathcal{C}(\mathcal{H})$ such that $\mathcal{F} \in \mathcal{C}'$ and $\mathcal{F}' \subset \bar{\mathcal{F}}$ implies $\mathcal{F}' \in \mathcal{C}'$. With the set of combinatorial roots of $C(W,S)$ corresponds the following set of subcomplexes of $\mathcal{C}(\mathcal{H})$.

Definition 8.27
*Given $H \in \mathcal{H}$ write $\bar{D}_H^+$ (resp. $\bar{D}_H^-$) for the closed half-space of $\mathbb{R}^{(S)}$ defined by H and containing C^0 (resp. not containing C^0). Let $\mathcal{C}_H^+$ (resp. $\mathcal{C}_H^-$) be the subcomplex of $\mathcal{C}(\mathcal{H})$ given by the set of facets $\mathcal{F}$ contained in $\overline{D}_H^+$ (resp. $\overline{D}_H^-$). Then there is $\overline{D}_H^+ = \underset{\mathcal{F} \in \mathcal{C}_H^+}{\cup} \mathcal{F}$ (resp. $\overline{D}_H^- = \underset{\mathcal{F} \in \mathcal{C}_H^-}{\cup} \mathcal{F}$), and $\mathcal{C}_H = \mathcal{C}_H^+ \cap \mathcal{C}_H^-$, where $\mathcal{C}_H$ is the subcomplex given by the facets contained in H. A **closed half space subcomplex** of $\mathcal{C}(\mathcal{H})$ is a subcomplex of the form $\mathcal{C}_H^+$ (resp. $\mathcal{C}_H^-$). If $H \in \mathcal{H}$ and $A \neq \emptyset$ is a set contained in $\mathbb{R}^{(S)} - H$ write $\bar{D}_H(A)$ for the closed half space subcomplex defined by H and A.*

The bijection $C(W,S) \xrightarrow{\sim} \mathcal{C}(\mathcal{H})$ induces a complex isomorphism

$$A^+(t) \simeq \mathcal{C}_H^+ \ (\text{resp. } A^-(t) \simeq \mathcal{C}_H^-) \text{ where } t = s_H.$$

$\mathcal{C}_H^+$ (resp. $\mathcal{C}_H^-$) is called the **positive root** (resp. **negative root**) defined by H. If no confusion arises write H (resp. $\bar{D}_H^+, \bar{D}_H^-$) for $\mathcal{C}_H$ (resp. $\mathcal{C}_H^+, \mathcal{C}_H^-$).

8.6.4 Simplicial decomposition of a Convex Hull

Definition 8.28
*A **convex subcomplex** K of $\mathcal{C}(\mathcal{H})$ is the intersection subcomplex $K = \underset{i \in I}{\cap} \bar{D}_i$, defined by a family of closed half space subcomplexes $(\bar{D}_i)_{i \in I}$, i.e. an intersection of root subcomplexes of $\mathcal{C}(\mathcal{H})$.*

The class of convex subcomplexes of $\mathcal{C}(\mathcal{H})$ and the class of convex subcomplexes of $C(W,S)$ are clearly in bijection.

Definition 8.29
*The **Convex Envelope** (resp. **Convex Hull**) $Env(\mathcal{F}, \mathcal{F}') \subset \mathcal{C}(\mathcal{H})$ of the facets $\mathcal{F}$ and $\mathcal{F}' \in \mathcal{C}(\mathcal{H})$ is the smallest convex subcomplex of $\mathcal{C}(\mathcal{H})$ containing $\mathcal{F}$ and $\mathcal{F}'$, i.e. the intersection (subcomplex) of all the roots of $\mathcal{C}(\mathcal{H})$ containing $\mathcal{F}$ and $\mathcal{F}'$. Alternatively $Env(\mathcal{F}, \mathcal{F}')$ is the set of all facets $\mathcal{F} \in \mathcal{C}(\mathcal{H})$ contained in the intersection $\cap \bar{D}$ of all the closed half spaces $\bar{D}$ of $\mathbb{R}^{(S)}$ defined by $H \in \mathcal{H}$ and containing $\mathcal{F}$ and $\mathcal{F}'$. As $Env(\mathcal{F}, \mathcal{F}')$ is an intersection of roots, the underlying set of points of $Env(\mathcal{F}, \mathcal{F}')$ is given by*

$$Env(\mathcal{F}, \mathcal{F}') = \underset{\mathcal{F}'' \in \mathcal{C}(\mathcal{H}), \mathcal{F}'' \subset Env(\mathcal{F}, \mathcal{F}')}{\cup} \mathcal{F}''.$$

Let $L = \bigcap_{i \in I} H_i$ be the intersection subspace of $\mathbb{R}^{(S)}$ defined by a family of hyperplanes $(H_i)_{i \in I} \subset \mathcal{H}$. Write $\mathcal{C}_L = \{\mathcal{F} \in \mathcal{C}(\mathcal{H}) \mid \mathcal{F} \subset L\}$ for the set of facets contained in L. Clearly $L = \bigcup_{\mathcal{F} \in \mathcal{C}_L} \mathcal{F}$. Let $\mathcal{F}' \subset \overline{\mathcal{F}}$ be a facet contained in the closure $\overline{\mathcal{F}}$ of a facet $\mathcal{F} \in \mathcal{C}_L$ then $\mathcal{F}' \in \mathcal{C}_L$. Thus $\mathcal{C}_L$ is a subcomplex of $\mathcal{C}(\mathcal{H})$. Let

$$\mathcal{H}_L = \{H \in \mathcal{H} \mid L \subset H\} \text{ and } \mathcal{H}'_L = \mathcal{H} - \mathcal{H}_L .$$

$\mathcal{C}_L$ may be interpreted as the set of equivalence classes given by the equivalence relation "$\sim$" defined, following [4], *Ch V*, by the family of hyperplanes $\mathcal{H}'_L$ on L as follows:
"$x \sim y$ if for all $H \in \mathcal{H}'_L$ either $x, y \in H \cap L$ or both x and y belong to a same open half space of the two open half spaces defined by $H \cap L$."

   The subcomplex $\mathcal{C}_L \subset \mathcal{C}(\mathcal{H})$ is endowed with a natural building structure. The set of chambers $Ch \; \mathcal{C}_L$ is given by the set of connected components of $L - \bigcup_{H \in \mathcal{H}'_L} H \cap L$. The chambers of $\mathcal{C}_L$ may be characterized as the maximal dimension facets $\mathcal{F}$ (resp. minimal codimension facets $\mathcal{F}$) of $\mathcal{C}$ contained in L. There is then $L - \bigcup_{H \in \mathcal{H}'_L} H \cap L = \bigcup_{C \in Ch \; \mathcal{C}_L} C$ and $L = \bigcup_{C \in Ch \; \mathcal{C}_L} \bar{C}$. This means that every facet in $\mathcal{C}_L$ is contained in the closure of some chamber. In other terms every facet in $\mathcal{C}_L$ is incident to a chamber. Given two chambers C' and C'' there is a gallery $\Gamma : C' = C_0, \dots, C_n = C''$ connecting them. If the points $x \in C'$ and $y \in C''$ are "generic enough" the segment $[x, y]$ intersects only chambers and codimension 1 facets of $\mathcal{C}_L$. Then $\Gamma(C', C'')$ is taken as the ordered set of chambers that $[x, y]$ intersects.

Definition 8.30 *The* **carrier** *subspace* $L = L_{(\mathcal{F}, \mathcal{F}')} \subset \mathbb{R}^{(S)}$ *of the convex hull (resp. convex envelope)* $Env(\mathcal{F}, \mathcal{F}')$ *is defined by* $L = \bigcap_{H \in \mathcal{H} \text{ and } \mathcal{F}, \mathcal{F}' \subset H} H$. *Let* $\mathcal{H}_L$ *be the subset of hyperplanes containing* L.

The sub-building structure of $Env(\mathcal{F}, \mathcal{F}') \subset \mathcal{C}_L$ is examined in detail.

Lemma 8.31 *The carrier* $L_{\mathcal{F}''}$ *of a facet of maximal dimension in* $Env(\mathcal{F}, \mathcal{F}')$ *is equal to the carrier* $L_{(\mathcal{F}, \mathcal{F}')}$ *of* $Env(\mathcal{F}, \mathcal{F}')$.

Proof *As* $L_{\mathcal{F}''} \subset L_{(\mathcal{F}, \mathcal{F}')}$ *it suffices to prove that* $L_{\mathcal{F}''} \supset Env(\mathcal{F}, \mathcal{F}')$. *Otherwise one has either* $\mathcal{F} \not\subset L_{\mathcal{F}''}$ *or* $\mathcal{F}' \not\subset L_{\mathcal{F}''}$. *Then there exists either* $p \in \mathcal{F} - L_{\mathcal{F}''}$ *or* $p \in \mathcal{F}' - L_{\mathcal{F}''}$, *and thus*

   dimension of the convex hull K *of* p *and* $\mathcal{F}'' >$ *dimension of* $\bar{\mathcal{F}}''$.

It follows that $K \subset \bigcup_{\tilde{F} \in Env(\mathcal{F}, \mathcal{F}')} \tilde{F}$ *with dimension* $\tilde{F} \leqslant \dim \mathcal{F}'' < \dim K$, *and this is impossible because* $\cup \tilde{F}$ *is a finite union.*

As a maximal dimensional facet $\mathcal{F}''$ of $\mathrm{Env}(\mathcal{F},\mathcal{F}')$ is open in $L_{(\mathcal{F},\mathcal{F}')}$ one deduces that the set of facets $\mathcal{F} \subset \mathrm{Env}(F,F')$ which are open in $L_{(\mathcal{F},\mathcal{F}')}$ is not empty, and thus that $\overset{\circ}{\mathrm{Env}}(F,F') \neq \emptyset$. On the other hand, one has that $x \in \overset{\circ}{\mathrm{Env}}(\mathcal{F},\mathcal{F}')$ and $y \in \mathrm{Env}(\mathcal{F},\mathcal{F}')$ implies that $[x,y[\subset \overset{\circ}{\mathrm{Env}}(\mathcal{F},\mathcal{F}')$. It results that

$$\text{``}\mathrm{Env}(\mathcal{F},\mathcal{F}') \text{ is the closure of } \overset{\circ}{\mathrm{Env}}(\mathcal{F},\mathcal{F}')\text{''}$$

It is easy to see that a facet $\mathcal{F}''$ contained in $\mathrm{Env}(\mathcal{F},\mathcal{F}')$, such that $\mathcal{F}'' \cap \overset{\circ}{\mathrm{Env}}(\mathcal{F},\mathcal{F}') \neq \emptyset$, is contained in $\overset{\circ}{\mathrm{Env}}(\mathcal{F},\mathcal{F}')$. Let $x \in \mathcal{F}'' \cap \overset{\circ}{\mathrm{Env}}(\mathcal{F},\mathcal{F}')$ and $y \in \mathcal{F}''$. Since $\mathcal{F}''$ is open in its carrier, it results that there is a segment $[x,z[\subset \mathrm{Env}(\mathcal{F},\mathcal{F}')$, with $z \in \mathcal{F}''$, such that $y \in [x,z[$, thus $y \in \overset{\circ}{\mathrm{Env}}(\mathcal{F},\mathcal{F}')$. Thus it is concluded that

$$\overset{\circ}{\mathrm{Env}}(\mathcal{F},\mathcal{F}') = \bigcup_{\mathcal{F}'' \cap \overset{\circ}{\mathrm{Env}}(\mathcal{F},\mathcal{F}') \neq \emptyset} \mathcal{F}'' .$$

Let

$$\mathcal{H}'_{(\mathcal{F},\mathcal{F}')} = \{ \partial\Phi \in \mathcal{H} \mid \mathrm{Env}(\mathcal{F},\mathcal{F}') \subset \Phi \},$$

where $\partial\Phi$ denotes the hyperplane defined by the root Φ, be the set of hyperplanes of $H \in \mathcal{H}$ which **do not separate** the facets $\mathcal{F}$ and $\mathcal{F}'$. One may write

$$\mathcal{H}'_{(\mathcal{F},\mathcal{F}')} = \mathcal{H}_L \coprod \mathcal{H}''_{(\mathcal{F},\mathcal{F}')} ,$$

where $\mathcal{H}''_{(\mathcal{F},\mathcal{F}')} = (\mathcal{H} - \mathcal{H}_L) \cap \mathcal{H}'_{(\mathcal{F},\mathcal{F}')}$ denotes the set of $H \in (\mathcal{H} - \mathcal{H}_L)$ which do not separate $\mathcal{F}$ and $\mathcal{F}'$. Given $H \in \mathcal{H}''_{(\mathcal{F},\mathcal{F}')}$ let $\overline{D}_{H \cap L} = \overline{D}_{H \cap L}(\mathrm{Env}(\mathcal{F},\mathcal{F}')) \subset \mathcal{C}_L$ be the closed half space of L defined by $H \cap L$ and containing $\mathrm{Env}(\mathcal{F},\mathcal{F}')$, and $D_{H \cap L} = D_{H \cap L}(\mathrm{Env}(\mathcal{F},\mathcal{F}')) \subset \overline{D}_{H \cap L}(\mathrm{Env}(\mathcal{F},\mathcal{F}'))$ be the corresponding open half space. There is then (cf. loc. cit.):

1) $\mathrm{Env}(\mathcal{F},\mathcal{F}') = \underset{H \in \mathcal{H}''_L}{\cap} \overline{D}_{H \cap L}$;

2) $\overset{\circ}{\mathrm{Env}}(\mathcal{F},\mathcal{F}') = \underset{H \in \mathcal{H}''_L}{\cap} D_{H \cap L}$;

3) The equality $\overline{\cap D_{H \cap L}} = \cap \overline{D}_{H \cap L}$ gives again **closure** of $\overset{\circ}{\mathrm{Env}}(\mathcal{F},\mathcal{F}') = \mathrm{Env}(\mathcal{F},\mathcal{F}')$.

Remark 8.32 *It results that $\overset{\circ}{\mathrm{Env}}(\mathcal{F},\mathcal{F}')$ is a facet defined by the family of hyperplanes $\mathcal{H}'_{(\mathcal{F},\mathcal{F}')}$ of $\mathbb{R}^{(S)}$, i.e. an equivalence class of the equivalence relation defined by the set of hyperplanes $\mathcal{H}'_{(\mathcal{F},\mathcal{F}')}$ instead of $\mathcal{H}$.*

From 2) it results that $\overset{\circ}{\mathrm{Env}}(\mathcal{F}, \mathcal{F}')$ is the union of facets $\mathcal{F}'' \subset L$ defined by the set of hyperplanes $(H \cap L)_{H \in \mathcal{H}_L''}$ in L $\big($resp. the union of facets $\mathcal{F}'' \in \mathcal{C}(\mathcal{H})$ contained in $\overset{\circ}{\mathrm{Env}}(\mathcal{F}, \mathcal{F}')\big)$. Since every facet $\mathcal{F}$ of $\mathcal{C}_L$ is contained in the closure $\bar{C}$ of some chamber C of $\mathcal{C}_L$, given $\mathcal{F}'' \subset \overset{\circ}{\mathrm{Env}}(\mathcal{F}, \mathcal{F}')$ there exists a maximal dimensional facet $C \subset \overset{\circ}{\mathrm{Env}}(\mathcal{F}, \mathcal{F}')$ such that $\mathcal{F}'' \subset \bar{C}$. It is deduced then that $\overset{\circ}{\mathrm{Env}}(\mathcal{F}, \mathcal{F}') \subset \bigcup\limits_{C \subset \overset{\circ}{\mathrm{Env}}(\mathcal{F}, \mathcal{F}')} \bar{C}$ $(C \in Ch\ \mathcal{C}_L)$. By 3) it is concluded that $\bigcup\limits_{C \subset \overset{\circ}{\mathrm{Env}}(\mathcal{F}, \mathcal{F}')} \bar{C} = \mathrm{Env}(\mathcal{F}, \mathcal{F}')$.

This proves that any facet $\mathcal{F}''$ of $Env(\mathcal{F}, \mathcal{F}')$ is incident to a chamber C of $\mathcal{C}_L$ contained in $Env(\mathcal{F}, \mathcal{F}')$ (resp. $\overset{\circ}{\mathrm{Env}}(F, F')$). If $C', C'' \subset \mathrm{Env}(\mathcal{F}, \mathcal{F}')$ the gallery Γ defined by a segment $[x, y]$, with $x \in C'$ and $y \in C''$, is contained in $\mathrm{Env}(\mathcal{F}, \mathcal{F}')$ as $[x, y] \subset \mathrm{Env}(\mathcal{F}, \mathcal{F}')$, because $\mathrm{Env}(\mathcal{F}, \mathcal{F}')$ is a convex set. We resume all these results in the following

Proposition 8.33 *The subcomplex $Env(\mathcal{F}, \mathcal{F}')$ of $\mathcal{C}_L$ is a sub-building. The chambers of $Env(\mathcal{F}, \mathcal{F}')$ are the facets $\mathcal{F}'' \subset Env(\mathcal{F}, \mathcal{F}')$ which are open in L (resp. whose carrier $L_{\mathcal{F}''}$ is equal to the carrier L of $Env(\mathcal{F}, \mathcal{F}')$). We may characterize $proj_{\mathcal{F}'}\ \mathcal{F}$ as the unique chamber C of $Env(\mathcal{F}, \mathcal{F}')$ such that $C \supset \mathcal{F}'$.*

8.7 Combinatorial Representation of the Coxeter Complex of a root system

An Apartment $A(R)$ is associated with a root system R, formed by its **parabolics subsets** (cf. definition below) as its facets, and an isomorphism with the Coxeter Complex $C(W(R), S) \simeq A(R)$, given by the Coxeter system $(W(R), S)$, where $W(R)$ is the Weyl group of R, and S the reflexions defined by a system of simple roots of R. $A(R)$ is called the **Combinatorial Realization** of $C(W(R), S)$. It is remarked that a parabolic subset is characterized by the set of systems of positive roots (resp. simple roots systems) that it contains. On the other hand, for each type of root system a correspondence may be established between the simple systems of roots and some classical geometrical configurations, thus obtaining a geometrical description of $A(R)$ (cf. [2], [3], [28], and [50]).

It is explained now how the preceding construction of a geometric realization of a Coxeter complex specializes for a $C(W(R), S)$ given by a simple root system. Here [4], Chap. VI is followed.

Definition 8.34 *Let V be an $\mathbb{R}$-vector space, and $R \subset V$ a finite subset verifying:*

1) $0 \notin R$ and R generates V;

2) *there is a map $\alpha \mapsto \alpha^\vee$ from R to the dual space V^* of V, such that $< \alpha, \alpha^\vee >= 2$ ($< x, y >$ denotes the duality pairing on $V \times V^*$), and the reflection*

$$s_\alpha : x \mapsto x - < x, \alpha^\vee > \alpha$$

leaves R stable;

3) *(Crystallographic condition) for all $\alpha \in R$ we have $\alpha^\vee(R) \subset \mathbb{Z}$. It is said that (V, R) is a* **root system**.

4) *Let $Aut(R)$ be the subgroup of the group of automorphisms of V leaving R stable. By definition the Weyl group $W(R)$ of R is the subgroup of $Aut(R)$ generated by the reflections $\{s_\alpha \mid \alpha \in R\}$.*

The image $R^\vee$ in V^* of R by $\alpha \mapsto \alpha^\vee$ defines a root system in the dual space $(V^*, R^\vee)$ (cf. also [23], Definition 1.1.1., and 6.4 of this work). The map $u \mapsto {}^t u^{-1}$ (inverse of the transposed map) is an isomorphism from $W(R)$ to $W(R^\vee)$. These two groups are identified by this isomorphism. It follows from [4], Chap VI, §1, that V^* is endowed with a canonical bilinear form $\Phi_{R^\vee}$, positive, non-degenerate, and $Aut(R)$-invariant. The group $W(R) \subset Aut(R)$ acts on V^* as a group of $\Phi_{R^\vee}$-orthogonal transformations. Write $\Phi_{R^\vee}(x, y) = (x, y)$. The orthogonal transformation s_{H_α} given by the hyperplane $H_\alpha = Ker(\alpha)$ is equal to $s_{\alpha^\vee} : x - < \alpha, x > \alpha^\vee$. From the definition of $W(R)$ follows that $W(R)$ is the group of orthogonal transformations generated by the orthogonal reflections $(s_{H_\alpha})_{H_\alpha \in \mathcal{H}}$.

Let $Q(R) \subset V$ be the subgroup generated by R. It is easily seen that there is at least one ordered group structure on $Q(R)$ (cf. [23], Exp. XXI, Remarque 3.2.7, and [4], Chap VI, Corollaire 2). Thus there is a decomposition

$$R = R_+ \coprod R_-,$$

where R_+ (resp R_-) is the system of positive roots (resp. negative roots) defined by the order. A system of positive roots R_+ is characterized by the decomposition $R = R_+ \coprod -R_+$. To a system of positive roots is associated a particular basis of V.

PROPOSITION - DEFINITION 8.35 1) *There is a unique subset $B(R_+) = \{\alpha_1, \ldots, \alpha_l\} \subset R_+$ characterized by : every $\alpha \in R_+$ may be written in a unique way as $\alpha = \sum n_i \alpha_i$ ($1 \le i \le l$) ($n_i \in \mathbb{N}$). $B(R_+)$ is called a* **system of simple roots** *for R.*

2) *Two positive systems of roots $R_+, R'_+ \subset R$ are conjugate under $W(R)$, i.e. $\exists w \in W(R)$ such that $w(R_+) = R'_+$. Thus one has $w(B(R_+)) = B(R'_+)$.*

3) *The action of W on the set of R_+ is simply transitive (cf. [4], Ch VI, Théorème 2).*

Definition 8.36 *Write:*

$$S_{R_+} = \{s_1, \ldots, s_l\},$$

where $s_1 = s_{\alpha_1}, \ldots, s_l = s_{\alpha_l}$. It may be proved (cf. [4], Ch. VI, §1,5) that $(W(R), S_{R_+})$ is a Coxeter system. In view of the preceding proposition if $R'_+ \subset R$ is another system of positive roots, $(W(R), S_{R_+})$ and $(W(R), S_{R'_+})$ are canonically isomorphic Coxeter systems.

It is observed that if w is defined by $w(R_+) = R'_+$ then $S_{R'_+} = \left\{w \circ s_1 \circ w^{-1}, \ldots, w \circ s_l \circ w^{-1}\right\}$.

Definition 8.37 *A **parabolic set** $P \subset R$ is a closed subset of R, i.e. satisfying*

$$\alpha, \beta \in P \ and \ \alpha + \beta \in R \Longrightarrow \alpha + \beta \in P \ ,$$

*and so that $R = P \cup (-P)$ (cf. [4], Ch VI, Définition 4). Denote by $A(R)$ the set of parabolic subsets of R endowed with the opposed relation defined by the inclusion relation between subsets. By definition P and $P' \in A(R)$ are **incident** if $P \cap P'$ is a parabolic set, i.e. $P \cap P' \in A(R)$.*
Let

$Vert(A(R))$ *be the set of **maximal proper parabolic subsets** of R,*

*i.e. the set of $P \in A(R)$ such that : $P \neq R$, $P' \in A(R)$ and $P \subset P' \Rightarrow P = P'$ or $P' = R$, and $Ch(A(R))$ the set of positive systems of roots (resp. **minimal parabolic subsets**).*

Two chambers C_{R_+} and $C_{R'_+}$ in $A(R)$ are connected by a gallery (cf. [39], *Chap VIII, Th.* 2). The following proposition shows that the couple $(Vert(A(R)), Ch(A(R)))$ defines an apartment $A(R)$, whose set of vertices is $Vert(A(R))$ and whose set of chambers is $Ch(A(R))$.

PROPOSITION - DEFINITION 8.38 *There is a bijection between the class of parabolic sets $A(R)$ and the class of subsets $\{P_{i_1}, \ldots, P_{i_\lambda}\}$ of $Vert(A(R))$, whose elements are two by two incidents defined by*

$$\{P_{i_1}, \ldots, P_{i_\lambda}\} \mapsto P = P_{i_1} \cap \ldots \cap P_{i_\lambda},$$

*i.e. $A(R)$ is a flag complex. It is said that $P_{i_1}, \ldots, P_{i_\lambda}$ are the **vertices** of P.*

Let R_+ be a system of positive roots, $B(R_+) = \{\alpha_1, \ldots, \alpha_l\}$ be the simple roots defined by R_+, and $S = \{s_1, \cdots, s_l\} \subset W$ the corresponding reflections. Write $R_0 = B(R_+)$, and for $1 \leq i \leq l$:

$$R_+^{(i)} = \mathbb{Z}(R_0 - \{\alpha_i\}) \cap R \cup R_+.$$

The subsets $R_+^{(i)} \subset R$ are two by two incident maximal parabolic sets. With $X \subset S$ is associated the parabolic set $P_X = \underset{s_i \in X}{\cap} R_+^{(i)}$. The set $(R_+^{(i)})_{s_i \in X}$ is the set of vertices of P_X. The stabilizer of P_X in W is given by $Stab\, P_X = W_{S-X}$. From the simple transitive action of W on $Ch(A(R))$ it may be proved that given $P \in A(R)$ there exists a unique $X \subset S$ and a unique $\overline{w} \in W/W_{S-X}$ with $w(P_X) = P$.

Proposition 8.39

1) $Stab_W\, P_X = W_{S-X}$;

2) For every parabolic set P there exists $w \in W(R)$ and a unique $X \in S$ with $w(P_X) = P$, the class $\overline{w} \in W/W_{S-X}$ being unique.

It results easily from this proposition the following

Proposition 8.40

1) There is a unique W-equivariant incidence preserving map

$$C(W, S) \longrightarrow A(R)$$

defined by : $W^{(s_i)} \mapsto R_+^{(i)}$ and preserving the incidence relation. This map sends the facet $w(F_X) = \{w(W^{(s_i)}) \mid s_i \in X \subset S\}$, where $F_X = \{W^{(s_i)} \mid s_i \in X \subset S\}$, to the parabolic set $w(P_X) = \underset{i \in X}{\cap}\, w(R_+^{(i)})$. The set of chambers $Ch\, C(W, S)$ corresponds to the set of positive systems of roots of R. The chamber C_e corrresponds to R_+. It may be noted that the set of vertices of $w(P_X)$ is precisely $\{w(R_+^{(i)}) \mid s_i \in X\}$.

2) If $C(W, S) = \underset{Y \subset S}{\amalg} W/W_Y$ is written then $C(W, S) \longrightarrow A(R)$ is given by

$$\overline{w}(\in W/W_{S-X}) \mapsto w(P_X).$$

3) There is a bijection

$$R \xrightarrow{\sim} \text{ set of combinatorial roots } \Phi \text{ of } A(R)$$

associating to $\alpha \in R$ the combinatorial root $\Phi_\alpha \subset R$ given by the sub-complex

$$\Phi_\alpha = \{P \in A(R) \mid \alpha \in P\}.$$

The wall $\partial\Phi_\alpha$ defined by Φ_α is thus equal to $\{P \in A(R) \mid \alpha, -\alpha \in P\}$.

8.8 Geometric representation of the Coxeter Complex of a root system

A simplicial decomposition of V^* is obtained in terms of the hyperplanes defined by R and corresponding to the geometrical representation of $C(W(R), S)$. This simplicial decomposition is a specialization of the one associated to a general Coxeter complex.

Definition 8.41 *Let the indexed set* $\mathcal{H} = (H_\alpha)_{(\alpha \in R)}$ *of hyperplanes in* V^* *be defined by:*

$$H_\alpha = \{x \in V^* \mid \alpha(x) = 0\}.$$

(obviously $H_\alpha = H_{-\alpha}$*). Write:*

$$D_\alpha^+ = \{ x \in V^* \mid \alpha > 0 \} \ (resp. \ \overline{D}_\alpha^+ = \{ x \in V^* \mid \alpha \geq 0 \}) \,,$$

and $D_\alpha^- = -D_\alpha^+$ *(resp.* $\overline{D}_\alpha^- = -\overline{D}_\alpha^+$*).*

The set $\mathcal{H}$ defines an equivalence relation $\sim$ on V^*, as in [4], Ch. V, §1, whose equivalence classes are simplicial cones of V^* which are also called facets. Denote by $\mathcal{C}_R(\mathcal{H})$ the quotient set $V^*/\sim$, and by $\mathcal{C}_\alpha^+$ the set of facets (resp. simplicial cones) contained in the closed half space $\overline{D}_\alpha^+$. Write $\mathcal{C}_\alpha^- = -\mathcal{C}_\alpha^+$. The set of chambers $Ch\ \mathcal{C}_R(\mathcal{H})$ is given by $Ch\ \mathcal{C}_R(\mathcal{H}) = (\mathcal{C}_{R_+})$, where $\mathcal{C}_{R_+} = \underset{\alpha \in R_+}{\cap}\ D_\alpha^+$ and R_+ runs on the positive root systems of R.

There is a natural mapping $A(R) \longrightarrow \mathcal{C}_R(\mathcal{H})$, $P \mapsto F_P$ defined as follows. Given a parabolic set $P \in A(R)$, let the $"\sim"$ - equivalence class $F_P \in \mathcal{C}_R(\mathcal{H})$ be defined as follows. Write

$$P^0 = P \cap (-P) \ (resp. \ P_+ = P - P \cap (-P)).$$

There is then $P = P^0 \coprod P_+$. Define

$$F_P = \left(\bigcap_{\alpha \in P^0} H_\alpha \right) \cap \left(\bigcap_{\alpha \in P_+} D_\alpha^+ \right)$$

(The facet of $\mathcal{C}_R(\mathcal{H})$ associated to P) (cf. [4], Ch V, §1, 2.). It is easy to see that F_P is an equivalence class for $\sim$. Actually the correspondence

$$P \mapsto F_P$$

is a bijection between $A(R)$ and $\mathcal{C}_R(\mathcal{H})$. The reciprocal map associates to $F \in \mathcal{C}_R(\mathcal{H})$ the parabolic subset

$$P_F = \{\alpha \in R \mid \forall x \in F\ (\alpha, x) \geq 0\}.$$

$P_F \subset R$ is a closed subset containing a positive system of roots and thus a parabolic set. The following duality formula holds:

$$P = \{\alpha \in R \mid \forall x \in F_P \ (\alpha, x) \geq 0\}.$$

The root Φ_α corresponds to the set $\mathcal{C}_\alpha^+$ by this bijection.

The Theorem 1 of [4], Ch. V, §3 applies to $\mathcal{H}$ and one obtains :

Proposition 8.42

1) *Let C be a chamber of V^* defined by $\mathcal{H}$, S the set of reflections defined by the walls of C (= bounding hyperplanes of C), and W the group generated by S. Then (W, S) is a Coxeter system. If C corresponds to C_{R_+} then the set of bounding hyperplanes is given by $(H_\alpha)_{\alpha \in B(R_+)}$, in fact $(W, S) = (W(R^\vee), S^\vee)$ where $S^\vee = (s_{\alpha^\vee})_{\alpha \in B(R_+)}$.*

2) *The set of hyperplanes H of V^* so that the orthogonal reflection $s_H \in W$ is given by $\mathcal{H}$.*

Definition 8.43 *It follows from (2)- that the set of hyperplanes $\mathcal{H}$ such that $H \in \mathcal{H} \Rightarrow s_H \in W(R)$ is canonically indexed by the set T of conjugates of S in $W(R)$. It is concluded that $C_R(\mathcal{H})$ in V^* may be looked at as the geometrical representation of $C(W(R), S(R_+))$. $C_R(\mathcal{H})$ is called the* **canonical geometrical representation** *of $C(W(R), S(R_+))$.*

Remark 8.44 *1) Let P be a parabolic set of R, i.e. $P \in A(R)$, and $F_P \in C_R(\mathcal{H})$ the corresponding facet (resp. simplicial cone) given by $A(R) \longrightarrow C_R(\mathcal{H})$. Then the carrier $L_{F_P} \subset V^*$ of F_P is given by*

$$L_{F_P} = \bigcap_{\alpha \in P^0} H_\alpha.$$

2) The combinatorial root Φ_α of $A(R)$ corresponds by $A(R) \longrightarrow C_R(\mathcal{H})$ to the closed half space

$$\overline{D}_\alpha^+ = \{x \mid \alpha(x) \geq 0\}$$

of V^*, and the wall $\partial \Phi_\alpha$ to the hyperplane H_α.

3) The facet $F_P \subset L_F$ may be seen as the chamber of L_F given by

$$F_P = L_F \cap \left(\bigcap_{\alpha \in P_+} D_\alpha^+ \right).$$

One has $D_\alpha^+ = \overline{D}_\alpha^+ - H_\alpha$.

The following construction plays a role in the study of a minimal generalized gallery between two facets F and F'.

## 8.9	The opposition involution in a Coxeter complex

Let $C(W, S)$ be a finite Coxeter complex. Given $C \in \mathrm{Ch}\, C(W, S))$, C^{opp} denotes the only chamber C' of $C(W, S)$ so that there is no root Φ of $C(W, S)$ with $C, C' \in \mathrm{Ch}\, \Phi$ (**The opposed chamber to** $C \in \mathrm{Ch}\, C(W, S)$). In other terms C^{opp} is the unique chamber C' satisfying $\mathcal{H}(C, C') = \mathcal{H}$.

Definition 8.45 *Following [4], Ex 22, Ch IV, there exists a unique **involutive automorphism** φ of $C(W, S)$ such that*

$$\varphi(C) = C^{opp}$$

for every chamber C of $C(W, S)$, and

$$\varphi(H) = H$$

*for every wall H of $C(W, S)$. Write $F^{opp} = \varphi(F)$ $(F \in C(W, S))$ (**The opposed facet to** F).*

Remark 8.46 *The following two properties characterizes F^{opp}:*

1) $L_{F^{opp}} = L_F$;

2) $\mathcal{H}(F^{opp}, F) = \mathcal{H}(F) = \{H \in \mathcal{H} \mid F \notin H\}$.

It is known that $\mathrm{St}_{F'}$ $(F' \in C(W, S))$ is isomorphic to some Coxeter complex $C(W', S')$, so the above definition applies to $\mathrm{St}_{F'}$. Denote by $\varphi_{F'} : \mathrm{St}_{F'} \longrightarrow \mathrm{St}_{F'}$ the involutive automorphism of $\mathrm{St}_{F'}$.

Definition 8.47 *Given $F \in \mathrm{St}_{F'}$, i.e. $F \supset F'$, write*

$$F^{opp}(F', F) = \varphi_{F'}(F)$$

(**The opposed facet to** F **relatively to** F' (**resp. of** F **in** $\mathrm{St}_{F'}$)).
As $\mathrm{St}_{F_\emptyset} = C(W, S)$ given $F \in C(W, S)$ one has

$$F^{opp} = F^{opp}(F_\emptyset, F),$$

where $F_\emptyset$ denotes the facet incident to every facet indexed by the empty set.

### 8.9.1	Opposed parabolic set in a star subcomplex

If the Coxeter system (W, S) is associated to a root system $R \subset V$, $R^\vee \subset V^*$, endowed with a system of positive roots R_+ (resp. simple roots $R_0 = \{\alpha_1, \dots, \alpha_l\}$), and $S = (s_\alpha)_{\alpha \in R_0}$, the above definitions specialize as follows.

Remark 8.48

It is clear that the mapping $g_{A(R)} : \alpha \mapsto -\alpha$ is an involutive automorphism of the root system R, thus $g_{A(R)}$ gives an involutive automorphism $\varphi_{A(R)}$ of $A(R)$. It is immediate that $\varphi_{A(R)}(\partial\Phi_\alpha) = \partial\Phi_\alpha$ and that $\varphi_{A(R)}(\Phi_\alpha) = \Phi_{-\alpha}$. It is concluded that $\varphi_{A(R)}$ corresponds to the involutive isomorphism φ of $C(W, S)$ by the isomorphism

$$C(W, S) \longrightarrow A(R),$$

Definition 8.49 *Define P^{opp} ($P \in A(R)$) following the pattern of the definition of F^{opp}, i.e. $P^{opp} = -P$. More precisely let $F \mapsto P$ by the above isomorphism. Write $P^{opp} = P_{F^{opp}}$. Let $P' = P_{F'}$ (resp. $P = P_F$), and $F' \subset F$. Write $P^{opp}(P', P) = P_{F^{opp}(F', F)}$.*

8.9.2 Determination of the opposed parabolic set in a star subcomplex

Proposition 8.50 *Let $R' \subset R$, and $V_{R'} = Vect_{\mathbb{R}}(R')$. The following statements are equivalent:*

1) R' is a closed and symmetric subset of R, i.e. $R' = -R'$.

2) R' is a closed subset of R and R' is a root system in the vector space $V_{R'}$.

For all $\alpha \in R'$ let $\alpha_1^\vee$ be the restriction of $\alpha^\vee$ to $V_{R'}$. Then the mapping $\alpha \to \alpha_1^\vee$ is the canonical bijection $R' \simeq R'^\vee$. The Weyl group of the root system R' is the subgroup $W_{R'} \subset W$ generated by $(s_\alpha)_{\alpha \in R'}$.

(cf. [4], *Ch. VI, n^0 1.1, Prop. 4*)
Clearly the above proposition applies to $R_X = P_X \cap (-P_X) \subset R$ and one has:

Proposition 8.51 *A simple system of roots of R_X is given by $\{\alpha \in R_0 \mid s_\alpha \in S - X\}$, and $(W(R_X), S - X)$ is a Coxeter system. One has $W(R_X) = W_{S-X}$. Thus $(R_X, R_X^\vee, V_{R_X}, V_{R_X}^*)$ defines a system of roots endowed with the system of simple roots $\{\alpha \in R_0 \mid s_\alpha \in S - X\}$, and $(W_{S-X}, S - X)$ as its associated Coxeter system.*

The content of this section is motivated by the correspondence between parabolic subsets of R contained in P_X and the parabolic subsets of R_X. The facet $F_X = \{W^{(s_i)} \mid s_i \in X\} \in C(W, S)(X \in \mathcal{P}(S))$ is associated with a star complex in $C(W, S)$(resp. a Coxeter system, a Coxeter complex):

$$St_{F_X} \subset C(W, S) \ (\text{resp. } (W_{S-X}, S - X), C(W_{S-X}, S - X)),$$

where $W_{S-X} = \text{Stab}\, F_X$.

Let $A(R_X)$ be the apartment defined by the system of roots R_X. One knows that there is a canonical isomorphism $C(W_{S-X}, S - X) \simeq A(R_X)$. On the other hand, there is a building isomorphism:

$$St_{P_X} \simeq A(R_X) \,,$$

where St_{P_X} denotes the star complex of P_X in $A(R)$, defined by $P \mapsto P \cap [P_X \cap (-P_X)] = P \cap R_X$. Its reciprocal mapping $A(R_X) \xrightarrow{\sim} St_{P_X}$ is defined by:

$$Q \mapsto Q \coprod [P_X - P_X \cap (-P_X)].$$

It is recalled that $St_{P_X} \subset A(R_X)$ is given by the set of of parabolic sets P' of $A(R_X)$ with $P' \subset R_X$, and that P' may be completed into a parabolic set of R by addition of the set $P_X - P_X \cap (-P_X) \subset R_+$.

On the other hand, the isomorphism $C(W, S) \simeq A(R)$ induces an isomorphism

$$(C(W_{S-X}, S - X) \xrightarrow{\sim}) \, St_{F_X} \xrightarrow{\sim} St_{P_X}.$$

By this isomorphism the building involution $\varphi_{F_X} : St_{F_X} \longrightarrow St_{F_X}$ corresponds to the involution φ_{P_X} of St_{P_X} described as follows.

Proposition 8.52 *The involutive isomorphism* $\varphi_{P_X} : St_{P_X} \longrightarrow St_{P_X}$ *is given by*

$$\varphi_{P_X} : P = [P' \cap (-P') \coprod (P' - P' \cap (-P'))] \coprod (P_X - P_X \cap (-P_X)) \longrightarrow$$
$$[P' \cap (-P') \coprod -(P' - P' \cap (-P'))] \coprod (P_X - P_X \cap (-P_X)),$$

where $P' = P \cap R_X$. *Observe that* $[P' \cap (-P') \coprod (P' - P' \cap (-P'))] = [P \cap (-P) \coprod (P - P \cap (-P))] \cap R_X$

Corollary 8.53 *The opposed parabolic* $P^{opp}(P_X, P)$ *is given by:*

$$P^{opp}(P_X, P) = [P' \cap (-P') \coprod -(P' - P' \cap (-P'))] \coprod (P_X - P_X \cap (-P_X))$$

in $A(R)$, *and by*

$$P'^{opp}(P_X, P) = [P' \cap (-P') \coprod -(P' - P' \cap (-P'))]$$

in $A(R_X)$.

8.10 Opposed parabolic set in a star subcomplex simplicial representation

Retain the notation of the preceding section. One obtains a geometric representation $St_{F_X} \simeq \mathcal{C}_{R_X}(\mathcal{H}_X)$ as follows. The set of root hyperplanes in $Vect_{\mathbb{R}}(R_X)$ is given by $\mathcal{H}_X = (H_\alpha \cap Vect_{\mathbb{R}}(R_X))_{\alpha \in R_X}$, where $H_\alpha \in \mathcal{H}_{F_X} \subset \mathcal{H}$. On the other hand, there are the following direct sum decompositions

$$V^* = Vect_{\mathbb{R}}(R_X^\vee) \oplus \bigcap_{\alpha \in R_X} H_\alpha \ (\text{resp. } H_{\alpha'} = H_{\alpha'} \cap Vect_{\mathbb{R}}(R_X^\vee) \oplus \bigcap_{\alpha \in R_X} H_\alpha \ (\alpha' \in R_X)),$$

thus establishing a bijection $\mathcal{H}_X \simeq \mathcal{H}_{F_X}$. By this bijection the set of chambers $\mathcal{C}_{R_X}(\mathcal{H}_X)$ corresponds to the set of chambers of $\mathcal{C}_R(\mathcal{H})$ incident to F_X, and more generally the set of of facets $\mathcal{C}_{R_X}(\mathcal{H}_X)$ to the set of facets in $\mathcal{C}_R(\mathcal{H})$ incidents to F_X. On the other hand, from the bijection $St_{P_X} \simeq A(R_X)$, $P \mapsto P' = P \cap R_X$ one obtains the expressions of $P^{opp}(P_X, P) = P'^{opp} \coprod (P_X - R_X)$. The following proposition makes explicit how this expression translates by this geometric representation.

From the above discussion it results easily the following

Proposition 8.54 *Let F' (resp. $F^{opp}(F_X, F)$) be the facet in $\mathcal{C}_{R_X}(\mathcal{H}_x)$ corresponding to P' (resp. to $P^{opp}(P_X, P)$ in $\mathcal{C}_R(\mathcal{H})$), and $L_{F'}$ be the carrier of F' in $\mathcal{C}_{R_X}(\mathcal{H})$. There is the following representation of $L_{F'}$ in $\mathcal{C}_{R_X}(\mathcal{H})$:*

$$L_{F'} = L_{F'^{opp}} = \left(\bigcap_{\alpha \in P' \cap (-P')} H_\alpha \right) \cap Vect_{\mathbb{R}}(R_X^\vee)$$

and

$$F'^{opp} = L_{F'} \cap \left(\bigcap_{\alpha \in -(P' - P' \cap (-P'))} D_\alpha^+ \right).$$

The representation of $F^{opp}(F_X, F)$ in $\mathcal{C}_R(\mathcal{H})$ is given by:

$$F^{opp}(F_X, F) = L_F \cap \left(\bigcap_{\alpha \in -(P' - P' \cap (-P'))} D_\alpha^+ \right) \cap \left(\bigcap_{\alpha \in (P_X - P_X \cap (-P_X))} D_\alpha^+ \right).$$

Observe that

$$P \cap (-P) = P' \cap (-P') \ (\text{resp. } P - P \cap (-P) = P' - P' \cap (-P') \coprod P_X - P_X \cap (-P_X),$$

and thus $L_F = L_{F'}$.

Corollary 8.55 *The set $\mathcal{H}_{F_X}(F, F^{opp}(F_X, F))$ of walls $\partial \Phi_\alpha \supset F_X$ separating F and $F^{opp}(F_X, F)$ is given by*

$$\mathcal{H}_{F_X}(F, F^{opp}(F_X, F)) = \mathcal{H}_{F_X}(F) = \mathcal{H}_{F_X}(F^{opp}(F_X, F)).$$

In fact this property characterizes $F^{opp}(F_X, F)$, i.e. $F^{opp}(F_X, F)$ is the unique facet incident to F_X satisfying this property.

Chapter 9

Minimal Generalized Galleries in a Coxeter Complex

The **minimal generalized galleries** in the setting of Coxeter complexes, are introduced. The existence of minimal generalized galleries is proven by means of the geometrical realization of a Coxeter Complex as a decomposition of an euclidean space by simplicial cones (chambers). The correspondence between minimal generalized galleries and **block decompositions** of elements in W is established. It is specifically explained how geometric constructions with generalized galleries translate into special decompositions of elements of W. A unicity result about generalized galleries with associated minimal galleries of types, i.e. galleries of types defined by minimal generalized galleries, is proven. The **Convex hull subcomplex of two facets** which plays an important role in this work is introduced. This follows from the result obtained here: a Minimal Generalized Gallery is contained in the Convex Envelope of its extremities. Thus generalizing a result proven for the Flag Complex.

In this chapter $\mathcal{A}$ denotes a finite apartment with a selected chamber C endowed with a building morphism

$$\text{typ} : \mathcal{A} \to \text{typ } \mathcal{A} \ ,$$

so that the restriction of typ to the sub-building $\Delta(C)$, formed by the facets incident to C, induces a building isomorphism $\Delta(C) \simeq \text{typ}$. Let $W_{\mathcal{A}} \subset Aut(\mathcal{A})$ be the subgroup of automorphisms of $\mathcal{A}$ commuting with typ, i.e. preserving the type of a facet. It is supposed that there is an isomorphism of $\mathcal{A}$ with a Coxeter Complex, $C(W, S) \simeq \mathcal{A}$ sending the chamber

C_e, given by the identity of W, to C, so that the induced isomorphism $Aut(C(W,S)) \simeq Aut(\mathcal{A})$ sends $W \subset Aut(C(W,S))$ to $W_{\mathcal{A}}$. There are building isomorphisms $C(W,S)/W \simeq \mathcal{A}/W_{\mathcal{A}} \simeq \Delta(C)$, and an identification $\mathcal{A}/W_{\mathcal{A}} \simeq \text{typ } \mathcal{A}$. Denote by S_C the image of the generating reflexions S by $W \simeq W_{\mathcal{A}}$. There is a canonical identification $C(W_{\mathcal{A}}, S_C) \simeq \mathcal{A}$. Given a chamber C of $\mathcal{A}$ there is a building isomorphism $\Delta(C) \simeq \text{typ } \mathcal{A}$ given by typ. We define the **Roots**, the **Hyperplanes**,... etc of $\mathcal{A}$ as the images of the corresponding objects in $C(W,S)$. Denote by $\mathcal{H}_{\mathcal{A}}$ (resp. $\mathcal{D}_{\mathcal{A}}$) the set of Hyperplanes (resp. Roots). The subindex $\mathcal{A}$ is omitted if no confusion arises. With $F, F' \in \mathcal{A}$ are associated the subset $\mathcal{H}_F \subset \mathcal{H}_{\mathcal{A}}$ of Hyperplanes containing F, $\mathcal{H}_F(F') = \mathcal{H}_F - \mathcal{H}_{F'}$, the Hyperplanes separating F and F', $\mathcal{H}(F, F') \subset \mathcal{A} - (\mathcal{H}_F \cup \mathcal{H}_{F'})$, and $\mathcal{H}_F(F', F'') = \mathcal{H}_F \cap \mathcal{H}(F', F'')$. Φ^{opp} denotes the opposed root to $\Phi \in \mathcal{D}_{\mathcal{A}}$ and by $\partial\Phi = \Phi \cap \Phi^{opp}$ its associated Hyperplane. With $F \in \mathcal{A}$ and $H \in \mathcal{H}_{\mathcal{A}} - \mathcal{H}_F$ is associated the Root $\Phi_H(F)$ so that $\partial\Phi_H(F) = H$ and $F \in \Phi_H(F) = H$.

As an example of an abstract Coxeter Complex as defined above, consider $A(R)$ the apartment defined by a root system, with $C - C_{R_0}$, where C_{R_0} denotes the chamber defined by the simple system of roots R_0, and typ : $A(R) \longrightarrow \mathcal{P}(R_0)$ associates with a parabolic subset, the corresponding set of simple roots in $\mathcal{P}(R_0)$.

The building typ $\mathcal{A}$ **(the typical simplex of $\mathcal{A}$)** of types of facets of $\mathcal{A}$, is by definition the quotient set $\mathcal{A}/W_{\mathcal{A}}$, endowed with the relation induced by the inclusion of facets. Given $F \in \mathcal{A}$, the type $t = \text{typ } F$ of the facet F is by definition the image of F by the quotient mapping typ : $\mathcal{A} \to \text{typ } \mathcal{A}$. Write:

$$\mathcal{A} = \coprod_{t \in typ\ \mathcal{A}} \mathcal{A}_t \ ,$$

where $\mathcal{A}_t = (\text{typ})^{-1}(t)$. Given $t \in \text{typ } \mathcal{A}$ and $C \in \text{Ch } \mathcal{A}$ let $F_t(C) \subset C$ be the unique facet of type t incident to C. The reciprocal isomorphism of $\Delta(C) \simeq \text{typ } \mathcal{A}$ is defined by $t \mapsto F_t(C)$.

Define the set *Relpos $\mathcal{A}$* of types of relative positions of facets in $\mathcal{A}$ by

$$Relpos\ \mathcal{A} := \mathcal{A} \times \mathcal{A}/W_{\mathcal{A}} \ .$$

Denote by $\tau : \mathcal{A} \times \mathcal{A} \longrightarrow Relpos\ \mathcal{A}$ the quotient mapping. Write

1) $Relpos_{(t,s)}\ \mathcal{A} = \mathcal{A}_t \times \mathcal{A}_s/W_{\mathcal{A}}$ $((t,s) \in typ\ \mathcal{A} \times typ\ \mathcal{A})$;

2) $(\mathcal{A} \times \mathcal{A})_{\tau_0} = (\tau)^{-1}(\tau_0)$ $(\tau_0 \in Relpos\ \mathcal{A})$;

3) $(Ch\ \mathcal{A} \times \mathcal{A})_{\tau_0} = \{ (C, F) \mid \tau(F_s(C), F) = \tau_0 \}$.

If $\mathcal{A} = C(W,S)$ the quotient set $\mathcal{A}/W$ may be identified with the combinatorial simplex $\mathcal{P}(S)$ given by the set of subsets of S, endowed with relation of inclusion of subsets of S. Given a facet by its set of vertices

$$F = \{wW^{(s)} \mid s \in X\}\ (w \in W,\ X \subset S,\ W^{(s)} = W_S - \{s\})$$

of $C(W, S)$ there is

$$\text{typ } F = X \quad (\text{resp. Stab } F = w(W_{S-X})w^{-1}).$$

Let $Y \in \mathcal{P}(S) = \text{typ } C(W, S)$, and let $C_e = \{W^{(s)}|\ s \in S\} \in \text{Ch } C(W, S)$. One thus obtains

$$F_Y(C_e) = \{W^{(s)}|\ s \in Y\} \quad (\text{resp. Stab } F_Y(C_e) = W_{S-Y}) \ .$$

It is recalled that there is a natural bijection

$$\coprod_{X \in \mathcal{P}(S)} W/W_X \xrightarrow{\sim} C(W, S)$$

described as follows. Let $X \in \mathcal{P}(S)$. Then $W/W_X \to C(W, S)$ is given by $\overline{w} \mapsto w(F_{S-X}(C_e))$ where $F_X(C_e) = \{W^{(s)}|\ s \in X\}$.

PROPOSITION - DEFINITION 9.1 *In each class* $\overline{w} \in W/W_X$ *there is a unique minimal length element* $w^m \in \overline{w}$, *given by*

$$w^m = w(C_e, proj_{w(F_{S-X})}C_e).$$

Write $C_{\overline{w}} = proj_{w(F_{S-X})}C_e$.

Proof *This is immediate from the characterization of* $proj_{w(F_{S-X})}C_e$ *as the unique chamber of* $\text{St}_{w(F_{S-X})}$ *satisfying*

$$d(C_e, proj_{w(F_{S-X})}C_e) = \min_{C \in \text{Ch } \text{St}_{w(F_{S-X})}} d(C_e, C) \ .$$

$\text{Gall}_\mathcal{A}$ (resp. $\text{Gall}_\mathcal{A}(F)$, $\text{Gall}_\mathcal{A}(F, F')$) is defined as the set of generalized galleries of $\mathcal{A}$ (resp. issued from F, with extremities (F, F')) (cf. 5.3). Given a gg γ of $\mathcal{A}$ let

$$g = \text{typ } \gamma \subset \text{typ } \mathcal{A}$$

be the **gg of types** (resp. typ $\mathcal{A}$) whose facets are given by the images of the facets of γ by typ. There is a mapping

$$\text{Gall}_\mathcal{A} \to \text{gall}_\mathcal{A} = \text{Gall}_{\text{typ } \mathcal{A}}$$

defined by $\gamma \to \text{typ } \gamma$, where typ γ denotes the gallery of typ $\mathcal{A}$ defined by the types of the facets of γ. This mapping can also denoted by "typ" when there is no confusion. It is observed that the restriction of typ to $Gall_{\Delta(C)}$ gives a bijection $Gall_{\Delta(C)} \simeq gall_\mathcal{A}$.

Let $g \in \text{gall}_\mathcal{A}$. Write

$$\text{Gall}_\mathcal{A}(g) := \text{typ}^{-1}(g) \quad (\text{resp. } \text{Gall}_\mathcal{A}(g, F) := \text{typ}^{-1}(g) \cap \text{Gall}_\mathcal{A}(g))$$

(The **gg**'s of type g (resp. of type g and extremity F)).

The following terminology adapted from [23], Exp. XXVI, 4. (cf. also [4], *Ch. IV*, 1., *Ex.* 22) is introduced.

Definition 9.2 *It is said that a couple of facets* $(F, F') \in \mathcal{A} \times \mathcal{A}$ *is in transversal position or simply transversal if there exists a chamber* $C \in \mathrm{Ch}\,\mathcal{A}$ *satisfying:* $F_{typ(F)}(C) = F$ *and* $F_{typ(F')}(C^{opp}) = F'$.

It is clear that C and F are transversal if and only if

$$d(C, F) = |\mathcal{H}(C, F)| \geqslant |\mathcal{H}(C, F')| = d(C, F'),$$

for $F' \in \mathcal{A}_t$, thus there is

Proposition 9.3 *A chamber* C *and a facet* F *of* $\mathcal{A}$ *are in* **transversal position** *if the length* n *of a minimal gallery*

$$\Gamma : C = C_0, \cdots, C_n \supset F$$

is maximal on the set of all minimal galleries between C *and a facet* $F' \in \mathcal{A}_t$ *of the same type* t *as* F.

The following lemma allows defining the center of a Bruhat cell.

Lemma 9.4 *Given* $C \in \mathrm{Ch}\,\mathcal{A}$, *and* $t \in \mathrm{typ}\,\mathcal{A}$ *there is only one facet* $F_t^{tr}(C) \in \mathcal{A}_t$ *such that* C *and* $F_t^{tr}(C)$ *are transversal in* $\mathcal{A}$.

Proof *Let* C^{opp} *be the opposed chamber to* C *in* $\mathcal{A}$. *Then* $F_t^{tr}(C)$ *may be characterized as the only facet* $F \in \mathcal{A}_t$ *incident* $F \subset C^{opp}$ *to* C^{opp}. *This proves the lemma.*

Let $\delta := \{(t, t) \in \mathrm{typ}\,\mathcal{A} \times \mathrm{typ}\,\mathcal{A} | \ t \in \mathrm{typ}\,\mathcal{A}\}$ and write

$$\mathrm{typ}^{(2)}\,\mathcal{A} := \{(t, s) \in \mathrm{typ}\,\mathcal{A} \times \mathrm{typ}\,\mathcal{A} - \delta | \ t \subset s\}.$$

Given $(t, s) \in \mathrm{typ}^{(2)}\,\mathcal{A}$, and $C \in \mathrm{Ch}\,\mathcal{A}$, denote by

$$F_{(t,s)}^{tr} := F_{(t,s)}^{tr}(C) \in \mathrm{St}_{F_t(C)},$$

the unique facet of type s in $\mathcal{A}$ is such that:

" C and $F_{(t,s)}^{tr}(C)$ are transversal in the Star Complex $\mathrm{St}_{F_t(C)}$".

If $\mathcal{A} = C(W, S)$ one obtains

$$\mathrm{typ}^{(2)}\,\mathcal{A} = \{(X, Y) \in \mathcal{P}(S) \times \mathcal{P}(S) | \ X \neq Y, \ X \subset Y\}.$$

An indexed set of elements of W associated with a chamber C in $C(W, S)$ is introduced so that with minimal generalized galleries issued from C are associated words in these elements.

Definition 9.5 *Let $C \in \mathrm{Ch}\ \mathcal{A}$ (resp. $(t,s) \in typ^{(2)}\ \mathcal{A}$). Write $F = F_{(t,s)}^{tr}(C)$ and define*

$$w(t,s) = w_C(t,s) = w(C, proj_F C).$$

The mapping $(t,s) \mapsto w(t,s)$ (resp. $(X,Y) \mapsto w(X,Y)$) defines an indexed family of elements of W

$$\mathscr{S} := \mathscr{S}_C = (w_C(t,s))_{(t,s)\in typ^{(2)}\ \mathcal{A}}$$

$$(resp.\ \mathscr{S} := \mathscr{S}_{C_e} = (w(X,Y))_{(X,Y)\in typ^{(2)}\ C(W,S)}).$$

9.1 Minimal Generalized Galleries issued from a chamber

Definition 9.6 *Let $C \in \mathrm{Ch}\ \mathcal{A}$ and $F \in \mathcal{A}$. A generalized gallery (gg) of the form*

$$\gamma = \gamma(C,F):\ \ C = F_r \supset F_r' \subset F_{r-1} \cdots F_0 \supset F_0' = F \text{ (right open gallery)}$$

(resp. $\gamma = \gamma(C,F):\ \ C = F_r \supset F_r' \subset F_{r-1} \cdots \subset F_0 = F$) (right closed gallery),

or in abbreviated notation

$$\gamma = \gamma(C,F):\ \ (F_i \supset F_i')\ (r \geqslant i \geqslant 0),\ \text{and}\ (F_i' \subset F_{i-1})\ (r \geqslant i \geqslant 1),$$

$$F_r = C,\ F_0' = F$$

(resp. $\gamma = \gamma(C,F):\ \ (F_i \supset F_i')\ (r \geqslant i \geqslant 1),\ \text{and}\ (F_i' \subset F_{i-1})\ (r \geqslant i \geqslant 1),$

$$F_r = C,\ F_1' \subset F_0 = F)$$

with $F_i \neq F_i'$ (resp. $F_i' \neq F_{i-1}$), is a **Minimal Generalized Gallery (MGG)** *between C and F, or with extremities C and F, if:*

(1) The sets $\mathcal{H}_{F_i'}(F_{i-1})\ (r \geqslant i \geqslant 1)$ are two by two disjoint, i.e. $i \neq j \Rightarrow \mathcal{H}_{F_i'}(F_{i-1}) \cap \mathcal{H}_{F_j'}(F_{j-1}) = \emptyset$;

(2) $\mathcal{H}(C,F) = \bigcup \mathcal{H}_{F_i'}(F_{i-1}) = \coprod \mathcal{H}_{F_i'}(F_{i-1})$ for $(r \geqslant i \geqslant 1)$.

Denote by $Gall^m(C,F)$ the set of Minimal Generalized Galleries between C and F.

Remark 9.7   *1) Observe that if $\gamma(C,F)$ is a Minimal Generalized Gallery right open then $F_0 \subset proj_F\ C$.*

2) To a minimal gallery $\Gamma:\ \ C = C_r \cdots C_0 \supset F$ corresponds the MGG:

$$(C_i \supset F_i)\ (r \geqslant i \geqslant 0)\ \ \ \ (resp.\ (F_{i+1} \subset C_i)\ (r-1 \geqslant i \geqslant 0)),$$

$$C_r = C,\ F_0 = F,$$

where $F_i = C_i \cap C_{i-1}$ ($r \geqslant i \geqslant 1$) denotes the common codimension 1 facet to C_i and C_{i-1}. This result comes from the following characterization of the minimal galleries of $\mathcal{A}$:
An injective gallery $\Gamma : C = C_r \cdots C_0 \supset F$ between C and F is minimal if Γ crosses only once each wall L it encounters, and $L \in \mathcal{H}(C, F)$ (cf. [4] Chap. 4, §1.4, Lemma 2).
As in 8.6 we denote by

$$\Psi^*(\Gamma) = (L_r, L_{r-1}, \cdots, L_1)$$

the sequence of walls that Γ crosses, and by

$$\Psi(\Gamma) = (t_r, t_{r-1}, \cdots, t_1)$$

the sequence of reflections defined by $\Psi^*(\Gamma)$. Remark that L_i ($r \geqslant i \geqslant 1$) is the carrier of the codimension 1 facet $F_i = C_i \cap C_{i-1}$.

9.2 Adapted minimal galleries to a Minimal Generalized Gallery

Assume the notation of the preceding section. A family of minimal galleries "majorating" a Minimal Generalized Gallery is defined. Write:

$$C_r = C \quad \text{and} \quad C_{i-1} = \text{proj}_{F_{i-1}} C_i \quad (r \geqslant i \geqslant 1).$$

Remark that C_i, $C_{i-1} \in \text{Ch St}_{F_i'}$ ($r \geqslant i \geqslant 1$), as C_i, $F_{i-1} \in \text{St}_{F_i'}$ and $\text{Ch St}_{F_i'}$ is a convex set of chambers of $\mathcal{A}$, i.e. if a minimal gallery $\Gamma(C, C')$ satisfies $C, C' \in \text{St}_{F_i'}$ then $\Gamma(C, C') \subset \text{St}_{F_i'}$. For each $r \geqslant i \geqslant 1$ a minimal gallery $\Gamma^i = \Gamma^i(C_i, C_{i-1})$ between C_i and C_{i-1} is chosen. Thus $\Gamma^i \subset \text{St}_{F_i'}$. Define

$$\Gamma = \Gamma(C, C_0) = \Gamma^r \circ \Gamma^{r-1} \circ \cdots \circ \Gamma^1,$$

i.e. Γ is the composed gallery defined by (Γ^i). Let $\Gamma' = \Gamma'(C, F)$ be the gallery obtained as "the composition of $\Gamma(C, C_0)$ and the inclusion $C_0 \supset F$".

Lemma 9.8 *$\Gamma'(C, F)$ is a minimal gallery between C and F.*

Proof *As set $\Psi^*(\Gamma^i) \subset \mathcal{H}_{F_i}(F_{i-1})$, it is deduced that*

$$\Psi^*(\Gamma) \subset \coprod \mathcal{H}_{F_i'}(F_{i-1}).$$

By (1) of Definition 9.6 it is concluded that Γ is injective, and each wall L it encounters crosses only once, and by (2) of Definition 9.6 that this wall satisfies $L \in \mathcal{H}(C, F)$. It is deduced that Γ' is a minimal gallery between C and F.

Definition 9.9 *It is said that a gallery* $\Gamma(C, F)$ *is a* **minimal gallery adapted** *to the generalized gallery* $\gamma(C, F)$ *if it is obtained from* $\gamma(C, F)$ *according to the pattern of the construction of* $\Gamma'(C, F)$.

Remark 9.10   *a) From the above lemma it follows that the set* $\Psi^*(\Gamma)$ *is equal to* $\mathcal{H}(C, F)$. *As set* $\Psi^*(\Gamma) = \coprod \mathcal{H}(C_i, C_{i-1})$, *and* $\mathcal{H}(C_i, C_{i-1}) \subset \mathcal{H}_{F_i'}(F_{i-1})$, *it is deduced by (2) of Definition 9.6 that* $\mathcal{H}(C_i, C_{i-1}) = \mathcal{H}_{F_i'}(F_{i-1})$. *One also has* $\mathcal{H}(C_i, C_{i-1}) = \mathcal{H}(C_i, F_{i-1})$ *as* $C_{i-1} = \mathrm{proj}_{F_{i-1}} C_i$.

   b) The equality $\mathcal{H}(C_i, F_{i-1}) = \mathcal{H}_{F_i'}(F_{i-1})$ *means that the distance* $d(C_i, F_{i-1})$ *is the maximal* $d(C, F)$ *between a chamber* C *and a facet* F *of* $\mathrm{St}_{F_i'}$ *of the same type as* F_{i-1}. *It is then deduced that* C_i *and* F_{i-1} *are transversal in* $\mathrm{St}_{F_i'}$ *(cf. Definition 9.3).*

The following reformulation of Definition 9.6 is then given:

Definition 9.11 *The same notation as in Definition 9.6 is kept. A gg*

$$\gamma(C, F): \ (F_i \supset F_i') \ (r \geqslant i \geqslant 0), \ (F_i' \subset F_{i-1}) \ (r \geqslant i \geqslant 1), \ F_r = C, \ F_0' = F$$

$$(\text{resp. } \gamma(C, F): \ (F_i \supset F_i') \ (r \geqslant i \geqslant 1), \ (F_i' \subset F_{i-1}) \ (r \geqslant i \geqslant 1),$$

$$F_r = C, \ F_0 = F)$$

is a **Minimal Generalized Gallery (MGG)** *if:* $F_i \neq F_i'$ *(resp.* $F_i' \neq F_{i-1}$*) and*

(1)-bis $\mathcal{H}_{F_i'}(F_{i-1}) \cap \mathcal{H}_{F_j'}(F_{j-1}) = \emptyset$ *for* $i \neq j$;

(2)-bis $|\mathcal{H}(C, F)| = \sum |\mathcal{H}_{F_i'}(F_{i-1})|$.

Proof *It is proved that (1) and (2) in Definition 9.6 is equivalent to (1)-bis and (2)-bis of Definition 9.11.*
The implication " $\Rightarrow$ " being immediate, let it be seen " $\Leftarrow$ ". The gallery $\Gamma'(C, F)$ *constructed as in Lemma 9.8 from* $\gamma(C, F)$ *satisfies:*

$$(*) \qquad |\mathcal{H}(C, F)| \leqslant |\Psi^*(\Gamma'(C, F))| \leqslant \sum |\mathcal{H}_{F_i'}(F_{i-1})| = |\mathcal{H}(C, F)|.$$

This establishes that $\Gamma'(C, F)$ *is a minimal gallery, and thus set (defined by)* $\Psi^*(\Gamma'(C, F)) = \mathcal{H}(C, F)$. *On the other hand, by construction of* $\Gamma'(C, F)$, *there is*

$$\text{set } \Psi^*(\Gamma'(C, F)) = \coprod \text{set } \Psi^*(\Gamma(C_i, C_{i-1})) \subset \coprod \mathcal{H}_{F_i'}(F_{i-1}),$$

which in view of $(*)$ *implies*

$$\mathcal{H}(C, F) = \text{set } \psi^*(\Gamma'(C, F)) = \coprod \mathcal{H}_{F_i'}(F_{i-1}).$$

Without proof the following reciprocal of Lemma 9.8 is given.

Lemma 9.12 *Given a gg $(s_\alpha \supset t_\alpha)$ $(r \geqslant \alpha \geqslant 0)$ $(t_\alpha \subset s_{\alpha-1})$ $(r \geqslant \alpha \geqslant 1)$ of types of $\mathcal{A}$ with $t_\alpha \neq s_{\alpha-1}$ (resp. $t_\alpha \neq s_\alpha$), a minimal gallery*

$$\Gamma = \Gamma(C, F): \ C = C_n, \cdots, C_0 \supset F$$

with types typ $C_n = s_r$ *and* typ $F = t_0$, *and a strictly increasing function*

$$j: \ [0, r] \to [0, n]$$

with $j(0) = 0$ and $j(r) = n$, one writes

$$F_\alpha = F_{s_\alpha}(C_{j(\alpha)}), \ F'_\alpha = F_{t_\alpha}(C_{j(\alpha)}).$$

It is supposed for $1 \leqslant \alpha \leqslant r$:

1. $\forall \ j(\alpha - 1) \leqslant i \leqslant j(\alpha)$, $F'_\alpha = F_{t_\alpha}(C_{j(\alpha)})$;

2. $C_{j(\alpha)}$ *and $F_{\alpha-1}$ are at maximal distance (resp. transversal) in* $\mathrm{St}_{F'_\alpha}$, *and* $\mathrm{proj}_{F_{\alpha-1}} C_{j(\alpha)} = C_{j(\alpha-1)}$.

Then the gg $(F_\alpha \supset F'_\alpha)$ $(r \geqslant \alpha \geqslant 0)$, $(F'_\alpha \subset F_{\alpha-1})$, $(r \geqslant \alpha \geqslant 1)$ is a MGG of $\mathcal{A}$.

9.3 Reduced words corresponding to Minimal Generalized Galleries

Definition 9.13 *Let g be a generalized gallery in* typ $\mathcal{A}$ *given by*

$$g: \ t_r \subset s_{r-1} \cdots t_1 \subset s_0$$

$$(\text{resp. } g: \ t_r \subset s_{r-1} \cdots t_1 \subset s_0 \supset t_0) .$$

A $\mathscr{S}_C$-reduced expression of type g is the product $w = \prod w_i$ $(r \geqslant i \geqslant 1)$ of the elements of the r-uple $(w_r, \cdots, w_1) \in \prod^r \mathscr{S}_C$ defined by

$$w_i = w(t_i, s_{i-1}) \quad (r \geqslant i \geqslant 1),$$

if

$$l_S(w) = l_S(\prod w_i) = l_S(w_r) + \cdots + l_S(w_1).$$

Then it is said that $w = \prod w_i$ is a $\mathscr{S}_C$-reduced expression of w of type g.

Let $w \in W$ (resp. $\overline{w} \in W/W_t$). It is known that the set of minimal galleries in the usual sense $\mathrm{Gall}'^m(C_e, C_w)$ (resp. $\mathrm{Gall}'^m(C_e, F_{\overline{w}})$) of $C(W, S)$, between C_e and $C_w = w(C_e)$ (resp. $F_{\overline{w}} = w(F_t(C_e)))$, is in bijection with the set of S-reduced expressions $\mathrm{Red}_S(w)$ (resp. $\mathrm{Red}_S(w^m)$) of w (resp. of the minimal

length element w^m of $\overline{w}$) (cf. [4], Ch. IV, Ex 16).
The set of Minimal Generalized Galleries $Gall^m(C_e, w^m(F_t(C)))$, between C_e and $w^m(F_t(C))$, is in bijection with the set of $\mathscr{S}_C$-reduced expressions of w^m. Retain the notation of Lemma 9.8. It is shown how a minimal generalized gallery $\gamma = \gamma(C_e, w^m(F_t(C)))$ gives rise to a $\mathscr{S}_C$-reduced expression of w^m of type typ γ.

Write $F_r = C_e$, and $s_i = \mathrm{typ}\, F_i$ (resp. $t_i = \mathrm{typ}\, F_i'$). Let $C_r, C_{r-1}, \cdots, C_0$ be the sequence of chambers, and $\Gamma(C_e, F) = \Gamma^r \circ \cdots \circ \Gamma^1$ the minimal gallery of $\mathcal{A}$ defined as in Lemma 9.8. Define for $r \geqslant i \geqslant 1$:

1. $w_i = w(C_i, C_{i-1})$;

2. $v_i = w_i \cdots w_r$;

3. $u_i = w(t_i, s_{i-1})$.

Clearly

$$v_1 = w(C_e, C_0) = w(C_e, \mathrm{proj}_F C_e)$$

(this results from the fact that C_e is the first term of a minimal gallery between C_e and F), and more generally

$$v_i = w(C_e, C_{i-1}).$$

There is:

$$w_i = w(C_i, C_{i-1}) = v_{i+1} w(t_i, s_{i-1}) v_{i+1}^{-1} = v_{i+1} u_i v_{i+1}^{-1},$$

as $v_{i+1}^{-1}(C_i) = C_e$, and $v_{i+1}^{-1}(\mathrm{proj}_{F_{i-1}} C_i) = \mathrm{proj}_{\overline{F}} C_e$, where $\overline{F} = F_{(t_i, s_{i-1})}^{\mathrm{tr}}(C_e)$. Let it be proved:

a) $w_i w_{i+1} \cdots w_r = u_r \cdots u_{i+1} u_i \ (r \geqslant i \geqslant 1)$;

b) $u_r \cdots u_1 = w(C_e, \mathrm{proj}_F C_e)$.

It is clear that b) follows from a). Thus it is proved a) by induction on i. It is clear that $w_r = u_r$. Let it be supposed that $i < r$ and $v_{i+1} = w_{i+1} \cdots w_r = u_r \cdots u_{i+1}$. One thus obtains

$$\begin{aligned} w_i w_{i+1} \cdots w_r &= (v_{i+1} u_i v_{i+1}^{-1}) u_r \cdots u_{i+1} = (v_{i+1} u_i v_{i+1}^{-1}) v_{i+1} \\ &= v_{i+1} u_i = u_r \cdots u_{i+1} u_i. \end{aligned}$$

Let it be proved now that $v_1 = u_r \cdots u_1$ is a $\mathscr{S}$-reduced expression. As Γ^i is a minimal gallery between C_i and $C_{i-1} = \mathrm{proj}_{F_{i-1}} C_i$ the transformed gallery $\Gamma_e^i := v_{i+1}^{-1}(\Gamma^i) \subset \mathrm{St}_{F_{t_i}}$ is also a minimal gallery between C_e and $\mathrm{proj}_{F_{(t_i, s_{i-1})}} C_e$. To Γ_e^i corresponds a reduced expression, relatively to S,

$$u_i = w(C_e, \mathrm{proj}_{F_{(t_i, s_{i-1})}} C_e) = \prod_{1 \leqslant \alpha_i \leqslant l_S(u_i)} u_i(\alpha_i) ,$$

with $u_i(\alpha_i) \in S$. Remark that

$$l_S(u_i) = \text{length } \Gamma_e^i = \text{length } \Gamma^i = |\mathcal{H}_{F_i'}(F_{i-1})|.$$

From Remark 9.10 a) it is found

$$l_S(u_r \cdots u_1) = l_S(w(C_e, \text{proj}_F C_e)) = \sum |\mathcal{H}_{F_i'}(F_{i-1})| = |\mathcal{H}(C_e, F)|.$$

It is then concluded that

$$w(C_e, \text{proj}_F C_e) = \prod_{r \geqslant i \geqslant 1} \left(\prod_{1 \leqslant \alpha_i \leqslant l_S(u_i)} u_i(\alpha_i) \right)$$

is an S-reduced expression, and finally that

$$w(C_e, \text{proj}_F C_e) = u_r \cdots u_1 = w(t_{r+1}, s_r) \cdots w(t_1, s_0)$$

is an $\mathscr{S}$-reduced expression.

Reciprocally to an $\mathscr{S}_C$-reduced expression $w(C_e, proj_F C_e) = \prod_{r \geqslant i \geqslant 1} w(t_i, s_{i-1})$
of type g corresponds a minimal generalized gallery $\gamma(C_e, F)$ of type g. Let $(w_i)_{r \leqslant i \leqslant 1}$ be the sequence of elements associated with $(w(t_i, s_{i-1}))_{r \geqslant i \geqslant 1}$ as above. Define a sequence of chambers $(C_i)_{r \geqslant i \geqslant 1}$ by $C_{i-1} = (w_i \cdots w_r)(C_e)$. Let $\gamma(C_e, F)$ be generalized gallery whose type is given by $g = (t_i, s_{i-1})_{r \geqslant i \geqslant 1}$ and its set of facets by $(F_{t_i}(C_i), F_{s_{i-1}}(C_i))_{r \geqslant i \geqslant 1}$. It is easy to see that $\gamma(C_e, F)$ is minimal (cf. loc. cit.). Thus a bijection between the set of $\mathscr{S}_C$-reduced expressions and the Minimal Generalized Galleries $Gall^m(C_e)$ issued from C_e is obtained.

9.4 Minimal Generalized Galleries

Notation 9.14 *The following notation is introduced to distinguish the four classes of galleries to which the following definition applies:*

i) $\gamma_2(F, F') = (F = F_r \supset F_r' \subset F_{r-1} \ldots F_1 \supset F_1' \subset F_0 = F')$ *a closed gallery;*

ii) $\gamma_2'(F, F') = (F = F_r \supset F_r' \subset F_{r-1} \supset F_{r-1}' \ldots F_1 \supset F_1' = F')$ *a right open gallery;*

iii) $\gamma_1(F, F') = (F = F_{r+1}' \subset F_r \supset F_r' \subset F_{r-1} \ldots F_1' \subset F_0 = F')$ *a left open gallery;*

iv) $\gamma_1'(F, F') = (F = F_{r+1}' \subset F_r \supset F_r' \subset F_{r-1} \ldots F_0 \supset F_0' = F')$ *an open gallery.*

Given $F'' \supset F$ one denotes by $\gamma_1(F''; F, F')$ (resp. $\gamma_1'(F''; F, F')$) the gg between F'' and F' given by the "composition" of $F'' \supset F$ and $\gamma_1(F, F')$ (resp. $\gamma_1'(F, F')$), and by $\gamma_2(F''; F, F')$ (resp. $\gamma_2'(F''; F, F')$), the gg $\gamma_2(F''; F, F)$: $F'' \supset F_r' \cdots F_1' \subset F_0 = F'$ (resp. $\gamma_2'(F''; F, F')$: $F'' \supset F_r' \cdots F_0 \supset F_0' = F'$) obtained by substitution of $F_r = F$ by F''.

Definition 9.15 *Given a couple of facets $(F, F') \in \mathcal{A} \times \mathcal{A}$ and $C \in Ch \, \mathrm{St}_F$ it is said that C is at maximal distance from F' if it satisfies $d(C, F') = \max_{C' \in Ch \, \mathrm{St}_F} d(C', F')$.*

The following definition is suggested by proposition 6.16.

Definition 9.16 *1) Let $(F, F') \in \mathcal{A} \times \mathcal{A}$ a* **Minimal Generalized Gallery (MGG)**, *$\gamma(F, F')$ is a gg of the form $\gamma(F, F') = \gamma_1(F, F')$ (resp. $\gamma_1'(F, F')$, $\gamma_2(F, F')$, $\gamma_2'(F, F')$), with $F_r \neq F_r'$ such that for every $C \in Ch \, \mathrm{St}_F$ at maximal distance from F' the gg $\gamma(C; F, F') = \gamma_1(C; F, F')$ (resp. $\gamma_1'(C; F, F')$, $\gamma_2(C; F, F')$, $\gamma_2'(C; F, F')$) is a MGG between C and F' (cf. Definition 9.6).*

9.4.1 Reformulation of the definition

It is recalled that the *carrier* L of $\mathrm{Env}(F, F')$ in $\mathcal{A}$ is obtained as the intersection of all the walls H of $\mathcal{A}$ containing F and F', i.e. $L = \bigcap_{\substack{H \in \mathcal{H}, \\ F, F' \in H}} H$. It is known that L is a building (but not a sub-building of $\mathcal{A}$ in general) whose chambers are the facets $F \subset L$ of minimal codimension, and that $\mathrm{Env}(F, F')$ is a sub-building of L.

The facet $\mathrm{proj}_F F' \in \mathrm{Env}(F, F')$ is characterized as the only facet containing F so that every facet of $\mathrm{Env}(F, F')$ containing F is contained in $\mathrm{proj}_F F'$. Thus $L_{\mathrm{proj}_F F'} = L$, where $L_{\mathrm{proj}_F F'}$ denotes the carrier of $\mathrm{proj}_F F'$, i.e. the maximal facet contained in the convex hull $Env(F, F')$ and incident to F.

Definition 9.17 *If $F \neq \mathrm{proj}_F F'$ (resp. $F = \mathrm{proj}_F F'$) it is written*

$$c(F, F') := F^{opp}(F, \mathrm{proj}_F F')$$

$$(\text{resp. } c(F, F') := \mathrm{proj}_F F' = F).$$

(cf. §2. e)). By definition the facet $c(F, F')$ satisfies the property

$$\mathcal{H}(c(F, F'), \mathrm{proj}_F F') = \mathcal{H}_F(\mathrm{proj}_F F')$$

as results from the Corollary 8.55. If $c(F, F') = F$ then $\mathcal{H}_F(\mathrm{proj}_F F') = \emptyset$. The facet $c(F, F')$ is a chamber of the building L. This means that the carrier $L_{c(F, F')}$ is equal to L.

Write
$$\mathcal{M} := \mathcal{H}(F, F') \coprod \mathcal{H}_F(F')$$

(the **set of walls separating** F **and** F' **or containing** F **without containing** F'), and $c := c(F, F')$.

Proposition 9.18 *With the above notation one has:*

1) If C is a chamber of St_F then $\mathcal{H}(C, F') \subset \mathcal{M}$.

2)

$$
\begin{aligned}
\text{Ch St}_c &= \{C \in \text{Ch St}_F | \ \mathcal{H}(C, F') = \mathcal{M}\} \\
&= \{C \in \text{Ch St}_F | \ d(C, F') = \max_{C' \in \text{Ch St}_F} d(C', F')\}.
\end{aligned}
$$

Proof *1) Let $H \in \mathcal{H}(C, F')$, it is first supposed that $H \notin \mathcal{H}_F$. Thus if $C \supset F \Rightarrow H \in \mathcal{H}(F, F') \subset \mathcal{M}$. If $H \in \mathcal{H}_F$ then $H \in \mathcal{H}_F(F')$ (F' is not contained in H as $H \in \mathcal{H}(C, F')$) thus $H \in \mathcal{M}$.*

To prove 2) it suffices to show that the first equality is satisfied. Let it first be seen that
$$C \supset c = c(F, F') \Rightarrow \mathcal{H}(C, F) = \mathcal{M}.$$
It is certain that in this case $\mathcal{H}(F, F') \subset \mathcal{H}(C, F')$. If $H \in \mathcal{H}_F(F') = \mathcal{H}_F(\text{proj}_F F')$, then H does not separate F and F'. Thus the root Φ determined by H and F' contains $\text{Env}(F, F')$, and a fortiori H does not separate $\text{proj}_{F'}F$ and F'. Then if $H \in \mathcal{H}_F(\text{proj}_F F')$ ($\neq \emptyset$) one has $\text{proj}_F F' \notin H$, and by definition of $c(F, F')$ the wall $H \supset F$ separates $c(F, F')$ and $\text{proj}_F F'$. Thus $C \supset c(F, F')$ and $H \in \mathcal{H}_F(F') \Rightarrow H \in \mathcal{H}(C, \text{proj}_F F')$. One finally has $H \in \mathcal{H}(C, F') = \mathcal{H}(C, \text{proj}_F F')$. It was proved that $C \supset c(F, F') \Rightarrow \mathcal{H}_F(F') \subset \mathcal{H}(C, F)$, and finally that $\mathcal{M} = \mathcal{H}(F, F') \coprod \mathcal{H}_F(F') \subset \mathcal{H}(C, F')$. Let $C \in \text{Ch St}_F$ be such that $\mathcal{H}(C, F') = \mathcal{M}$. It is first shown that $C \in \text{St}_c$ under the assumption $c \neq \text{proj}_F F'$. If $C \notin \text{St}_c$, i.e. $C \not\supset c$, there exists $H \in \mathcal{H}_F$ (= walls of St_F) separating C and c, i.e. $H \in \mathcal{H}(C, c)$. As $c \notin H$ there is $H \not\supset L_c$(carrier of c) $= L$(carrier of $\text{Env}(F, F')$). Thus by definition of c, H separates c and $\text{proj}_F F'$, i.e. $H \in \mathcal{H}(\text{proj}_F F', c)$. It is concluded that H does not separate C and $\text{proj}_F F'$. On the other hand, $H \supset F$ implies that H does not separate $\text{proj}_F F'$ and F', as it does not separate F and F'. It is concluded that $H \in \mathcal{H}_F(F')$ does not separate C and F'. This being contrary to the assumption $\mathcal{H}(C, F') = \mathcal{M} = \mathcal{H}(F, F') \coprod \mathcal{H}_F(F')$, it has been proved that $\mathcal{H}(C, F') = \mathcal{M} \Rightarrow C \in \text{St}_c$. If $c = \text{proj}_{F'}F$ then $F = \text{proj}_F F$ and there is nothing to prove.

$\Big($ *To prove that if $C \in St_F$ realizes the maximal distance between a chamber $C \in \text{Ch St}_F$ and F', then C is incident to $c(F, F')$, one may argue as follows. It may be assumed that $F \neq \text{proj}_F F'$. As a hyperplane $H \in \mathcal{H}_F$ does* *not separate F and F' and $c(F,F')^{opp} = proj_F\ F'$ in St_F, one has that H separates $c(F,F')$ from F'. Let $C' \in Ch\ St_F$ so that $c(F,F') \not\subseteq C'$. Thus for a chamber $c(F,F') \subset C$ there exists $H \in \mathcal{H}_F$ separating C and C' (resp. F'). It is concluded that $d(C',F') < d(C,F')$.*$\Big)$

Corollary 9.19 *The set of chambers $C \in St_F$ at maximal distance from F' is principal homogeneous under $W_{c(F,F')} = Stab\ c(F,F') \subset W(resp.\ W_F)$.*

The definition 9.16 may be reformulated as follows:

Definition 9.20 *1) Given $F, F' \in \mathcal{A}$ with $c(F,F') \neq F$ a **Minimal Generalized Gallery (MGG)**, $\gamma(F,F')$ is a gg of the form $\gamma(F,F') = \gamma_1(F,F')$ (resp. $\gamma'_1(F,F')$, $\gamma_2(F,F')$, $\gamma'_2(F,F')$), with $F_r \neq F'_r$ so that for every $C \in Ch\ St_{c(F,F')}$ the gg $\gamma(C;F,F') = \gamma_1(C;F,F')$ (resp. $\gamma'_1(C;F,F')$, $\gamma_2(C;F,F')$, $\gamma'_2(C;F,F')$) is a MGG between C and F' according to Definition 9.6.*

2) Given $F, F' \in \mathcal{A}$ with $c(F,F') = F$ is a Minimal Generalized Gallery (MGG) $\gamma(F,F')$ is a gg of the form $\gamma(F,F') = \gamma_2(F,F')$ (resp. $\gamma'_2(F,F')$) with $F_r \neq F'_r$ such that for every $C \in St_{c(F,F')}$ $\gamma(C;F,F') = \gamma_2(C;F,F')$ (resp. $\gamma'_2(C;F,F')$) is a MGG.

Proposition 9.21 *Let $\gamma(F,F')$ be a generalized gallery. Then $\gamma(F,F')$ is a MGG if and only if it satisfies the following conditions:*

1) $\gamma(F,F') \subset Env(F,F')$;

2) $\exists\ C \in St_{c(F,F')}$ so that $\gamma(C;F,F')$ is a MGG between C and F'.

Proof *It will be seen in the next sub-paragraphs, that: "$\gamma(F,F')$ is a MGG $\Rightarrow$ 1)". The condition 2) is verified by definition. If $\gamma(F,F')$ as in 1) satisfies $\gamma(F,F') \subset Env(F,F')$ then:*

$$\forall\ w \in W_{c(F,F')} \Rightarrow w(\gamma(F,F')) = \gamma(F,F').$$

Thus it is deduced that if $C \in St_{c(F,F')}$ and $w \in W_{c(F,F')}$, there is:

$$w(\gamma(C;F,F')) = \gamma(w(C);F,F'),$$

as $L_{c(F,F')} = carrier\ of\ Env(F,F')$. On the other hand, if it is supposed that $\gamma(C;F,F')$ is a MGG it is concluded that

$$\forall\ w \in W_{c(F,F')},\quad \gamma(w(C);F,F')\ \text{is a MGG},$$

(If $\gamma(c,F)$ is a MGG then: $\forall w \in W\ w(\gamma(c,F))$ is a MGG) and, as $Ch\ St_{c(F,F')}$ is principal homogeneous under $W_{c(F,F')}$, finally that $\forall\ C \in St_{c(F,F')}\ \gamma(C;F,F')$ is a MGG, i.e. $\gamma(F,F')$ is a MGG.

Proposition 9.22 *Let $C \in Ch \; \mathcal{A}$ and $(F, F') \in (\mathcal{A} \times \mathcal{A})_{\tau_0}$, with $\tau_0 \mapsto (s, t)$, so that $F_s(C) = F$ satisfying*

$$d(C, F') = max \; \{d(C, F'') \mid (F, F'') \in (\mathcal{A} \times \mathcal{A})_{\tau_0}\}$$

$$\left(resp. \; d(C, F') = \max_{\tau(F, F'') = \tau_0} d(C, F'') \right) .$$

Then $F' \in \mathcal{A}_t$ is unique with this property. Moreover

$$d(C, F') = \max_{\overline{C} \supset F} d(\overline{C}, F'),$$

i.e. F' is at maximal distance from the chamber $F \subset C$, and thus $C \supset c(F, F')$.

Proof *First observe that $d(C, F') = \max_{\tau(F, F'') = \tau_0} d(C, F'')$ implies $d(C, F') = \max_{\overline{C} \supset F} d(\overline{C}, F')$. Let $(C, F'') \in (Ch \; \mathcal{A} \times \mathcal{A})_{\tau_0}$ satisfying $d(C, F'') = d(C, F')$. From $\tau(F, F'') = \tau(F, F')$ it results that there exists $w \in W_F$ so that $F'' = w(F')$, and thus that $Env(F, F'') = w(Env(F, F'))$. Assume that $c(F, F') \neq F$. One has $d(C, F'') = \max_{\overline{C} \supset F} d(\overline{C}, F'')$ and $\text{proj}_F F'' = w(\text{proj}_F F')$. The couples $(C, \text{proj}_F F'')$ and $(C, \text{proj}_F F')$ are both in transversal position in St_F as $C \supset c(F, F') = (\text{proj}_F F')^{opp}$ (resp. $C \supset c(F, F'') = (\text{proj}_F F'')^{opp}$). It results $\text{proj}_F F'' = \text{proj}_F F'$ and $\text{proj}_F F'' = w(\text{proj}_F F')$, and w leaves $L_{(F,F')}$ pointwise fixed (cf. [4], Ch. V, §23, Prop. 1), and thus $F'' = F'$. The case $c(F, F') = F$ follows immediately from this last remark.*

Let the facet F' of Proposition 9.22 be determined in terms of W. Given $C \in Ch \; \mathcal{A}$ denote as usual by $S = S_C$ the set of generators of W given by the reflections defined by the walls of C. Let $i_C : C(W, S) \to \mathcal{A}$ be the building isomorphism corresponding to C. Under this isomorphism the relative position mapping

$$\tau : \; \mathcal{A} \times \mathcal{A} \to \text{Relpos} \; \mathcal{A},$$

corresponds to the mapping

$$\tau_{C(W,S)} : \; C(W, S) \times C(W, S) \to \coprod_{(s,t) \in \text{typ} \; \mathcal{A} \times \text{typ} \; \mathcal{A}} W_s \setminus W / W_t ,$$

defined by:

$$\tau_{C(W,S)} : (\overline{w}, \overline{w'}) \mapsto \text{double class of } w^{-1} w' \text{ in } W_s \setminus W / W_t ,$$

where $(\overline{w}, \overline{w'}) \in (W/W_s) \times (W/W_t)$. Clearly $\tau_{C(W,S)}$ is a well defined mapping. Given $\tau \in \text{Relpos} \; \mathcal{A}$ it is denoted by $\tilde{w}_\tau \in W_s \setminus W / W_t$ the double class corresponding to τ, i.e.

$$\tau(F_s(C), w_\tau(F_t(C))) = \tau.$$

Let $\tilde{w} \in W_s \backslash W/W_t$ be the double class defined by w. Observe that $\tilde{w}$ is a subset of W/W_t. Thus one has

$$\tilde{w} = \tilde{w}_\tau \iff \overline{w} \in \tilde{w} \subset W/W_t.$$

The isomorphism i_C induces a bijection between the elements in a double class and a class of facets:

$$\{\overline{w} \in W/W_t| \; \tilde{w} = \tilde{w}_\tau \; (\text{resp. } \overline{w} \in \tilde{w}_\tau)\} \to (\mathcal{A} \times \mathcal{A})_{(\tau, F_s(C))} =$$

$$\{F' \in \mathcal{A}_t \mid \tau(F_s(C), F') = \tau_0\},$$

defined by $\overline{w} \mapsto (F_s(C), w(F_t(C)))$. Let $\overline{w}_{F'} \in W/W_t$ be the class corresponding to F' as in Lemma 9.22, i.e. $w_{F'}(F_t(C)) = F'$. The elements $\overline{w} \in W/W_t$ such that $\overline{w} \in (w^{-1}w')$, where $(\overline{w}, \overline{w}') \mapsto (F, F')$, correspond to the facets $F' \in \mathcal{A}_t$ with $\tau(F_s(C), F') = \tau_0$. Thus there is only one "maximal length class" $\overline{w}^{max} \in (w^{-1}w')$ satisfying $(C, \overline{w}^{max}(F_t(C))$ are at maximal distance. Thus one has the following

Proposition 9.23 *With the above notation* $\overline{w}_{F'} \in W/W_t$ *is characterized by the following properties:*

1) $\tilde{w}_{F'} = \tilde{w}_\tau$ *(resp.* $\overline{w}_{F'} \in \tilde{w}_\tau$*);*

2) $l_S(\overline{w}_{F'})$ *is maximal on the set* $(w^{-1}w') \subset W/W_t$.

9.5   An expression for the convex Hull of two facets

An expression of the convex hull of two facets is obtained. It allows one to prove that a Minimal Generalized Gallery is contained in the convex hull of its extremities.

Given $F, F' \in \mathcal{A}$, write $\mathcal{H}_{F'}(F)$ for the set of hyperplanes containing F' but not F, and

$$E(F, F') := \{\Phi \in \mathcal{D}_\mathcal{A}| \; F, F' \in \Phi\}$$

(The set of Roots containing F and F'). By definition of the convex envelope one has $\mathrm{Env}(F, F') = \bigcap_{\Phi \in E} \Phi$. Define four subsets of $\mathcal{D}_\mathcal{A}$ in terms of F and F' as follows:

1) $\overline{\mathcal{D}}(F, F') = \{\Phi \in \mathcal{D}_\mathcal{A}| \; F, F' \notin \partial\Phi\} = (\partial)^{-1}(\mathcal{H} - (\mathcal{H}_F \cup \mathcal{H}_{F'}))$.

2) $\overline{\mathcal{D}}_{F'}(F) = \{\Phi \in \mathcal{D}_\mathcal{A}| \; \partial\Phi \in \mathcal{H}_{F'}(F)\} = (\partial)^{-1}(\mathcal{H}_{F'}(F))$.

3) $\overline{\mathcal{D}}_F(F') = \{\Phi \in \mathcal{D}_\mathcal{A}| \; \partial\Phi \in \mathcal{H}_F(F')\} = (\partial)^{-1}(\mathcal{H}_F(F'))$.

4) $\overline{\mathcal{D}}_L = \{\Phi \in \mathcal{D}_\mathcal{A}| \; F, F' \in \partial\Phi\} = \{\Phi \in \mathcal{D}_\mathcal{A}| \; L \subset \partial\Phi\} = (\partial)^{-1}(\mathcal{H}(L)$ $(L =$ carrier of $\mathrm{Env}(F, F'))$.

One has the partition of the set of roots $\mathcal{D}_A$ given by (F, F').:

$$\mathcal{D}_A = \overline{\mathcal{D}}(F, F') \coprod \overline{\mathcal{D}}_{F'}(F) \coprod \overline{\mathcal{D}}_F(F') \coprod \overline{\mathcal{D}}_L.$$

This partition induces a partition of the se $E(F, F')$. Write $E = E(F, F')$ and $c = c(F, F')$.

Write:

1)' $\mathcal{D}(F, F') = \overline{\mathcal{D}}(F, F') \cap E = \{ \Phi \mid F, F' \in \Phi \};$

2)' $\mathcal{D}_{F'}(F) = \overline{\mathcal{D}}_{F'}(F) \cap E = \{ \Phi \mid F' \in \partial\Phi, \ F \in \Phi - \partial\Phi \};$

3)' $\mathcal{D}_F(F') = \overline{\mathcal{D}}_F(F') \cap E = \{ \Phi \mid F' \in \partial\Phi, \ F \in \Phi - \partial\Phi \};$

4)' $\mathcal{D}_L = \overline{\mathcal{D}}_L \cap E = \overline{\mathcal{D}}_L$

Retain the above notation. Observe that if $C \in Ch \ St_c$ then $C \supset c \supset F$.

Proposition 9.24

a) $E = \mathcal{D}(F, F') \coprod \mathcal{D}_{F'}(F) \coprod \mathcal{D}_F(F') \coprod \mathcal{D}_L;$

b) $E' = \{\Phi \in \mathcal{D}_A \mid c, F' \in \Phi\} = \mathcal{D}(c, F') \coprod \mathcal{D}_{F'}(c) \coprod \mathcal{D}_L$ and $\mathrm{Env}(c, F') = \bigcap_{\Phi \in E'} \Phi;$

c) $E'' = \{\Phi \in \mathcal{D}_A \mid C, F' \in \Phi\} = \mathcal{D}(C, F') \coprod \mathcal{D}_{F'}(C)$ and $\mathrm{Env}(C, F') = \bigcap_{\Phi \in E''} \Phi;$

d) $\mathcal{D}_{F'}(C) = \mathcal{D}_L(C) \coprod \mathcal{D}_{F'}(F)$ $(= \mathcal{D}_{F'}(c))$, where $\mathcal{D}_L(C) = \{\Phi \mid L \subset \partial\Phi, \ C \in \Phi\}.$

Proof *The above partition of $\mathcal{D}_A$ gives rise to the following one of the subset $E \subset \mathcal{D}_A$:*

$$E = \mathcal{D}_A \cap E = \overline{\mathcal{D}}(F, F') \cap E \coprod \overline{\mathcal{D}}_{F'}(F) \cap E \coprod \overline{\mathcal{D}}_F(F') \cap E \coprod \overline{\mathcal{D}}_L \cap E,$$

which may be re-written as $E = \mathcal{D}(F, F') \coprod \mathcal{D}_{F'}(F) \coprod \mathcal{D}_F(F') \coprod \mathcal{D}_L$. The disjoint union of a) follows from

$$E = \mathcal{D}_A \cap E = \overline{\mathcal{D}}(F, F') \cap E \coprod \overline{\mathcal{D}}_{F'}(F) \cap E \coprod \overline{\mathcal{D}}_F(F') \cap E \coprod \overline{\mathcal{D}}_L \cap E.$$

One obtains b) (resp. c)), as particular cases of a), by writing $F = c = c(F, F')$ (resp. $F = C$) in the equality of a). Remark that $E' = E(c, F')$ and $E'' = E(C, F')$. Thus b) follows from a) by writing $F = c(F, F')$ and c) by writing $F = C$. From the inclusion $\mathcal{D}_{F'}(C) \subset E =$

$\mathcal{D}(F, F') \coprod \mathcal{D}_{F'}(F) \coprod \mathcal{D}_F(F') \coprod \mathcal{D}_L$, and on account of $\mathcal{D}_{F'}(C) \cap \mathcal{D}_{F'}(F) = \mathcal{D}_{F'}(F)$, $\mathcal{D}_{F'}(C) \cap \mathcal{D}(F, F') = \emptyset$, and $\mathcal{D}_{F'}(C) \cap \mathcal{D}_F(F') = \emptyset$ one obtains

$$
\begin{aligned}
\mathcal{D}_{F'}(C) &= \{\Phi \in \mathcal{D}_A| \ F' \in \partial\Phi, \ (F \subset)C \in \Phi\} \\
&= \{\Phi \in \mathcal{D}_A| \ \underbrace{F', F \in \partial\Phi}_{L \subset \partial\Phi}, \ (F \subset) \ C \in \Phi\} \\
&\coprod \ \{\Phi \in \mathcal{D}_A| \ F' \in \partial\Phi, \ F \notin \partial\Phi, \ C \in \Phi\} \ (= \mathcal{D}_{F'}(F), \text{ as } F \subset C) \\
&= \mathcal{D}_L(C) \coprod \mathcal{D}_{F'}(F) \ (= \mathcal{D}_{F'}(c)).
\end{aligned}
$$

Remark that if $\Phi \in \mathcal{D}_{F'}(F)$ then $C \in \Phi$ as $F \subset C$.

Notation 9.25 *The following convention is used in: if $\mathcal{F} = (E_i)_{i \in I}$ is a family of sets one writes $\bigcap \mathcal{F} = \bigcap_{i \in I} E_i$.*

From a), b), c), and "c)+d)" it is respectively obtained:

Proposition 9.26

a)' $\mathrm{Env}(F, F') = (\bigcap \mathcal{D}(F, F'))(\bigcap \mathcal{D}_{F'}(F))(\bigcap \mathcal{D}_F(F'))(\bigcap \mathcal{D}_L) \ (= L);$

b)' $\mathrm{Env}(c, F') = (\bigcap \mathcal{D}(c, F'))(\bigcap \mathcal{D}_{F'}(c))(\bigcap \mathcal{D}_L)(= L);$

c)' $\mathrm{Env}(C, F') = (\bigcap \mathcal{D}(C, F'))(\bigcap \mathcal{D}_{F'}(C));$

d)' $\mathrm{Env}(C, F') = (\bigcap \mathcal{D}(C, F'))(\bigcap \mathcal{D}_L(C))(\bigcap \mathcal{D}_{F'}(c)).$

Proof *The four statements result respectively from a), b), c), and "c)+d)"*.

Remark 9.27 *One has $\Phi \in \mathcal{D}_L \Leftrightarrow \mathrm{Env}(F, F') \subset \partial\Phi$. Thus $\Phi \in \mathcal{D}_L \Leftrightarrow \Phi^{opp}$ (opposed root to Φ) $\in \mathcal{D}_L$, as $\partial\Phi = \partial\Phi^{opp}$. It is then deduced that:*

$$
\bigcap \mathcal{D}_L = \bigcap_{\Phi \in \mathcal{D}_L} (\Phi \cap \Phi^{opp}) = \bigcap_{\Phi \in \mathcal{D}_L} \partial\Phi = \bigcap_{H \in \mathcal{H}_L} H = L.
$$

Proposition 9.28

e)' $\mathrm{Env}(c, F') = (\bigcap(\mathcal{D}_F(F') \cap \mathcal{D}(c, F'))) \bigcap (\cap \mathcal{D}(F, F')) \bigcap (\cap \mathcal{D}_{F'}(F))L.$

f)' $\mathrm{Env}(F, F') = (\bigcap \mathcal{D}_F(F')) \bigcap \mathrm{Env}(c(F, F'), F').$

Proof *As $\mathcal{D}(c, F') = \{\Phi \ | \ \Phi \supset \mathrm{Env}(c, F'), \ c, F' \notin \Phi\}$ and in view $c \supset F$ one has*

$$
\mathcal{D}(c, F') = \mathcal{D}(F, F') \coprod \mathcal{D}_F(F') \bigcap \mathcal{D}(c, F') \ .
$$

By substitution of $\mathcal{D}(c, F')$ by the second member of the above equality in b)', and by remarking that $\mathcal{D}_{F'}(c) = \mathcal{D}_{F'}(F)$, one obtains e').

By a)', $\mathrm{Env}(F, F') = (\bigcap \mathcal{D}(F, F'))(\bigcap \mathcal{D}_{F'}(F))(\bigcap \mathcal{D}_F(F'))(\bigcap \mathcal{D}_L) \ (= L) =;$

$(\bigcap \mathcal{D}_{F'}(F))(\bigcap(\mathcal{D}_F(F') \cap \mathcal{D}(c, F')))(\bigcap \mathcal{D}(F, F'))(\bigcap \mathcal{D}_F(F'))L$

by remarking that the inclusion $\mathcal{D}_F(F') \cap \mathcal{D}(c, F') \subset \mathcal{D}_F(F')$ *gives* $\bigcap \mathcal{D}_F(F') \subset \bigcap(\mathcal{D}_F(F') \cap \mathcal{D}(c, F'))$. *One concludes by* e') *that* $\mathrm{Env}(F, F') = (\bigcap \mathcal{D}_F(F')) \bigcap \mathrm{Env}(c(F, F'), F')$.

Let $c \subset C$. By definition of $\mathcal{D}_L(C)$: $\mathcal{D}_L(C) = \{\Phi \in \mathcal{D}_\mathcal{A} | \partial \Phi \supset L, C \in \Phi\}$, one has $\mathcal{D}_L(C) = \mathcal{D}_c(C)$, as "carrier of c" $= L$.

Lemma 9.29 *The intersection of the family of convex subcomplexes* $(\bigcap \mathcal{D}_L(C))_{C \in \mathrm{Ch}\ \mathrm{St}_c}$ *is given by*

$$\bigcap_{C \in \mathrm{Ch}\ \mathrm{St}_c} (\bigcap \mathcal{D}_L(C)) = L.$$

Proof *Let* $C \in \mathrm{Ch}\ \mathrm{St}_c$ *and* $C^{opp} = F^{opp}(c, C)$ *the chamber opposed to* C *relatively to* c *(resp. in* St_c*). By definition of* C^{opp} *one has:*

$$\Phi \in \mathcal{D}_c(C) \Leftrightarrow \Phi^{opp} \in \mathcal{D}_c(C^{opp}).$$

It is concluded that

$$(\bigcap \mathcal{D}_c(C)) \bigcap (\bigcap \mathcal{D}_c(C^{opp})) = \bigcap_{\Phi \in \mathcal{D}_c(C)} (\Phi \cap \Phi^{opp})$$

$$= \bigcap_{\Phi \in \mathcal{D}_c(C)} \partial \Phi = \bigcap_{H \in \mathcal{H}_c} = L.$$

The equality of the lemma follows from:

$$\bigcap_{C \in \mathrm{St}_c} (\bigcap \mathcal{D}_c(C)) = \bigcap_{C \in \mathrm{St}_c} ((\bigcap \mathcal{D}_c(C)) \bigcap (\bigcap \mathcal{D}_c(C^{opp}))) = L.$$

Proposition 9.30 *One has:*

1)

$$\bigcap_{C \in \mathrm{St}_c} \mathrm{Env}(C, F') = \mathrm{Env}(c, F').$$

2)

$$\mathrm{Env}(F, F') = (\bigcap \mathcal{D}_F(F')) \bigcap (\bigcap_{C \in \mathrm{St}_c} \mathrm{Env}(C, F')).$$

Proof *As 2) follows immediately from* f)$'$ *and 1) let it be proved 1). From* d)$'$ *it is obtained:*

$$\bigcap_{C \in \mathrm{St}_c} \mathrm{Env}(C, F') = \bigcap_{C \in \mathrm{St}_c} [(\bigcap \mathcal{D}(C, F'))(\bigcap \mathcal{D}_L(C))(\bigcap \mathcal{D}_{F'}(c))]$$

$$= \Big[\bigcap_{C \in \mathrm{St}_c} (\bigcap \mathcal{D}(C, F')) \Big] \Big[\bigcap_{C \in \mathrm{St}_c} (\bigcap \mathcal{D}_L(C))][\bigcap \mathcal{D}_{F'}(c) \Big].$$

On the other hand, the inclusion $(C \supset c)$ $\mathcal{D}(c, F') \subset \mathcal{D}(C, F')$, implies

$$\bigcap \mathcal{D}(C, F') \subset \bigcap \mathcal{D}(c, F')$$

and gives

$$\bigcap_{C \in \mathrm{St}_c} \left(\bigcap \mathcal{D}(C, F') \right) \subset \bigcap \mathcal{D}(c, F').$$

Using the preceding lemma finally the inclusion is obtained:

I) $\displaystyle\bigcap_{\substack{C \in \mathrm{St}_c \\ c \subset C}} \mathrm{Env}(C, F') \subset \left(\bigcap \mathcal{D}(c, F') \right) L \left(\bigcap \mathcal{D}_{F'}(c) \right) = \mathrm{Env}(c, F')$ *by b)'. As one has that:*

II) $\mathrm{Env}(c, F') \subset \displaystyle\bigcap_{C \in \mathrm{St}_c} \mathrm{Env}(C, F')$

The equality of 1) results from I) and II).

9.6   The convex envelope of a Minimal Generalized Gallery

Here it is proved the

Proposition 9.31 *Let $\gamma(F, F')$ be a MGG between F and F' then*

$$\gamma(F, F') \subset \mathrm{Env}(F, F'),$$

i.e. all the facets of $\gamma(F, F')$ belong to $\mathrm{Env}(F, F')$.

Let it first be shown that:

Proposition 9.32

$$\gamma(F, F') \subset \mathrm{Env}(c(F, F'), F').$$

Proof   *Let $\gamma(C; F, F')$, for $C \supset c(F, F')$, be defined as in Proposition 9.21. Let $\Gamma'(C, F')$ be the minimal gallery of Lemma 9.8 given by $\gamma(C; F, F')$. One has*

$$\Gamma'(C, F') \subset \mathrm{Env}(C, F').$$

By construction of $\Gamma'(C, F')$ every facet of $\gamma(C; F, F')$, and a fortiori of $\gamma(F, F')$, is incident to some chamber of $\Gamma'(C, F')$. It is thus obtained by the reformulation of Definition 9.16

$$C \in \mathrm{St}_{c(F, F')} \Rightarrow \gamma(C; F, F') \subset \mathrm{Env}(C, F')$$

(resp. $\gamma(F, F') \subset \mathrm{Env}(C, F')$). It is concluded that

$$\gamma(F, F') \subset \bigcap_{C \in \mathrm{St}_{c(F, F')}} \mathrm{Env}(C, F') = \mathrm{Env}(c(F, F'), F') \ (cf. \ \text{Proposition 9.30}).$$

Observe that the inclusion is evident in the case $\gamma(F, F') = \gamma_1(F, F')$ (resp. $\gamma_1'(F, F')$). To obtain this inclusion in the case $\gamma(F, F') = \gamma_2(F, F')$ (resp. $\gamma_2'(F, F')$) remark that $F \subset c(F, F')$.

Remark 9.33 *If $F = \mathrm{proj}_F F' = c(F, F')$ then*

$$\mathrm{Env}(F, F') = \mathrm{Env}(c(F, F'), F'),$$

and in this case one has

$$\gamma(F, F') \subset \mathrm{Env}(F, F'),$$

i.e. the assertion of Proposition 9.31 is verified.

9.6.1 Proof of Proposition 9.31

The case $F = c(F, F')$ follows immediately as pointed in the above remark. Let one consider $F \neq c(F, F')$. The case $c(F, F') \neq F$, $\gamma(F, F') = \gamma_1(F, F')$ (resp. $\gamma_1'(F, F')$) is first treated.

By f)' above one has

$$\mathrm{Env}(F, F') = \left(\bigcap \mathcal{D}_F(F')\right) \bigcap \mathrm{Env}(c(F, F'), F').$$

Thus in order to prove that $\gamma(F, F') \subset \mathrm{Env}(F, F')$ one needs, in view of $\gamma(F, F') \subset \mathrm{Env}(c(F, F'), F')$, only to see that:

$$\Phi \in \mathcal{D}_F(F') \Rightarrow \gamma(F, F') \subset \Phi.$$

Let C be a chamber so that $C \supset c(F, F')$, i.e. at maximal distance from F' (thus transversal to $\mathrm{proj}_F F'$ in St_F), and $\Gamma'(C, F')$ the minimal gallery of Lemma 9.8 adapted to $\gamma(C; F, F')$. It is recalled that $\Gamma'(C, F')$ is defined by the sequence of chambers

$$C_{r+1} = C, \quad C_i = \mathrm{proj}_{F_i} C_{i+1} \quad (r \geqslant i \geqslant 0),$$

where $F_i = F_i(\gamma(F, F'))$, i.e. F_i denotes the i-th facet of $\gamma(F, F') = (F = F_{r+1}' \subset F_r \supset F_r' \subset F_{r-1} \dots F_1' \subset F_0 = F')$ and a sequence of choices of minimal galleries

$$\Gamma^{i+1} = \Gamma^{i+1}(C_{i+1}, C_i) \subset \mathrm{St}_{F_{i+1}'} \quad (F_{i+1}' = F_{i+1}'(\gamma(F, F'))) \quad (r \geqslant i \geqslant 0).$$

The minimal gallery $\Gamma'(C, F')$ is obtained as the composed gallery

$$\Gamma'(C, F') = \Gamma^{r+1} \circ \cdots \circ \Gamma^1.$$

Lemma 9.34 *Let $H \in \mathcal{H}_{F_{i+1}'}(F_i)$. Denote by $\Phi_H(C_0) =$ the root of $\mathcal{A}$ defined by $\partial \Phi_H(C_0) = H$ and $C_0 \subset \Phi_H(C_0)$. There are the inclusions:*

$$\Phi_H(C_0) \supset C_i, C_{i-1}, \cdots, C_0$$

and a fortiori

$$\Phi_H(C_0) \supset F_i, F_{i-1}, \cdots, F_0$$

(resp. $\Phi_H(C_0) \supset F'_{i+1}, \cdots, F'_1$).

Proof *The proof follows from the following remark. By Remark 9.10 b) it is known that*

$$\mathcal{H}(C_{i+1}, C_i) = \mathcal{H}_{F'_{i+1}}(F_i).$$

Thus $\mathcal{H}(C, F') = \coprod_{r \geqslant i \geqslant 0} \mathcal{H}_{F'_{i+1}}(F_i)$, and the uple $\Psi^*(\Gamma^{i+1})$ of walls that Γ^{i+1} crosses, gives the set $\mathcal{H}_{F'_{i+1}}(F_i)$ in a certain order. It is concluded that a hyperplane $H \in \mathcal{H}_{F'_{i+1}}(F_i)$ cannot separate the chambers C_i and C_0, and a fortiori the chambers C_j and C_0 with $0 < j < i$ which are contained in $Env(C_i, C_0)$.

    Let
$$\Phi \in \mathcal{D}_F(F') = \{\Phi \in \mathcal{D}_{\mathcal{A}} |\ F, F' \in \Phi,\ \partial\Phi \in \mathcal{H}_F(F')\}\ .$$

Then $Env(F, F') \subset \Phi$ and

$$H = \partial\Phi \in \mathcal{H}_F(\text{proj}_F F') = \mathcal{H}_F(F')\ .$$

Thus H separates $c(F, F')$ and $\text{proj}_F F'$ but does not separate F and F'. It is recalled that $c(F, F')$ being the opposed facet to $\text{proj}_F F'$ in St_F implies that every $H \in \mathcal{H}_F(\text{proj}_F F')$ separates $c(F, F')$ and $\text{proj}_F F'$. On the other hand, from $C \supset c(F, F')$, and $\text{proj}_F F' \in Env(F, F')$ it follows that:

- H separates C and $\text{proj}_F F'$;

- H separates C and $Env(F, F')$ as $\text{proj}_F F' \in Env(F, F')$.

It is concluded that $\Phi \supset Env(F, F')$, and "$\partial\Phi$ separates C and F'". Thus

$$H = \partial\Phi \in \mathcal{H}(C, F') = \coprod \mathcal{H}_{F'_{i+1}}(F_i).$$

Let $r \geqslant i_0 = i_0(H) \geqslant 0$ be defined by

$$H \in \mathcal{H}_{F'_{i_0+1}}(F_{i_0}).$$

Lemma 9.35 *Retain the above notation and assume $i_0(H) < r$. There is then $F_r, F_{r-1} \cdots F_{i_0+1} \in H$ $(H \in \mathcal{H}_F(F'))$, in other terms:*

$$\text{"If } H \notin \mathcal{H}_{F'_{r+1}}(F_r) \text{ then } i_0(H) = \min\{i|\ F_r, \cdots, F_i \in H\} - 1\text{"}.$$

Proof *Clearly if $H \in \mathcal{H}_{F'_{r+1}}$ and $H \notin \mathcal{H}_{F'_{r+1}}(F_r)$ one has $F_r \in H$. Let it be seen that:*

$$F_r, \cdots, F_{i+1} \in H \text{ and } H \notin \mathcal{H}_{F'_{i+1}}(F_i) \Rightarrow F_i \in H.$$

As $F_{i+1} \in H$, and $F'_{i+1} \subset F_{i+1}$ it is deduced that $H \in \mathcal{H}_{F'_{i+1}}$. Thus if it is supposed $H \notin \mathcal{H}_{F'_{i+1}}(F_i)$ it is concluded that $F_i \in H$.

Thus at each step there are two issues $F_i \in H$ or $F_i \notin H$, and the process stops if $F_i \notin H$. The hypothesis $H \in \mathcal{H}_F(F')$, with $F = F'_{r+1}$ (resp. $F' = F_0$), gives $\min\{i|\ F_r, \cdots F_i \in H\} \geqslant 1$. It follows from the above reasoning that the integer $i_0(H)$ satisfies:

1. $i_0(H) \geqslant 1$;

2. $F_r, \cdots, F_{i_0(H)+1} \in H$, and $F_{i_0(H)} \notin H$.

Clearly these properties characterize $i_0(H)$.

The proof of Proposition 9.31 in the case $c(F, F') \neq F$ and $\gamma(F, F')$ a left open gallery, thus let $\gamma(F, F') = \gamma_1(F, F')$ (resp. $\gamma'_1(F, F')$) runs as follows. Let $H \in \mathcal{H}_F(F')$. If $H \in \mathcal{H}_{F'_{r+1}}(F_r)$ on the basis of Lemma 9.34 one has that

$$\Phi_H(F') \supset \Phi_H(C_0) \supset F_r, \cdots, F_0$$

and a fortiori $\gamma(F, F') \subset \Phi_H(C_0)$. Otherwise on the basis of Lemma 9.35 there exists an integer $i_0(H)$ so that:

$$F_r, \cdots, F_{i_0(H)+1} \in H$$

(resp. $F'_{r+1}, \cdots, F'_{i_0(H)+1} \in H$), and $H \in \mathcal{H}_{F'_{i_0(H)+1}}(F_{i_0(H)})$. On the basis of Lemma 9.34 it is concluded that

$$\Phi_H(F_0) = \Phi_H(C_0) \supset F_{i_0(H)}, \cdots, F_0$$

(resp. $F'_{i_0(H)}, \cdots, F'_1$), and finally that $\gamma(F, F') \subset \Phi_H(F')$.

Proof of Proposition 9.31 in the case $c(F, F') \neq F$, and $\gamma(F, F')$ a left closed gallery $\gamma(F, F') = \gamma_2(F, F')$ (resp. $\gamma(F, F') = \gamma'_2(F, F')$)

Given $C \supset c(F, F') \supset F = F_r$, and $H \in \mathcal{H}_F(F')$, let $\gamma(C; F, F')$ be obtained from $\gamma(F, F')$ as in Definition 9.16. One proceeds to construct $\Gamma'(C, F')$ as in Lemma 9.8. It is deduced then using Lemmas 9.34 and 9.35 as above that

$$\Phi_H(F') \supset F'_r, F_{r-1}, \cdots, F'_1, F_0.$$

By hypothesis $F_r \in H (\in \mathcal{H}_F(F'))$ thus $F \in H$ and consequently $\gamma(F, F') \subset \Phi_H(F')$.

9.7 Minimal Generalized Galleries unicity properties

Given a Minimal Generalized Gallery:

$$\gamma(C, F'): \ C = F_r \supset F'_r \cdots F_0 \supset F'_0 = F'$$

between a chamber C of $\mathcal{A}$ and $F' \in \mathcal{A}$, write:

$$s_i = \text{typ } F_i \quad (\text{resp. } t_i = \text{typ } F'_i) \quad (r \geqslant i \geqslant 0).$$

Let $g = \text{typ } \gamma : s_r \supset t_r \cdots s_0 \subset t_0$ be the corresponding gallery of types defined by $\gamma = \gamma(C, F')$.

Recall that given a chamber C of $\mathcal{A}$ and a couple of types $t \subset s$ ($t \neq s$) of $\mathcal{A}$ it is denoted by $F^{tr}_{(t,s)}(C)$ the unique facet of $\text{St}_{F_t(C)}$, with typ $F^{tr}_{(t,s)}(C) = s$, transversal to $C \in \text{Ch } \text{St}_{F_t(C)}$ in the star complex $\text{St}_{F_t(C)}$, i.e. $F^{tr}_{(t,s)}(C) =$ facet of type s of $\text{St}_{F_t(C)}$ at maximal distance from C.

Proposition 9.36 *Let* $\gamma(C, F') = \gamma'_2(C, F')$ *(resp.* $\gamma(C, F') = \gamma_2(C, F')$*) be a right open (resp. closed) Minimal Generalized Gallery of type* $g = \text{typ } \gamma(C, F')$*. Then* $\gamma(C, F')$ *is the unique Minimal Generalized Gallery issued from* C *of type* g*.*

Proof *By induction on* $r \geqslant i \geqslant 0$ *one defines a sequence of triples of facets* $(\overline{C}_i, \overline{F}_i, \overline{F'}_i)$ *of* $\mathcal{A}$ *such that:*

$$\overline{C}_i \in \text{Ch } \mathcal{A} \quad (\text{resp. typ } \overline{F}_i = s_i, \text{ typ } \overline{F'}_i = t_i),$$

as follows. Write

$$C_r = C, \quad \overline{F}_r = C, \quad \overline{F'}_r = F_{t_r}(C) = F_{t_r}(C_r),$$

and if $r > i$

$$\overline{F'}_i = F_{t_i}(C_{i+1}), \quad \overline{F}_{i+1} = F^{tr}_{(t_i, s_i)}(C_{i+1}), \quad C_i = \text{proj}_{\overline{F}_{i+1}} C_{i+1}.$$

Let $(C_i)_{r \geqslant i \geqslant 0}$ *be the sequence of chambers associated with* $\gamma(C, F')$ *as in Lemma 9.8, and defined by:*

$$C_r = C, \quad C_i = \text{proj}_{F_i} C_{i+1} \quad (r > i \geqslant 0).$$

Clearly C *and sequence of facets* $(F_i)_{r \geqslant i \geqslant 0}$ *(resp.* $(F'_i)_{r \geqslant i \geqslant 0}$*) define a generalized gallery:*

$$\gamma_g(C, F') : \quad C = \overline{F}_r \supset \overline{F'}_r \cdots \overline{F}_0 \supset \overline{F'}_0 = F'$$

By an easy induction one has $\overline{C}_i = C_i$ ($r \geqslant i \geqslant 0$). It is clear that $\overline{C}_r = C_r$, $\overline{F}_r = F_r$, $\overline{F'}_r = F'_r$. It is supposed for $r \geqslant i \geqslant 1$ that $\overline{C}_i = C_i$, $\overline{F}_i = F_i$, $\overline{F'}_i = F'_i$. It is deduced that $\overline{F}_{i-1} = F^{tr}_{(t_i, s_{i-1})}(C_i) = F_{i-1}$, as $\mathcal{H}(C_i, F_{i-1}) = \mathcal{H}_{F'_i}(F_{i-1})$, and this proves that $\overline{C}_i = C_i$ and F_{i-1} are transversal (by Lemma 9.4 there is only one facet of type s_{i-1} transversal to C_i relatively to $\overline{F'}_i = F_i$). It is concluded that $\overline{C}_{i-1} = C_{i-1} = \text{proj}_{F_i} C_i$. It *has thus been proved that with $C \in$ Ch and $g \in$ gall$_A$ there is associated a generalized gallery $\gamma_g(C, F')$ satisfying:*

$$g \in \text{gall}_A \text{ and } \gamma(C, F') \text{ a mgg of type } g \Longrightarrow \gamma(C, F') = \gamma_g(C, F') \text{ .}$$

This implies that if $\gamma(C, F')$ is a MGG of type g then it is unique to this property.

It is clear that the preceding argument proves *mutatis mutandis* the proposition in the case $\gamma(C, F') = \gamma_2(C, F')$.

Corollary 9.37 *Let $\gamma(F, F')$ and $\overline{\gamma}(F, F')$ be two Minimal Generalized Galleries with extremities F and F' of the same type, i.e. so that* typ $\gamma(F, F') =$ typ $\overline{\gamma}(F, F')$ *between F and F'. Then $\gamma(F, F') = \overline{\gamma}(F, F')$.*

Proof *Let $C \supset c(F, F')$. By Definition 9.16 $\gamma(C; F, F')$ (resp. $\overline{\gamma}(C; F, F')$) is a MGG between C and F' and* typ $\gamma(C; F, F') =$ typ $\overline{\gamma}(C; F, F')$. *On the basis of Proposition 9.36 it is concluded that $\gamma(C; F, F') = \overline{\gamma}(C; F, F')$ and a fortiori $\gamma(F, F') = \overline{\gamma}(F, F')$.*

The following proposition allows assigning to a type of relative position a set of minimal galleries of types. This set furnishes a family of smooth resolutions of a Schubert variety defined by this type.

Proposition 9.38 *Let $\gamma(F, F')$ (resp. $\overline{\gamma}(F, F')$) be a Minimal Generalized Gallery (resp. a generalized gallery) between F and F' of type g. Then*

$$\overline{\gamma}(F, F') = \gamma(F, F') \text{ ,}$$

i.e. if there exists a MGG $\gamma(F, F')$ between F and F' of type g, this is the unique gg of type g between F and F'. In this case one writes $\gamma_g(F, F') = \gamma(F, F')$.

Proof *For the sake of briefness the proof in the case $g = g'_1$, $l(g) = r + 1$ (see i) below) is carried out. Thus write*

$$\gamma(F, F') : \quad F = F'_{r+1} \subset F_r \cdots F_0 \supset F'_0 = F'$$

$$(\text{resp. } \overline{\gamma}(F, F') : \quad F = \overline{F'}_{r+1} \subset \overline{F}_r \cdots \overline{F}_0 \supset \overline{F'}_0 = F').$$

Choose a chamber $C \supset c(F, F')$ (cf. Proposition 9.18) and define:

$$\overline{C}_{r+1} = C, \quad \overline{C}_i = \text{proj}_{\overline{F}_i} \overline{C}_{i+1} \quad (r \geqslant i \geqslant 0).$$

Observe that $\overline{C}_{i+1} \supset \overline{F'}_{i+1} \subset \overline{F}_i$, and choose a Minimal Gallery $\overline{\Gamma}^{i+1} = \overline{\Gamma}^{i+1}(\overline{C}_{i+1}, \overline{C}_i)$, between $\overline{C}_{i+1}$ and $\overline{C}_i$. The set Ch St$_{\overline{F'}_{i+1}}$ being a convex set

of chambers, one then has $\overline{\Gamma}^{i+1} \subset \text{Ch St}_{\overline{F}'_{i+1}}$ *and thus "set of walls given by*
$\Psi^*(\overline{\Gamma}^{i+1}) \subset \mathcal{H}_{\overline{F}'_{i+1}}(\overline{F}_i)$". *Write*

$$\overline{\Gamma}(C, F') = \overline{\Gamma}^{r+1} \circ \cdots \circ \overline{\Gamma}^1$$

(the composed gallery given by $(\overline{\Gamma}^i)_{r+1 \geqslant i \geqslant 1}$ *which is looked at as a gg between*
C *and* F'), *and* $|\Psi^*(\overline{\Gamma}(C, F'))| = $ *number of components of* $\Psi^*(\overline{\Gamma}(C, F'))$ *(resp.*
$|\Psi^*(\overline{\Gamma}^i(C_i, C_{i-1}))| = $ *number of components of* $\Psi^*(\overline{\Gamma}^i(C_i, C_{i-1})))$. *There is*

$$|\Psi^*(\overline{\Gamma}^i(C_i, C_{i-1}))| \leqslant |\mathcal{H}_{\overline{F}'_i}(\overline{F}_{i-1})|$$

$$(\text{resp. } |\Psi^*(\overline{\Gamma}(C, F'))| = \sum_{r+1 \geqslant i \geqslant 1} |\Psi^*(\overline{\Gamma}^i(C_i, C_{i-1}))|).$$

On the other hand, the hypothesis $\gamma(F, F') = MGG$ *(cf. Definition 9.11) gives*
the equality

$$\sum |\mathcal{H}_{\overline{F}'_i}(\overline{F}_{i-1})| = \sum |\mathcal{H}_{F'_i}(F_{i-1})| = |\mathcal{H}(C, F')|.$$

Thus it is deduced
$$|\Psi^*(\overline{\Gamma}(C, F'))| \leqslant |\mathcal{H}(C, F')|.$$

This proves that $\overline{\Gamma}(C, F')$ *is a Minimal Gallery (adapted to* $\overline{\gamma}(F, F')$); *and*
thus that "set $\Psi^*(\overline{\Gamma}^i) \cap$ *set* $\Psi^*(\overline{\Gamma}^j) = \emptyset$" *for* $i \neq j$, *and* $|\Psi^*(\overline{\Gamma}(C, F'))| = $
$|\mathcal{H}(C, F')|$. *From the inclusions* $\Psi^*(\overline{\Gamma}^i) \subset \mathcal{H}_{\overline{F}'_i}(\overline{F}_{i-1})$ *and* $|\Psi^*(\overline{\Gamma}(C, F'))| = $
$\sum |\mathcal{H}_{F'_i}(F_{i-1})|$ *it is deduced*

$$\Psi^*(\overline{\Gamma}^i) = \mathcal{H}_{\overline{F}'_i}(\overline{F}_{i-1}).$$

It is finally obtained

$$\mathcal{H}(C, F) = \coprod \Psi^*(\overline{\Gamma}^i) \text{ (abusive language)} = \coprod \mathcal{H}_{F'_i}(F_{i-1}),$$

i.e. $\overline{\gamma}(F, F')$ *is a MGG between* F *and* F'. *On the other hand,* typ $\gamma(F, F') = $
$g = $ typ $\overline{\gamma}(F, F')$ *by hypothesis proves that* $\gamma(F, F') = \overline{\gamma}(F, F')$, *as results from*
Corollary 9.37.

9.8 The Type of Relative Position associated to Minimal Generalized Galleries of types

Definition 9.39

1) *Recall* $\text{gall}_{\mathcal{A}} := \text{Gall}_{\text{typ }\mathcal{A}}$, *i.e.* $\text{gall}_{\mathcal{A}} = $ *set of generalized galleries of the typical simplex* typ $\mathcal{A}$, *and define*

$$\text{gall}_{\mathcal{A}}^{m} := \{\text{typ } \gamma \mid \gamma \text{ a MGG of } \mathcal{A}\} \subset \text{gall}_{\mathcal{A}}$$

(the set of Minimal Generalized Galleries of types of $\mathcal{A}$). *Thus*

$$\text{gall}_{\mathcal{A}}^{m} = \textit{image by } \text{typ} : \text{Gall}_{\mathcal{A}} \rightarrow \text{gall}_{\mathcal{A}} \textit{ of } \text{Gall}_{\mathcal{A}}^{m} \subset \text{Gall}_{\mathcal{A}}.$$

Given $g \in \text{gall}_{\mathcal{A}}$ *(resp.* $F \in \mathcal{A}$ *) write:*

2) $\text{Gall}_{\mathcal{A}}(g) = \text{typ}^{-1}(g)$ *(resp.* $\text{Gall}_{\mathcal{A}}^{m}(g) = \text{Gall}_{\mathcal{A}}^{m} \cap \text{Gall}_{\mathcal{A}}(g)$*).* **(The Minimal Generalized Galleries of type g)**

3) $\text{Gall}_{\mathcal{A}}^{m}(g, F) = \text{Gall}_{\mathcal{A}}^{m}(g) \cap \text{Gall}_{\mathcal{A}}(F)$ **(The Minimal Generalized Galleries of type g with left extremity F)**

Let $\gamma(F, F')$ be a gg of $\mathcal{A}$. Given an automorphism $f : \mathcal{A} \rightarrow \mathcal{A}$ of $\mathcal{A}$, it is denoted by

$$f(\gamma(F, F')) = (f.\gamma)(f(F), f(F'))$$

the transformation of the gallery $\gamma = \gamma(F, F')$ under f. If $f \in W_{\mathcal{A}}$ then $f.\gamma = f(\gamma(F, F'))$ is a generalized gallery of the same type as $\gamma(F, F')$. The mapping $\text{Gall}_{\mathcal{A}} \rightarrow \text{Gall}_{\mathcal{A}}$, defined by $\gamma \mapsto f.\gamma$, induces a mapping of the set of Minimal Generalized Galleries $\text{Gall}_{\mathcal{A}}^{m} \rightarrow \text{Gall}_{\mathcal{A}}^{m}$. It follows from its definition that if γ is a Minimal Generalized Gallery then $f.\gamma$ is a minimal too and thus $\text{Gall}_{\mathcal{A}}^{m}$ is stable under automorphisms of $\mathcal{A}$. If $f \in W_{\mathcal{A}}$, i.e. f is admissible, one has that $\forall \gamma \in \text{gall}_{\mathcal{A}}$

$$\text{typ } f.\gamma = \text{typ } \gamma.$$

Thus

$$\text{typ} : \text{Gall}_{\mathcal{A}} \rightarrow \text{gall}_{\mathcal{A}}$$

factors through the quotient map

$$q_{\mathcal{A}} : \text{Gall}_{\mathcal{A}} \rightarrow \text{Gall}_{\mathcal{A}}/W_{\mathcal{A}} .$$

Then there exists a mapping

$$\overline{\text{typ}} : \text{Gall}_{\mathcal{A}}/W_{\mathcal{A}} \rightarrow \text{gall}_{\mathcal{A}}$$

satisfying $\text{typ} = \overline{\text{typ}} \circ q_{\mathcal{A}}$. Denote by

$$q_{\mathcal{A}}^{m} : \text{Gall}_{\mathcal{A}}^{m} \rightarrow \text{Gall}_{\mathcal{A}}^{m}/W_{\mathcal{A}} \subset \text{Gall}_{\mathcal{A}}/W_{\mathcal{A}}$$

the mapping induced by $q_\mathcal{A}$, which is in fact the quotient mapping defined by the induced action of $W_\mathcal{A}$ on $\text{Gall}_\mathcal{A}^m$. Let

$$\overline{\text{typ}}^m : \text{Gall}_\mathcal{A}^m / W_\mathcal{A} \to \text{gall}_\mathcal{A}^m$$

be the mapping induced by $\overline{\text{typ}}$. One has thus

$$\text{typ}^m = \overline{\text{typ}}^m \circ q_\mathcal{A}^m.$$

From the following proposition it follows that $\overline{\text{typ}}^m$ is a bijection.

Proposition 9.40 *Let $\gamma(F, F')$ and $\overline{\gamma}(\overline{F}, \overline{F'})$ be two MGG of $\mathcal{A}$ with typ $\gamma(F, F') = $ typ $\overline{\gamma}(\overline{F}, \overline{F'})$. Then $\gamma(F, F')$ and $\overline{\gamma}(\overline{F}, \overline{F'})$ are conjugate under $W = W_\mathcal{A}$, i.e. $\exists\, w \in W$ so that $w(\gamma(F, F')) = \overline{\gamma}(\overline{F}, \overline{F'})$ (resp. $w.\gamma = \overline{\gamma}$).*

Proof *Let $C \in \text{St}_{c(F,F')}$ (resp. $\overline{C} \in \text{Ch St}_{c(\overline{F}, \overline{F'})}$). Then $\gamma(C; F, F')$ (resp. $\overline{\gamma}(\overline{C}; \overline{F}, \overline{F'}))$ is a MGG between C and F' (resp. $\overline{C}$ and $\overline{F'}$). Let $w \in W$ such that $w(\overline{C}) = C$. By Remark 9.10, b), one has (C_i, F_{i-1}) is a couple in transversal position thus it follows, by induction, that $w(\overline{F'}) = F'$. Thus by hypothesis $w(\overline{\gamma}(\overline{C}; \overline{F}, \overline{F'}))$ and $\gamma(C; F, F')$ are two MGG issued from C and of the same type and the same extremities, i.e. with typ $w(\overline{\gamma}(\overline{C}; \overline{F}, \overline{F'})) = $ typ $\gamma(C; F, F')$. From Proposition 9.36 it is deduced $w(\overline{\gamma}(\overline{C}; \overline{F}, \overline{F'})) = \gamma(C; F, F')$, and a fortiori $w(\overline{\gamma}(\overline{F}, \overline{F'})) = \gamma(F, F')$.*

The following corollary states that Minimal Generalized Galleries under the action of W are classified by $\text{gall}_\mathcal{A}^m$.

Corollary 9.41 *The mapping $\overline{\text{typ}}^m : \text{Gall}_\mathcal{A}^m / W_\mathcal{A} \to \text{gall}_\mathcal{A}^m$ is a bijection.*

Let $\gamma(F, F') \in \text{Gall}_\mathcal{A}^m$ be a representative of the class of $\text{Gall}_\mathcal{A}^m / W_\mathcal{A} = \text{gall}_\mathcal{A}^m$ indexed by $g \in \text{gall}_\mathcal{A}^m$. The class $\tau(F, F') \in (\mathcal{A} \times \mathcal{A})/W$ of (F, F') is independent of the choice of the representative $\gamma(F, F')$ of the class g. This results from the evident fact that the mapping $e : \text{Gall}_\mathcal{A} \to \mathcal{A} \times \mathcal{A}$ defined by

$$E = (E_1, E_2) : \gamma(F, F') \mapsto (F, F') \quad \text{(the extremities mapping)}$$

is $W = W_\mathcal{A}$-equivariant, by definition of the action of W on $\text{Gall}_\mathcal{A}$ (resp. $\mathcal{A} \times \mathcal{A}$).

A type of relative position correspondant to a Minimal Generalized Gallery of types $g \in \text{gall}_\mathcal{A}^m$ is defined as the type of relative position of the extremities of a representative gallery of the class g.

Definition 9.42 *Let $\tau_\bullet : \text{Gall}_\mathcal{A}^m / W_\mathcal{A} = \text{gall}_\mathcal{A}^m \to \text{Relpos } \mathcal{A} = (\mathcal{A} \times \mathcal{A})/W$ be the mapping defined by $\tau_\bullet : g \mapsto \tau_g = \tau_g(F, F')$, where $\gamma(F, F') \in \text{gall}_\mathcal{A}^m$ is a representative MGG of the class defined by g. If $E^m : \text{Gall}_\mathcal{A}^m \to \mathcal{A} \times \mathcal{A}$ denotes the restriction of E to $\text{Gall}_\mathcal{A}^m \subset \text{Gall}_\mathcal{A}$, one obtains the relation:*

$$\tau \circ E^m = \tau_\bullet \circ q_\mathcal{A}^m.$$

It is known that the quotient $(C(\mathfrak{S}_{r+1}, S) \times C(\mathfrak{S}_{r+1}, S))/\mathfrak{S}_{r+1}, S)$ is identified with the set of relative position matrices. Let M be a matrix defining a type of relative position. One has defined a minimal generalized gallery $g(M)$. By its construction it is clear that to $g(M)$ it corresponds by the above definition the type of relative position defined by M. In fact one has thus defined a section of $\tau_\bullet : gall^m_{C(\mathfrak{S}_{r+1}, S))} \longrightarrow Relpos\ C(\mathfrak{S}_{r+1}, S))$, i.e. $\tau_\bullet(g(M)) = M$. Let $\tau_0 \in$ Relpos $\mathcal{A} = (\mathcal{A} \times \mathcal{A})/W_\mathcal{A}$. Recall that $(\mathcal{A} \times \mathcal{A})_{\tau_0} = \tau^{-1}(\tau_0)$.

Definition 9.43 *Write*
$$gall^m_\mathcal{A}(\tau_0) = \tau_\bullet^{-1}(\tau_0) = \{g \subset gall^m_\mathcal{A} \mid \tau_\bullet(g) = \tau_0\}$$
(resp. $Gall^m_\mathcal{A}(\tau_0) = (q^m_\mathcal{A})^{-1}(\overline{typ}^m)^{-1}gall^m_\mathcal{A}(\tau_0) = (typ^m)^{-1}(gall^m_\mathcal{A}(\tau_0))).$

There is the following disjoint union:

$$Gall^m_\mathcal{A}(\tau_0) = \coprod_{g \in gall^m_\mathcal{A}(\tau_0)} Gall^m_\mathcal{A}(g) \ .$$

The set of MGG $\gamma(F, F')$ with $\tau(F, F') = \tau_0$ decomposes in a disjoint union indexed by the set of types of MGG $gall^m(\tau_0)$. The following proposition proves that the restriction of E^m to $Gall^m_\mathcal{A}(g)$ induces a bijection $E^m|_{Gall^m_\mathcal{A}(g)} : Gall^m_\mathcal{A}(g) \simeq (\mathcal{A} \times \mathcal{A})_{\tau_0}$. In other words given a couple $((F, F'), g)$ with $\tau(F, F') = \tau_0$ and $g \in gall^m_\mathcal{A}(\tau_0)$ there exists a unique Minimal Generalized Gallery $\gamma_g(F, F')$ of type g characterized by $E^m(\gamma_g(F, F')) = (F, F')$. Proposition 9.38 implies:

Proposition 9.44 *The mapping*

$$E^m_g : Gall^m_\mathcal{A}(g) \to (\mathcal{A} \times \mathcal{A})_{\tau_g},$$

induced by E^m, *admits a* W-*equivariant section* $\sigma_g : (\mathcal{A} \times \mathcal{A})_{\tau_g} \to Gall^m_\mathcal{A}(g)$ *defined by*
$$\sigma_g : (F, F') \mapsto \gamma_g(F, F') \ ,$$

where $\tau(F, F') = \tau_g$ *and* $\gamma_g(F, F')$ *denotes the unique MGG of type* g *(cf. Proposition 9.38) between* F *and* F'. *In fact* σ_g *is a* $W_\mathcal{A}$-*equivariant bijection.*

Let $(\mathcal{A} \times \mathcal{A}) \times_{\text{Relpos } \mathcal{A}} gall^m_\mathcal{A}$ be the fiber product set defined by the couple $(\tau, \tau_\bullet)$. Given $\tau \in$ Relpos $\mathcal{A}$ the fiber over τ is given by

$$((\mathcal{A} \times \mathcal{A}) \times_{\text{Relpos } \mathcal{A}} gall^m_\mathcal{A})_\tau = (\mathcal{A} \times \mathcal{A})_\tau \times gall^m_\mathcal{A}(\tau).$$

Thus the following decomposition holds

$$(\mathcal{A} \times \mathcal{A}) \times_{\text{Relpos } \mathcal{A}} gall^m_\mathcal{A} = \coprod_{\tau \in \text{Relpos } \mathcal{A}} (\mathcal{A} \times \mathcal{A})_\tau \times gall^m_\mathcal{A}(\tau)$$

$$= \coprod_{\substack{(\tau, g) \in (\text{Relpos } \mathcal{A}) \times gall^m_\mathcal{A}, \\ \tau = \tau_g}} (\mathcal{A} \times \mathcal{A})_\tau \times \{g\}.$$

Consider also the decomposition defined by the types of galleries

$$\mathrm{Gall}_{\mathcal{A}}^m = \coprod_{g \in \mathrm{gall}_{\mathcal{A}}^m} \mathrm{Gall}_{\mathcal{A}}^m(g).$$

The natural mapping $E^m \times_{Relpos_{\mathcal{A}}} \mathrm{typ} : \mathrm{Gall}_{\mathcal{A}}^m \longrightarrow \sigma : (\mathcal{A} \times \mathcal{A}) \times_{\mathrm{Relpos}} {}_{\mathcal{A}} \mathrm{gall}_{\mathcal{A}}^m \to \mathrm{Gall}_{\mathcal{A}}^m$ admits a section

$$\sigma : (\mathcal{A} \times \mathcal{A}) \times_{\mathrm{Relpos}} {}_{\mathcal{A}} \mathrm{gall}_{\mathcal{A}}^m \to \mathrm{Gall}_{\mathcal{A}}^m$$

defined by

$$\sigma = \coprod_{g \in \mathrm{gall}_{\mathcal{A}}^m} \overline{\sigma}_g$$

where

$$\overline{\sigma}_g : (\mathcal{A} \times \mathcal{A})_{\tau_g} \times \{g\} \to \mathrm{Gall}_{\mathcal{A}}^m(g),$$

is defined by $\overline{\sigma}_g : ((F, F'), g) \mapsto \gamma_g(F, F')$. From the fact that $\overline{\sigma}_g$ ($g \in \mathrm{gall}_{\mathcal{A}}^m$) is a $W_{\mathcal{A}}$-equivariant bijection, it follows that σ is a $W_{\mathcal{A}}$-equivariant bijection. For example observe that $\overline{\sigma}_{g(M)} : (\Delta^{(r)'} \times (\Delta^{(r)'})_M \longrightarrow \mathrm{Gall}_{\Delta^{(r)'}}^m(g(M))$ is given by $\overline{\sigma}_{g(M)} : (D, D') \mapsto \gamma_{g(M)}(D, D')$, where $\gamma_{g(M)}(D, D') = \gamma(M)$.

9.9   The Contracted Product defined by a Gallery of Types

The correspondence between Minimal Generalized Galleries issued from a chamber C and a set of words in $\mathscr{S}_C$ is extended to a correspondence between the generalized galleries issued from C and words in a set $\mathcal{W}_C^m$ larger than $\mathscr{S}_C$.

It is recalled that there is a building isomorphism $\mathrm{typ}\, \mathcal{A} \xrightarrow{\sim} \Delta(C)$ defined by $t \mapsto F_t(C)$. Let $g \in \mathrm{gall}_{\mathcal{A}}$. Define $\gamma_{g,C} \subset \Delta(C)$ as the image of g by the preceding isomorphism. It is called $\gamma_{g,C}$ the **basical gg of type** g **defined by** C.

Definition 9.45 *With the notation of Definition 9.16* **the length** $l(\gamma_1(F, F'))$ **is defined** *(resp.* $l(\gamma_1'(F, F'))$, $l(\gamma_2(F, F'))$, $l(\gamma_2'(F, F')))$ *by:*

$$l(\gamma_1(F, F')) = r + 1 \quad (\text{resp. } l(\gamma_1'(F, F')) = r + 1, \; l(\gamma_2(F, F'))$$

$$= r + 1, \; l(\gamma_2'(F, F')) = r + 1).$$

It may be represented thus $\gamma_1(F, F')$ *(resp.* $\gamma_1'(F, F')$, $\gamma_2(F, F')$, $\gamma_2'(F, F'))$ *by:*

$$\gamma_1(F, F') : (F_i)_{r \geqslant i \geqslant 0} \quad , \quad (F_i')_{r+1 \geqslant i \geqslant 1}$$
$$(\text{resp. } \gamma_1'(F, F') : (F_i)_{r \geqslant i \geqslant 0} \quad , \quad (F_i')_{r+1 \geqslant i \geqslant 0}$$
$$\gamma_2(F, F') : (F_i)_{r \geqslant i \geqslant 0} \quad , \quad (F_i')_{r \geqslant i \geqslant 1}$$
$$\gamma_2'(F, F') : (F_i)_{r \geqslant i \geqslant 0} \quad , \quad (F_i')_{r \geqslant i \geqslant 0}).$$

Write typ $F_i = s_i$ *and typ* $F_i' = t_i$.

Let $g_1 = \mathrm{typ}\, \gamma_1(F, F')$ *(resp.* $g_1' = \mathrm{typ}\, \gamma_1'(F, F')$, $g_2 = \mathrm{typ}\, \gamma_2(F, F')$, $g_2' = \mathrm{typ}\, \gamma_2'(F, F'))$. *Write:*

2) $l(g_1)$ *(length of g_1)* $= l(\gamma_1(F, F'))$ *(resp.* $l(g_1') = l(\gamma_1'(F, F'))$ *(length of g_1'), $l(g_2) = l(\gamma_2(F, F'))$ (length of g_2), $l(g_2') = l(\gamma_2'(F, F'))$ (length of g_2'))*

3) $s(g_1) = (s_i) = (\mathrm{typ}\, F_i)_{r \geqslant i \geqslant 0}$, $t(g_1) = (t_i) = (\mathrm{typ}\, F_i')_{r+1 \geqslant i \geqslant 1}$ *(resp.* $s(g_1') = (s_i) = (\mathrm{typ}\, F_i)_{r \geqslant i \geqslant 0}$, $t(g_1') = (t_i) = (\mathrm{typ}\, F_i')_{r+1 \geqslant i \geqslant 0}$, $s(g_2) = (s_i) = (\mathrm{typ}\, F_i)_{r \geqslant i \geqslant 0}$, $t(g_2) = (t_i) = (\mathrm{typ}\, F_i')_{r \geqslant i \geqslant 1}$, $s(g_2') = (s_i) = (\mathrm{typ}\, F_i)_{r \geqslant i \geqslant 0}$, $t(g_2') = (t_i) = (\mathrm{typ}\, F_i')_{r \geqslant i \geqslant 0})$;

4) $s(g) = (s_i)$, $t(g) = (t_i)$ *we write* $t_i(g) := t_i$, $s_i(g) := s_i$.

Denote by $gall_1$ *(resp.* $gall_1'$, $gall_2$, $gall_2'$) *the set of* $g \in gall_\mathcal{A}$ *which are of the form:* $g_1 = \mathrm{typ}\, \gamma_1(F, F')$ *(resp.* $g = g_1' = \mathrm{typ}\, \gamma_1'(F, F')$, $g = g_2 = \mathrm{typ}\, \gamma_2(F, F')$, $g = g_2' = \mathrm{typ}\, \gamma_2'(F, F'))$. *Let* $\gamma \in Gall_\mathcal{A}$ *with* $\mathrm{typ}\, \gamma \in gall_1$ *(resp.* $\mathrm{typ}\, \gamma \in gall_1'$, $\mathrm{typ}\, \gamma \in gall_2$, $\mathrm{typ}\, \gamma \in gall_2'$). *Write* $\gamma = \gamma_1$ *(resp.* $\gamma = \gamma_1'$, $\gamma = \gamma_2$, $\gamma = \gamma_2'$), *and denote by*

$$F_{t_i}(\gamma) = F_{t_i(\mathrm{typ}\, \gamma)}(\gamma) \quad (resp.\ F_{s_i}(\gamma) = F_{s_i(\mathrm{typ}\, \gamma)})\ ,$$

the corresponding facet of γ.

Remark 9.46 *With this convention the generalized gallery* γ *associated with an injective gallery* $\Gamma = (C_r, \cdots, C_0)$, *as in the commentary that follows Definition 9.6, is of length $r + 1$, while Γ is a gallery of length r.*

Definition 9.47 *Let* $s \in \mathrm{typ}\, \mathcal{A}$ *be the type of a chamber, i.e.* $s = \mathrm{typ}\, C$, $C \in \mathrm{Ch}\, \mathcal{A}$.

Define a mapping $gall_\mathcal{A} \longrightarrow gall_\mathcal{A}$ *by* $g \mapsto g^* \in gall_\mathcal{A}$ *where g^* is defined as follows. If* $g = g_1$, g_1' *(resp.* $g = g_2$, g_2') *with* $l(g) = r + 1$, *let* $g^* \in gall_\mathcal{A}$ *be given by:*

$$l(g^*) = r + 2 \quad (resp.\ l(g^*) = r + 1), \quad and$$

$$g^* : (s_i^*)_{r+1 \geqslant i \geqslant 0},\ t(g) \quad (resp.\ g^* : (s_i^*)_{r \geqslant i \geqslant 0},\ t(g)),$$

with $s_i^* = s$ $(r + 1 \geqslant i \geqslant 0)$ *(resp.* $s_i^* = s$ $(r \geqslant i \geqslant 0)$). *Thus g^* is obtained from g by changing each type $s_i(g)$ into the type of a chamber $s_i(g^*)$.*

Write $W_t = W_{F_t(C)}$ ($t \in \mathrm{typ}\, \mathcal{A}$). Let $C(W_\mathcal{A}, S_C)$ be the Coxeter system defined by the set of reflexions S_C defined by the walls of C, and $C(W_\mathcal{A}, S_C) \simeq \mathcal{A}$ the corresponding building isomorphism, and $\mathscr{S}_C := (w(t, s))$ $((t, s) \in \mathrm{typ}^{(2)}\mathcal{A})$, where $w(t, s) = w(C, \mathrm{proj}_F C)$, with $F = F_{(t,s)}^{tr}(C)$. There is then $w(t, s) \in W_t$. Let $X \in \mathcal{P}(S_C)$ correspond to t under the identification

$$\mathcal{P}(S_C) \simeq \mathrm{typ}\, \mathcal{A},$$

i.e. $F_t \subset C$ corresponds to the facet of $C(W_\mathcal{A}, S_C)$ whose set of vertices is $X \subset S_C$. Then $W_t = W_{S_C - X}$.

One defines for each gallery of types g a couple of groups $W_C(t(g))$ (resp. $W_C(s(g))$) and an action of $W_C(s(g))$ on $W_C(t(g))$.

Definition 9.48 *Let $g \in \text{gall}_A$ be characterized by $t(g) = (t_i(g))$ (resp. $t(s) = (s_j(g))$). Define the group $W_C(t(g))$ (resp. $W_C(s(g))$) in the four different cases by:*

1) $g = g_1$

$$W_C(t(g)) = W_C(t(g)) \quad = \prod_{r+1 \geqslant i \geqslant 1} W_{t_i(g)}$$

$$(\text{resp. } W_C(s(g)) = W_C(s(g)) \quad = \prod_{r \geqslant i \geqslant 0} W_{s_i(g)});$$

2) $g = g_1'$

$$W_C(t(g)) = W_C(t(g)) \quad = \prod_{r+1 \geqslant i \geqslant 1} W_{t_i(g)}$$

$$(\text{resp. } W_C(s(g)) = W_C(s(g)) \quad = \prod_{r \geqslant i \geqslant 0} W_{s_i(g)});$$

3) $g = g_2$

$$W_C(t(g)) = W_C(t(g)) \quad = \prod_{r \geqslant i \geqslant 1} W_{t_i(g)}$$

$$(\text{resp. } W_C(s(g)) = W_C(s(g)) \quad = \prod_{r-1 \geqslant i \geqslant 0} W_{s_i(g)});$$

4) $g = g_2'$

$$W_C(t(g)) = W_C(t(g)) \quad = \prod_{r \geqslant i \geqslant 1} W_{t_i(g)}$$

$$(\text{resp. } W_C(s(g)) = W_C(s(g)) \quad = \prod_{r-1 \geqslant i \geqslant 0} W_{s_i(g)}).$$

Remark 9.49 *The above construction is a particular case of a general one. Let $\gamma \in \text{Gall}_A$, and $g = \text{typ } \gamma$. One associates with γ two groups as follows. Write $t(\gamma) = (F'_{t_i(g)}(\gamma))$ (resp. $s(\gamma) = (F_{s_j(g)}(\gamma))$), and define:*

$$W(t(\gamma)) = \prod W_{F'_{t_i(g)}} \quad (\text{resp. } W(s(\gamma)) = \prod W_{F_{s_j(g)}}),$$

where i (resp. j) runs over a convenient set of indices which depends on the class of g. Clearly there is:

$$W_C(t(g)) = W(t(\gamma_{g,C})) \quad (\text{resp. } W_C(s(g)) = W(s(\gamma_{g,C}))).$$

Definition 9.50 *Define a right action $W_C(t(g)) \times W_C(s(g)) \to W_C(t(g))$, of the product group $W_C(s(g))$, on the underlying set of $W_C(t(g))$, by $(x, y) \mapsto x'$, where:*

1) *if $g = g_1$, $x = (x_i)_{r+1 \geqslant i \geqslant 1}$ (resp. $x' = (x'_i)_{r+1 \geqslant i \geqslant 1}$), $y = (y_i)_{r \geqslant i \geqslant 0}$, then*
   $x'_{r+1} = x_{r+1} y_r$, $x'_i = y_i^{-1} x_i y_{i-1}$ $(r \geqslant i \geqslant 1)$;

2) *if $g = g'_1$, $x = (x_i)_{r+1 \geqslant i \geqslant 1}$ (resp. $x' = (x'_i)_{r+1 \geqslant i \geqslant 1}$), $y = (y_i)_{r \geqslant i \geqslant 0}$, then*
   $x'_{r+1} = x_{r+1} y_r$, $x'_i = y_i^{-1} x_i y_{i-1}$ $(r \geqslant i \geqslant 1)$;

3) *if $g = g_2$, $x = (x_i)_{r \geqslant i \geqslant 1}$ (resp. $x' = (x'_i)_{r \geqslant i \geqslant 1}$), $y = (y_i)_{r-1 \geqslant i \geqslant 0}$, then*
   $x'_r = x_r y_{r-1}$, $x'_i = y_i^{-1} x_i y_{i-1}$ $(r - 1 \geqslant i \geqslant 1)$;

4) *if $g = g'_2$, $x = (x_i)_{r \geqslant i \geqslant 1}$ (resp. $x' = (x'_i)_{r \geqslant i \geqslant 1}$), $y = (y_i)_{r-1 \geqslant i \geqslant 0}$, then*
   $x'_r = x_r y_{r-1}$, $x'_i = y_i^{-1} x_i y_{i-1}$ $(r - 1 \geqslant i \geqslant 1)$;

The quotient set $W_C(t(g))/W_C(s(g))$ is called **the contracted product de-fined by the basical gallery** $\gamma_{g,C}$.

Remark 9.51 *Two groups $W(s(\gamma))$ and $W(t(\gamma))$ may be similarly defined and a right action of $W(s(\gamma))$ on $W(t(\gamma))$ for a generalized gallery γ in $\mathcal{A}$, and correspondingly define the quotient set*

$$W_C(t(\gamma))/W_C(s(\gamma))$$

(the contracted product associated with the gg γ).

9.9.1   Sets of Generalized Galleries as Contracted Products

Notation 9.52 *Given $\gamma = \gamma(F, F') \in \mathrm{Gall}_A$ of type $\mathrm{typ}\, \gamma = g$ and length $l(\gamma)$ we write:*

$$\gamma : (F_j)_{j \in I(g)} \; ; \; (F'_i)_{i \in I'(g)} ,$$

where the set of indices $I(g)$(resp. $I'(g)$) depend on the class of g. It is understood that:

1) *the inclusions $F_i \supset F'_i$, $F'_{i+1} \subset F_i$ hold whenever they are defined;*

2) *this same convention applies mutatis mutandis to a $g \in \mathrm{gall}_A$, it is written:*

$$g : (s_j)_{j \in I(g)} , \; (t_j)_{i \in I'(g)} .$$

Let the gg $\gamma = \gamma(F, F')$ with $l(\gamma) = r + 1$ be given, according to the above notation, by $\gamma = \gamma(F, F') : (F_i), (F'_j)$, where $\gamma = \gamma_i$ or γ'_i $(i = 1, 2)$. Write $\mathrm{typ}\, \gamma = g : (s_i), (t_j)$. Thus

$$F_i = F_{s_i}(\gamma), \; F_j = F_{t_j}(\gamma) .$$

The gallery $\gamma_{g,C} \subset \Delta(C)$ is given by

$$\gamma_{g,C} : (F_{s_i}(C)), (F_{t_j}(C)).$$

Let $\mathrm{Gall}_{\mathcal{A}}(g, F)$ denote the set of gg of type g issued from F, and $x = (x_i)_{r+1 \geqslant i \geqslant 1}$ (resp. $x = (x_i)_{r \geqslant i \geqslant 1}) \in W_C(t(g))$. If $g = g_1, g_1'$ (resp. $g = g_2, g_2'$). Write

$$z_i = \prod_{r+1 \geqslant \alpha \geqslant i} x_\alpha, \quad z_{r+2} = 1$$

$$(\text{resp. } z_i = \prod_{r \geqslant \alpha \geqslant i} x_\alpha, \quad z_{r+1} = 1),$$

and define, by means of $\gamma_{g,C}$, the mapping:

$$W_C(t(g)) \quad \to \quad \mathrm{Gall}_{\mathcal{A}}(g, F_{t_{r+1}}(C))$$
$$(\text{resp. } W_C(t(g)) \quad \to \quad \mathrm{Gall}_{\mathcal{A}}(g, F_{s_r}(C))),$$

if $g = g_1, g_1'$ (resp. $g = g_2, g_2'$), by

$$x \mapsto \gamma_{g,C}(x): \ (z_{i+1}(F_{s_i}(C))), \ (z_{j+1}(F_{t_j}(C))).$$

Also write $x \cdot \gamma_{g,C} = \gamma_{g,C}(x)$ with $x \in W_C(t(g))$. Remark that:

$$F_{s_i}(C) \supset F_{t_i}(C) \quad \Rightarrow \quad z_{i+1}(F_{s_i}(C)) \supset z_{i+1}(F_{t_i}(C))$$
$$(\text{resp. } F_{t_{i+1}}(C) \subset F_{s_i}(C) \quad \Rightarrow \quad z_{i+2}(F_{t_{i+1}}(C)) \subset z_{i+1}(F_{s_i}(C))),$$

as $z_{i+1} = z_{i+2} x_{i+1}$ with $x_{i+1} \in W_{F_{t_{i+1}}}$ gives $z_{i+2}(F_{t_{i+1}}(C)) = z_{i+1}(F_{t_{i+1}}(C))$. Thus $\gamma_{g,C}(x)$ is a gg of the same type as $\gamma_{g,C}$, i.e. $\mathrm{typ}\,\gamma_{g,C}(x) = g$. Given $C \in \mathrm{Ch}\,\mathcal{A}$ one writes:

$$F_g(C) = F_{t_{r+1}(g)}(C), \quad \text{if } g = g_1, g_1'$$
$$(\text{resp. } F_g(C) = F_{s_r(g)}(C), \quad \text{if } g = g_2, g_2').$$

Definition 9.53 *The above defined mapping*

$$W_C(t(g)) \to \mathrm{Gall}_{\mathcal{A}}(g, F_g(C))$$

factors through the quotient mapping:

$$W_C(t(g)) \to W_C(t(g))/W_C(s(g)),$$

as it is easily verified. Denote by

$$i_{g,C}: \ W_C(t(g))/W_C(s(g)) \to \mathrm{Gall}_{\mathcal{A}}(g, F_g(C))$$

the induced mapping.

9.9.2 Bijectivity of this mapping

Let the isomorphism $C(W, S_C) \simeq \mathcal{A}$ be made explicit. The choice of $C \in \mathrm{Ch}\,\mathcal{A}$ gives rise to a building isomorphism

$$i_C = \coprod_{t \in \mathrm{typ}\,\mathcal{A}} i_{t,C} : C(W, S_C) = \coprod_{X \subset \mathcal{P}(S_C)} W/W_{S_C - X} \to \mathcal{A} = \coprod_{t \in \mathrm{typ}\,\mathcal{A}} \mathcal{A}_t$$

defined as follows. Let $t \in \mathrm{typ}\,\mathcal{A}$ correspond to $Y \in \mathcal{P}(S_C) = \mathrm{typ}\,C(W, S_C)$ under the induced bijection $\mathrm{typ}\,C(W, S_C) \simeq \mathrm{typ}\,\mathcal{A}$, given by $X \mapsto F_X(C)$ where $F_X(C) \subset C$ denotes the facet invariant under $W_{S_C - X}$, one writes $t = t_X$. Then there is: $W_t = \mathrm{Stab}\,F_t(C) = W_{S_C - X}$. Define $i_{t,C} : W/W_t \to \mathcal{A}_t$ by:

$$i_{t,C} : \overline{w} \mapsto w(F_t(C)) \,.$$

Proposition 9.54 *Let $g \in \mathrm{gall}_{\mathcal{A}}$. It is supposed $g = g'_1$, $l(g) = r + 1$. Write, according to the preceding convention (cf. Notation 9.52)*

$$g : (s_i)_{r \geqslant i \geqslant 0}, \ (t_j)_{r+1 \geqslant j \geqslant 0}.$$

Then, with the above notation (cf. Definition 9.48), the mapping

$$W_C(t(g)) \to \mathrm{Gall}_{\mathcal{A}}(g, F_{t_{r+1}}(C))$$

defined by $x \mapsto \gamma_{g,C}(x)$ $(x = (x_j) \in W_C(t(g)))$ where

$$\gamma_{g,C}(x) : \ (z_{i+1}(F_{t_i}(C)))_{r \geqslant i \geqslant 0}, \ (z_{j+1}(F_{s_j}(C)))_{r+1 \geqslant j \geqslant 0},$$

and $z_\alpha = \prod_{r+1 \geqslant j \geqslant \alpha} x_j$ (resp. $z_{r+2} = 1$), satisfies:

$$\gamma_{g,C}(x) = \gamma_{g,C}(u) \ \ (u = (u_i)_{r+1 \geqslant j \geqslant 1} \in W_C(t(g))) \Rightarrow$$

$$\exists \, y = (y_i)_{r \geqslant i \geqslant 0} \in W_C(s(g))$$

so that: $x.y = (x_{r+1}y_r, \ y_r^{-1}x_ry_{r-1}, \cdots, y_1^{-1}x_1y_0) = u$.

Proof *Denote by $\overline{z}_{i+1}$ (resp. $\overline{z}'_{j+1}$) the class of z_{i+1} (resp. z'_{j+1}) in W/W_{s_i} (resp. W/W_{t_j}). Let $\overline{g} \in \mathrm{gall}_{C(W,S_C)}$ correspond to g by $\mathrm{typ}\,\mathcal{A} \xrightarrow{\sim} \mathcal{P}(S_C)$, and $Y \subset S_C$ to t_{r+1}, i.e. $t_Y = t_{r+1}$. The composed mapping of $x \mapsto \gamma_{g,C}(x)$ with the bijection $i_C^{-1} : \mathcal{A} \xrightarrow{\sim} C(W, S_C)$, induces the following mapping:*

$$\mathrm{Gall}_{\mathcal{A}}(g, F_{t_{r+1}}(C)) \xrightarrow{\sim} \mathrm{Gall}_{C(W,S_C)}(\overline{g}, F_Y(C_e))$$

transforming $\gamma_{g,C}(x)$ into $\overline{\gamma}_g(x) : (\overline{z}_{i+1}), (\overline{z}'_{j+1})$. Write $v_\alpha = \prod_{r+1 \geqslant i \geqslant \alpha} u_i$. The hypothesis $\gamma_{g,C}(x) = \gamma_{g,C}(u)$ translates into:

$$z_{\alpha+1} \equiv v_{\alpha+1} \ (\mathrm{mod}\ W_{s_\alpha})$$

$$(\text{resp. } z_{\alpha+1} \equiv v_{\alpha+1} \ (\mathrm{mod}\ W_{t_\alpha})) \ \ (r \geqslant \alpha \geqslant 0).$$

Then

$$(*) \qquad\qquad \forall\, r \geqslant \alpha \geqslant 0 \; \exists y_\alpha \in W_{s_\alpha} / \; v_{\alpha+1} = z_{\alpha+1} y_\alpha.$$

Let it be proved by induction that $\forall\, r \geqslant \alpha \geqslant 1$ *the equality*

$$E(\alpha) : (x_{r+1} y_r, \; y_r^{-1} x_r y_{r-1}, \cdots, x_\alpha y_{\alpha-1}) = (u_{r+1}, \cdots, u_\alpha)$$

holds.

For $\alpha = r + 1$ *this is just* $(*)$. *Let then* $r \geqslant \alpha$, *and suppose that* $E(\alpha + 1)$ *holds. Thus* $v_{\alpha+1} = z_{\alpha+1} y_\alpha$. *By* $(*)$ *one obtains*

$$v_{\alpha+1} u_\alpha = v_\alpha = z_\alpha y_{\alpha-1} = z_{\alpha+1} x_\alpha y_{\alpha-1}.$$

By substitution $v_{\alpha+1} = z_{\alpha+1} y_\alpha$ *one obtains*

$$z_{\alpha+1} y_\alpha u_\alpha = z_{\alpha+1} x_\alpha y_{\alpha-1},$$

i.e. $u_\alpha = y_\alpha^{-1} x_\alpha y_{\alpha-1}$. *This equality joint to* $E(\alpha+1)$ *gives* $E(\alpha)$. *The equality* $E(1)$ *is that of the proposition.*

Remark 9.55 *The proof of proposition 9.54 may be easily adapted, mutatis mutandis, to that of the corresponding statement obtained by supposing* $g = g_1$ *(resp.* $g = g_2$, $g = g_2'$*), instead of* $g = g_1'$.

The proposition gives immediatly the following

Corollary 9.56 *The induced mapping* $i_{g,C}$: $W_C(t(g))/W_C(s(g)) \to \mathrm{Gall}_A(g, F_g(C))$ *defined by* $x \mapsto \gamma_{g,C}(x)$ $(x \in W_C(t(g)))$ *is* **injective**.

Definition 9.57   I) *Let* $\gamma \in \mathrm{Gall}_A(g)$, *where* $g = g_1'$ *(resp.* $g = g_1$*), and* $l(g) = r + 1$, *be given by*

$$\gamma : (F_i)_{r \geqslant i \geqslant 0}, \; (F_j')_{r+1 \geqslant j \geqslant 0} \quad (resp. \; (F_j')_{r+1 \geqslant j \geqslant 1}).$$

Given $r \geqslant \alpha \geqslant 0$ *one defines the* α-*truncated gallery* $\gamma^{(\alpha)}$ *of* γ *by:*

$$\gamma^{(\alpha)} : (F_i)_{r \geqslant i \geqslant \alpha}, \; (F_j')_{r+1 \geqslant j \geqslant \alpha+1}.$$

Given $r \geqslant \alpha \geqslant 0$ *(resp.* $r \geqslant \alpha \geqslant 1$*) one defines the* α-*truncated gallery* $\gamma^{(\alpha)'}$ *of* γ *by:*

$$\gamma^{(\alpha)'} : (F_i)_{r \geqslant i \geqslant \alpha}, \; (F_j')_{r+1 \geqslant j \geqslant \alpha}.$$

Thus $\mathrm{typ}\,\gamma^{(\alpha)} \in gall_1$ *(resp.* $\mathrm{typ}\,\gamma^{(\alpha)'} \in gall_1'$*).*

II) *Let* $\gamma \in \mathrm{Gall}_A(g)$, *where* $g = g_2'$ *(resp.* $g = g_2$*), and* $l(g) = r + 1$, *be given by*

$$\gamma : (F_i)_{r \geqslant i \geqslant 0}, \; (F_j')_{r \geqslant j \geqslant 0} \quad (resp. \; (F_j')_{r \geqslant j \geqslant 1}).$$

Given $r > \alpha \geqslant 0$ one defines the α-truncated gallery $\gamma^{(\alpha)}$ of γ by:

$$\gamma^{(\alpha)} : \ (F_i)_{r \geqslant i \geqslant \alpha}, \ (F'_j)_{r \geqslant j \geqslant \alpha+1}.$$

Given $r \geqslant \alpha \geqslant 0$ (resp. $r \geqslant \alpha \geqslant 1$) one defines the α-truncated gallery $\gamma^{(\alpha)'}$ of γ by

$$\gamma^{(\alpha)'} : \ (F_i)_{r \geqslant i \geqslant \alpha}, \ (F'_j)_{r \geqslant j \geqslant \alpha}.$$

Thus typ $\gamma^{(\alpha)} \in gall_2$ *(resp.* typ $\gamma^{(\alpha)'} \in gall'_2$*).*

The following proposition is interesting by itself and does not play any role in the proof of the bijectivity of $i_{g,C}$.

Proposition 9.58 *Let $\gamma = \gamma(C, F')$ a MGG between the chamber C and the facet F' of $\mathcal{A}$. It is supposed that* typ $\gamma = g'_2$, *and $l(\gamma) = r + 1$. Write*

$$\gamma = \gamma(C, F') : \ (F_i)_{r \geqslant i \geqslant 0}, \ (F'_j)_{r \geqslant j \geqslant 0}.$$

*One then has that the α-**truncated galleries***

$$\gamma^{(\alpha)} = \gamma^{(\alpha)}(C, F_\alpha) \ \ (r \geqslant \alpha \geqslant 0) \quad (resp. \ \gamma^{(\alpha)'}(C, F'_\alpha) \ \ (r \geqslant \alpha \geqslant 0))$$

are Minimal Generalized Galleries.

Proof *As in Lemma 9.8 a minimal gallery adapted to $\gamma(C, F')$ is constructed. Thus one defines the sequence of chambers*

$$C_r = C, \ \ C_\alpha = \mathrm{proj}_{F_\alpha} C_{\alpha+1} \ \ (r > \alpha \geqslant 0),$$

and a sequence of minimal galleries:

$$(\ \Gamma^\alpha (= \Gamma^\alpha (C_\alpha, C_{\alpha-1}))) \ _{r \geqslant \alpha \geqslant 1}.$$

It is known (cf. Subsection 9.2 Construction of a minimal gallery adapted to $\gamma(C, F')$) that the gallery obtained by composition of (Γ^α):

$$\Gamma^{(1)} = \Gamma^{(1)}(C_r, C_0) = \Gamma^r \circ \cdots \circ \Gamma^1$$

is a minimal gallery adapted to $\gamma(C, F')$. A fortiori the α-truncated gallery $\Gamma^{(\alpha)} = \Gamma^r \circ \cdots \circ \Gamma^\alpha$ is also a minimal gallery. Let it be proved that $\Gamma^{(\alpha)}$ is in fact a minimal gallery between C and F'_α. One must then see that for every wall

$$H \in \mathcal{H}_{F'_r}(F_{r-1}) \coprod \cdots \coprod \mathcal{H}_{F'_{\alpha+1}}(F_\alpha) \quad (= \mathcal{H}(C, C_\alpha))$$

that $\Gamma^{(\alpha)}$ crosses one has $F'_\alpha \notin H$. It is clear that if this property holds, there is

$$\mathcal{H}(C, F'_\alpha) = \mathcal{H}_{F'_{r-1}}(F_{r-1}) \coprod \cdots \coprod \mathcal{H}_{F'_{\alpha+1}}(F_\alpha),$$

and consequently that $\Gamma^{(\alpha)}(C, F'_\alpha)$ is a minimal gallery between C and F'_α. It is supposed that there exists $j \geqslant \alpha$ so that

$$(*) \qquad H \in \mathcal{H}_{F'_{j+1}}(F_j) \ \text{ and } \ F'_\alpha \in H.$$

Let it be proved that necessarily $j > \alpha$. Otherwise one must have $H \in \mathcal{H}_{F'_{\alpha+1}}(F_\alpha)$ and $F'_\alpha \in H$. It is known that $H \in \mathcal{H}_{F'_{\alpha+1}}(F_\alpha) \Rightarrow H \in \mathcal{H}(C_r, F'_0)$, thus $F'_0 \notin H$.
On the other hand,

$$H \in \mathcal{H}_{F'_{\alpha+1}}(F_\alpha) \ \text{ and } \ \mathcal{H}_{F'_{\alpha+1}}(F_\alpha) \cap \mathcal{H}_{F'_\alpha}(F_{\alpha-1}) = \emptyset$$

implies $F_{\alpha-1} \in H$. Thus as $F'_{\alpha-1} \subset F_{\alpha-1}$ it is deduced that $H \in \mathcal{H}_{F'_{\alpha-1}}$, and finally that $F_{\alpha-2} \in H$, otherwise one should have $H \in \mathcal{H}_{F'_{\alpha-1}}(F_{\alpha-2})$. Pursuing this same reasoning one must certainly obtain the conclusion $F_0 \in H$ (and a fortiori $F'_0 \in H$), which contradicts the hypothesis $F'_0 \notin H$.
Let now $j > \alpha$ so that $(*)$ holds for j. One then has $H \in \mathcal{H}_{F'_\alpha} \cap \mathcal{H}(C_r, F'_0)$. As $\mathcal{H}_{F'_{j+1}}(F_j) \cap \mathcal{H}_{F'_\alpha}(F_{\alpha-1}) = \emptyset$, one must have $F_{\alpha-1} \in H$. By the same argument as above, the contradiction $F'_0 \in H$ is obtained. Thus no H satisfies $(*)$. This proves that $\Gamma^{(\alpha)}$ is a minimal gallery and finally that $\gamma^{(\alpha)}(C, F'_\alpha)$ is a MGG.

Corollary 9.59 *[of the proof]* One has $\mathrm{proj}_{F'_\alpha} C = C_\alpha$ $(r > \alpha \geqslant 0)$.

Remark 9.60 *In fact it is easily seen that, more generally, the truncation $\gamma^{(\alpha)}(F, F')$ of a MGG, $\gamma(F, F')$ is also a MGG.*

Proposition 9.61 *The mapping $i_{g,C}$ is* **surjective**.

Proof *It suffices to see that $x \mapsto \gamma_{g,C}(x)$ is a surjective mapping in the case $g = g'_1$ and $l(g) = r+1$. The cases $g = g_1, g_2, g'_2$ may be handled in essentially the same way. Write $t_j := t_j(g)$ (resp. $s_i := s_i(g)$). Let it be proved that given*

$$\gamma : \ (F_i)_{r \geqslant i \geqslant 0}, \ \ (F'_j)_{r+1 \geqslant j \geqslant 0},$$

with $F'_{r+1} = F_{t_{r+1}}(C)$ there exists $x = (x_i) \in W_C(t(g))$, so that, with the notation of proposition 9.54, one has:

$$F_i = z_{i+1}(F_{s_i}(C)) \quad (\text{resp.} \ \ F'_j = z_{j+1}(F_{t_j}(C))).$$

In this case one writes $\gamma = x.\gamma_{g,C}$.
One proceeds to prove by induction that:

$$(*) \qquad \forall \, r \geqslant \alpha \geqslant 0, \ \exists \, x^{(\alpha+1)} = (x_j) \in \prod_{r+1 \geqslant j \geqslant \alpha+1} W(t_j(g)),$$

so that:

$$F_i = z_{i+1}(F_{s_i}(C)) \quad (r \geqslant i \geqslant \alpha) \quad (\text{resp. } F'_j = z_{j+1}(F_{t_j}(C)) \quad (r \geqslant j \geqslant \alpha)).$$

Let $\alpha = r$. One has

$$\gamma_{g^{(r)'},C} : F_{t_{r+1}}(C) \subset F_{s_r}(C) \supset F_{t_r}(C).$$

There is a bijection $W_{t_{r+1}}/W_{s_r} \to (\mathrm{St}_{F_{t_{r+1}}(C)})_{s_r}$, defined by: $\overline{x} \mapsto x(F_{s_r}(C))$ ($\overline{x} \in W_{t_{r+1}}/W_{s_r}$). Thus given $F_{t_{r+1}} \subset F_r$, with typ $F_r = s_r$, there exists $x \in W_{t_{r+1}}$ with $x(F_{s_r}(C)) = F_r$, i.e. $\exists\, x^{(r+1)} \in W(t(g^{(r)'})) = W_{t_{r+1}}$ so that $x^{(r+1)}.\gamma_{g^{(r)'},C} = \gamma^{(r)'}$.

Let $r > \alpha \geqslant 0$. Reasoning by induction, it is supposed that there exists

$$x^{(\alpha+2)} = (x_j) \in \prod_{r+1 \geqslant j \geqslant \alpha+2} W(t_j(g)),$$

satisfying: $x^{(\alpha+2)}.\gamma_{g^{(\alpha+1)'},C} = \gamma^{(\alpha+1)'}$. Then one has $F'_{\alpha+1} = z_{\alpha+2}(F_{t_{\alpha+1}}(C)$. Thus there is a surjection

$$W_{F'_{\alpha+1}} \to (\mathrm{St}_{F'_{\alpha+1}})_{s_\alpha},$$

defined by: $w \mapsto w(z_{\alpha+2}.F_{s_\alpha}(C))$, as $(\mathrm{St}_{F'_{\alpha+1}})_{s_\alpha}$ is a $W_{F'_{\alpha+1}}$-homogeneous set. On the other hand,

$$W_{F'_{\alpha+1}} = z_{\alpha+2} W_{t_{\alpha+1}} z_{\alpha+2}^{-1}.$$

It is concluded that there exists $x_{\alpha+1} \in W_{t_{\alpha+1}}$, so that

$$F_\alpha = z_{\alpha+2} x_{\alpha+1} z_{\alpha+2}^{-1}(z_{\alpha+2}(F_{s_\alpha}(C))) = z_{\alpha+2} x_{\alpha+1}(F_{s_\alpha}(C)).$$

Write $x^{(\alpha+1)} = (x'_j)_{r+1 \geqslant j \geqslant \alpha+1}$, with $x'_{\alpha+1} = x_{\alpha+1}$ (resp. $(x'_j) = x^{(\alpha+2)}$ ($r+1 \geqslant j \geqslant \alpha+2$)). It is finally obtained that $x^{(\alpha+1)}.\gamma_{g^{(\alpha)'},C} = \gamma^{(\alpha)'}$. This concludes the proof of $(*)$, and consequently

$$i_{g,C} : W_C(t(g)) \to \mathrm{Gall}_{\mathcal{A}}(g, F_g(C))$$

defined by $i_{g,C} : \quad x \mapsto \gamma_{g,C}(x)$, where $F_g(C) = F_{e_1(g)}(C)$, is a surjective mapping.

From Corollary 9.56 and Proposition 9.61 one deduces the

Proposition 9.62 *The mapping*

$$i_{g,C} : W_C(t(g))/W_C(s(g)) \to \mathrm{Gall}_{\mathcal{A}}(g, F_g(C))$$

*is a **bijection**.*

One has associated with $g \in \text{gall}_A$ a gallery $g^* \in \text{gall}_A$ (cf. Definition 9.47). Observe that

$$W(t(g^*))/W(s(g^*)) = W(t(g^*)).$$

Thus, in this case, there is a bijection $i_{g^*,C} : W(t(g^*)) \to \text{Gall}_A(g^*, C)$. On the other hand, there is a natural mapping

$$q_{g^*} : \text{Gall}_A(g^*) \to \text{Gall}_A(g),$$

defined by:

$$q_{g^*} : \gamma^* \mapsto F_g(\gamma^*) \ (\gamma^* \in \text{Gall}_A(g^*)) \,,$$

where $F_g(\gamma^*)$ is the gg of type g defined by $F_{s_i}(F_g(\gamma^*)) = F_{s_i}(C_i) \ (r \geqslant i \geqslant 0)$, and $F'_{t_j}(F_g(\gamma^*)) = F_{t_j}(C_j)$, with $(t_j) = t(g) \ (resp. \ (s_i) = s(g))$.

From the inclusions $t_{i+1}(g) \subset s_i(g) \supset t_i(g)$ it results that $F_{t_{i+1}}(C_{i+1}) \subset F_{s_i}(C_i) \supset F_{t_i}(C_i)$. i.e. the sets of facets $(F_{s_i}(C_i))$, $(F_{t_j}(C_j))$ satisfy the inclusions defining a generalized gallery $F_g(\gamma^*)$.

Denote by

$$q_{g^*,C} : \text{Gall}_A(g^*, C) \to \text{Gall}_A(g, F_g(C))$$

the restriction of q_{g^*} to the set of g^*-galleries (resp. galleries of type g^*) issued from C, $\text{Gall}_A(g^*, C) \subset \text{Gall}_A(g^*)$. The set $\text{Gall}_A(g, C)$ thus appears as quotient set of $\text{Gall}_A(g^*, C)$.

Let $W(q_{g^*,C}) : W_C(t(g)) \to W_C(t(g))/W_C(s(g))$ be the mapping defined by

$$W(q_{g^*,C}) = i_{g,C}^{-1} \circ q_{g^*,C} \circ i_{g*,C}.$$

The following lemma shows that $q_{g^*,C}$ corresponds to the quotient mapping $W(t(g)) \longrightarrow W(t(g))/W(s(g))$ by the isomorphisms $i_{g,C} \ (resp. \ i_{g*,C})$. It is easy to see (cf. Proposition 9.54) the

Lemma 9.63 *The mapping $W(q_{g^*,C})$ coincides with the quotient mapping defined by the right action of $W_C(s(g))$ on $W_C(t(g))$ (cf. Definition 9.50).*

Definition 9.64

A) *Let $g \in \text{gall}_A$ with $l(g) = r + 1$, and $C \in \text{Ch } A$. Define a mapping $s_{g^*,C} : \text{Gall}_A(g, F_g(C)) \to \text{Gall}_A(g^*, C)$ by*

$$s_{g^*,C} : \gamma \mapsto \gamma^*,$$

(The canonical section of $q_{g^*,C}$) *where γ^* is defined as follows.*

1) *Let $g = g_1$ (resp. g'_1), and $\gamma : \ (F_i)_{r \geqslant i \geqslant 0}, \ (F'_j)_{r+1 \geqslant j \geqslant 1}$ (resp. $(F'_j)_{r+1 \geqslant j \geqslant 0}$) . Define γ^* by*

$$\gamma^* : \ (C_i)_{r+1 \geqslant i \geqslant 0}, \ (F'_j)_{r+1 \geqslant j \geqslant 1} \ (resp. \ (F'_j)_{r+1 \geqslant j \geqslant 0}),$$

where $C_{r+1} = C, \ C_i = \text{proj}_{F_i} C_{i+1} \ (r \geqslant i \geqslant 0)$.

2) *Let* $g = g_2$ *(resp.* g_2'*), and* $\gamma :$ $(F_i)_{r \geqslant i \geqslant 0}$, $(F_j')_{r \geqslant j \geqslant 1}$ *(resp.* $(F_j')_{r \geqslant j \geqslant 0}$*). Define* γ^* *by*

$$\gamma^* : (C_i)_{r \geqslant i \geqslant 0}, \ (F_j')_{r \geqslant j \geqslant 1} \ (\text{resp. } (F_j')_{r \geqslant j \geqslant 0}),$$

where $C_r = C$, $C_i = \text{proj}_{F_i} C_{i+1}$ $(r > i \geqslant 0)$.

Denote by $s_{g^*,C}^m : \text{Gall}_{\mathcal{A}}^m(g, F_g(C)) \to \text{Gall}_{\mathcal{A}}(g^*, C)$ *the restriction of* $s_{g^*,C}$ *to* $\text{Gall}_{\mathcal{A}}^m(g, F_g(C)) \subset \text{Gall}_{\mathcal{A}}(g, F_g(C))$.

B) *Let* $\mathcal{G} \subset \text{Gall}_{\mathcal{A}}(g)$. *Write:*

$$\text{Ch } \mathcal{A} * \mathcal{G} := \{(C, \gamma(F, F')) \in \text{Ch } \mathcal{A} \times \mathcal{G} | \ F_{typ \ F}(C) = F\},$$

i.e. $\text{Ch } \mathcal{A} * \mathcal{G} = \text{Ch } \mathcal{A} \times_{\mathcal{A}} \mathcal{G}$ *where the second member is the fiber product defined by the couple of mappings* $\text{Ch } \mathcal{A} \to \mathcal{A}$ *(resp.* $\mathcal{G} \to \mathcal{A}$*), given by* $C \mapsto F_{e_1(g)}(C)$ *(resp.* $E_1|_{\mathcal{G}} : \gamma(F, F') \mapsto F$*). The section* $s_{g^*,C}$ *may be extended to a section on* $\text{Ch } \mathcal{A} * \text{Gall}_{\mathcal{A}}(g)$.
Let $s_{g^*,ch} : \text{Ch } \mathcal{A} * \text{Gall}_{\mathcal{A}}(g) \to \text{Gall}_{\mathcal{A}}(g^*)$ *be the mapping defined by*

$$s_{g^*,ch} : \ (C, \gamma(F, F')) \mapsto \gamma^*(C, F').$$

Thus given $(C, \gamma) \in \text{Ch } \mathcal{A} * \text{Gall}_{\mathcal{A}}(g)$ *one has* $s_{g^*,ch}((C, \gamma)) = s_{g^*,C}(\gamma)$.
Let

$$s_{g^*,ch}^m : \ \text{Ch } \mathcal{A} * \text{Gall}_{\mathcal{A}}^m(g) \to \text{Gall}_{\mathcal{A}}(g^*)$$

be the restriction of $s_{g^*,ch}$ *to* $\text{Ch } \mathcal{A} * \text{Gall}_{\mathcal{A}}^m(g) \subset \text{Ch } \mathcal{A} * \text{Gall}_{\mathcal{A}}(g)$.

Let $E = (E_1, E_2) : \text{Gall}_{\mathcal{A}} \to \mathcal{A} \times \mathcal{A}$ be defined by $E : \gamma(F, F') \mapsto (F, F')$ (the extremities mapping). Let

$$E_1 * q_{g^*} : \ \text{Gall}_{\mathcal{A}}(g^*) \to \text{Ch } \mathcal{A} * \text{Gall}_{\mathcal{A}}(g)$$

be the fiber product mapping

$$E_1 * q_{g^*} = E_1 \times_{\mathcal{A}} q_{g^*}.$$

It results immediately from Definition 9.64 and the definition of q_{g^*} that:

$$(\forall \ \gamma \in \text{Gall}_{\mathcal{A}}(g, F_g(C))) \quad , \quad (q_{g^*,C} \circ s_{g^*,C})(\gamma) = \gamma$$
$$(\text{resp. } \forall \ (C, \gamma) \in \text{Ch } \mathcal{A} * \text{Gall}_{\mathcal{A}}(g) \quad , \quad ((E_1 * q_{g^*}) \circ s_{g^*,ch})((C, \gamma)) = (C, \gamma) \ ).$$

Then one has

$$q_{g^*,C} \circ s_{g^*,C} = 1_{\text{Gall}_{\mathcal{A}}(g, F_g(C))}$$
$$(\text{resp. } (E_1 * q_{g^*}) \circ s_{g^*,ch} = 1_{\text{Ch } \mathcal{A} * \text{Gall}_{\mathcal{A}}(g)}).$$

9.10   The Representatives Set of a Contracted Product defined by a gallery of types

With the natural section $s_{g^*,C}$ of the quotient mapping $q_{g^*,C}$ corresponds a section of the contracted product $W(t(g))/W(s(g)) \longrightarrow W(t(g))$ whose values gives a set of canonical representatives of the quotient. These representatives are calculated in terms of the minimal length elements in the classes $(W/W_F)_{F \subset C}$.

Define
$$W(s_{g^*,C}) : \; \overline{W}(g) = W_C(t(g))/W_C(s(g)) \to W_C(t(g))$$

by
$$W(s_{g^*,C}) = i_{g^*,C}^{-1} \circ s_{g^*,C} \circ i_{g,C}.$$

One thus has

$$
\begin{aligned}
W(q_{g^*,C}) \circ W(s_{g^*,C}) &= (i_{g,C}^{-1} \circ q_{g^*,C} \circ i_{g^*,C}) \circ (i_{g^*,C}^{-1} \circ s_{g^*,C} \circ i_{g,C}) \\
&= i_{g,C}^{-1} \circ 1_E \circ i_{g,C} = 1_{\overline{W}(g)}.
\end{aligned}
$$

It is concluded that:

Proposition 9.65 *The mapping $W(s_{g^*,C})$ is a section of the quotient mapping $W(q_{g^*,C}) : W_C(t(g)) \to \overline{W}(g) = W_C(t(g))/W_C(s(g))$.*

Notation 9.66 *Given a class $\overline{x} \in \overline{W}(g)$ one denotes by $w_{g^*,C}(\overline{x})$ the element which makes $W(s_{g^*,C})$ correspond to $\overline{x}$. It is thus concluded that the set*

$$(w_{g^*,C}(\overline{x}))_{\overline{x} \in \overline{W}(g)} \subset W_C(t(g))$$

is a set of representatives of $W(q_{g^*,C}) : \; W_C(t(g)) \to W_C(t(g))/W_C(s(g))$.

9.10.1   Explicit calculation of the Representatives set

Write:

$$g: \; (s_i)_{r \geqslant i \geqslant 0}, \; (t_j)_{r+1 \geqslant j \geqslant 0}; \; x = (x_i)_{r+1 \geqslant i \geqslant 0} \in W_C(t(g)); \; z_i =$$

$$\prod_{r+1 \geqslant \alpha \geqslant i} x_\alpha \quad (\text{resp. } z_{r+2} = 1).$$

Thus one obtains $i_{g,C}(\overline{x}) = \gamma_{g,C}(x)$ for $\overline{x} \in \overline{W}(g)$ cf. Definition 9.53, with

$$\gamma_{g,C}(x): \; (z_{i+1}(F_{s_i}(C)))_{r \geqslant i \geqslant 0}, \; (z_{j+1}(F_{t_j}(C)))_{r+1 \geqslant j \geqslant 0}.$$

Let $\gamma_{g,C}^*(x)$ be the image of $\gamma_{g,C}(x)$ by $s_{g^*,C}$ (cf. Definition 9.63). Write

$$\gamma_{g,C}^*(x): \; (C_i(x))_{r+1 \geqslant i \geqslant 0}, \; (z_{j+1}(F_{t_j}(C)))_{r+1 \geqslant j \geqslant 0}.$$

The determination of $\gamma^*_{g,C}(x)$ amounts to that of the sequence of chambers

$$C_{r+1} = C, \; C_i = C_i(x) = \text{proj}_{F_i(x)} C_{i+1}(x) \quad (r \geqslant i \geqslant 0)$$

where $F_i = F_i(x) = z_{i+1}(F_{s_i}(C)) \; (r \geqslant i \geqslant 0)$. Clearly the following inclusions holds

$$C_i \supset F_i \supset F'_i = z_{i+1}(F_{t_i}(C)).$$

Definition 9.67 *Define* $x^m = x^m_i = $ *by the following recursive definition. Let* $x^m_{r+1} = $ *the minimal length* $l_{S_C}(x^m_{r+1})$ *representative of the class* $\overline{x}_{r+1} \in W_{t_{r+1}}/W_{s_r}$, *and* $y_r \in W_{s_r}$ *be the element defined by* $x^m_{r+1} = x_{r+1} y_r$, *i.e.* $y_r = x^{-1}_{r+1} x^m_{r+1}$. *For* $r \geqslant i \geqslant 1$ *one writes* $x^m_i = $ *the minimal length* $l_{S_C}(x^m_i)$ *representative of the class* $y^{-1}_i x_i \in W_{t_i}/W_{s_{i-1}}$, *and let* $y_{i-1} \in W_{s_{i-1}}$ *be the element defined by* $x^m_i = y^{-1}_i x_i y_{i-1}$, *i.e.* $y_{i-1} = x^{-1}_i y_i x^i_m$.

Let it be written $w_{g^*,C}(\overline{x}) = x'^m = (x'^m_i)$. The characteristic property of x'^m being

$$x'^m \cdot \gamma_{g^*,C} = \gamma_{g^*,C}(x'^m) = s_{g^*,C}(\gamma_{g,C}(\overline{x})) = \gamma^*_{g,C}(x),$$

or equivalently

$$(*) \qquad z^m_{i+1}(C) = C_i \quad (r \geqslant i \geqslant 0),$$

where $z^m_{i+1} = \displaystyle\prod_{r+1 \geqslant \alpha \geqslant i+1} x'^m_\alpha$.

Proposition 9.68 *The following identity holds:* $x^m = w_{g^*,C}(\overline{x})$.

Proof *Let* $w'_i = w(C_i, C_{i-1}) \; (r+1 \geqslant i \geqslant 1)$ *(resp.* $z_{r+2} = y_{r+1} = 1$*). One then has* $w'_{r+1} = w(C_{r+1}, C_r) = x^m_{r+1}$. *Define for* $r+1 \geqslant i \geqslant 1$

$$w^m_i = (z_{i+1} y_i)(y^{-1}_i x_i y_{i-1})(z_{i+1} y_i)^{-1} = (z_{i+1} y_i) x^m_i (z_{i+1} y_i)^{-1}.$$

From the equality $x^m = (x^m_i) = (y^{-1}_i x_i y_{i-1})_{r+1 \geqslant i \geqslant 1}$ *it is obtained*

$$z^m_{i+1} = \prod_{r+1 \geqslant \alpha \geqslant i+1} x^m_\alpha = \prod_{r+1 \geqslant \alpha \geqslant i+1} (y^{-1}_\alpha x_\alpha y_{\alpha-1}) = (\prod_{r+1 \geqslant \alpha \geqslant i+1} x_\alpha) y_i = z_{i+1} y_i.$$

Then it follows that $w^m_i = z^m_{i+1} x^m_i (z^m_{i+1})^{-1}$. *Consequently it is obtained, as may be easily verified by induction, that*

$$(**) \qquad z_{i+1} y_i = x^m_{r+1} \cdots x^m_{i+1} = w^m_{i+1} \cdots w^m_{r+1}.$$

Let it now be proved by induction $w'_i = w(C_i, C_{i-1}) = w^m_i$. *It is known that* $w'_{r+1} = w(C_{r+1}, C_r) = x^m_{r+1} = x_{r+1} y_r = w^m_{r+1}$. *Let* $r \geqslant i \geqslant 1$. *It is supposed by recursive hypothesis that* $w'_{\alpha+1} = w^m_{\alpha+1}$ *for* $\alpha \geqslant i$. *Thus one obtains*

$$w'_{i+1} \cdots w'_{r+1} = w^m_{i+1} \cdots w^m_{r+1} = z^m_{i+1} = z_{i+1} y_i,$$

then $(z_{i+1}y_i)(C) = w'_{i+1}\cdots w'_{r+1}(C) = C_i \supset F_i \ (\supset F'_i)$. *On the other hand,* $C'_{i-1} = z_i(C) \supset z_i(F_{s_{i-1}}(C)) = F_{i-1}$, *i.e.* C'_{i-1} *is a chamber incident to* F_{i-1}. *As*

$$z_i = z_{i+1}x_i = (z_{i+1}y_i)(y_i^{-1}x_i)(z_{i+1}y_i)^{-1}(z_{i+1}y_i) = z_{i+1}^m(y_i^{-1}x_i)(z_{i+1}^m)^{-1}z_{i+1}^m.$$

It is recalled that $z_{i+1}^m(C) = C_i$, *thus it is deduced that*

$$C'_{i-1} = z_i(C) = z_{i+1}^m(y_i^{-1}x_i)(z_{i+1}^m)^{-1}(C_i).$$

Let u_i = *the minimal length* $l_{S(C_i)}(u_i)$ *element of the class of* $z_{i+1}^m(y_ix_i)(z_{i+1}^m)^{-1}$ *in* $W_{F'_i}/z_{i+1}^m W_{s_{i-1}}(z_{i+1}^m)^{-1} = z_{i+1}^m(W_{t_i}/W_{s_{i-1}})(z_{i+1}^m)^{-1}$. *It is concluded that* $C_{i-1} = \mathrm{proj}_{F_{i-1}}C_i = u_i(C_i)$, *i.e.* $u_i = w(C_i, C_{i-1})$. *On the other hand, it is known that by definition*

$$x_i^m = y_i^{-1}x_iy_{i-1} \in W_{t_i}/W_{s_{i-1}},$$

is the minimal length $l_{S_C}(x_i^m)$ *element of the class* $\overline{y_i^{-1}x_i} \in W_{t_i}/W_{s_{i-1}}$, *and finally one has* $w'_i = z_{i+1}^m x_i^m(z_{i+1}^m)^{-1} = w_i^m$. *From* (∗∗) *it is concluded that* (∗) *holds in view of* $(w'_{i+1}\cdots w'_{r+1})(C) = C_i$. *This proves that* x^m *gives* $w_{g^*,C}(\overline{x})$, *and thus achieves the proof of the proposition.*

9.11　Minimal length class representatives in facet stabilizer subgroups

Write

$$\mathrm{St}_{F_t}^{\Delta(C)} := \{F \in \Delta(C)|\ F_t \subset F\}$$

(**Star Complex of** F_t **in** $\Delta(C)$), and $\mathrm{St}_{F_t}^{\mathcal{A}}$ for the Star Complex of F_t in $\mathcal{A}$, where $F_t = F_t(C)$ ($t \in$ typ $\mathcal{A}$). On the other hand, the isomorphisms $i_C : C(W_{\mathcal{A}}, S_C) \simeq \mathcal{A}$ (resp. typ $C(W, S_C) = \mathcal{P}(S_C) \simeq$ typ $\mathcal{A} \simeq \Delta(C)$) induce the isomorphisms

$$i_C^{.\mathrm{St}_{F_t}} : C(W_{S_C - X_t}, S_C - X_t) \to \mathrm{St}_{F_t} = \mathrm{St}_{F_t}^{\mathcal{A}},$$

where $X_t \subset S_C$ corresponds to t by $\mathcal{P}(S_C) \simeq$ typ $\mathcal{A}$; thus X_t indexes the vertices of F_t, and

$$\mathrm{typ}\, C(W_{S_C - X_t}, S_C - X_t) = \mathcal{P}(S_C - X_t) \simeq \mathrm{typ}\, \mathrm{St}_{F_t}^{\mathcal{A}} \simeq \mathrm{St}_{F_t(C)}^{\Delta(C)}.$$

One has that

$$C(W_{S_C - X_t}, S_C - X_t) = \coprod_{F \in \mathrm{St}_{F_t(C)}^{\Delta(C)}} W_{S_C - X_t}/W_F$$

and given $s \supset t$ the restriction

$$i_{C,s}^{.\mathrm{St}_{F_t}} : W_{S_C - X_t}/W_{F_s} \simeq (\mathrm{St}_{F_t}^{\mathcal{A}})_s \ (F_s \supset F_t),$$

is defined by $\overline{w} \mapsto w(F_s) \in (\mathrm{St}_{F_t}^{\mathcal{A}})_s$.

Definition 9.69

Given $F \in (\mathrm{St}_{F_t(C)})_s$, let $\overline{w}_F \in W_{S_C - X_t}/W_{F_s}$, where $W_{F_s} = W_{S - X_s}$, be the class of an element w_F satisfying $w_F(F_s(C)) = F$. In other terms $\overline{w}_F = (i_C^{\mathrm{St}_{F_t(C)}})^{-1}(F)$. Let

$$w_F^m = w(C, \mathrm{proj}_F C) \in W_{S_C - X_t}.$$

One has that $w_F^m \in \overline{w}_F$ is the minimal length element in the class $\overline{w}_F$, i.e. $l_{S_C}(w_F^m)$ is a minimum of the function $w \mapsto l_{S_C}(w)$ ($w \in \overline{w}_F$). As $w_F^m \in W_{S_C - X_t}$ one then has

$$l_{S_C - X_t}(w_F^m) = l_{S_C}(w_F^m).$$

Thus $w_F^m \in \overline{w}_F$ is characterized as the minimal length element, relatively to $S_C - X_t$, of this class $\overline{w}_F \subset W_{S_C - X_t}$.

Write $W_t = W_{S_C - X_t}$ thus $W_{F_s} = W_s$.

Definition 9.70

Let $(t, s) \in \mathrm{typ}^{(2)} \mathcal{A}$ (cf. Definition 9.5). Write:

$$\mathcal{W}_C^m(t, s) := \{w_F^m \in W_t | \ F \in (\mathrm{St}_{F_t(C)})_s \simeq W_t/W_s\}$$

(The set of minimal length representatives of the classes W_t/W_s), and

$$\mathcal{W}_C^m = \coprod_{(t,s) \in \mathrm{typ}^{(2)} \mathcal{A}} \mathcal{W}_C^m(t, s) .$$

Given $g \in \mathrm{gall}_{\mathcal{A}}$ so that $g = g_1, g_1'$ (resp. $g = g_2, g_2'$) with $l(g) = r + 1$, let:

$$\mathcal{W}_C^m(g) \ := \prod_{r+1 \geqslant i \geqslant 1} \mathcal{W}_C^m(t_i(g), s_{i-1}(g))$$

$$(\text{resp. } \mathcal{W}_C^m(g) \ := \prod_{r \geqslant i \geqslant 1} \mathcal{W}_C^m(t_i(g), s_{i-1}(g))).$$

Clearly one has the inclusion $\mathcal{W}_C^m(g) \subset W_C(t(g))$.

Remark that the family of classes of representatives $(\mathcal{W}_C^m(t, s))_{(t,s) \in \mathrm{typ}^{(2)} \mathcal{A}}$ is indexed by the same set of indices as $\mathscr{S}_C \subset W$ (cf. loc. cit.). The set $\mathcal{W}_C^m(t, s) \subset W_t$ (cf. Definition 9.70) is in fact a set of representatives of the quotient set W_t/W_s, by means of the following correspondence. Given $\overline{w} \in W_t/W_s$, one writes $F_{\overline{w}} := w(F_s(C))$. Define a bijection $W_t/W_s \to \mathcal{W}_C^m(t, s)$, by $\overline{w} \mapsto w_{F_{\overline{w}}}^m$ (cf. loc. cit.). The set $\mathcal{W}_C^m$ plays the same role with respect to generalized galleries issued from C as $\mathcal{S}_C$ with respect to Minimal Generalized Galleries, i.e. Generalized Galleries issued from C correspond to words in $\mathcal{W}_C^m$.

Definition 9.71

Define the **length of a class** $\overline{w} \in W_t/W_s$ relatively to $S = S_C$ by $l_S(\overline{w}) = l_S(w^m)$, where $w^m \in \overline{w}$ is the minimal length $l_S(w^m)$ element of this class. If $F = F^{tr}_{(t,s)}(C)$ then the element w^m_F (cf. loc. cit.) satisfies the following characteristic property:

$$(\forall\ \overline{w} \in W_t/W_s)\ \ l_S(\overline{w^m_F}) \geqslant l_S(\overline{w}).$$

By definition $F = F^{tr}_{(t,s)}(C)$ is the facet of type s at maximal distance from C in $\mathrm{St}_{F_t(C)}$, i.e. C and F are in transversal position in $\mathrm{St}_{F_t}(C)$ and $w^m_F = w(C, \mathrm{proj}_F C)$, thus

$$(\forall \overline{w} \in W_t/W_s)\ \ \ l_S(w^m_F) = d(C, F) \geqslant d(C, F_{\overline{w}}) = l_S(\overline{w}).$$

By definition of $w(t, s)$ it is obtained

$$w(t, s) = w(C, \mathrm{proj}_F C) = w^m_F.$$

Thus $w(t, s)$ is the representative of the maximal length class in W_t/W_s.

Proposition 9.72 With the notation of Proposition 9.54 one has: The mapping

$$i^m_{g,C} :\ \mathcal{W}^m_C(g) \to \mathrm{Gall}_{\mathcal{A}}(g, F_g(C))$$

defined by $i^m_{g,C} :\ x \mapsto \gamma_{g,C}(x)$ $(x \in \mathcal{W}^m_C(g))$ is bijective.

Observe that $i^m_{g,C} = $ restriction of $q_{g^*,C} \circ i_{g^*,C}$ (cf. 9.62) to $\mathcal{W}^m_C(g)$ (resp. the composition of the natural mapping $\mathcal{W}^m_C(g) \longrightarrow W_C(t(g))/W_C(s(g))$ followed by $i_{g,C}$).

Proof It suffices to carry out the proof in the case $g = g_1, g'_1$. In this case $F_g(C) = F_{t_{r+1}(g)}(C)$. Remark that

(*) $x = (x_i) \in \mathcal{W}^m_C(g) \Leftrightarrow x_i$ is the minimal length element of its own class

$$\overline{x}_i \in W_{t_i(g)}/W_{s_{i-1}(g)}.$$

As $i_{g,C} :\ \overline{W}(g) \to \mathrm{Gall}_{\mathcal{A}}(g, F_g(C))$ is bijective given $\gamma \in \mathrm{Gall}_{\mathcal{A}}(g, F_g(C))$, there exists $x = (x_i)$ so that $\gamma_{g,C}(x) = \gamma$. Let $y = (y_i) \in W_C(s(g))$ be defined as in Subsection 9.10.1. Then $x^m = (x_j) \in W_C(t(g))$ given by

$$x^m_{r+1} = x_{r+1}y_r, \quad x^m_i = y_i^{-1}x_iy_{i-1} \quad (r \geqslant i \geqslant 1),$$

satisfies:

1) x^m_i is the minimal length element of its own class $\overline{x^m_i} \in W_{t_i(g)}/W_{s_{i-1}(g)}$.

2) $\gamma_g(x^m) = \gamma_{g,C}(x)$,

as follows from loc. cit. This proves that $i_{g,C}^m$ is a surjective mapping. Let it be proved that given $x = (x_i)$, and $t = (t_i) \in \mathcal{W}_C^m(g)$, then $\gamma_{g,C}(x) = \gamma_g(t) \Rightarrow x = t$, i.e. $i_{g,C}^m$ is an injective mapping. By $(*)$ it is deduced as in loc. cit. $x_{r+1}^m = x_{r+1}$, and thus $y_r = 1$. One determines x_i^m (resp. y_i) by induction. Let it be supposed $y_i = 1$. Following the recursive pattern of definition of x^m one obtains $x_{i-1}^m = x_{i-1}$, and $y_{i-1} = 1$ (cf. loc. cit.). Thus $x^m = x$. In the same way it is proved that $t^m = t$ and thus $x = x^m = t^m = t$.

Corollary 9.73 *The restriction to $\mathcal{W}_C^m(g)$*

$$W(q_{g^*,C})|_{\mathcal{W}_C^m(g)} : \mathcal{W}_C^m(g) \to W_C(t(g))/W_C(s(g)),$$

of the quotient mapping $W(q_{g^*,C}) : W_C(t(g)) \to \overline{W}(g)$ (cf. Lemma 9.63) satisfies

$$W(q_{g^*,C})|_{\mathcal{W}_C^m(g)} = i_{g,C}^{-1} \circ i_{g,C}^m$$

(cf. Proposition 9.62). Thus the set $\mathcal{W}_C^m(g) \subset W_C(t(g))$ is a set of representatives of $W_C(t(g))/W_C(s(g))$, and one has

$$\text{Im } W(s_{g^*,C}) = (w_{g^*,C}(\overline{x}))_{\overline{x} \in W_C(t(g))/W_C(s(g))} = \mathcal{W}_C^m(g)$$

(cf. Subsection 9.10.1).

Proof *By definition one has*

$$W(q_{g^*,C}) = i_{g,C}^{-1} \circ q_{g^*,C} \circ i_{g^*,C},$$

and by the remark following Proposition 9.72 one obtains

$$i_{g,C}^m = (q_{g^*,C} \circ i_{g^*,C})|_{\mathcal{W}_C^m(g)}.$$

Finally one gets

$$W(q_{g^*,C})|_{\mathcal{W}_C^m(g)} = i_{g,C}^{-1} \circ i_{g,C}^m.$$

From Proposition 9.62 and Proposition 9.72 it follows that $W(q_{g^*,C})|_{\mathcal{W}_C^m(g)}$ is a bijective mapping. Thus it is concluded that $\mathcal{W}_C^m(g)$ is a set of representatives of $\overline{W}(g)$. The construction of 9.10.1 proves the last assertion.

9.12 Minimal length class representatives sequence associated with a minimal generalized gallery

Define $w_C^m(g) \in \mathcal{W}_C^m(g)$ by

$$w_C^m(g) := (w(t_{i+1}, s_i)) \quad (r \geqslant i \geqslant 0)$$
$$(\text{resp. } w_C^m(g) := (w(t_{i+1}, s_i)) \quad (r - 1 \geqslant i \geqslant 0)).$$

according to $g = g_1, g_1'$ (resp. $g = g_2, g_2'$). Let $F, F' \in \mathcal{A}$, and $g \in \text{gall}_{\mathcal{A}}^m$ the gallery of types of a MGG between F and F'. With the minimal generalized gallery $\gamma_g(F, F')$ of type g, and a chamber $C \supset F$ at maximal distance from F', i.e. so that $C \supset c(F, F')$, is associated a $\mathscr{S}_C$-reduced expression of type g of $w_{F'}^m = w(C, \mathrm{proj}_{F'}C)$, as follows.

Let $\gamma_g(C; F, F')$ be the MGG between C and F' associated with $\gamma_g(F, F')$ as in Definition 9.16. It has been seen how to associate a $\mathscr{S}$-reduced expression of $w(C_e, \mathrm{proj}_F C_e)$ with a MGG $\gamma(C_e, F)$(cf. Subsection 9.3). These considerations apply here and one obtains, in the case $g = g_1, g_1'$ $l(g) = r + 1$:

$$w_{F'}^m = w(C, \mathrm{proj}_{F'}C) = w(t_{r+1}, s_r) \cdots w(t_1, s_0).$$

It is concluded that the element of $W_C(t(g))$ defined by $\gamma_g(C; F, F')$ is precisely $w_C^m(g)$. It is easy to verify that

$$i_{g,C}^m(w_C^m(g)) = \gamma_g(F, F').$$

Thus one has proved the

Proposition 9.74

Let $\gamma_g(F, F')$ be the MGG of type $g \in \mathrm{gall}_{\mathcal{A}}^m$ between F and F', and $C \supset F$ a chamber at maximal distance from F'. Then

$$w_C^m(g) = (w(t_i, s_{i-1})) \in \mathcal{W}_C^m(g),$$

is an $\mathscr{S}_C$-reduced expression of $w(C, \mathrm{proj}_{F'}C)$, and moreover

$$\gamma_g(w_C^m(g)) = \gamma_g(F, F').$$

Proposition 9.75

Assume $g \in \mathrm{gall}_{\mathcal{A}}^m$ and that $w = w_C(t_r, s_{r-1}) \cdots w_C(t_1, s_0)$ is a $\mathscr{S}_C$-reduced expression. Denote by $w_C^m(g) \in \mathcal{W}_C^m(g)$ its corresponding element. Write $F = w(F_{s_0}(C))$ (resp. $F = w(F_{t_0}(C))$), according to $g = g_2$ (resp. $g = g_2'$). Let $\gamma(F_s, F)$ be the generalized gallery constructed from $w_C^m(g)$. Then $\gamma(F_s, F)$ is a Minimal Generalized Gallery, $\gamma(F_s, F) = \gamma_g(F_s, F)$, and C realizes the maximal distance between a chamber $C \supset F_s$ and F.

Proof *It is supposed $g = g_2'$, and one can write:*

$$w_i = w(t_i, s_{i-1}) \ (r \geqslant i \geqslant 1), \qquad z_{i+1} = w_r \cdots w_{i+1} \ (r > i \geqslant 0)$$
$$(\text{resp. } z_{r+1} = 1), \quad u_i = z_{i+1} w_i z_{i+1}^{-1} \ (r \geqslant i \geqslant 1).$$

Then $u_i \cdots u_r = w_r \cdots w_i$ $(r \geqslant i \geqslant 1)$ (resp. $w(C, \mathrm{proj}_F C) = w = u_1 \cdots u_r$). Let $C_i = z_{i+1}(C)$ $(r \geqslant i \geqslant 1)$. One chooses, for $r \geqslant i \geqslant 1$ a minimal gallery $\Gamma^i = \Gamma^i(C_i, C_{i-1})$. Observe that $u_i = w(C_i, C_{i-1})$. (One has: $u_i(C_i) = z_{i+1} w_i z_{i+1}^{-1}(z_{i+1}(C)) = z_{i+1} w_i(C) = z_i(C) = C_{i-1}$). Write:

$$\gamma_g(w_C^m(g)): \ (F_i)_{r \geqslant i \geqslant 0}, \ (F_j')_{r \geqslant j \geqslant 0}.$$

By definition one has: $F_i = z_{i+1}(F_{s_i}(C))$ *(resp.* $F_i' = z_{i+1}(F_{t_i}(C))$*). As* $z_{i+1}(F_{s_i}(C)) = F_{s_i}(z_{i+1}(C)) = F_{s_i}(C)$ *(resp.* $z_{i+1}(F_{t_i}(C)) = F_{t_i}(z_{i+1}(C)) = F_{t_i}(C_i)$*), one obtains immediately* $C_i \supset F_i \supset F_i'$. *On the other hand,* $w_i = w(t_i, s_{i-1}) \in W_{t_i}$ *gives*

$$F_{t_i}(C) \subset F_{s_{i-1}}(C) \Rightarrow F_{t_i}(C) \subset w_i(F_{s_{i-1}}(C)),$$

thus $F_i' = z_{i+1}(F_{t_i}(C)) \subset z_{i+1}w_i(F_{s_{i-1}}(C)) = z_i(F_{s_{i-1}}(C))$, *and it is deduced that* $F_i' \subset F_{i-1} \subset C_{i-1}$.

Thus $\Gamma^i(C_i, C_{i-1}) \subset \mathrm{St}_{F_i'}$. *Apply the following lemma 9.76 to* $(C, w_i(C))$, *and* $(z_{i+1}(C), z_{i+1}w_i(C)) = (C_i, C_{i-1})$, *and* $w_i(F_{s_{i-1}}(C)) \subset w_l(C) = \mathrm{proj}_{F_{s_{i-1}}(C)}C$. *One obtains*

$$z_{i+1}(\mathrm{proj}_{w_i(F_{s_{i-1}}(C))}C) = \mathrm{proj}_{w_i(F_{i-1}(C))}C_i = C_{i-1}$$

$$(\ C_i = z_{i+1}(C),\ \ z_{i+1}w_i(F_{s_{i-1}}(C)) = F_{i-1}).$$

Thus $\Gamma^i(C_i, C_{i-1})$ *is a Minimal Gallery between* C_i *and* F_{i-1}, *so that*

$$\mathrm{set}\ \Psi^*(\Gamma^i(C_i, C_{i-1})) = \mathcal{H}_{F_i'}(F_{i-1}),$$

and thus

$$(*) \quad \mathrm{length}\ \Gamma^i(C_i, C_{i-1}) = |\mathcal{H}_{F_i'}(F_{i-1})| = l_S(w_i) \quad (\mathcal{S} = \mathcal{S}(C)).$$

Let $\Gamma(C, C_0) = \Gamma^r \circ \cdots \circ \Gamma^1$. *The hypothesis gives:*

$$d(C, F) = l_S(w_F^m) = l_S(w_r) + \cdots + l_S(w_1),$$

and $(*)$ *implies:*

$$\mathrm{length}\ \Gamma(C, C_0) = l_S(w_r) + \cdots + l_S(w_1),$$

thus

$$\mathrm{length}\ \Gamma(C, C_0) = d(C, F).$$

It is concluded that $\Gamma(C, C_0)$ *is a Minimal Gallery between* C *and* F, *such that*

$$\mathrm{set}\ \Psi^*(\Gamma(C, C_0)) = \coprod_{r \geqslant i \geqslant 1} \mathcal{H}_{F_i'}(F_{i-1}),$$

and consequently

$$\mathcal{H}(C, F) = \coprod \mathcal{H}_{F_i'}(F_{i-1}).$$

By a step by step comparison of this gallery with a Minimal Generalized Gallery of type g *one obtains that* $\tau(F_s, F) = \tau_g$ *and thus that* $\gamma(F_s, F)$ *is minimal of type* g. *Thus* $\gamma_g(w_C^m(g)) = \gamma_g(F, F')$.

To see that $C \supset F_s$ *is at maximal distance from* F *let objects and their properties introduced in the next chapter be used. Observe that* $\dim \Sigma(F_s, F) \leqq \dim \hat{\Sigma}(F_s, F) = \dim \hat{\Sigma}(C, F)$. *On the other hand one has* $\Sigma(C, F) \subset \Sigma(F_s, F)$, *and* $\dim \hat{\Sigma}(C, F) = \dim \Sigma(C, F)$. *Thus* $\dim \Sigma(C, F) = \dim \Sigma(F_s, F)$. *(Where* $\hat{\Sigma}(F_s, F) = \hat{\Sigma}(g, F_s)$ *and* $\hat{\Sigma}(C, F) = \hat{\Sigma}(\bar{g}, C)$*). This achieves the proof.*

The construction of the projection of a facet over another one (cf. Definition 8.15) commutes with the action of W.

Lemma 9.76 *Let (C, C'), $(\overline{C}, \overline{C'}) \in \mathrm{Ch}\ \mathcal{A} \times \mathrm{Ch}\ \mathcal{A}$, and $F \subset C'$ so that $C' = \mathrm{proj}_F C$. Suppose that there exists $w \in W$ so that $(w(C), w(C')) = (\overline{C}, \overline{C'})$, and one writes $\overline{F} = w(F)$. Then*

$$w(C') = w(\mathrm{proj}_F C) = \mathrm{proj}_{\overline{F}} \overline{C}.$$

Proof *One has $w(\overline{C}, \overline{C'}) = ww(C, C')w^{-1}$, and thus*

$$l_{S_{\overline{C}}}(w(\overline{C}, \overline{C'})) = l_{S_C}(w(C, C')).$$

The hypothesis $C' = \mathrm{proj}_F C$ gives:

$$l_{S_C}(w(C, C')) = d(C, C') = |\mathcal{H}(C, F)|.$$

As clearly $w(\mathcal{H}(C, F)) = \mathcal{H}(\overline{C}, \overline{F})$ it is concluded that $l_{S_C}(w(C, C')) = |\mathcal{H}(\overline{C}, \overline{F})|$. As $\overline{F} \subset \overline{C'}$ it is deduced that

$$w(\mathrm{proj}_F C) = w(C') = \overline{C'} = \mathrm{proj}_{\overline{F}} \overline{C}.$$

Recall that $\mathrm{proj}_{\overline{F}} \overline{C}$ is the chamber in $St_{\overline{F}}$ at minimal distance from $\overline{C}$.

9.13 Couples of facets and Minimal Generalized Galleries correspondence

Let $\tau_g \in \mathrm{Relpos}\ \mathcal{A}$, be the type of relative position given by $g \in \mathrm{gall}_{\mathcal{A}}^m$. Define

$$\sigma_g : (\mathcal{A} \times \mathcal{A})_{\tau_g} \to \mathrm{Gall}_{\mathcal{A}}^m(g)\ ,$$

as the mapping induced by

$$\sigma : (\mathcal{A} \times \mathcal{A}) \times_{\mathrm{Relpos}\ \mathcal{A}} \mathrm{gall}_{\mathcal{A}}^m \to \mathrm{Gall}_{\mathcal{A}}^m$$

(cf. Notation 9.43), i.e.

$$\sigma_g : (F, F') \mapsto \gamma_g(F, F'), \text{ where } \tau(F, F') = \tau_g\ .$$

Thus by σ_g to a couple of facets (F, F'), whose type of relative position is τ_g, corresponds the unique minimal generalized gallery of type g, $\gamma_g(F, F')$ joining F and F'. Observe that σ_g is in fact a bijection.

Denote by

$$\sigma_{g,ch} : (\mathrm{Ch}\ \mathcal{A} \times \mathcal{A})_{\tau_g} \to \mathrm{Ch}\ \mathcal{A} * \mathrm{Gall}_{\mathcal{A}}^m(g) = \mathrm{Ch}\ \mathcal{A} \times_{\mathcal{A}} \mathrm{Gall}_{\mathcal{A}}^m(g)$$

(cf. Definition 9.64) the mapping defined by

$$\sigma_{g,ch} : (C, F) \mapsto (C, \gamma_g(F_{e_1(g)}(C), F')),$$

and write

$$\sigma_{(g,F),ch} = \sigma_{g,ch}\big|(\text{Ch } \mathcal{A} \times \mathcal{A})_{\tau_g,F}$$

for the restriction of $\sigma_{g,ch}$ to $(\text{Ch } \mathcal{A} \times \mathcal{A})_{\tau_0,F}$.

Let

$$(\text{Ch } \mathcal{A} \times \mathcal{A})'_{\tau_0,F} \subset (\text{Ch } \mathcal{A} \times \mathcal{A})_{\tau_g,F}$$

be the subset formed by the couples (C, F') satisfying the hypothesis of Proposition 9.22 (resp. $C \supset F$ is at maximal distance from F'). The following proposition characterizes the image of $(\text{Ch } \mathcal{A} \times \mathcal{A})'_{\tau_g,F} \subset (\text{Ch } \mathcal{A} \times \mathcal{A})_{\tau_g,F}$ by $\sigma_{(g,F),ch}$ as a subset of $\mathcal{A} * \text{Gall}^m_{\mathcal{A}}(g, F) = \text{Ch } \mathcal{A} \times_{\mathcal{A}} \text{Gall}^m_{\mathcal{A}}(g, F)$.

Proposition 9.77 $(C, \gamma_g(F, F')) \in \mathcal{A} * \text{Gall}^m_{\mathcal{A}}(g, F) = \text{Ch } \mathcal{A} \times_{\mathcal{A}} \text{Gall}^m_{\mathcal{A}}(g, F)$ *belongs to* $Im\ \sigma_{(g,F),ch}$ *if and only if* $\gamma_g(C; F, F')$ *is a MGG.*

Proof Let $d(C, F')$ be maximal on the set of F' with $\tau(F, F') = \tau_g$. Then $d(C, F')$ is maximal on the set of chambers C incident to F. Thus it follows by definition of MGG that $\gamma_g(C; F, F')$ is a MGG. Reciprocally if $\gamma_g(C; F, F')$ is a MGG it results from the definition that $\mathcal{H}(C, F') = \coprod \mathcal{H}_{F'_i}(F_{i-1})$, where the i's run on a set depending on the class of g. It is concluded from this that $d(C, F')$ is maximal on the set of chambers C incident to F, and thus that it is maximal on the set of F' such that $\tau(F, F') = \tau_g$. This achieves the proof.

9.14 Weyl group Minimal Generalized Galleries characterization

Taking into account that $Ch\ \text{St}_{c(F,F')}$ is homogeneous under $W_{c(F,F')} = $ *Stabilizer of* $L_{(F,F')}$ and Proposition 9.21 one obtains:

Proposition 9.78 Let $\gamma(F, F')$ be a generalized gallery. Then $\gamma(F, F')$ is a MGG if and only if it satisfies the following 2 conditions:

1) $\gamma(F, F') \subset L_{(F,F')}$ (*The carrier of* F *and* F');

2) $\exists\ C \in \text{St}_{c(F,F')}$ such that $\gamma(C; F, F')$ is a MGG between C and F'.

One now proceeds to characterize a MGG between $F_s(C)$ and $w(F_t(C))$ in terms of W by translating the above proposition in terms of W.

Let $F \in \mathcal{A}$ (resp. $C \in \text{Ch } \mathcal{A}$). One supposes $F = F_s(C)$. Let L be the carrier of $\text{Env}(F, F')$, and $F'' \in \mathcal{A}$. The following two assertions are equivalent:

1) $F'' \in L$;

2) F'' is stable under the set of reflections $\{s_H|\ H \in \mathcal{H}_F \cap \mathcal{H}_{F'}\}$ defined by the set of walls $\mathcal{H}_F \cap \mathcal{H}_{F'}$.

Remark that given $\overline{w} \in W/W_t$ the set of reflections $\{s_H \mid H \in \mathcal{H}_{F_s} \cap \mathcal{H}_{w(F_t)}\}$ $(F_s = F_s(C), F_t = F_t(C))$ may be characterized as the set of reflections of W common to the subgroups W_s and wW_tw^{-1}.
If $F'' = w(F_t)$ then one immediately obtains

$$s_H(F'') = F'' \iff s_H w \in \overline{w} \in W/W_t.$$

Let $w^m \in \overline{w}$ be defined by $w^m = w(C, \mathrm{proj}_{w(F_t)}C)$ (the minimal length $l_S(w^m)$ element of the class $\overline{w}$). As in definition 9.70 one writes

$$l_S(\overline{w}) = l_S(w^m).$$

Notation 9.79 *Let $\tilde{w} \in W_s \setminus W/W_t$. Denote by $\overline{w'} \in \tilde{w}$ the maximal length class $l_S(\overline{w'})$ element of $\tilde{w}$, and by $w^m \in \overline{w'}$ the minimal length element of the class $\overline{w'}$. Thus one has by definition of w^m, $l_S(w^m) = l_S(\overline{w'})$ (resp. $\tilde{w}^m = \tilde{w}$, $l_S(\overline{w^m}) \geqslant l_S(\overline{w''})$ for every $\overline{w''} \in \tilde{w}$).*

Observe that $d(C, \overline{w'}(F_t))$ realizes the maximal distance $d(C, \overline{w}(F_t))(\overline{w} \in \tilde{w})$.

One fixes a gallery of types $g = g_1'$ given by

$$g: \ (s_i)_{r \geqslant i \geqslant 0}, \ (t_j)_{r+1 \geqslant j \geqslant 0}.$$

Consider the element

$$w_C^m(g) = (w(t_{i+1}, s_i))_{r \geqslant i \geqslant 0} \in \mathcal{W}_C^m(g).$$

Let $w^m = w(t_{r+1}, s_r) \cdots w(t_1, s_0)$. Write

$$F = F_{t_{r+1}}(C) \quad (\text{resp.} \ F' = w^m(F_{t_0}(C))),$$

and $z_{i+1} = w(t_{r+1}, s_r) \cdots w(t_{i+1}, s_i) \ (r \geqslant i \geqslant 0)$.

Let $\overline{z}_{i+1} \in W/W_{s_i}$ be the class defined by z_{i+1}. Remark that with the notation of Subsection 9.5 one has $\mathcal{H}_F \cap \mathcal{H}_{F'} = \mathcal{D}_L$, i.e. $\mathcal{H}_F \cap \mathcal{H}_{F'} = $ set of walls containing F and F'. One has the following criterion to decide if $\gamma_g(w_C^m(g))$ is a MGG between F and F'.

Proposition 9.80 *With the above notation $\gamma_g(w_C^m(g))$ is a MGG between F and F' if the following two conditions are satisfied by $w_C^m(g)$:*

1) $w^m = w(t_{r+1}, s_r) \cdots w(t_1, s_0)$ *is a $\mathcal{S}_C$-reduced expression;*

2) *for all $H \in \mathcal{H}_F \cap \mathcal{H}_{F'}$, $s_H z_{i+1} \in \overline{z}_{i+1} \ (r \geqslant i \geqslant 0)$.*

Let $\tilde{w} \in W_{t_{r+1}} \setminus W/W_{t_0}$ *(resp.* $\overline{w}' \in W/W_{t_0}$*) be the double class defined by* w^m *(resp. the maximal length class defined by* $\tilde{w}$*). Then* w^m *is the minimal length element of* $\overline{w}'$*.*

Proof *The first condition assures that* $\gamma(C; F, F')$ *defined by* $\gamma(F, F') = \gamma_g(w_C^m(g))$ *is a MGG. By 2) we have that* $\gamma(F, F') \subset L = $ *carrier of* $\mathrm{Env}(F, F')$*. Thus* $\gamma(F, F')$ *is a MGG between* F *and* F' *of type* g*, i.e.* $\gamma_g(w_C^m(g)) = \gamma_g(F, F')$*.*

9.15 Existence proof of Minimal Generalized Galleries between two given facets

Let $\mathcal{C}(\mathcal{H})$ be the geometric realization of $\mathcal{A}$ obtained from some isomorphism $\mathcal{A} \simeq C(W, S)$. The carrier L of $\mathrm{Env}(F, F')$ is defined as the subspace of $\mathbb{R}^{(S)}$ defined by $L = \bigcap_{H \in \mathcal{H},\ F, F' \subset H} H$. In this section $\mathcal{H}$ denotes the set of hyperplanes of $\mathbb{R}^{(S)}$ given by the set of conjugates of the reflections, associated with the walls of the chamber (resp. simplicial cone) $\subset \mathbb{R}^{(S)}$, corresponding to C_e in $C(W, S)$.

It is known that L has a building structure. The set Ch L of chambers of L is given by

$$\mathrm{Ch}\ L = \{F \in \mathcal{C}(\mathcal{H}) |\ L_F = L\}.$$

Let $\mathcal{H}_L := \{H \in \mathcal{H} |\ L \subset H\}$ and $\mathcal{H}'_L := \{H \cap L|\ H \in \mathcal{H} - \mathcal{H}_L\}$. Then the set of hyperplanes of L is given by $\mathcal{H}'_L$. One has $\mathcal{H}_L = \mathcal{H}_{c(F, F')}$. One chooses $C = C_{c(F, F')} \in \mathrm{Ch}\ \mathrm{St}_{c(F, F')}$, and considers the sub-building of $\mathcal{A}$

$$\mathcal{A}(c(F, F'), C) := \bigcap_{\substack{\Phi \text{ root of } \mathcal{A}, \\ C \in \Phi,\ c(F, F') \in \partial\Phi}} \Phi.$$

$\left(\text{Observe that according to [50]},\right.$

$$\mathrm{proj}_{c(F, F')}^{-1} C = \bigcap_{\substack{\Phi \text{ root of } \mathcal{A}, \\ C \in \Phi,\ c(F, F') \in \partial\Phi}} \mathrm{Ch}\ \Phi,$$

where $\mathrm{proj}_{c(F, F')}^{-1} C = \{C' \in \mathrm{Ch}\ \mathcal{A} |\ \mathrm{proj}_{c(F, F')} C' = C\}$. One has that $\mathrm{proj}_{c(F, F')}^{-1} C$ is a convex set of chambers (cf. loc. cit.).$\Big)$ It is known that $c(F, F')$ is a chamber of L, then there exists one and only one chamber $C_{c(F, F')}$ in $\mathcal{A}(c(F, F'), C)$ such that $c(F, F')$ is incident to $C_{c(F, F')}$, i.e. $\overline{C}_{c(F, F')} \supset c(F, F')$, since the intersection $\left(\bigcap_{C_{c(F, F')} \in \Phi,\ c(F, F') \in \partial\Phi} \Phi\right) \cap \left(\mathrm{Ch}\ \mathrm{St}_{c(F, F')}\right)$ defines one and only one chamber C of $St_{c(F,F')}$ namely $C = C_{c(F,F')}$.

Observe that the set of facets F of L may be defined as the equivalence classes, given in the usual way, by the set of hyperplanes $\mathcal{H}'_L$ of L. It may thus be denoted by $C(\mathcal{H}'_L)$ the building defined by L.

Definition 9.81 *A segment $[x,y] \subset \mathbb{R}^{(S)}$ (resp. $[x,y] \subset L$) is called a generic segment of $C(\mathcal{H})$ (resp. $C(\mathcal{H}'_L)$) if it satisfies:*

(1) $x \in C$ and $y \in C'$, with $C, C' \in \mathrm{Ch}\, C(\mathcal{H})$ (resp. $\mathrm{Ch}\, C(\mathcal{H}'_L)$);

(2) $[x,y]$ intersects only chambers and codimension 1 facets of $C(\mathcal{H})$ (resp. $C(\mathcal{H}'_L)$).

Let
$$\Gamma([x,y]): \quad C = C_n, \cdots, C_0 = C'$$

be the minimal gallery of $C(\mathcal{H})$ (resp. $C(\mathcal{H}'_L)$) defined by the ordered set of chambers $C = C_n \cdots C_1 = C'$ that $[x,y]$ intersects.
Let

$$\gamma([x,y]): \quad C = C_n \supset \cdots C_i \subset C_i \cap C_{i-1} \supset C_{i-1} \cdots \subset C_0 = C'$$

be the corresponding generalized gallery, where $C_i \cap C_{i-1}$ denotes the codimension 1 facet F_i, incident to C_i and C_{i-1} (i.e. $F_i \subset \overline{C}_i$, $\overline{C}_{i-1}$), which $[x,y]$ intersects.

One now constructs a generalized gallery $\gamma(F, F')$ between F and F' as follows. Recall first that $\mathrm{proj}_F F'$ (resp. $\mathrm{proj}_{F'} F$) defines a chamber of $C(\mathcal{H}'_L)$. Choose $x \in \mathrm{proj}_F F'$ and $y \in \mathrm{proj}_{F'} F$ so that the segment $[x,y]$ is a generic segment of $C(\mathcal{H}'_L)$. Assume $c(F, F') \neq \mathrm{proj}_F F'$, and denote by

$$\Gamma(\mathrm{proj}_F F', \mathrm{proj}_{F'} F): \quad F_r, F_{r-1}, \cdots, F_0$$

$(F_r = \mathrm{proj}_F F', F_0 = \mathrm{proj}_{F'} F)$ the minimal gallery of $C(\mathcal{H}'_L)$: $\Gamma([x,y])$. Write $F'_i = F_i \cap F_{i-1}$, and define

$$\gamma(F, F'): \quad F = F'_{r+1} \subset F_r = \mathrm{proj}_F F' \supset F'_r \cdots F'_1 \subset F_0 = \mathrm{proj}_{F'} F \supset F'_0 = F'.$$

The first and last inclusions may not be strict, i.e. one may have $F' = \mathrm{proj}_{F'} F$ (resp. $F = \mathrm{proj}_F F'$).
It is now proved that $\gamma(F, F')$ is a MGG. Let $u \in c(F, F')$ so that $[u,x]$ is a generic segment of $C(\mathcal{H}'_L)$. It may be supposed that after suitable choices of u, x and y, these three points are on the same line and $[u,x] \cup [x,y] = [u,y]$. One then has
$$\gamma([u,y]) = \gamma([x,y]) \circ \gamma([u,x]).$$

From the definition of $c(F, F')$ it is clear that $\gamma([u, x]) \subset \mathrm{St}_F$ "crosses" all the hyperplanes $H \in \mathcal{H}_F(\mathrm{proj}_F F')$. On the other hand, one has

$$\mathcal{H}(F_r, F_0) = \mathcal{H}(F_r, F_{r-1}) \coprod \cdots \coprod \mathcal{H}(F_1, F_0),$$

$F_r = \mathrm{proj}_F F'$ (resp. $F_0 = \mathrm{proj}_{F'} F$), as $[x, y]$ intersects precisely these hyperplanes, as easily seen. One finally gets

$$\mathcal{H}(c(F, F'), F_0) = \mathcal{H}_{F'_{r+1}}(F_r) \coprod \mathcal{H}(F_r, F_{r-1}) \coprod \cdots \coprod \mathcal{H}(F_1, F_0)$$

$(F'_{r+1} = F)$.
Observe now that as $L = L_{F_i} = L_{F_{i-1}}$, and $F'_i = $ codimension 1 facet of $\mathcal{C}(\mathcal{H}'_L)$ then:

$$\mathcal{H}(F_i, F_{i-1}) = \{H \in \mathcal{H} - \mathcal{H}_L | \ H \supset F'_i\} = \mathcal{H}_{F'_i}(F_{i-1}).$$

Thus one has

$$\mathcal{H}(c(F, F'), F_0) = \mathcal{H}_{F'_{r+1}}(F_r) \coprod \mathcal{H}_{F'_r}(F_{r-1}) \coprod \cdots \coprod \mathcal{H}_{F'_1}(F_0).$$

Let C_{F_0} be the unique chamber with $C_{F_0} \supset F_0 = \mathrm{proj}_{F'} F$ and $C_{F_0} \in \mathcal{A}(c(F, F'), C)$. Choose $\bar{u} \in C = C_{c(F, F')}$ and $\bar{y} \in C_{F_0}$ so that:

" $[\bar{u}, \bar{y}]$ is near enough to $[u, y]$ as to intersect the same hyperplanes as $[u, y]$, i.e.
$$H \cap [u, y] \neq \emptyset \ \Leftrightarrow \ H \in \mathcal{H}(c(F, F'), F_0) \text{ ''}.$$

Thus the minimal gallery $\Gamma(C, F_0) = \Gamma([\bar{x}, \bar{y}])$ satisfies:

$$\text{set } \Psi^*(\Gamma(C, F_0)) = \mathcal{H}_{F'_{r+1}}(F_r) \coprod \cdots \coprod \mathcal{H}_{F'_1}(F_0).$$

As $\mathcal{H}_F \cap \mathcal{H}_{F'_i}(F_{i-1}) = \emptyset$ $(r + 1 \geqslant i \geqslant 1)$ it is deduced that $\Gamma(C, F_0)$ is a minimal gallery between C and F'_0. It is concluded that $\gamma(F, F')$ is a mgg between F and F'. The case $F = \mathrm{proj}_F F'$ is similarly handled.

Chapter 10

Minimal Generalized Galleries in a Reductive Group Building

The reader is referred to [50] for details about the Building $I(G)$ of a reductive algebraic group G over an algebraically closed field k, and to [6], [9] and [23] for the basic definitions and properties of a parabolic subgroup of G. The reader may follow the constructions of this section by considering those of the Flag complex as a guiding example. The role of Flags there is played here by the Parabolic subgroups. The developments of this chapter are reconsidered in more detail in the next ones, in the setting of Group Schemes. The **Building** $I(G)$ of a reductive algebraic group G, over an algebraically closed field k, is the simplicial complex whose simplices are the parabolic subgroups of G. The incidence relation is defined as the symmetric inclusion relation between parabolics subgroups. It thus extends the definition of the Flag Complex of $Gl(k^{r+1})$ to the setting of k-reductive groups. The **Minimal Generalized Galleries in the Building** $I(G)$ are introduced. A minimal generalized gallery is contained in the Convex Envelope of its extremities, in $I(G)$ and in all the Apartments containing these extremities. With a minimal generalized gallery of types g is associated a type of relative position τ_g, satisfying the important property: Given a couple of facets (F, F') with a type of relative position τ_g, there is a unique Minimal Generalized Gallery $\gamma(F, F')$ with associated gallery of types g and extremities (F, F'). The set of generalised galleries of type g, issued from a fixed facet F admits a canonical "Cell Decomposition" indexed by generalized galleries of type g in the Weyl complex, which generalizes Bruhat cell decomposition. This set is described in terms of a contracted product along a gallery of types g in the Weyl Complex. Natural Parametrizations of the Cells are given.

10.1 Building of a reductive group

Let k be an algebraically closed field, and G a k-reductive group. Denote by $G = G(k)$ the group given by the set of k-points of G. By a parabolic subgroup $P \subset G(k)$ (resp. maximal torus $T \subset G(k)$) it can be understood in this section the set of k-points $P(k)$ (resp. $T(k)$) of some parabolic subgroup $P \subset G$ (resp. maximal torus $T \subset G$). Let $\mathcal{P}\mathrm{ar}(G)$ (resp. $\mathcal{B}\mathrm{or}(G)$, $\mathcal{T}\mathrm{or}(G)$) be the set of parabolic subgroups (maximal torus) of G defined over k.

The building $I(G)$ of G may be described as the set of parabolic subgroups $\mathcal{P}\mathrm{ar}(G)$ of G endowed with the relation opposed to the inclusion of parabolic subgroups of G. The group G acts on $I(G)$ by conjugation, and thus as a group of automorphisms of $I(G)$. Given $F \in I(G)$ we denote by P_F the corresponding parabolic subgroup. The set of apartments $\mathcal{A}\mathrm{pp}\, I$ of I is indexed by $\mathcal{T}\mathrm{or}(G)$, in fact there is a bijection

$$\mathcal{T}\mathrm{or}(G) \simeq \mathcal{A}\mathrm{pp}(I)$$

which associates with the maximal torus T the sub-building

$$\mathcal{A}_T = \{F \in I(G) \mid T \subset P_F\}.$$

There is also a bijection

$$\mathcal{B}\mathrm{or}(G) \simeq \mathrm{Ch}\, \mathrm{I(G)}$$

between the set of Borel subgroups and the set of chambers $\mathrm{Ch}\, \mathrm{I(G)}$ of the Tits building $I(G)$.

Given a maximal torus T we denote by $R = R_T$ the system of roots defined by T. Let $\mathcal{A}(R_T)$ be the apartment whose facets F are given by the parabolic subsets of R_T, and the inclusion $(F' \subset F)$ of facets by the opposed relation to the inclusion of parabolic subsets.

There is a building isomorphism

$$\mathcal{A}(R_T) \simeq \mathcal{A}_T$$

which associates with the facet $F \in \mathcal{A}(R_T)$ the R-subgroup P_F of G defined by the maximal torus T and the parabolic set R_F given by F (cf. [23], Exp. XXII).

The reciprocal isomorphism

$$\mathcal{A}_T \simeq \mathcal{A}(R_T)$$

associates to $F \in \mathcal{A}_T$ the set R_P of roots given by the action of T on the Lie algebra $\mathrm{Lie}(P_F)$. Let $B \supset T$ be a minimal parabolic subgroup containing T, and $C = C_B \in \mathcal{A}_T$ the corresponding chamber. Write

$$W_T = N(T)/T$$

where $N(T)$ denotes the normalizer of T in G. Then there is an isomorphism:

$$C(W_T, S(C)) \simeq \mathcal{A}(R_T) \ .$$

A frame $E \in Ep(G)_k$ (cf. [23], Exp. XXII) (cf. also the definition of a frame of G given in next *chapters*) gives rise to a killing couple:

$$E \mapsto (T \subset B)_E \ .$$

Given E, $E' \in Ep(G)_k$, let $\alpha_{EE'} : G \longrightarrow G$ be the automorphism defined by $\alpha_{EE'}(E) = E'$. Define the Weyl complex of G as the building given by the inductive limit

$$C(W, S) = \varinjlim_{E} C(W_T, S(C_B)) \ ,$$

the transition isomorphisms being induced by the $(\alpha_{EE'})$.

Given a frame E the associated Killing couple $(T \subset B)_E$ gives rise to the building isomorphisms:

$$C(W, S) \simeq C(W_T, S_{C_B}) \simeq \mathcal{A}(R_T) \simeq \mathcal{A}_T \subset I(G)$$

inducing a bijection

$$\mathrm{typ}\, C(W, S) \simeq \mathrm{typ}\, I(G) = I(G)/G$$

(resp. $\mathrm{Relpos}\, C(W, S) \simeq \mathrm{Relpos}\, I(G) = I(G) \times I(G)/G$ (diagonal action)).

Two parabolic subgroups P and Q are **incident** if: their intersection $P \cap Q$ is a parabolic subgroup. (resp. there exists a minimal parabolic subgroup $B \subset P, Q$). The building $I(G)$ may also be obtained from the set

$$\mathrm{Vert}\, I(G) \subset I(G)$$

of the maximal parabolic subgroup of G, which is endowed with the relation induced by the incidence relation, as follows. The facets of $I(G)$ are given by the subsets

$$\sigma \subset \mathrm{Vert}\, I(G)$$

of maximal parabolics two by two incidents. The set of chambers of $I(G)$ thus corresponds to the set $(\sigma_B)_{B \in \mathcal{B}or(G)}$ where

$$\sigma_B = \{P \in \mathrm{Vert}\, I(G) \mid P \supset B\}.$$

The class of subsets σ of $\text{Vert } I(G)$ defines a building structure $\mathfrak{I}(G)$ on $\text{Vert } I(G)$, according to the general definition, with $(\sigma_B)_{B\in\mathcal{B}\text{or}(G)}$ as the set of chambers. There is a building isomorphism

$$I(G) \simeq \mathfrak{I}(G)$$

defined by

$$F \mapsto \sigma_F$$

where

$$\sigma_F = \{P \in \text{Vert } I(G) \mid P_F \subset P\}.$$

10.2 Minimal Generalized Galleries in a reductive group building

It has been seen that given two flags $\mathcal{D}$ and $\mathcal{D}'$ in k^{r+1} there exists a basis e of k^{r+1} adapted to $\mathcal{D}$ and $\mathcal{D}'$. The result generalizes to a building $I(G)$. Given a couple of parabolics (P, Q) of G there exists a maximal torus T contained in both P and Q. This property translates into the building setting as follows, the facets F_P and F_Q are contained in an apartment $\mathcal{A}_T$. With a fixed Killing couple $B \supset T$ is associated a Coxeter system $C(W, S_C)$, where $W = N(T)/T$ and S_C is the set of reflexions defined by the system of simple roots R_0 given by the system of positive roots R_+ defined by B.
Define

$$\text{typ } I(G) := I(G)/G \ (\textit{resp. Relpos } I(G) := I(G) \times I(G)/G \ (\text{diagonal action})) \,.$$

Write $I = I(G)$ and denote by:

$$\text{typ}_I : I \longrightarrow \text{typ } I = I/G$$

(resp. $\tau_I : I \times I \longrightarrow \text{Relpos } I = I \times I/G$) the quotient mapping. Let

$$\text{typ}_I^g : \text{Gall}_I \longrightarrow \text{Gall}_{\text{typ } I}$$

be the mapping induced by typ_I assigning to a generalized gallery its gallery of types.

From the conjugation of maximal torus in G it results that the natural inclusion $C(W, S_C) \hookrightarrow I(G)$ induces the natural identifications:

$$\text{typ } C(W, S_C) = \text{typ } I \ (\textit{resp. Relpos } C(W, S_C) = \textit{Relpos } I \ ) \,.$$

Write:

$$\text{gall}_I = \text{Gall}_{\text{typ } I}$$

and remark that we may identify gall_I with $\text{gall}_{C(W,S)}$. Denote by

$$\text{gall}_I^m \subset \text{gall}_I$$

the subset of gall_I corresponding to the subset $\text{gall}_{C(W,S)}^m \subset \text{gall}_{C(W,S)}$ of types of minimal generalized galleries in $C(W,S))$, and by $E_I : \text{Gall}_I \longrightarrow I \times I$ the extremities mapping. Write $E_I = (E_1, E_2)$. Denote by

$$\boldsymbol{\tau} : \text{gall}_I^m \longrightarrow \text{Relpos } I$$

the mapping corresponding to $\boldsymbol{\tau} : \text{gall}_{C(W,S)}^m \longrightarrow \text{Relpos } C(W,S)$ which associates with a Minimal Generalized Gallery of types a type of relative position. Let $F \in I$ (resp. $g \in \text{gall}_I$). Write

$$\text{Gall}_I(g) = (\text{typ}_I^g)^{-1}(g)$$

(resp. $\text{Gall}_I(g, F) = \text{Gall}_I(g) \cap E_1^{-1}(F)$).

Definition 10.1 (Minimal Generalized Galleries in I)

*A generalized gallery $\gamma \in \text{Gall}_I$ is minimal (**MGG**) if*

$$\text{typ}_I^g(\gamma) \in \text{gall}_I^m$$

and

$$\bar{\tau}(E_I(\gamma)) = \boldsymbol{\tau}(\text{typ}_I^g(\gamma))$$

i.e. if $\gamma = \gamma(F, F')$, and we write $g = \text{typ}_I^g(\gamma)$, then $g \in \text{gall}_I^m$ and $\tau(F, F') = \bar{\tau}(= \boldsymbol{\tau}(g))$.

A generalized gallery γ is minimal if its corresponding gallery of type g is minimal and the type of relative position of its extremities $E_I(\gamma) = (F, F')$ is precisely the associated one to g, $\tau_\bullet(g) = \tau(F, F')$.

Let Gall_I^m denote the set of Minimal Generalized Galleries of I. Write for $g \in \text{gall}_I^m$

$$\text{Gall}_I^m(g) = \text{Gall}_I^m \cap \text{Gall}_I(g)$$

(resp. for $F \in \mathcal{A}$, $\text{Gall}_I^m(g, F) = \text{Gall}_I^m \cap \text{Gall}_I^m(g, F)$). Write E_I^m for the restriction of E_I to $\text{Gall}_I^m \subset \text{Gall}_I$.

The convex hull $\text{Env}^{\mathcal{A}_T}(F, F')$ of a couple of facets contained in an apartment $\mathcal{A}_T$ may be described as the set of parabolic subgroups P containing the intersection subgroup $P_F \cap P_{F'} \subset P$. More precisely as the set of facets F'' such that $P_F \cap P_{F'} \subset P_{F''}$. On the other hand, two maximal tori T, T' of G contained in $P_F \cap P_{F'}$ are conjugate in $P_F \cap P_{F'}$, thus

$$\text{Env}^{\mathcal{A}_T}(F, F') = \text{Env}^{\mathcal{A}_{T'}}(F, F') .$$

Given (F, F') define $\text{Env}^I(F, F') = \text{Env}^{\mathcal{A}_T}(F, F')$ (**Convex Hull** (resp. **Envelope) in** I of (F, F')) for some apartment $F, F' \subset \mathcal{A}_T$. The convex hull of two facets in I may be defined in the setting of general buildings (cf. [50]). The projection $\mathbf{proj}^I_{\mathbf{F'}}$ $\mathbf{F}$ in I is defined following the same pattern. The existence of a Minimal Generalized Gallery $\gamma_g(F, F')$ with $\tau_\bullet(g) = \tau(F, F')$ results from the following remark.

Remark 10.2 *If* $\mathcal{A}$ *is an apartment containing* F *and* F' *it is known that* $\mathcal{A} \simeq C(W, S)$ *Thus there exists such a* **MGG** $\gamma_g(F, F')$ *between* F *and* F' *in* $\mathcal{A}$, *and moreover* $\gamma_g(\Gamma, \Gamma') \subset Env^I(F, F')$.

A minimal gallery $\Gamma(C, C')$, in the usual sense, i.e. of minimal length, is contained in $\text{Env}^I(C, C')$ (cf. loc. cit.). From this property and the unicity of a minimal generalized gallery $\gamma_g(F, F')$ of fixed type g in an apartment $\mathcal{A}$ one obtains the corresponding unicity property in I.

Let
$$\gamma(F, F') : (F_i)(r \geqslant i \geqslant 0), (F'_j)(r + 1 \geqslant j \geqslant 0)$$

with $F = F'_{r+1}$, $F' = F'_0$, be a MGG in I of length $l(\gamma(F, F')) = r + 1$, and of the form $\gamma(F, F') = \gamma'_1(F, F')$. Let $\mathcal{A}$ be an apartment containing F and F'. Write
$$g = \text{typ}^g_I \, \gamma(F, F') \in \text{gall}^m_I = \text{gall}^m_{\mathcal{A}}.$$

By Definition 10.1 one has
$$\tau_I(E_I(\gamma(F, F'))) = \tau_I(F, F') = \tau_\bullet(g) \, (= \tau_g).$$

Let
$$\gamma^*(F, F') : (F^*_i) \ (r \geqslant i \geqslant 0), (F^{*'}_j) \ (r + 1 \geqslant j \geqslant 0)$$

with $F^{*'}_{r+1} = F$, $F^{*'}_0 = F'$, be the unique MGG in $\mathcal{A}$ between F and F' of type g, i.e. $\gamma^*(F, F') = \gamma_g(F, F')$ (cf. Proposition 9.38).
Let it be proved that $\gamma(F, F') = \gamma^*(F, F')$. Choose a chamber in $\mathcal{A}$ so that $C \supset F$ and at maximal distance $d(C, F')$ from F', i.e. $C \supset c(F, F')$ (cf. Definition 9.17, Proposition 9.18). Define a sequence of chambers
$$C^*_{r+1} = C, \quad C^*_i = \text{proj}^{\mathcal{A}}_{F_i} C^*_{i+1} \ (r \geqslant i \geqslant 0)$$

$$\left(\text{resp. } C_{r+1} = C, \quad C_i = \text{proj}^I_{F_i} C_{i+1} \ (r \geqslant i \geqslant 0) \right).$$

Choose a sequence of minimal galleries
$$\Gamma^{*i+1} = \Gamma^{*i+1}(C^*_{i+1}, C^*_i) \subset \text{St}^{\mathcal{A}}_{F^{*'}_{i+1}} \ (r \geqslant i \geqslant 0)$$

$$\left(\text{resp. } \Gamma^{i+1} = \Gamma^{i+1}(C_{i+1}, C_i) \subset \text{St}^I_{F'_{i+1}} \ (r \geqslant i \geqslant 0) \right).$$

Recall that $\text{Ch St}^I_{F'_{i+1}}$ is a convex set of chambers in I. Remark that

$$\text{length of } \Gamma^{*i+1} = \left| \mathcal{H}_{F'^*_{i+1}}(F^*_i) \right|$$

Let

$$\Gamma^* = \Gamma^*(C, C^*_0) = \Gamma^{*r+1} \circ \cdots \circ \Gamma^{*1}$$

(resp. $\Gamma = \Gamma(C, C_0) = \Gamma^{r+1} \circ \cdots \circ \Gamma^1$) be the composite gallery. By Remark 3.7,a) one has

$$\text{length of } \Gamma^*(C, C^*_0) = \Sigma^r_{i=0} \left| \mathcal{H}_{F'^*_{i+1}}(F^*_i) \right| \ ,$$

and by Lemma 9.8 that $\Gamma^*(C, C^*_0)$ gives rise to a minimal gallery in $\mathcal{A}$, thus in I, between C and $F' \subset C^*_0$. On the other hand, as $\Gamma^{i+1}(C_{i+1}, C_i)$ is a minimal gallery in $\text{St}^I_{F'_{i+1}}$ between a chamber C_{i+1} and the facet $F_i \supset F'_{i+1}$, with

$$\text{typ}_I F'_{i+1} = t_{i+1}(g) = \text{typ}_{\mathcal{A}} F^{*'}_{i+1}$$

(resp. $\text{typ}_I F_i = s_i(g) = \text{typ}_{\mathcal{A}} F^*_i$), it is deduced that:

$$\text{length } \Gamma^{i+1}(C_{i+1}, C_i) \leqslant \left| \mathcal{H}_{F'^*_{i+1}}(F^*_i) \right| (C_i = \text{proj}_{F'_{i+1}} F_i) \ ,$$

and finally one obtains

$$\text{length } \Gamma(C, C_0) = \Sigma^r_{i=0}\text{length } \Gamma^{i+1}(C_{i+1}, C_i) \leqslant \Sigma^r_{i=0} \left| \mathcal{H}_{F'^*_{i+1}}(F^*_i) \right| =$$

$$\text{length } \Gamma^*(C, C^*_0) \ .$$

As $\Gamma^*(C, C^*_0)$ (resp. $\Gamma(C, C_0)$) gives rise to a minimal gallery (resp. gallery) between C and F' in I (Observe that a minimal gallery in $\mathcal{A}$ is also minimal in I), it is concluded that:

"$\Gamma(C, C_0)$ is a minimal gallery between C and F' in I".

It results from this that

$$\Gamma(C, C_0) \subset \text{Env}^I (C, F') \subset \mathcal{A}$$

as C, $F' \in \mathcal{A}$. On the other hand, the inclusions

$$F'_i \subset F_i \subset C_i \quad (r \geqslant i \geqslant 0)$$

(resp. $F'_{r+1} \subset C_{r+1}$), give

$$\gamma(F, F') \subset \text{Env}^I (C, F') \subset \mathcal{A}$$

By unicity of the MGG of type $g \in \text{gall}^m_{\mathcal{A}}$, between the facets F and F' with $\tau(F, F') = \mathfrak{z}$, in $\mathcal{A}$, one obtains that $\gamma(F, F') = \gamma^*(F, F')$. It has thus been proved the

Proposition 10.3 *Given $(F, F') \in I \times I$, and $g \in \mathrm{gall}_I^m$ so that $\mathfrak{T} = \tau(F, F')$, there exists one and only one MGG $\gamma_g(F, F')$ of type g in I with extremities F and F'. Furthermore one has*

$$\gamma_g(F, F') \subset Env^I (F, F') .$$

$$\gamma_g(F, F') \subset \mathrm{Env}^I (F, F') .$$

Remark 10.4

It follows from this proposition that a generalized gallery $\gamma(F, F')$ is **minimal** *if and only if it is contained in the convex hull of its extremities $\gamma(F, F') \subset Env^I(F, F')$ and $\gamma(F, F')$ is minimal in an apartment $Env^I(F, F') = Env^{\mathcal{A}}(F, F') \subset \mathcal{A}$ containing $Env^I(F, F')$.*

The action of G on $I(G)$ induces an action on Gall_I. It results from the definition of a mgg the subset $\mathrm{Gall}_I^m \subset \mathrm{Gall}_I$ is stable under this action. The mapping $\mathrm{typ}_I^g : \mathrm{Gall}_I \longrightarrow \mathrm{gall}_I$ being G-equivariant, if gall_I is endowed with the trivial action of G, induces a G-equivariant mapping

$$\mathrm{q}_I^m : \mathrm{Gall}_I^m \longrightarrow \mathrm{gall}_I^m .$$

Let

$$\overline{\mathrm{q}}_I^m : \mathrm{Gall}_I^m / G \longrightarrow \mathrm{gall}_I^m$$

denote the quotient mapping. Given an apartment $\mathcal{A} = \mathcal{A}_T$ of I there is a natural mapping

$$\mathrm{j}_{\mathcal{A}} : \mathrm{Gall}_{\mathcal{A}}^m / W_{\mathcal{A}} \longrightarrow \mathrm{Gall}_I^m / G$$

Lemma 10.5 *The mapping $\mathrm{j}_{\mathcal{A}}$ is bijective.*

Proof *Given $(F, F') \in I \times I$ there exists an apartment $\mathcal{A}'$ such that $(F, F') \in \mathcal{A}' \times \mathcal{A}'$. On the other hand, there exists $x \in G$ with $x(\mathcal{A}') = \mathcal{A}$. Thus*

$$(x(F), x(F')) \in \mathcal{A} \times \mathcal{A}$$
$$\text{and} \qquad \tau_I(F, F') = \tau_I(x(F), x(F')) .$$

Let $\gamma(F, F') \in \mathrm{Gall}_I^m$, $g = \mathrm{typ}_I^g(\gamma(F, F')) \in \mathrm{gall}_I^m$, thus $\tau(F, F') = \mathfrak{T} (= \mathfrak{T}_*(g))$ by definition of a MGG. By proposition 9.31 one has

$$\gamma(F, F') \subset \mathrm{Env} (F, F') \subset \mathcal{A}'$$

By Proposition 9.36 and Proposition 9.38 it is obtained:

$$\gamma(F, F') = \gamma_g(F, F')$$

where $\gamma_g(F, F')$ *is the unique MGG in* $\mathcal{A}$ *with* $E(\gamma_g(F, F')) = (F, F')$ ($\tau_I(F, F') = \tilde{g}$). *It is then deduced that:*

$$x(\gamma_g(F, F')) = \gamma_g(x(F), x(F')) \subset \mathcal{A}$$

again by loc.cit., and thus that $\gamma(F, F')$ *is equivalent to* $\gamma_g(x(F), x(F')) \subset \mathcal{A}$. *This proves that* $j_{\mathcal{A}}$ *is a surjective mapping.*

Given

$$\gamma_a = \gamma_a(F_a, F'_a), \ \gamma_b = \gamma_b(F_b, F'_b) \in \mathrm{Gall}^m_{\mathcal{A}}$$

such that there exists $x \in G$ *with:*

$$x(\gamma_a) = \gamma_b$$

it is deduced that

$$g = \mathrm{typ}_{\mathcal{A}} \gamma_a = \mathrm{typ}_{\mathcal{A}} \gamma_b$$

and

$$\tilde{g} = \tau_I(F_a, F'_a) = \tau_I(F_b, F'_b)$$

Thus $\gamma_a = \gamma_g(F_a, F'_a)$ *(resp.* $\gamma_b = \gamma_g(F_b, F'_b)$*) (cf. Corollary 9.37 and Proposition 9.38). Let* $w \in W$ *so that*

$$(w(F_a), w(F_a)) = (F_b, F'_b)$$

then

$$w(\gamma_a) = \gamma_b \ (cf. \ Proposition \ 9.38)$$

i.e. $j_{\mathcal{A}}$ *is injective, and thus bijective.*

The bijective mapping

$$\overline{\mathrm{typ}}^m_{\mathcal{A}} : \mathrm{Gall}^m_{\mathcal{A}} /W \longrightarrow \mathrm{gall}^m_{\mathcal{A}} = \mathrm{gall}^m_I$$

of Corollary 9.37 factors as

$$\overline{\mathrm{typ}}^m_{\mathcal{A}} = \overline{\mathrm{q}}^m_I \circ j_{\mathcal{A}}$$

Corollary 10.6 *The mapping*

$$\overline{\mathrm{q}}^m_I : \mathrm{Gall}^m_I /G \longrightarrow \mathrm{gall}^m_I$$

is bijective. (Compare with the mapping $\mathrm{q}^m_{\mathcal{A}}$ *in section 9.8.)*

10.3   A combinatorial cartesian square

Let it be verified that the following commutative square

$$
\begin{array}{ccc}
\mathrm{q}_I^m : \mathrm{Gall}_I^{\,m} & \longrightarrow & \mathrm{gall}_I^{\,m} \\
\Big\downarrow {\scriptstyle E_I^m} & & \Big\downarrow {\scriptstyle \bullet} \\
\tau_I : I \times I & \longrightarrow & \mathrm{Relpos}\, I
\end{array}
$$

is cartesian. One defines a bijective G-equivariant mapping

PROPOSITION - DEFINITION 10.7
   There is a bijective G-equivariant mapping

$$
\Theta : (I \times I) \times_{Relpos\, I} gall_I^{\,m} \longrightarrow Gall_I^m
$$

which is defined following the same pattern as in the definition of the W-equivariant mapping

$$
\sigma : (\mathcal{A} \times \mathcal{A}) \times_{Relpos\, \mathcal{A}} gall_{\mathcal{A}}^{\,m} \longrightarrow Gall_{\mathcal{A}}^{\,m} \quad (cf.\ Notation\ 9.43).
$$

Proof   *There is*

$$
Gall_I^m = \coprod_{g \in gall_I^m} Gall_I^m(g),
$$

and

$$
(I \times I) \times_{Relpos\, I} gall_I^{\,m} = \coprod_{\tau \in Relpos\, I} (I \times I)_\tau \times \tau_\bullet^{-1}(\tau)
$$

$$
= \coprod_{(\tau,g)\in Relpos\, I \times gall_I^m \ with\ \tau_g = \tau} (I \times I)_\tau \times \{g\}.
$$

Let

$$
\Theta = \coprod_{g \in gall_I^m} \overline{\Theta}_g,
$$

where

$$
\overline{\Theta}_g : (I \times I)_\tau \times \{g\} \longrightarrow Gall_I^m(g) \qquad (\tau = \tau_g)
$$

is defined by

$$
\overline{\Theta}_g : ((F, F'), g) \longrightarrow \gamma_g(F, F') \qquad (\tau(F, F') = \tau_g) .
$$

It follows from Proposition 10.3 that the mapping

$$(I \times I)_{\bar{\tau}} \longrightarrow \mathrm{Gall}_I^m(g) \qquad (g \in \mathrm{gall}_I^m)$$

associating with (F, F') with $\tau(F, F') = \bar{\tau}$, the unique MGG $\gamma_g(F, F')$ between F and F' in I of type $g \in \mathrm{gall}_I^m$ is a G-equivariant bijection, thus $\overline{\Theta}_g$ also is. It is deduced that $\Theta = \coprod\limits_{g \in \mathrm{gall}_I^m} \overline{\Theta}_g$ is a G-equivariant bijection.

Remark 10.8 *The equivariant mapping Θ may be seen as a section of the natural mapping*

$$E_I^m \times \mathrm{q}_I^m : \mathrm{Gall}_I^m \longrightarrow (I \times I) \times_{\mathrm{Relpos}\, I} \mathrm{gall}_I^m$$

which in fact is a bijection. For every $g \in \mathrm{gall}_I^m$ and $(F, F') \in I \times I$, with $\tau_I(F, F') = \bar{\tau}$, one has

$$\begin{aligned}\left((E_I^m \times \mathrm{q}_I^m) \circ \Theta\right)((F, F'), g) &= \left(E_I^m \times \mathrm{q}_I^m\right)\left(\gamma_g(F, F')\right) \\ &= \left(E_I^m(\gamma_g(F, F')), \mathrm{q}_I^m(\gamma_g(F, F'))\right) \\ &= ((F, F'), g)\,.\end{aligned}$$

Thus

$$(E_I^m \times \mathrm{q}_I^m) \circ \Theta = \text{identity of } (I \times I) \times_{\mathrm{Relpos}\, I} \mathrm{gall}_I^m\,.$$

NOTATION 10.9 *Let $F \in I$, $\tau_0 \in \mathrm{Relpos}\, I$, $g \in \mathrm{gall}_I^m$ one writes:*

1. $\Sigma(\tau_0) = (I \times I)_{\tau_0} \qquad$ (resp. $\Sigma = \coprod\limits_{\tau \in \mathrm{Relpos}} \Sigma(\tau) = I \times I$) ;

2. $\Sigma(\tau_0, F) = \Sigma(\tau_0) \cap (\{F\} \times I)$;

3. $\Sigma(g) = \mathrm{Gall}_I^m(g) \qquad$ (resp. $\Sigma^m = \coprod\limits_{g \in \mathrm{gall}_I^m} \Sigma(g) = \mathrm{Gall}_I^m$);

4. $\Sigma(g, F) = \mathrm{Gall}_I^m(g, F) = \mathrm{Gall}_I^m(g) \cap E_1^{-1}(F)$;

5. $\hat{\Sigma}(g) = \mathrm{Gall}_I(g) \qquad$ (resp. $\hat{\Sigma} = \coprod\limits_{g \in \mathrm{gall}_I^m} \hat{\Sigma}(g)$);

6. $\hat{\Sigma}(g, F) = \mathrm{Gall}_I(g, F)$.

It follows from its definition that $\Sigma(\tau_0, F)$ is P_F-homogeneous. Actually $\Sigma(\tau_0, F)$ is a P_F-orbit in $\{F\} \times I$. Observe that $\Sigma^m \subset \hat{\Sigma}$.

Definition 10.10 *For $g \in \mathrm{gall}_I^m$ one defines*

$$\Theta_g : \Sigma(\mathfrak{f}) = (I \times I)_{\mathfrak{f}} \longrightarrow \Sigma(g) \subset \hat{\Sigma}(g)$$

as the co-restriction to $\Sigma(g) = \mathrm{Gall}_I^m(g) \subset \Sigma^m = \mathrm{Gall}_I^m$ of the composed mapping of

$$\Sigma(\mathfrak{f}) = (I \times I)_{\mathfrak{f}} \simeq (I \times I)_{\mathfrak{f}} \times \{g\} \hookrightarrow (I \times I) \times_{\mathrm{Relpos}\, I} \mathrm{gall}_I^m$$

with Θ.

Observe that $\Theta_g(\Sigma(\mathfrak{f}, F)) \subset \Sigma(g, F)$. Let

$$\Theta_{g,F} : \Sigma(\mathfrak{f}, F) \longrightarrow \Sigma(g, F) \qquad (F \in I)$$

be the mapping induced by Θ_g. Denote by

$$\pi^m : \Sigma^m \longrightarrow \Sigma \qquad (\text{resp. } \pi : \hat{\Sigma} \longrightarrow \Sigma)$$

the mapping given by the restriction of the extremities mapping $E_I : \mathrm{Gall}_I \longrightarrow I \times I$ to $\Sigma^m = \mathrm{Gall}_I^m$ (resp. $\hat{\Sigma}$) $\subset \mathrm{Gall}_I$, i.e. $\pi^m = E_I^m$ (cf. Definition 10.1). By Definition 10.1 of MGG of type g in I one has

$$\Sigma(g) = \mathrm{Gall}_I^m(g) = \{\gamma \in \mathrm{Gall}_I(g) \mid \mathfrak{f}(E_I(\gamma)) = \mathfrak{f}\}$$

thus

$$\pi^m(\Sigma(g)) = E_I^m(\mathrm{Gall}_I^m(g)) \subset (I \times I)_{\mathfrak{f}} = \Sigma(\mathfrak{f}).$$

On the other hand, $\Theta_g : (I \times I)_{\mathfrak{f}} \longrightarrow \mathrm{Gall}_I^m(g)$ being a bijective section of $E|_{\mathrm{Gall}_I^m(g)} = \pi^m|_{\Sigma(g)}$, it is deduced that the restriction

$$\pi^m(g) = \pi^m|_{\Sigma(g)} : \Sigma(g) \longrightarrow \Sigma(\mathfrak{f})$$

(resp. $\pi^m(g, F) = \pi^m|_{\Sigma(g,F)} : \Sigma(g, F) \longrightarrow \Sigma(\mathfrak{f}, F)$) is bijective. Given $\tau \in \mathrm{Relpos}\, I$ and $g \in \mathrm{gall}_I^m$, with $\mathfrak{f} = \tau$, the results of the next chapters prove that the image

$$\pi(\hat{\Sigma}(g)) \subset I \times I \qquad (\text{resp. } \pi(\hat{\Sigma}(g, F) \subset I))$$

is independent of the choice of g with $\mathfrak{f} = \tau$.

Definition 10.11 *Define* **the combinatorial closure** $\Sigma(\tau) \subset \overline{\Sigma}^c(\tau)$ *(resp.*
$\Sigma(\tau, F) \subset \overline{\Sigma}^c(\tau, F))$ *of* $\Sigma(\tau)$ *(resp.* $\Sigma(\tau, F))$ *in* $I \times I$ *(resp. I), by*

$$\overline{\Sigma}^c(\tau) = \pi(\hat{\Sigma}(g)) \qquad (resp. \ \overline{\Sigma}^c(\tau, F) = \pi(\hat{\Sigma}(g, F)))$$

with $g \in \underset{\bullet}{\tau}^{-1}(\tau) \subset \text{gall}_I^m$. *It follows from Remark 10.8 that* Θ_g *(resp.*
$\Theta_{g,F} = \Theta(g)|_{\Sigma(g,F)}$) *is a G-equivariant section (resp. P_F-equivariant sec-*
tion) of $\pi^m(g)$ *(resp.* $\pi^m(g, F))$ *which is bijective.*

10.4   Cell decomposition of the set of galleries

For a building $I = I(G)$ of a k-reductive group G let

$$\text{Relpos}_{(s',s)} I = I_{s'} \times I_s / G \qquad ((s', s) \in \text{typ} I \times \text{typ} I) \ .$$

Thus $\text{Relpos}_{(s',s)} I$ denotes the set of types of relative positions of couples of
facets of types s and s'. If $s' = \text{typ}_I C$ $(C \in \text{Ch} I)$, i.e. if s' is the type of a
chamber, we simply write:

$$\text{Relpos}_s I = \text{Relpos}_{(s',s)} I \quad (resp. \ \text{Relpos}' I = \coprod_{s \in \text{typ} I} \text{Relpos}_s I).$$

Given $\tau_0 \in \text{Relpos}' I$, a chamber C and an apartment $\mathcal{A}$ containing C, let
$F_{\tau_0}(C)$ denote the unique facet satisfying $\tau_I(C, F_{\tau_0}(C)) = \tau_0$. There is a
bijection $\text{Relpos}' I \simeq \mathcal{A}$, given by $\tau_0 \mapsto F_{\tau_0}(C)$. The composition of $F \mapsto$
$\tau_I(C, F)$ with the preceding bijection is a building morphism $\rho_{\mathcal{A},C} : I \longrightarrow \mathcal{A}$
(**Retraction of I on $\mathcal{A}$ with center C**).

Definition 10.12 *Define for* $t, s, s' \in \text{typ} I$ *with* $t \subset s, s'$

$$\overset{t}{\text{Relpos}}_{(s',s)} \ = \ \overset{t}{\text{Relpos}}_{(s',s)} I$$

$$(resp. \ \overset{t}{\text{Relpos}}_s \ = \ \overset{t}{\text{Relpos}}_s I)$$

as the image of

$$\text{Relpos}_{(s',s)} \text{St}_F \longrightarrow \text{Relpos} I$$

$$(resp. \qquad \text{Relpos}_s \text{St}_F \longrightarrow \text{Relpos} I \ )$$

where $F \in I_t$ $(t \in \text{typ} I)$. *This image is independant of the choice of F and*
is the set of types of relative positions of couples of facets of types s and s'
which are incident to a facet of type t.
For $g \in \text{gall}_I$ *with* $l(g) = r + 1$ *such that* $g = g_1, g_1'$ *(resp.* $g = g_2, g_2'$*) one*
writes:

$$\overset{gall}{\text{Relpos}}_I(g) = \prod \text{Relpos}_{s_{i-1}(g)}^{t_i(g)} \qquad (r + 1 \geqslant i \geqslant 1) \ (resp. \ r \geqslant i \geqslant 1)$$

(the set of relative position types galleries of type g), *and* $Relpos_I^{gall} =$ $\coprod_{g \in gall_I} Relpos_I^{gall}(g).$

It is recalled that given $g \in gall_I = gall_A$, $g^* \in gall_I$ has been defined by:

$$s_i(g^*) = \operatorname{typ} C \qquad (C \in \operatorname{ChI}) \ (r+1 \geqslant i \geqslant 0), \ t_j(g^*) = t_j$$

(resp. $\qquad s_i(g^*) = \operatorname{typ} C \qquad (r \geqslant i \geqslant 0), \ t_j(g^*) = t_j$)

and $F_{e_1(g)}(C)$ by:

$$F_{e_1(g)}(C) = F_{t_{r+1}(g)}(C) \qquad (\text{resp. } F_{e_1(g)}(C) = F_{t_r}(g))$$

according to $g = g_1, g_1'$ (resp. $g = g_2, g_2'$) (cf. §9.9.2, i)).

Definition 10.13 *Let*

$$s_{g^*,C}^I : \operatorname{Gall}_I(g, F_{e_1(g)}(C)) \longrightarrow \operatorname{Gall}_I(g^*, C)$$

be defined according to the same pattern as

$$s_{g^*,C} : \operatorname{Gall}_A(g, F_{e_1(g)}(C)) \longrightarrow \operatorname{Gall}_A(g^*, C)$$

in Definition 9.64 with the big building I instead of the apartment $\mathcal{A}$. In this case $\operatorname{proj}_{F_i}^I C_{i+1}$ *plays the role of* $\operatorname{proj}_{F_i}^{\mathcal{A}} C_{i+1}$ *in the building I. If no confusion arises one writes* $s_{g^*,C}^{\mathcal{A}} = s_{g^*,C}.$

Let

$$\tau_{g^*,C}^I : \operatorname{Gall}_I(g^*, C) \longrightarrow Relpos_I^{gall}(g)$$

denote the mapping

$$\tau_{g^*,C}^I : \gamma^* \mapsto \underline{\tau}(\gamma^*) = \left(\tau(C_i, F_{s_{i-1}}(C_{i-1}))\right)$$

where $\gamma^* \in \operatorname{Gall}_I(g^*, C)$, and $C_i = F_i(\gamma^*)$ is the chamber of γ^*. Finally let

$$\tau_{g,C}^I : \operatorname{Gall}_I(g, F_{e_1(g)}(C)) \longrightarrow Relpos_I^{gall}$$

be the composed mapping:

$$\tau_{g,C}^I = \tau_{g^*,C}^I \circ s_{g^*,C}^I .$$

Let $\mathcal{A} = \mathcal{A}_T$ be an apartment containing $C = C_B$, i.e. $(\mathcal{A}, C)$ corresponds to the killing couple (T, B). The following commutative diagram

$$s^{\mathcal{A}}_{g^*,C} : \mathrm{Gall}_{\mathcal{A}}(g, F_{e_1(g)}(C)) \longrightarrow \mathrm{Gall}_{\mathcal{A}}(g^*, C)$$

$$\downarrow \qquad\qquad\qquad\qquad\qquad\qquad \downarrow$$

$$s^I_{g^*,C} : \mathrm{Gall}_I(g, F_{e_1(g)}(C)) \longrightarrow \mathrm{Gall}_I(g^*, C)$$

where the vertical arrows are the inclusions induced by $\mathcal{A} \hookrightarrow I$, expresses the compatibility between the mappings $s^{\mathcal{A}}_{g^*,C}$, and $s^I_{g^*,C}$.

Denote by $\tau^{\mathcal{A}}_{g,C}$ (resp. $\tau^{\mathcal{A}}_{g^*,C}$) the composed mapping of

$$\mathrm{Gall}_{\mathcal{A}}(g, F_{e_1(g)}(C)) \hookrightarrow \mathrm{Gall}_I(g, F_{e_1(g)}(C))$$

(resp. $\mathrm{Gall}_{\mathcal{A}}(g^*, C) \hookrightarrow \mathrm{Gall}_I(g^*, C)$) followed by $\tau_{I,C}(g)$ (resp. $\tau_{I,C}(g^*)$), i.e. the restriction of $\tau^I_{g,C}$ (resp. $\tau^{\mathcal{A}}_{g^*,C}$) to $\mathrm{Gall}_{\mathcal{A}}(g, F_{e_1(g)}(C))$ (resp. $\mathrm{Gall}_{\mathcal{A}}(g^*, C)$).

Definition 10.14 (Cell decomposition of $\mathrm{Gall}_I(g, F)$ defined by a chamber $C \supset F$).
Write

$$\mathscr{C}_C(g, \underline{\tau}) = (\tau^I_{g,C})^{-1}(\underline{\tau}) \subset \mathrm{Gall}_I(g, F)$$

where $\underline{\tau} \in \mathrm{Relpos}^{\mathrm{gall}}_I(g)$, and $F = F_{e_1(g)}(C)$. Clearly

$$\mathrm{Gall}_I(g, F) = \coprod_{\underline{\tau} \in \mathrm{Relpos}^{\mathrm{gall}}_I(g)} \mathscr{C}_C(g, \underline{\tau}) \ .$$

*One calls $\mathscr{C}_C(g, \underline{\tau})$ the **Cell** defined by the **gallery of relative position types** $\underline{\tau}$ and $C \supset F$.*

The next aim is to give a canonical parametrization in terms of a block decomposition of $\mathscr{C}_C(g, \underline{\tau})$ once an apartment $\mathcal{A}$ containing C is chosen.

10.5 Galleries of relative position types and Galleries in an apartment

For the sake of briefness it is supposed $g = g'_1$, and $l(g) = r+1$ in the following developments.

With a gallery of relative position types is associated a sequence of elements in W. Let $\mathcal{W}^m_C(t, s)$ with $C \in \mathcal{A}$ be as in Definition 9.70. There is a bijection

$$\mathcal{W}^m_C(t, s) \longrightarrow \mathrm{Relpos}^t_s I$$

defined by

$$w \longrightarrow \tau\left(C, w(F_s(C))\right) \ .$$

There is thus an induced bijection

$$\mathcal{W}_C^m(g) = \prod \mathcal{W}_C^m(t_i(g), s_{i-1}(g)) \longrightarrow \prod \text{Relpos}_{s_{i-1}(g)}^{t_i(g)} = \text{Relpos}_g^{\text{gall}} I .$$

On the other hand, by Proposition 9.72 there is a bijection

$$i_{g,C}^m : \mathcal{W}_C^m(g) \longrightarrow \text{Gall}_{\mathcal{A}}(g, F_{e_1(g)}(C))$$

given by:

$$x \mapsto \gamma_g(x) \qquad (x = (x_i)_{r+1 \geqslant i \geqslant 1} \in \mathcal{W}_C^m(g))$$

where $\gamma_g(x) = x \cdot \gamma_g(C)$ is defined as in Proposition 9.54, i.e. $\gamma_g(x) =$ the translation by x of the basical gallery $\gamma_g(C) \subset \Delta(C)$.

Definition 10.15 *Let* $\underline{\tau} = (\tau_{r+1}, \ldots, \tau_1) \in \text{Relpos}_g^{\text{gall}} I$ $(g = g_1, g_1')$. *One associates with* $\underline{\tau}$ *a sequence of chambers, and two sequences of facets of* $\mathcal{A}$ *by the following recursive pattern. Write* $C_{r+1} = C$ *(resp.* $F_{r+1}' = F_{t_{r+1}(g)}(C_{r+1}) = F_{e_1(g)}(C)$, $F_r = F_{\tau_{r+1}}(C)$, $C_r = \text{proj}_{F_r} C_{r+1})$. *Given* (F_i, C_i) $(r \geqslant i \geqslant 1)$, *with* $F_i \subset C_i$ *we define, for* $g = g_1'$:

$$F_i' = F_{t_i(g)}(C_i) , \quad F_{i-1} = F_{\tau_i}(C_i) , \quad C_{i-1} = \text{proj}_{F_{i-1}} C_i.$$

Let $\gamma_{\underline{\tau}} \in \text{Gall}_{\mathcal{A}}(g, F)$ *(resp.* $\gamma_{\underline{\tau}}^* \in \text{Gall}_{\mathcal{A}}(g^*, C))$ *be defined by*

$$\gamma_{\underline{\tau}} : (F_i) \ (r \geqslant i \geqslant 0) , \ (F_j') \ (r+1 \geqslant j \geqslant 0)$$

(resp. $\gamma_{\underline{\tau}}^* : (C_i) \ (r \geqslant i \geqslant 0) , \ (F_j') \ (r+1 \geqslant j \geqslant 0)).$

Denote by

$$\tau_{g,C}'^{\mathcal{A}} : \text{Relpos}_g^{\text{gall}} I \longrightarrow \text{Gall}_{\mathcal{A}}(g, F_{e_1(g)}(C))$$

(resp. $\tau_{g^*,C}'^{\mathcal{A}} : \text{Relpos}_g^{\text{gall}} I \longrightarrow \text{Gall}_{\mathcal{A}}(g^*, C))$

the mapping defined by

$$\tau_{g,C}'^{\mathcal{A}} : \underline{\tau} \mapsto \gamma_{\underline{\tau}}$$

(resp. $\tau_{g^*,C}'^{\mathcal{A}} : \underline{\tau} \mapsto \gamma_{\underline{\tau}}^*).$

From the construction of $\gamma_{\underline{\tau}}$ (resp. $\gamma_{\underline{\tau}}^*$) it results that

$$s_{g^*,C}(\gamma_{\underline{\tau}}) = \gamma_{\underline{\tau}}^* \quad \text{(cf. Definition 9.64)}$$

and thus by definition of $\tau_{g,C}^{\mathcal{A}}$ that

$$\tau_{g,C}^{\mathcal{A}}(\gamma_{\underline{\tau}}) = \tau_{g^*,C}^{\mathcal{A}}\left(s_{g^*,C}(\gamma_{\underline{\tau}})\right) = \tau_{g^*,C}^{\mathcal{A}}(\gamma_{\underline{\tau}}^*) = \underline{\tau}$$

as by construction of $\gamma_{\underline{\tau}}^*$ and definition of $\tau_{g*,C}^I$ one has that

$$\tau_{g*,C}^I(\gamma_{\underline{\tau}}^*) = \underline{\tau}.$$

Finally one obtains

$$\forall\,\underline{\tau} \in \mathrm{Relpos}_g^{gall}\ \left(\tau_{g,C}^{\mathcal{A}} \circ \tau_{g,C}'^{\mathcal{A}}\right)(\underline{\tau}) = \underline{\tau}.$$

This gives that $\tau_{g,C}^{\mathcal{A}}$ is **surjective**. To prove the injectivity of $\tau_{g,C}^{\mathcal{A}}$ one needs the following

Lemma 10.16 *The triangle*

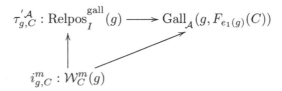

commutes. Where the vertical arrow is defined as above. (Remark that as two of its arrows are bijections it follows that $\tau_{g,C}'^{\mathcal{A}}$ also is a bijection, and that a fortiori $\tau_{g,C}^{\mathcal{A}}$ is a bijection).

Proof *Let $\underline{\tau}(x) = (\underline{\tau}_{r+1},\dots,\underline{\tau}_1) \in \mathrm{Relpos}_g^{gall}\,I$ be the gallery of relative positions corresponding to $x = (x_{r+1},\dots,x_1) \in W_C^m(g)$. Let $(C_{r+1},\dots,C_0)$ the sequence of chambers given by $\gamma_{\underline{\tau}}^*$. Write $w_i = w(C_i, C_{i-1})$ $(r+1 \geqslant i \geqslant 1)$. Let*

$$v_{i+1} = w_{r+1}\dots w_{i+1} \qquad (r \geqslant i \geqslant 0).$$

Then it may be seen by induction that the following relations holds between $(w_{r+1},\dots,w_1)$ and $x = (x_{r+1},\dots,x_1)$:

1. $x_{r+1} = w_{r+1}$, $x_i = v_{i+1}^{-1}\, w_i\, v_{i+1}$ $(r \geqslant i \geqslant 1)$;

2. $x_{r+1}\dots x_i = w_i\dots w_{r+1}$.

From (2) it follows immediately that

$$(x_{r+1}\dots x_i)(C) = C_{i-1} \qquad (r+1 \geqslant i \geqslant 1).$$

By definition of $\gamma_g(x)$ (cf. Proposition 9.54) it follows that $\gamma_g(x) = \gamma_{\underline{\tau}}$, and a fortiori that $i_g^m(x) = \gamma_{\underline{\tau}}$. This proves the commutativity of the triangle.

From the Lemma one obtains:

Proposition 10.17 $\tau_{g,C}^{\mathcal{A}} : \mathrm{Gall}_{\mathcal{A}}(g, F_{e_1(g)}(C)) \longrightarrow \mathrm{Relpos}_g^{gall}\,\mathcal{A}$ *is a bijective mapping whose inverse is given by*

$$\left(\tau_{g,C}^{\mathcal{A}}\right)^{-1} = \tau_{g,C}'^{\mathcal{A}}.$$

10.6   Galleries cells parametrizations

Let $g : (s_i)_{r \geqslant i \geqslant 0}$, $(t_j)_{r+1 \geqslant i \geqslant 0}$ be a gg $g \in \text{gall}_I = \text{gall}_\mathcal{A}$, with $l(g) = r + 1$, and $g = g_1'$ (resp. $\gamma_g(C) : (F_{s_i}(C))_{r \geqslant i \geqslant 0}$, $(F_{t_j}(C))_{r+1 \geqslant j \geqslant 0}$ the corresponding basical gg defined by C). It is recalled that given $F \in I(G)$ one has $\text{Stab}_G F = P_F = $ the parabolic subgroup of G given by F. Following the pattern of Definition 9.48 one writes:

$$\text{Stab}_{t(g)} \gamma_g(C) = \prod \text{Stab}_G F_{t_j}(C) \qquad (r + 1 \geqslant j \geqslant 1)$$

(resp. $\text{Stab}_{s(g)} \gamma_g(C) = \prod \text{Stab}_G F_{s_i}(C) \qquad (r \geqslant i \geqslant 0)$)

This definition corresponds to the case 2) of Definition 9.48. The definition of $\text{Stab}_{t(g)} \gamma_g(C)$ (resp. $\text{Stab}_{s(g)} \gamma_g(C)$) in the cases 1), 3), and 4) follows the corresponding patterns 1), 3), and 4). One develops the case 1), i.e. $g = g_1'$, in detail. It is easy to see that 1), 3) and 4) are similarly handled.

Definition 10.18 *Given $x \in \text{Stab}_{t(g)} \gamma_g(C)$ one writes*

$$z_\alpha = \prod x_i \qquad (r + 1 \geqslant i \geqslant \alpha) \ (resp. \ z_{r+2} = 1) \ .$$

     Let

$$i_g^I = i_{g,C}^I : \text{Stab}_{t(g)} \gamma_g(C) \longrightarrow \text{Gall}_I(g, F_{e_1(g)}(C))$$

$\left(F_{e_1(g)}(C) = F_{t_{r+1}}(C)\right)$ *be given by:*

$$i_{g,C}^I : x \mapsto \gamma_g(x) = x \cdot \gamma_g(C) \ ,$$

where

$$\gamma_g(x) = x \cdot \gamma_g(C) : (z_{i+1}(F_{s_i}(C))) \ (r \geqslant i \geqslant 0) \ , \ (z_{j+1}(F_{t_j}(C))) \ (r+1 \geqslant j \geqslant 1)$$

(see Proposition 9.54).

To avoid repetitions define a right action

$$\text{Stab}_{t(g)} \gamma_g(C) \times \text{Stab}_{s(g)} \gamma_g(C) \longrightarrow \text{Stab}_{t(g)} \gamma_g(C)$$

according to Definition 9.50, by writing $\text{Stab}_{t(g)} \gamma_g(C)$ (resp. $\text{Stab}_{s(g)} \gamma_g(C)$) instead of $W(t(g))$ (resp. $W(s(g))$).

One calls the quotient set

$$\text{Stab}_{t(g)} \gamma_g(C)/\text{Stab}_{s(g)} \gamma_g(C)$$

the **contracted product associated to the gg** $\gamma_g(C)$ **in I.**

The mapping $i_g^I : \text{Stab}_{t(g)} \gamma_g(C) \longrightarrow \text{Gall}_I(g, F_{e_1(g)}(C))$ *factors through the quotient mapping*

$$\text{Stab}_{t(g)} \gamma_g(C) \longrightarrow \text{Stab}_{t(g)} \gamma_g(C)/\text{Stab}_{s(g)} \gamma_g(C) .$$

Let

$$\bar{i}_g^I = \bar{i}_{g,C}^I : \text{Stab}_{t(g)} \gamma_g(C)/\text{Stab}_{s(g)} \gamma_g(C) \longrightarrow \text{Gall}_I(g, F_{e_1(g)}(C))$$

be the induced mapping.

Remark 10.19 *The mapping* $\bar{i}_{g,C}^I$ *corresponds to the mapping*

$$i_{g,C} : W(t(g))/W(s(g)) \longrightarrow \text{Gall}_A(g, F_{e_1(g)}(C))$$

of Corollary 9.56.

Proposition 10.20 *The mapping* $\bar{i}_{g,C}^I$ *is a bijection.*

The proof of this Proposition follows mutatis mutandis from that of Corollary 9.56 (which gives that $\bar{i}_{g,C}^I$ is injective) and Proposition 9.61 (which gives that $\bar{i}_{g,C}^I$ is surjective).

Let
$$q_{g^*} : \text{Gall}_I(g^*) \longrightarrow \text{Gall}_I(g)$$

denote the mapping
$$q_{g^*} : \gamma^* \mapsto F_g(\gamma^*)$$

where $F_g(\gamma^*)$ is the gg of type g defined by $\gamma^* \in \text{Gall}_I(g^*)$. By

$$q_{g^*,C}^I : \text{Gall}_I(g^*, C) \longrightarrow \text{Gall}_I(g, F_{e_1(g)}(C))$$

the induced mapping is denoted.

It is easy to see that the mapping

$$s_{g^*,C}^I : \text{Gall}_I(g, F_g(C)) \longrightarrow \text{Gall}_I(g^*, C)$$

is a section of the surjective mapping $q_{g^*,C}^I$, i.e. that

$$q_{g^*,C}^I \circ s_{g^*,C}^I = \text{identity of the set } \text{Gall}_I(g, F_{e_1(g)}(C)).$$

See Definition 9.64 and what follows.

10.6.1 Cells parametrizations

Given a cell $\mathscr{C}(g, \underline{\tau})$ a set of class representatives $U(w_{\underline{\tau}})w_{\underline{\tau}} \subset Stab_{t(g)}\, \gamma_g(C)$ is defined such that $i_{g,C}^I$ induces a bijection $U(w_{\underline{\tau}})w_{\underline{\tau}} \simeq \mathscr{C}(g, \underline{\tau})$.

We suppose $\mathcal{A} = \mathcal{A}_T$, T a maximal torus of G. Let $R = R_T$ be the system of roots defined by T. Denote by B_C the Borel subgroup corresponding to the chamber $C \in \mathrm{Ch}\,\mathcal{A}$, and by $R_+(C) \subset R$ the system of positive roots defined by C.

The natural action of $W = W_{\mathcal{A}} = N(T)/T$ on $\mathrm{Ch}\,\mathcal{A}$ corresponds to the action of W on $(R_+(C))$ $(C \in \mathrm{Ch}\,\mathcal{A})$, i.e.

$$w(R_+(C)) = R_+(w(C)).$$

It is recalled that by definition (cf. Notation 10.9,2.)

$$\Sigma(\tau, F) = (I \times I)_\tau \cap (\{F\} \times I) .$$

Let $\tau \in \mathrm{Relpos}_s = (\mathrm{Ch}\,I \times I_s)/W$. In this section one writes (by abuse of language):

$$\Sigma(\tau, C) = \{F \in I_s \mid (C, F) \in (I \times I)_\tau\} .$$

Let it be recalled that $F_\tau(C) \in \mathcal{A}_s$ is the unique facet such that

$$\tau(C, F_\tau(C)) = \tau$$

i.e. $\mathcal{A} \cap \Sigma(\tau, C) = \{F_\tau(C)\}$. Write

$$w_\tau^m = w_{F_\tau(C)}^m = w(C, \mathrm{proj}_{F_\tau(C)}\, C).$$

$F_\tau(C)$ may be seen as the image of $\Sigma(\tau, C)$ by the retraction $\rho_{\mathcal{A},C} : I \longrightarrow \mathcal{A}$ (cf. [50], 3.3). Thus $F_\tau(C) = w_\tau^m(F_s(C))$ (cf. Definition 9.70). Given $w \in W$ one considers a closed subset of the positive roots $R_+(C)$

$$R_+(w) = R_+(C) - R_+(C) \cap w(R_+(C)).$$

Define the subgroup

$$U(w) \subset B_C$$

as the image of the mapping

$$\prod \mathcal{X}_\alpha \, (\alpha \in R_+(w)) \longrightarrow B_C$$

induced by the multiplication of G, where by $\mathcal{X}_\alpha$ $(\alpha \in R)$ one denotes the root subgroup corresponding to α. Remark that

$$\mathcal{H}(C, w(C)) = \{\partial \Phi_\alpha \mid \alpha \in R_+(C)\} ,$$

where $\partial\Phi_\alpha$ denotes the hyperplane of $\mathcal{A}$ defined by the root α. It is well known that the mapping

$$U(w_\tau^m) \longrightarrow \Sigma(\tau, C) \qquad (\tau \in (\mathrm{ChI} \times \mathrm{I})/\mathrm{G})$$

defined by

$$u \mapsto u(F_\tau, C) = u\, w_\tau^m(F_s(C))$$

is a bijection, i.e. every $F \in \Sigma(\tau, C)$ is uniquely written in the form

$$F = u\, w_\tau^m(F_s(C)) \qquad (u \in U(w_\tau^m))$$

(cf. [23], Exp. XXVI, 4.5.3). Given $\underline{\tau} \in \mathrm{Relpos}_g^{gall}I$, $\underline{\tau} = (\tau_{r+1}, \ldots, \tau_1)$, it is known that to $\underline{\tau}$ corresponds a unique element (cf. Lemma 10.16, and Proposition 10.17)

$$w_{\underline{\tau}} = (w_{\tau_{r+1}}^m, \ldots, w_{\tau_1}^m) \in \mathcal{W}_C^m(g).$$

Write

$$U(w_{\underline{\tau}}) = U(w_{\tau_{r+1}}^m) \times \cdots \times U(w_{\tau_1}^m) \subset \mathrm{Stab}_{t(g)}\gamma_g(C).$$

Observe that $\tau_i \in \mathrm{Relpos}_{s_{i-1}}^{t_i} \implies w_{\tau_i}^m \in W_{F_{t_i}(C)}$, One has $\mathrm{proj}_{w_{\tau_i}^m(F_{s_{i-1}}(C))}C = w_{\tau_i}^m(C)$ and $F_{t_i}(C) \subset F_{\tau_i}(C)$, thus that $\mathcal{H}(C, w_{\tau_i}^m(C)) \subset \mathcal{H}_{F_{t_i}(C)}$ and $R_+(w_{\tau_i}^m) \subset R_{F_{\tau_i}(C)}$.

This gives

$$U(w_{\tau_i}^m) \subset P_{F_{t_i}}(C) = \mathrm{Stab}_G\, F_{t_i}(C)$$

and finally

$$U(w_{\underline{\tau}}) \subset \prod \mathrm{Stab}_G\, F_{t_i}(C) = \mathrm{Stab}_{t(g)}\gamma_g(C)$$

$(t_i = t_i(g))$.

Definition 10.21 *For each $w \in W = N(T)/T$ one denotes w a representative $w \in N(T)$ by the same letter if no confusion arises (see Definition 11.8).*

Let

$$U(w_{\underline{\tau}})w_{\underline{\tau}} \subset \mathrm{Stab}_{t(g)}\gamma_g(C)$$

be defined by

$$U(w_{\underline{\tau}})w_{\underline{\tau}} = U(w_{\tau_{r+1}}^m)w_{\tau_{r+1}}^m \times \cdots \times U(w_{\tau_1}^m)w_{\tau_1}^m$$

$$= \prod U(w_{\tau_i}^m)w_{\tau_i}^m.$$

Let

$$i_{\underline{\tau}, C}^m : U(w_{\underline{\tau}})w_{\underline{\tau}} \longrightarrow \mathrm{Gall}_I(g, F_{e_1(g}(C)) .$$

be the restriction of

$$i_{g,C}^I : x \mapsto \gamma_g(x) = x \cdot \gamma_g(C) \quad (cf. \; Definition \; 10.18)$$

to $U(w_{\underline{\tau}})w_{\underline{\tau}} \subset \mathrm{Stab}_{t(g)} \, \gamma_g(C).$

The following Proposition completes this definition.

Proposition 10.22

$$\mathrm{Im} \, \imath_{\underline{\tau},C}^m \subset \mathscr{C}_C(g,\underline{\tau}).$$

Proof *Write*

$$w_i = w_{\tau_i}^m.$$

Given

$$x = (u_{r+1}w_{r+1}, \ldots, u_1 w_1) \in U(w_{\underline{\tau}})w_{\underline{\tau}}$$

Define a sequence of chambers

$$C_{r+1} = C, \qquad C_i = (u_{r+1}w_{r+1} \ldots u_i w_i)(C) \quad (r \geqslant i \geqslant 1).$$

Let it be seen that

$$\tau_I\left(C_i, F_{s_{i-1}}(C_{i-1})\right) = \tau_i \qquad (resp. \; C_{i-1} = \mathrm{proj}_{F_{s_{i-1}}(C_{i-1})}(C_i))$$

One has

$$\tau_I((u_{r+1}w_{\tau_{r+1}}^m \cdots u_{i+1}w_{\tau_{i+1}}^m)(C), F_{s_{i-1}}((u_{r+1}w_{\tau_{r+1}}^m \cdots u_i w_{\tau_i}^m)(C)))$$
$$= \tau_I(C, F_{s_{i-1}}(w_{\tau_i}^m(C))$$

and

$$\tau_I(C, F_{s_{i-1}}(w_{\tau_i}^m(C))) = \tau_I(C, F_{\tau_i}(C)).$$

Write $w^{(i+1)} = u_{r+1}w_{\tau_{r+1}}^m \cdots u_{i+1}w_{\tau_{i+1}}^m$, *thus*

$$\tau_i = (w^{(i+1)}(C), w^{(i+1)}(F_{\tau_i}(C))) = (C_i, F_{s_{i-1}}(C_{i-1})),$$

and

$$w^{(i+1)}(\mathrm{proj}_{F_{\tau_i}(C)} C) = \mathrm{proj}_{w^{(i+1)}(F_{\tau_i}(C))} w^{(i+1)}(C) = \mathrm{proj}_{w^{(i+1)}(F_{\tau_i}(C))} C_i.$$

Observe that $F_{s_{i-1}}(\gamma_g(w^{(1)})) = w^{(i+1)}(F_{\tau_i}(C))$. *This achieves the proof of both equalities. It results from the above equalities that*

$$(\tau_{g,C}^I)(\gamma_g(x)) = \underline{\tau}$$

i.e. $\gamma_g(x) \in \mathscr{C}_C(g,\underline{\tau})$ *(cf. Definition 10.14), and this completes Definition 10.21.*

Lemma 10.23 $i^m_{\mathcal{I},C}$ *induces a bijection*

$$U(w_{\mathcal{I}})w_{\mathcal{I}} \simeq \mathscr{C}_C(g,\mathcal{I}) \, .$$

Proof   *Let it be proved that* $\gamma \in \mathscr{C}_C(g,\mathcal{I})$ *may be uniquely written*

$$\gamma = \gamma_g(x) = x \cdot \gamma_g(C) \, , \ x \in U(w_{\mathcal{I}})w_{\mathcal{I}}.$$

As $F_r(\gamma) \in \Sigma(\mathcal{T}_{r+1},C)$ *clearly one has,*

$$F_r(\gamma) = u_{r+1}\, w_{r+1}(F_{s_r}(C))$$

with $u_{r+1} \in U(w_{r+1})$ *uniquely determined. Now it is supposed that*

$$F_j(\gamma) = u_{r+1}\, w_{r+1} \ldots u_{j+1}\, w_{j+1}(F_{s_j}(C)) \qquad (r \geqslant j \geqslant i)$$

with $(u_{r+1}\, w_{r+1}, \ldots, u_{j+1}\, w_{j+1})$ *uniquely determined.*
One then has $v^{-1}_{i+1}(F_{i-1}(\gamma)) \in \mathrm{St}_{F_{t_i}(C)}$, *with*

$$v_{i+1} = u_{r+1}\, w_{r+1} \ldots u_{i+1}\, w_{i+1},$$

and $v^{-1}_{i+1}(F_{i-1}(\gamma)) \in \Sigma(\mathcal{T}_i,C)$, *as*

$$\tau_I\left(v^{-1}_{i+1}(C_i), v^{-1}_{i+1}(F_{i-1}(\gamma))\right) = \tau_I\left(C, v^{-1}_{i+1}(F_{i-1}(\gamma))\right) = \mathcal{T}_i.$$

(Define $C_r = u_{r+1}\, w_{r+1}(C)$, $C_{r-1} = u_{r+1}\, w_{r+1} u_r w_r(C) \ldots$ *recursively).*
Thus there exists $u_i\, w_i \in U(w_i)w_i$, *uniquely, determined so that*

$$v^{-1}_{i+1}\left(F_{i-1}(\gamma)\right) = u_i\, w_i\left(F_{s_{i-1}}(C)\right).$$

It is finally obtained

$$F_j(\gamma) = u_{r+1}\, w_{r+1} \ldots u_{j+1}\, w_{j+1}\left(F_{s_j}(C)\right) \qquad (r \geqslant j \geqslant i-1)$$

with $(u_{r+1}\, w_{r+1}, \ldots, u_{j+1}\, w_{j+1})$ *uniquely determined.*

It is clear that

$$\mathcal{I} \neq \mathcal{I}' \Longrightarrow U(w_{\mathcal{I}})w_{\mathcal{I}} \cap U(w_{\mathcal{I}'})w_{\mathcal{I}'} = \varnothing.$$

One may thus write

$$\bigcup_{\mathcal{I} \in \mathrm{Relpos}^{\mathrm{gall}}_g I} U(w_{\mathcal{I}})w_{\mathcal{I}} = \coprod_{\mathcal{I} \in \mathrm{Relpos}^{\mathrm{gall}}_g I} U(w_{\mathcal{I}})w_{\mathcal{I}} \subset \mathrm{Stab}_{t(g)}\gamma_g(C).$$

The bijective mapping

$$\mathcal{W}^m_C(g) \longrightarrow \mathrm{Relpos}^{\mathrm{gall}}_g I$$

is obtained as the composition of bijections

$$i_{g,C}^m : \mathcal{W}_C^m(g) \longrightarrow \mathrm{Gall}_{\mathcal{A}}(g, F_{e_1(g)}(C))$$

(cf. Proposition 9.72) followed by

$$\tau_{g,C}^{\mathcal{A}} : \mathrm{Gall}_{\mathcal{A}}(g, F_{e_1(g)}(C)) \longrightarrow \mathrm{Relpos}_g^{gall} I.$$

Thus the cell decomposition

$$\mathrm{Gall}_I(g, F) = \coprod_{\underline{\tau} \in \mathrm{Relpos}_g^{gall} I} \mathscr{C}_C(g, \tau)$$

may be indexed by the set $\mathcal{W}_C^m(g)$ (resp. $\mathrm{Gall}_{\mathcal{A}}(g, F_{e_1(g)}(C))$).

By Corollary 9.73 it is known that

$$\mathcal{W}_C^m(g) \subset \mathcal{W}_C(t(g))$$

is a set of representatives of the quotient set

$$W_C(t(g))/W_C(s(g)).$$

One now gives a set of representatives of $\mathrm{Stab}_{t(g)} \gamma_g(C)/\mathrm{Stab}_{s(g)} \gamma_g(C)$, closely tied to $\mathcal{W}_C^m(g)$, which generalizes the parametrization of double classes given by Bruhat decomposition.

Definition 10.24 *Write:*

$$U(\mathcal{W}_C^m(g)) = \coprod_{\underline{\tau} \in \mathrm{Relpos}_g^{gall} I} U(w_{\underline{\tau}})w_{\underline{\tau}} \subset \mathrm{Stab}_{t(g)}\gamma_g(C).$$

Define the bijective mapping:

$$i_{g,C}^{I,m} : U(\mathcal{W}_C^m(g)) \longrightarrow \mathrm{Gall}_I(g, F_{e_1(g)}(C)) = \coprod_{\underline{\tau} \in \mathrm{Relpos}_g^{gall} I} \mathscr{C}_C(g, \underline{\tau}) ,$$

by:

$$i_{g,C}^{I,m} = \coprod_{\underline{\tau} \in \mathrm{Relpos}_g^{gall} I} i_{\underline{\tau},C}^m .$$

The mapping $i_{g,C}^{I,m}$ being the restriction of $i_{g,C}^I$ (cf. Definition 10.18) it factors through the bijective mapping:

$$\overline{i}_{g,C}^I : \mathrm{Stab}_{t(g)} \gamma_g(C)/\mathrm{Stab}_{s(g)} \gamma_g(C) \longrightarrow \mathrm{Gall}_I(g, F_{e_1(g)}(C)).$$

It is deduced that the mapping

$$\left(\mathrm{Stab}_{t(g)} \gamma_g(C) \supset\right) U(\mathcal{W}_C^m(g)) \longrightarrow \mathrm{Stab}_{t(g)} \gamma_g(C)/\mathrm{Stab}_{s(g)} \gamma_g(C)$$

induced by the quotient mapping is also bijective. From this it results that $U(\mathcal{W}_C^m(g))$ may be viewed as a set of representatives of the $\mathrm{Stab}_{s(g)}\,\gamma_g(C)$-classes of $\mathrm{Stab}_{t(g)}\,\gamma_g(C)$.

The cells $\mathscr{C}_C(g,\underline{\tau})$ are obtained as the fibers of a retraction mapping.

Definition 10.25 *Define* **the retraction of galleries** *given by the couple* $(\mathcal{A},C)$.

$$\rho_{\mathcal{A},C}(g) : \mathrm{Gall}_I(g, F_{e_1(g)}(C)) \longrightarrow \mathrm{Gall}_{\mathcal{A}}(g, F_{e_1(g)}(C))$$

by:

$$\rho_{\mathcal{A},C}(g) := \left(\tau_{g,C}^{\mathcal{A}}\right)^{-1} \circ \tau_{g,C}^I.$$

(cf. Proposition 10.17)

The fibers of $\rho_{\mathcal{A},C}(g)$ are given by the set of cells $(\mathscr{C}_C(g,\underline{\tau}))_{\underline{\tau}\in\mathrm{Relpos}_g^{gall}I}$. More precisely, we have

$$\left(\rho_{\mathcal{A},C}(g)\right)^{-1}(\gamma_{\underline{\tau}}) = \mathscr{C}_C(g,\underline{\tau}),$$

where $\gamma_{\underline{\tau}} \in \mathrm{Gall}_{\mathcal{A}}(g,C)$ is the unique gg of $\mathcal{A}$ such that $(\tau_{g,C}^{\mathcal{A}})(\gamma_{\underline{\tau}}) = \underline{\tau}$. It is recalled that $\tau_{g,C}^{\mathcal{A}} : \mathrm{Gall}_{\mathcal{A}}(g, F_{e_1(g)}(C)) \longrightarrow \mathrm{Relpos}_g^{gall}I$ is bijective, and its inverse is given by $\tau_{g,C}^{'\mathcal{A}} : \underline{\tau} \mapsto \gamma_{\underline{\tau}}$ (cf. loc. cit.). On the other hand, there is a natural mapping

$$U(\mathcal{W}_C^m(g)) \longrightarrow \mathcal{W}_C^m(g),$$

defined by:

$$(u_{r+1}\,w_{r+1}, \ldots, u_1\,w_1) \mapsto (w_{r+1}, \ldots, w_1) \in W_C^m(g) .$$

The following commutative diagram expresses the compatibility between $i_{g,C}^{I,m}$ and $i_{g,C}^m$.

$$
\begin{array}{ccc}
i_{g,C}^{I,m} : U(\mathcal{W}_C^m(g)) & \longrightarrow & \mathrm{Gall}_I(g, F_{e_1(g)}(C)) \ . \\
\downarrow & & \downarrow {\scriptstyle \rho_{\mathcal{A},C}(g)} \\
i_{g,C}^m : \mathcal{W}_C^m(g) & \longrightarrow & \mathrm{Gall}_{\mathcal{A}}(g, F_{e_1(g)}(C))
\end{array}
$$

Given $g \in \mathrm{gall}^m = \mathrm{gall}_{\mathcal{A}}^m = \mathrm{gall}_I^m$, so that $g = g_1'$, one has associated with $C \in \mathrm{Ch}\,\mathcal{A}$ the bijections:

$$i_{g,C}^m : \mathcal{W}_C^m(g) \longrightarrow \mathrm{Gall}_I(g, F_{e_1(g)}(C)) \qquad (cf.\ Proposition\ 9.72)$$

$$(\text{resp. } \tau_{g,C}^{\mathcal{A}} : \mathrm{Gall}_{\mathcal{A}}(g, F_{e_1(g)}(C)) \longrightarrow \mathrm{Relpos}_g^{gall}I \ \ (cf.\ Proposition\ 10.17) \ .$$

It is easy to see that the composed mapping

$$\tau_{g,C}^{\mathcal{A}} \circ i_{g,C}^m : \mathcal{W}_C^m(g) \longrightarrow \mathrm{Relpos}_g^{gall} I$$

is the bijection induced by the mappings

$$\mathcal{W}_C^m(t,s) \longrightarrow \mathrm{Relpos}_s^t I$$

defined by: $w \mapsto \tau(C, w(F_s(C)))$ (cf. §9.3,c).

10.6.2 The generic gallery

Recall that

$$\mathscr{S}_C = (w(t,s))_{(t,s) \in \mathrm{typ}^{(2)}\mathcal{A}},$$

with $(\mathrm{typ}^{(2)}\mathcal{A} = \{(t,s) \in \mathrm{typ}\,\mathcal{A} \times \mathrm{typ}\,\mathcal{A} - \Delta \mid t \subset s\})$, where

$$w(t,s) = w(C, \mathrm{proj}_F\, C)$$

and $F = F_{(t,s)}^{tr}(C) = $ the facet of type s incident to C^{opp} in $\mathrm{St}_{F_t(C)}$. With the notation of §9.1, a), $F = F_{(t,s)}^{tr}(C)$ may be characterized as the facet of type s in $\mathrm{St}_{F_t(C)}$ at maximal distance from C, i.e. if $\tau = \tau(C, \mathrm{proj}_F\, C)$, then $w_F^m = w_\tau^m = $ the element of maximal length of $\{w_\tau^m \mid \tau \in \mathrm{Relpos}_s^t\}$, with respect to S_C. The couple $(C, F_{(t,s)}^{tr}(C))$ is in transversal position in $\mathrm{St}_{F_t(C)}$. Observe that

$$\mathscr{S}_C \cap \mathcal{W}_C^m(t_i(g), s_{i-1}(g)) = \{w(t_i(g), s_{i-1}(g))\}.$$

Definition 10.26
Let $\underline{\tau}^{tr} \in \mathrm{Relpos}_g^{gall} I$ denote the image of
$$\underline{w}^{tr} = (w_i^{tr}) = (w(t_i(g), s_{i-1}(g))) \in \mathcal{W}_C^m(g) = \prod \mathcal{W}_C^m(t_i(g), s_{i-1}(g)) \text{ by } \tau_{g,C}^{\mathcal{A}} \circ i_{g,C}^m.$$ *One calls the cell*

$$\mathscr{C}_C(g, \underline{\tau}^{tr}) \subset \hat{\Sigma}(g, F_{e_1(g)}(C)) = \mathrm{Gall}_I(g, F_{e_1(g)}(C))$$

corresponding to $\underline{\tau}^{tr}$, **the big cell** *of $\hat{\Sigma}(g, F_{e_1(g)}(C))$, with respect to C.*

The transversal relative position type τ_i^{tr} is the image of $w_i^{tr} = w(t_i(g), s_{i-1}(g)) \in \mathcal{W}_C^m(t_i(g), s_{i-1}(g))$ by the above defined mapping

$$\mathcal{W}_C^m(t_i(g), s_{i-1}(g)) \longrightarrow \mathrm{Relpos}_{s_{i-1}(g)}^{t_i(g)} I.$$

Thus

$$\underline{\tau}^{tr} = (\tau_i^{tr}) \qquad (r+1 \geqslant i \geqslant 1).$$

Remark 10.27

1. *Write*

$$w_g^{ch} = \prod w_i^{tr} \qquad (r+1 \geqslant i \geqslant 1).$$

Then $\underline{w}^{tr} = (w_i^{tr})$ *is a* $\mathscr{S}_C$-*reduced expression of type* g *of* $w_{g,C}^{ch}$, *according to Definition 9.13.*

2. *The image* $\gamma_g(\underline{w}^{tr}) = \underline{w}^{tr} \cdot \gamma_g(C)$ *by* $i_{g,C}^m$ *of* $\underline{w}^{tr} = (w_i^{tr})$ *is a MGG of type* g *between* $F = F_{e_1(g)}(C) = F_{t_{r+1}}(C)$ *and* $F' = w_g^{ch}(F_{t_0(g)}(C))$, *namely*

$$\gamma_g(\underline{w}^{tr}) = \gamma_g(F,F') \ .$$

3. *It is easy to verify that by definition of* $w_g^{ch}(F_{t_0(g)}(C))$ *one has: "$C \supset F$ is at maximal distance from* $w_g^{ch}(F_{t_0(g)}(C))$". *(w_g^{ch} is the maximal length element among those obtained as the product of the components of some* $\underline{w} \in \mathcal{W}_C^m(g)$). *It follows that* $\mathscr{C}_C(g, \underline{\tau}^{tr})$ *is the cell of maximal dimension.*

Write

$$\tau_g^{ch} = \tau(C, w_g^{ch}(F_{t_0(g)}(C))) \in (\mathcal{A} \times \mathcal{A})_{\tau_g}^{ch}/W_{\mathcal{A}} \qquad (\tau_g = \tau(F,F')).$$

Observe that $w_g^{ch} = w_{\tau_g^{ch}}^m$.

10.7 Minimal Generalized Gallery block decomposition of a Schubert cell parametrizing subgroup

One explicates the parametrization of the big cell of the galleries of type $g \in gall_{\mathcal{A}}^m$ and its relation with the corresponding Schubert cell.
By definition of MGG between F and F' one obtains

$$\mathcal{H}(C,F') = \coprod \mathcal{H}_{F_i'}(F_{i-1}) \ (r+1 \geqslant i \geqslant 1) \quad (\text{resp. } \mathcal{H}(C,F') = \mathcal{H}(C, \text{proj}_{F_0} C))$$

where $F_i = F_i(\gamma_g(F,F'))$ (resp. $F_j' = F_j(\gamma_g(F,F'))$).

One has defined a mapping

$$s_{g^*,C} : \mathrm{Gall}_{\mathcal{A}}(g, F_{e_1(g)}(C)) \longrightarrow \mathrm{Gall}_{\mathcal{A}}(g^*, C)$$

(cf.   Definition 9.64) which associates with $\gamma_g(F,F')$ a gg, namely, $s_{g^*,C}(\gamma_g(F,F'))$ characterized by the sequence of chambers $C_{r+1} = C$, $C_i = \text{proj}_{F_i} C_{i+1}$ ($r \geqslant i \geqslant 0$). Following a standard calculus (cf. §9.62, i)), one obtains

$$C_i = w_{r+1}^{tr} \dots w_{i+1}^{tr}(C) = v_{i+1}(C)$$

where

$$v_{i+1} = w_{r+1}^{tr} \ldots w_{i+1}^{tr}.$$

As $C_i \supset F_i \supset F_i'$, it is deduced that

$$
\begin{aligned}
F_i' &= F_i'(\gamma_g(F, F')) = v_{i+1}(F_{t_i(g)}(C)). \\
(\text{resp. } F_i &= F_i(\gamma_g(F, F')) = v_{i+1}(F_{s_i(g)}(C))) \;.
\end{aligned}
$$

On the other hand, by definition of τ_i^{tr} one has

$$\tau_i^{tr} = \tau(C, w_i^{tr}(F_{s_{i-1}(g)}(C))),$$

one then deduces that

$$\tau(C_i, F_{i-1}) = \tau_I(C, w_i^{tr}(F_{s_{i-1}(g)}(C))) = \tau_i^{tr},$$

(as $C_i = v_{i+1}(C), v_{i+1} \cdot w_i^{tr} = v_i$) and

$$v_{i+1}\left(\mathcal{H}(C, w_i^{tr}(F_{s_{i-1}(g)}(C)))\right) = \mathcal{H}(C_i, F_{i-1}).$$

From this last equality it follows that v_{i+1} transforms the set $R_+(C, w_i^{tr}(F_{s_{i-1}(g)}(C)))$ of roots $\alpha \in R_+(C)$, so that $\partial\Phi_\alpha \in \mathcal{H}(C, w_i^{tr}(F_{s_{i-1}(g)}(C)))$, into the set $R_+(C_i, F_{i-1})$. Remark that as $\gamma_g(\underline{w}^{tr})$ is a MGG one has

$$H \in \mathcal{H}(C_i, F_{i-1}) \Longrightarrow H \notin \mathcal{H}(C_{r+1}, C_i) \qquad (C_{r+1} = C)$$

(cf. Remark 3.6,a)), and that

$$v_{i+1} = w(C, C_i) \;, \; (C_i = v_{i+1}(C)) \;.$$

On the other hand, given $\alpha \in R_+(C)$ the following equivalence holds

$$v_{i+1}(\alpha) \in -R_+(C) \Longleftrightarrow \partial\Phi_\alpha \in \mathcal{H}(C, C_i) \;,$$

as $C_i = v_{i+1}(C)$. It is then deduced that

$$R_+(C_i, F_{i-1}) \subset R_+(C) \;.$$

As

$$\mathcal{H}_{F_i'}(F_{i-1}) = \mathcal{H}(C_i, F_{i-1})$$

one obtains

$$\mathcal{H}(C, F') = \coprod \mathcal{H}(C_i, F_{i-1}) \;,$$

and finally

$$R_+(C, F') = \coprod R_+(C_i, F_{i-1}) \;.$$

From the definition it follows that

$$R_+(w_g^{ch}) = R_+(C) - R_+(C) \cap w_g^{ch}(R_+(C))$$

and that

$$R_+(w_g^{ch}) = R_+(C, w_g^{ch}) = R_+(C, F').$$

Thus results the equality

$$R_+(C, F') = R_+(w_g^{ch}) = \coprod R_+(C_i, F_{i-1}).$$

One now remarks that a set of roots $R_+(C, F) = R_+(C) - R_+(C) \cap R_F$ is a closed set of roots. Then the image of the mapping

$$\prod \mathfrak{X}_\alpha(\alpha \in R_+(C, F)) \longrightarrow G$$

induced by the multiplication in G is a subgroup of B_C which is denoted by $U(C, F)$.

Lemma 10.28 *With the above notation*

$$U(C_i, F_{i-1}) = v_{i+1} U(w_i^{tr}) v_{i+1}^{-1}.$$

Proof *This equality results from the following facts:*

$$U(w_i^{tr}) = U(C, w_i^{tr}(F_{s_{i-1}(g)}(C)))$$

(resp. $R_+(C_i, F_{i-1}) = v_{i+1}\left(R_+(C, w_i^{tr}(F_{s_{i-1}(g)}(C)))\right)$), in view of the implication $(\bar{C}, \bar{F}) = (v(C), v(F)) \implies R(\bar{C}, \bar{F}) = v(R(C, F)) \implies U(\bar{C}, \bar{F}) = v\,U(C, F)v^{-1}.$

Proposition 10.29 *The equality*

$$R_+(w_g^{ch}) = \coprod R_+(C_i, F_{i-1})$$

gives a bijection immediately

$$\prod U(C_i, F_{i-1}) \longrightarrow U(w_g^{ch}) = U(C, F')$$

induced by the multiplication in G.

10.8   Minimal type galleries big cell

Let one establish the connection between $U(w_g^{ch}) = U(C, F')$ and

$$\Sigma(\tau_g^{ch}, C) \subset \Sigma(\tau_g, F_{e_1(g)}(C)),$$

where $\tau_g = \tau(F, F')$ and $\tau_g^{ch} = \tau(C, F')$, and $C \supset F$ is at maximal distance from F'.

Remark 10.30 *Observe that if* (C, F) *corresponds to* $(F_{e_1(g)}(C), F)$ *by the mapping*

$$(\mathcal{A} \times \mathcal{A})_{\tau_g}^{ch} \longrightarrow (\mathcal{A} \times \mathcal{A})_{\tau_g} \text{ , then}$$

$$\Sigma(\tau, C) \subset \Sigma(\tau_g, F_{e_1(g)}(C)),$$

where $\tau = \tau_I(C, F)$ *(cf. Lemma 9.22).*

Lemma 10.31 *The multiplication in* G *induces a bijection*

$$U(w_{\underline{\tau}^{tr}})w_{\underline{\tau}^{tr}} = \prod U(w_i^{tr})w_i^{tr} \simeq U(w_g^{ch})w_g^{ch} \text{ .}$$

Proof *One first observes that*

$$w_{r+1}^{tr} = w_{\tau_{r+1}^{tr}}^{m} \qquad (resp. \ w_{\underline{\tau}^{tr}} = (w_i^{tr})),$$

thus $U(w_{\underline{\tau}^{tr}})w_{\underline{\tau}^{tr}} = \prod U(w_i^{lr})w_i^{lr}.$

On the other hand, the mapping

$$U(w_{\underline{\tau}^{tr}})w_{\underline{\tau}^{tr}} \longrightarrow G$$

induced by the multiplication factors as follows. It is obtained as the composition of the bijection

$$\prod U(w_i^{tr})w_i^{tr} \longrightarrow \prod v_{i+1} U(w_i^{tr})v_{i+1}^{-1}$$

followed by the bijection

$$\prod U(C_i, F_{i-1}) = \prod v_{i+1} U(w_i^{tr})v_{i+1}^{-1} \longrightarrow U(w_g^{ch})$$

(cf. §10.7) induced by the multiplication in G, *and the right multiplication by* w_g^{ch}:

$$x \mapsto x \, w_g^{ch} \text{ .}$$

This follows immediately from the identity

$$u_{r+1} w_{r+1}^{tr} u_r w_r^{tr} \ldots u_1 w_1^{tr} = u_{r+1}(v_{r+1} u_r v_{r+1}^{-1})(v_r u_{r-1}v_r^{-1}) \cdots (v_1 uv_1^{-1})v_1.$$

$((u_{r+1} w_{r+1}^{tr}, \ldots, u_1 w_1^{tr}) \in U(w_{\underline{\tau}^{tr}})w_{\underline{\tau}^{tr}})$. *Observe that* $v_{i+1}^{-1} v_i = w_{i+1}^{tr}$

There is a bijective mapping

$$U(w_g^{ch})w_g^{ch} \longrightarrow \Sigma(\tau_g^{ch}, C) \subset \Sigma(\tau_g, F_{e_1(g)}(C))$$

defined by

$$u \mapsto u(F_{t_0(g)}(C)) \qquad (u \in U(w_g^{ch})w_g^{ch})$$

(cf. §10.8). On the other hand, the mapping

$$i^m_{\underline{\tau}^{tr},C} : U(w_{\underline{\tau}^{tr}})w_{\underline{\tau}^{tr}} \longrightarrow \mathscr{C}_C(g,\underline{\tau}^{tr}) \subset \Sigma(g, F_{e_1(g)}(C))$$

being induced by

$$i^I_{g,C} : \mathrm{Stab}_{t(g)}\,\gamma_g(C) \longrightarrow \hat{\Sigma}(g, F_{e_1(g)}(C)) = \mathrm{Gall}_I(g, F_{e_1(g)}(C))$$

defined by

$$i^I_{g,C} : x \mapsto \gamma_g(x) = x \cdot \gamma_g(C) \qquad (x \in \mathrm{Stab}_{t(g)}\,\gamma_g(C))\,.$$

(cf. Definition 10.21, and Definition 10.18), verifies

$$F'_0(\gamma_g(x)) = u_{r+1}\,w^{tr}_{r+1}\ldots u_1\,w^{tr}_1(F_{t_0(g)}(C))$$

where $x = (u_{r+1}\,w^{tr}_{r+1},\ldots,u_1\,w^{tr}_1) \in U(w_{\underline{\tau}^{tr}})w_{\underline{\tau}^{tr}} \subset \mathrm{Stab}_{t(g)}\,\gamma_g(C)$.

Putting together these facts from Lemma 10.31 it results that the composition mapping

$$E_2 \circ i^m_{\underline{\tau}^{tr},C} : U(w_{\underline{\tau}^{tr}})w_{\underline{\tau}^{tr}} \longrightarrow \Sigma(\bar{\tau}_g, F_{e_1(g)}(C))$$

sends $U(w_{\underline{\tau}^{tr}})w_{\underline{\tau}^{tr}}$ bijectively to

$$(U(w_g^{ch})w_g^{ch})(F_{t_0(g)}(C)) = \Sigma(\tau_g^{ch}, C) \subset \Sigma(\tau_g, F_{e_1(g)}(C))\,.$$

one has proved the

Proposition 10.32 *The following diagram*

$$
\begin{array}{ccccc}
i^m_{\underline{\tau}^m,C}: & U(w_{\underline{\tau}^{tr}})w_{\underline{\tau}^{tr}} & \longrightarrow & \mathscr{C}_C(g,\underline{\tau}^{tr}) \hookrightarrow & \Sigma(g, F_{e_1(g)}(C)) \\
 & \Big\downarrow & & & \Big\downarrow \pi^m_{g,F_{e_1(g)}(C)} \\
 & U(w_g^{ch})w_g^{ch} & \longrightarrow & \Sigma(\tau_g^{ch},C) \hookrightarrow & \Sigma(\bar{\tau}_g, F_{e_1(g)}(C))
\end{array}
$$

where $U(w_g^{ch})w_g^{ch} \longrightarrow \Sigma(\tau_g^{ch}, C)$ *is defined as above, and the left vertical arrow is given by Lemma 10.31, commutes.*

It is known that $\pi^m_{g,F_{e_1(g)}(C)} = E_2^m$ is a bijective mapping whose inverse mapping is given by the section

$$\Theta_{g,F_{e_1(g)}(C)} : \Sigma(\bar{g}, F_{e_1(g)}(C)) \longrightarrow \Sigma(g, F_{e_1(g)}(C))$$

defined by:

$$\Theta_{g,F_{e_1(g)}(C)} : F' \mapsto \gamma_g(F_{e_1(g)}(C), F')$$

$$(= \text{the MGG between } F_{e_1(g)}(C) \text{ and } F' \text{ of type g satisfying}$$
$$(\tau(F_{e_1(g)}(C), F') = \tau_g))$$

Thus one obtains that all the arrows of the above diagram, with the exception of the two inclusions, are bijective. The following is deduced

Corollary 10.33 *With the above notation*

$$\left(\Theta_{g,F_{e_1(g)}(C)}\right)\left(\Sigma(\tau_g^{ch}, C)\right) = \mathscr{C}_C(g, \underline{\tau}^{tr}).$$

10.9 Galleries of fixed type as fiber products

It is possible to write the set of galleries of type g, $Gall_I(g)$ as a fiber product of "closed universal Schubert cells" given by the set of type s facets incident to F_t, $(St_{F_t})_s$.

Recall $typ^{(2)}\mathcal{A} = \{(t, s) \in \operatorname{typ}\mathcal{A} \times \operatorname{typ}\mathcal{A} - \Delta \mid t \subset s\}$. There is a natural mapping

$$\tau^{(2)} : \operatorname{typ}^{(2)}\mathcal{A} \longrightarrow \operatorname{Relpos}\mathcal{A}$$

defined by

$$\tau^{(2)} : (t, s) \mapsto \tau_I(F', F) \quad ((t, s) \in \operatorname{typ}^{(2)}\mathcal{A})$$

where $(F', F) \in I_t \times I_s$ satisfies $F' \subset F$. Clearly the mapping $\tau^{(2)}$ is well-defined, injective, and $\tau^{(2)}(t, s) \in \operatorname{Relpos}^t_{(t,s)} I$.

Definition 10.34
    Given $(t, s) \in \operatorname{typ}^{(2)}\mathcal{A}$, *and* $F' \in I_t$ *write:*

$$\Sigma(t, s) = \Sigma(\tau^{(2)}(t, s)) \subset I_t \times I_s$$
$$(\textit{resp.} \qquad \Sigma(t, s; F') = \Sigma(\tau^{(2)}(t, s), F'))$$

One calls $\Sigma(t, s; F')$ **the** (t, s)**-elementary cell defined by** F' **(resp.** $\Sigma(t, s)$ **the** (t, s)**-universal elementary cell).** *One has* $\overline{\Sigma}^c(t, s) = \Sigma(t, s)$.

    Let $(t, s) \in \operatorname{typ}^{(2)}\mathcal{A}$, and a type $t' \subset s$. There is a natural mapping

$$\operatorname{pr}_{2,t'} = \operatorname{pr}^\Sigma_{2,t'} : \Sigma(t, s) \longrightarrow I_{t'} \qquad (\textit{resp.} \ \operatorname{pr}_1 = \operatorname{pr}^\Sigma_1 : \Sigma(t, s) \longrightarrow I_t)$$

associating with $(F', F) \in \Sigma(t, s)$ the unique facet $F_{t'}(F)$ of type t' incident to $F(\mathrm{typ}\, F = s \supset t')$, i.e.

$$\mathrm{pr}_{2,t'} : (F', F) \mapsto F_{t'}(F) .$$

(resp. defined by $\mathrm{pr}_1 : (F', F) \mapsto F'$).

Definition 10.35

1. *Let* $(t, s), (t', s') \in \mathrm{typ}^{(2)} \mathcal{A}$, *with* $t' \subset s$, *and* $F' \in I_t$. *Then it is said that* (t, s), *and* (t', s') *are* **composable**. *Write:*

$$\Sigma(t, s) * \Sigma(t', s') = \Sigma(t, s) \times_{I_{t'}} \Sigma(t', s')$$

(resp. $\Sigma(t, s; F') * \Sigma(t', s') = \Sigma(t, s; F') \times_{I_{t'}} \Sigma(t', s'))$

where the fiber product is defined relatively to the couple:

$$(\mathrm{pr}_{2,t'}, \mathrm{pr}_1) \qquad (\textit{resp.} \ (\mathrm{pr}_{2,t'}|_{\Sigma_{(t,s;F')}}, \mathrm{pr}_1)).$$

2. *Let* $F \in I$ *with* $\mathrm{typ}\, F \supset t$. *Define*

$$\Sigma(t, s; F) = \Sigma(t, s; F_t(F)) .$$

To $g \in \mathrm{Gall}_{\mathcal{A}}$, of length $r + 1$, with $g = g_1$, so that $(t_{i+1}(g), s_i(g)) \in \mathrm{typ}^{(2)} \mathcal{A}$ given by:

$$g : (s_i) \quad (r \geqslant i \geqslant 0), \qquad (t_j) \quad (r + 1 \geqslant j \geqslant 1)$$

one associates the star product $\Sigma^*(g)$ which is defined by induction from Definition 10.35:

$$\Sigma^*(g) = \Sigma(t_{r+1}, s_r) * \Sigma(t_r, s_{r-1}) \cdots * \Sigma(t_1, s_0).$$

Definition 10.36 *Given* $g = g_1'$ *we define*

$$\Sigma^*(g) = \Sigma^*(g^{(0)})$$

where $g^{(0)}$ *denotes the truncated gallery defined by* g *(cf. Definition 9.57). If* $g = g_2 : (s_i) \quad (r \geqslant i \geqslant 0), (t_j) \quad (r + 1 \geqslant j \geqslant 1)$ *one defines*

$$\Sigma^*(g) = \Sigma(t_r, s_{r-1}) * \cdots * \Sigma(t_1, s_0)$$

For $g = g_2'$ *one writes*

$$\Sigma^*(g) = \Sigma^*(g^{(0)}).$$

Proposition 10.37 *There is a canonical bijection*

$$\mathrm{Gall}_I(g) \simeq \Sigma^*(g)$$

for $g \in \mathrm{Gall}_{\mathcal{A}}$ verifying $(t_{i+1}(g), s_i(g)) \in \mathrm{typ}^{(2)}\mathcal{A}$.

Proof *Let $\gamma \in \mathrm{Gall}_I(g)$, with $g = g_1$ be given by:*

$$\gamma : (F_i) \quad (r \geqslant i \geqslant 0), \qquad (F'_j) \quad (r+1 \geqslant j \geqslant 1).$$

With γ one associates the element

$$((F'_{i+1}, F_i)) \in \prod \Sigma(t_{i+1}, s_i) \qquad (r \geqslant i \geqslant 0)$$

which clearly belongs to $\Sigma^*(g) \subset \prod \Sigma(t_{i+1}, s_i)$. Thus one defines a mapping

$$\mathrm{Gall}_I(g) \longrightarrow \Sigma^*(g) ,$$

which is immediately seen to be bijective. A bijection $\mathrm{Gall}_I(g) \simeq \Sigma^*(g)$ may be obtained in the case $g = g_2$ (resp. $g = g'_1$, $g = g'_2$) following the same pattern.

Given F with $\mathrm{typ}\, F = t_{r+1}(g)$ (resp. $\mathrm{typ}\, F = s_r(g)$) one defines

$$\Sigma^*(g, F) \subset \Sigma^*(g)$$

as the image of $\mathrm{Gall}_I(g, F)$, by the bijection $\mathrm{Gall}_I(g) \longrightarrow \Sigma^*(g)$, if $g = g_1, g'_1$ (resp. $g = g_2, g'_2$).

For $g \in \mathrm{gall}_{\mathcal{A}}$, and $r \geqslant \alpha \geqslant 0$ (resp. $r > \alpha \geqslant 0$) let

$$\pi_g^{(\alpha)} : \mathrm{Gall}_I(g) \longrightarrow \mathrm{Gall}_I(g^{(\alpha)})$$

be the mapping:

$$\pi_g^{(\alpha)} : \gamma \mapsto \gamma^{(\alpha)}$$

if $g = g_1, g'_1$ (resp. $g = g_2, g'_2$) (cf. Definition 9.57).

Let

$$\pi_{g,F}^{(\alpha)} : \mathrm{Gall}_I(g, F) \longrightarrow \mathrm{Gall}_I(g^{(\alpha)}, F)$$

denote the mapping induced by $\pi_g^{(\alpha)}$ by restriction to $\mathrm{Gall}_I(g, F) \subset \mathrm{Gall}_I(g)$, and co-restriction to $\mathrm{Gall}_I(g^{(\alpha)}, F) \subset \mathrm{Gall}_I(g^{(\alpha)})$.

From the definition of $\pi_{g^{(\alpha')}}^{(\alpha)}$ $(\alpha > \alpha')$ results the following proposition.

Proposition 10.38 *Let* $\gamma^{(\alpha)} \in \mathrm{Gall}_I(g^{(\alpha)})$. *The fiber* $\left(\pi_{g^{(\alpha')}}^{(\alpha)}\right)^{-1}(\gamma^\alpha)$ *of*

$$\pi_{g^{(\alpha')}}^{(\alpha)} : \mathrm{Gall}_I(g^{(\alpha')}) \longrightarrow \mathrm{Gall}_I(g^{(\alpha)}) \qquad (\alpha' \leqslant \alpha)$$

over γ *is given by:*

$$\left(\pi_{g^{(\alpha')}}^{(\alpha)}\right)^{-1}(\gamma^{(\alpha)}) = \Sigma(t_\alpha, s_{\alpha-1}; F_{t_\alpha}(F_{s_\alpha}(\gamma))) * \cdots * \Sigma(t_{\alpha'+1}, s_{\alpha'}).$$

If one introduces the notation

$$g^{[\alpha,\alpha']} : t_\alpha(g) \subset s_{\alpha-1}(g) \cdots t_{\alpha'+1}(g) \subset s_{\alpha'}(g)$$

the equality of the Proposition 10.38 may be written

$$\left(\pi_{g^{(\alpha')}}^{(\alpha)}\right)^{-1}(\gamma^{(\alpha)}) = \Sigma^*(g^{[\alpha,\alpha']}, F_{t_\alpha}(F_{s_\alpha}(\gamma^{(\alpha)}))) = \mathrm{Gall}_I(g^{[\alpha,\alpha']}, F_{t_\alpha}(F_{s_\alpha}(\gamma^{(\alpha)}))).$$

As a particular case of the Proposition 10.38 for $\alpha' = \alpha - 1$ one obtains the equality:

$$\left(\pi_{g^{(\alpha-1)}}^{(\alpha)}\right)^{-1}(\gamma^{(\alpha)}) = \Sigma(t_\alpha, s_{\alpha-1}; F_{t_\alpha}(F_{s_\alpha}(\gamma^{(\alpha)}))).$$

Definition 10.39 *Let* $g = g_1, g_1'$ *(resp.* $g = g_2, g_2'$*). It is called the sequence of mappings*

$$\left(\pi_{g^{(\alpha-1)}}^{(\alpha)}\right) \qquad (r+1 \geqslant \alpha \geqslant 1)$$

$$(resp. \qquad \left(\pi_{g^{(\alpha-1)}}^{(\alpha)}\right) \qquad (r \geqslant \alpha \geqslant 1))$$

the tower of fibrations associated with $\mathrm{Gall}_I(g)$. *If* $\mathrm{typ}\, F = t_{r+1}(g)$ *(resp.* $\mathrm{typ}\, F = s_r(g)$*) one has, corresponding to the preceding sequence, the sequence*

$$\left(\pi_{g^{(\alpha-1)},F}^{(\alpha)}\right) \qquad (r+1 \geqslant \alpha \geqslant 1)$$

$$(resp. \qquad \left(\pi_{g^{(\alpha-1)},F}^{(\alpha)}\right) \qquad (r \geqslant \alpha \geqslant 1))$$

(the tower of fibrations associated with $\mathrm{Gall}_I(g, F)$**).** *Where* $\pi_{g^{(r)}}^{(r+1)}$ *(resp.* $\pi_{g^{(r-1)}}^{(r)}$*) denotes the canonical mapping*

$$\mathrm{Gall}_I(g) \longrightarrow I_{t_{r+1}(g)}$$

$$(resp. \qquad \mathrm{Gall}_I(g) \longrightarrow I_{s_r(g)}).$$

10.10 Galleries cells tower fibration

Let $\tau, \tau' \in (\text{Ch}\,\mathcal{A} \times \mathcal{A})/W = (\text{Ch}\,\text{I} \times \text{I})/G$. Define

$$\text{pr}_2^\tau : \Sigma(\tau) \longrightarrow \text{Ch}\,\text{I}$$

$$(\textit{resp.} \qquad \text{pr}_1^\tau : \Sigma(\tau') \longrightarrow \text{Ch}\,\text{I} \,)$$

by

$$\text{pr}_2^\tau : (C, F) \mapsto \text{proj}_F\, C \qquad (\textit{resp.}\ \text{pr}_1^\tau : (C, F) \mapsto C\,).$$

Write

$$\Sigma(\tau) * \Sigma(\tau') = \Sigma(\tau) \times_{\text{Ch}\,\mathcal{A}} \Sigma(\tau')$$

where the fiber product is defined by the couple $(\text{pr}_2^\tau, \text{pr}_1^{\tau'})$. Given $C \in \text{Ch}\,\text{I}$ one writes: $\Sigma(C, \tau) * \Sigma(\tau') =$ fiber over C of the mapping

$$\Sigma(\tau) * \Sigma(\tau') \longrightarrow \Sigma(\tau) \overset{\text{pr}_1^\tau}{\longrightarrow} \text{Ch}\,\text{I} \ .$$

Definition 10.40 *For $\underline{\tau} \in \text{Relpos}_I^{\text{gall}}(g)$, and $C \in \text{Ch}\,\text{I}$ define*

$$\Sigma^*(\underline{\tau}) = \Sigma(\tau_{r+1}) * \cdots * \Sigma(\tau_1)$$

$$(\textit{resp.} \qquad \Sigma^*(\underline{\tau}, C) = \Sigma(\tau_{r+1}, C) * \cdots * \Sigma(\tau_1)).$$

where $\underline{\tau} = (\tau_{r+1}, \ldots, \tau_1)$. One supposes $g = g_1, g_1'$. The corresponding defini-tion in the case $g = g_2, g_2'$ is obtained by replacing $r + 1$ by r in the preceding equalities.

Suppose now $C \in \text{Ch}\,\mathcal{A}$ and $g = g_1$. The next aim is to obtain the cell $\mathscr{C}_C(g, \underline{\tau}) \subset \text{Gall}_I(g)$ as the star product $\Sigma^*(\underline{\tau}, C)$ along the gallery of relative positions

$$\underline{\tau} = (\tau_{r+1}, \ldots, \tau_1) \in \text{Relpos}_I^{\text{gall}}(g).$$

One represents an element $x \in \Sigma^*(\underline{\tau})$ in the form

$$x = ((C_{i+1}, F_i)) \qquad (r \geqslant i \geqslant 0)$$

and one defines

$$\mathrm{j}_C(g, \underline{\tau}) : \Sigma^*(\underline{\tau}, C) \longrightarrow \text{Gall}_I(g, F_{e_1(g)}(C))$$

by

$$\mathrm{j}_C(g, \underline{\tau}) : ((C_{i+1}, F_i)) \longrightarrow \gamma((C_{i+1}, F_i))$$

where this latter is given by

$$\gamma((C_{i+1}, F_i)) : (F_i) \ (r \geqslant i \geqslant 0), \qquad (F_{t_j(g)}(C_j)) \ (r + 1 \geqslant j \geqslant 1).$$

From the Definition 10.14 of $\mathscr{C}_C(g, \underline{\tau})$ one obtains the proposition:

Proposition 10.41 *The image* $\Im j_C(g, \underline{\tau})$ *is given by*

$$\Im j_C(g, \underline{\tau}) = \mathscr{C}_C(g, \underline{\tau}).$$

One identifies $\Sigma^*(\underline{\tau}, C)$ *with* $\mathscr{C}_C(g, \underline{\tau})$ *by means of* $j_C(g, \underline{\tau})$.

Let

$$\underline{\tau}^{(\alpha)} = (\tau_{r+1}, \ldots, \tau_\alpha) \in \mathrm{Relpos}_I^{\mathrm{gall}}(g^{(\alpha-1)})$$

be the α-truncated gallery of relative positions defined by $\underline{\tau}$. One has the inclusions:

$$\mathscr{C}_C(g^{(\alpha-1)}, \underline{\tau}^{(\alpha)}) = \Sigma^*(\underline{\tau}^{(\alpha)}, C) \subset \Sigma^*(g^{(\alpha-1)}, C) .$$

Let

$$\pi^{(\alpha)}_{\underline{\tau}^{(\alpha)}, C} : \Sigma^*(\underline{\tau}^{(\alpha)}, C) \longrightarrow \Sigma^*(\tau^{(\alpha+1)}, C)$$

denote the mapping obtained as the restriction to

$$\Sigma^*(\underline{\tau}^{(\alpha)}, C) \subset \hat{\Sigma}(g^{(\alpha-1)}, C)$$

and the co-restriction to

$$\Sigma^*(\underline{\tau}^{(\alpha+1)}, C) \subset \hat{\Sigma}(g^{(\alpha)}, C)$$

of $\pi^{(\alpha)}_{g^{(\alpha-1)}, C}$.

Given

$$\gamma \in \mathscr{C}_C(g^{(\alpha)}, \underline{\tau}^{(\alpha+1)}) = \Sigma^*(C, \underline{\tau}^{(\alpha+1)})$$

one associates with γ a sequence of chambers

$$C = C_{r+1} = C_{r+1}(\gamma), \qquad C_i(\gamma) = \mathrm{proj}_{F_{t_i(g)}(\gamma)} C_{i+1}(\gamma).$$

Proposition 10.42 *Let* $\gamma^{(\alpha)} \in \Sigma(\underline{\tau}^{(\alpha+1)}, C) \subset \text{Gall}_I(g^{(\alpha)}, C)$.
The fiber $\left(\pi_{\underline{\tau}^{(\alpha)},C}^{(\alpha)}\right)^{-1}(\gamma^{(\alpha)})$ *of* $\pi_{\underline{\tau}^{(\alpha)},C}^{(\alpha)}$ *over* $\gamma^{(\alpha)}$ *is given by:*

$$\left(\pi_{\underline{\tau}^{(\alpha)},C}^{(\alpha)}\right)^{-1}(\gamma^{(\alpha)}) = \Sigma(\tau_\alpha, C_\alpha(\gamma^{(\alpha)})).$$

$$
\begin{array}{ccc}
\Sigma^*(\underline{\tau}^{(\alpha)}, C) & \longrightarrow & \text{Gall}_I(g^{(\alpha-1)}, C) \\
\downarrow & & \downarrow \\
\Sigma^*(\underline{\tau}^{(\alpha+1)}, C) & \longrightarrow & \text{Gall}_I(g^{(\alpha)}, C)
\end{array}
$$

where the horizontal arrows are inclusions, and the vertical left arrows (resp. right arrow) is given by

$$\pi_{\underline{\tau}^{(\alpha)},C}^{(\alpha)} \qquad \left(\text{resp. } \pi_{g^{(\alpha-1)},C}^{(\alpha)}\right).$$

One has the inclusion of fibers over $\gamma^{(\alpha)}$:

$$\Sigma(\tau_\alpha, C_\alpha(\gamma^{(\alpha)})) \subset \Sigma(t_\alpha, s_{\alpha-1}; F_{t_\alpha}(F_{s_\alpha}(\gamma)))$$

(cf. Proposition 10.38).

Chapter 11

Parabolic Subgroups in a Reductive Group Scheme

One summarizes the essential definitions and results of reductive S-groups schemes which are useful for our purpose of defining Schubert Schemes and their associated canonical and functorial Smooth Resolutions and for extending to this setting the buildings constructions. In fact, it is remarked, that both Grothendieck reductive group schemes and Tits Buildings are both inspired by the fundamental Chevalley's Tohoku paper [10]. The principal objects associated with a building become in the schematic context twisted locally constant finite S-schemes. They are defined by etale descent.

One recalls the definition of a **Reductive S-Group scheme** G, of a **Splitting** (resp. Frame) of such a group, and of a Parabolic subgroup of G. According to [23], the **functor of Parabolics subgroups of** G, the **Types of Parabolics quotient scheme** of this functor by the adjoint action of G, i.e. the Dynkin scheme, the functor of **Couples Parabolics in Standard Position**, the **Types of Relative Positions quotient scheme** of this functor by the diagonal adjoint action of G, are introduced, along with **the class of (R)-subgroups of** G, which are of use to generalize buildings constructions to the relative setting. The representability of the (R)-**subgroups functor** implies the representability of functors of more restricted classes of subgroups allowing to introduce the analogues of Convex Hull subcomplexes of a Building in the relative frame. It implies in particular the representability of the Parabolics subgroups functor, the one of the Cartan subgroups giving the relative apartments $\cdots$ etc. Finally the construction of the **Weyl Complex scheme** of G is given. The reader is referred to *Chapters $XXII - XXVI$* of *SGA III* [23] for details.

11.1   Reductive group schemes

Let be stated first some important definitions.[1] In this number S denotes a base scheme. Given a point $s \in S$ and an S-scheme X, $X_{\bar{s}}$ denotes the geometric fiber of X on s, i.e.

$$X_{\bar{s}} = X \times_S \operatorname{Spec}(\bar{\kappa}(s)),$$

where $\bar{\kappa}(s)$ is the algebraic closure of the residual field

$$\kappa(s) = \mathcal{O}_{S,s}/m_{S,s}.$$

Definition 11.1   *1) By a **reductive** (resp.   **semi-simple**) group S-scheme G one understands an affine smooth S-group scheme G, such that $G_{\bar{s}}$ is a connected and reductive (resp. semi-simple) $\bar{\kappa}(s)$-group, for every $s \in S$.*

*2) A **maximal torus** T of G, is a torus $T \subset G$ so that $T_{\bar{s}} \subset G_{\bar{s}}$ is a maximal torus for every $s \in S$. A torus $T \subset G$ is **trivial** if there exists a finitely generated free $\mathbb{Z}$-module M satisfying:*

$$T \simeq \underline{Hom}_{S-gr}(M_S, \mathbb{G}_{mS}),$$

where M_S denotes the trivial sheaf defined by M.

Definition 11.2 *(cf. loc. cit., Exp. XIX, Definition 3.2.)*
Let G be a reductive S-group scheme, and $T \subset G$ an S-torus. Write $\mathcal{G} = Lie(G)$.
   *A character $\alpha \in Hom_{S-gr}(T, \mathbb{G}_{mS})$ is a **root** of G if for each $s \in S$, one has that the induced character $\alpha_{\bar{s}} \in Hom(T_{\bar{s}}, \mathbb{G}_{m\bar{s}})$, is a root of $G_{\bar{s}}$, i.e. if one considers the decomposition*

$$\mathcal{G}_{\bar{s}} = \mathcal{G}^0 \oplus \left(\underset{\alpha \in R}{\oplus} \mathcal{G}^\alpha \right) (R \subset Hom(T_{\bar{s}}, \mathbb{G}_{m\bar{s}})),$$

of $\mathcal{G}_{\underline{s}}$ under the action of $T_{\bar{s}}$, one has

$$\alpha_{\bar{s}} \in R.$$

   Let $W(\mathcal{G})$ be the S-**vector group defined by** $\mathcal{G} = Lie(G)$, i.e. the group S-functor defined by

$$S' \to \mathcal{O}_{S'} \otimes_{\mathcal{O}_S} \mathcal{G}.$$

   Define a subfunctor of $W(\mathcal{G})$ by

$$W(\mathcal{G})^\alpha(S') = \{x \in W(\mathcal{G})(S') \ / \ \mathrm{ad}(t)x = \alpha(t)x \text{ for every } t \in T(S''), \ S'' \to S'\}.$$

---

[1]For more details see loc. cit., Exp. XIX

There exists a rank 1 direct factor $\mathcal{O}_S$-submodule $\mathcal{G}^\alpha \subset \mathcal{G}$ so that

$$W(\mathcal{G})^\alpha = W(\mathcal{G}^\alpha) \text{ (cf. loc. cit., Exp. XIX, §4).}$$

If α is a root of G then $-\alpha$ is also a root of G.

Definition 11.3 *Let G be a reductive S-group scheme and T a maximal torus, and R a set of roots relatively to T. It is said that R is a root system of G relatively to T if*

$$\mathcal{G} = \mathcal{G}^0 \oplus \left(\bigoplus_{\alpha \in R} \mathcal{G}^\alpha \right) ,$$

Where $\mathcal{G}^0 = Lie(T)$. One also writes $Lie(T) = \mathcal{T}$.

Definition 11.4 *Let*

$$\mathscr{R} : (S' \longrightarrow S) \mapsto \text{ set of roots of } G_{S'} \text{ with respect to } T_{S'} ,$$

be **the root functor.** *The natural inclusion morphism $R_S \hookrightarrow Hom_{S-gr}(T, \mathbb{G}_m)$ gives rise to an isomorphism $R_S \simeq \mathscr{R}$.*

11.2 $\mathbb{Z}$-root systems

Definition 11.5 *(cf. loc. cit., Exp. XXI, Definition 1.1.1.)*
A **$\mathbb{Z}$-root system**
$$\mathcal{R} = (M, M^*, R, R^\vee)$$

is the data given by:

1. *A finitely generated $\mathbb{Z}$-module M, and its dual M^*. Denote by (x, y) the duality pairing*

$$M \times M^* \to \mathbb{Z}.$$

2. *A finite subset*

$$R \subset M \ (resp. \ R^\vee \subset M^*)$$

and a map

$$R \to R^\vee$$

given by

$$\alpha \mapsto \alpha^\vee \ (\alpha \in R).$$

Write for $\alpha \in R$, $x \in M$ (resp. $\alpha^\vee \in R$, $y \in M^*$), $s_\alpha(x) = x - (x, \alpha^\vee) \alpha$ (resp. $s_{\alpha^\vee}(y) = y - (\alpha, y) \alpha^\vee$).

This mapping satisfies:

$$(\alpha^\vee, \alpha) = 2 \ (resp. \ s_\alpha(R) \subset R, \ s_{\alpha^\vee}(R^\vee) \subset R^\vee).$$

Thus

$$s_\alpha : M \to M \text{ (resp. } s_{\alpha^\vee} : M^* \to M^*)$$

is a reflexion invariating R (resp. $R^\vee$).

Let

$$V = \mathbb{R} \otimes_{\mathbb{Z}} M \text{ (resp. } V^* = \mathbb{R} \otimes_{\mathbb{Z}} M^*).$$

Denote also by $R \subset V$ (resp. $R^\vee \subset V^*$) the image of R (resp. $R^\vee$) by $M \to \mathbb{R} \otimes_{\mathbb{Z}} M$ (resp. $M^* \to \mathbb{R} \otimes_{\mathbb{Z}} M$).

One has then

$$(V, V^*, R, R^\vee)$$

is a root system as defined in [4], Ch. VI, §1, Def. 1 (the **Root System** defined by $\mathcal{R}$).

11.3 $\mathbb{Z}$-root system defined by a splitting of a reductive group scheme

Let $T = \underline{\mathrm{Hom}}_{S-gr}(M_S, \mathbb{G}_{mS})$ be a maximal and trivial torus T of the S-reductive group scheme G. Write $\mathcal{G} = \mathrm{Lie}(G)$ (the Lie algebra of G). Let

$$\mathcal{G} = \mathcal{G}^0 \oplus (\oplus \mathcal{G}^\alpha (\alpha \in R))$$

be the **root decomposition** of $\mathcal{G}$ under the action of T (cf. loc. cit., Exp. XIX, Definition 3.6).

11.3.1 Co-roots defined by a splitting of a reductive group scheme

PROPOSITION - DEFINITION 11.6   *1) There exists a $\mathcal{O}_S$-module morphism:*

$$\mathcal{G}^\alpha \otimes_{\mathcal{O}_S} \mathcal{G}^{-\alpha} \longrightarrow \mathcal{O}_S,$$

*given by a **duality pairing**: $(X, Y) \longrightarrow < X, Y >$ identifying $\mathcal{G}^\alpha$ to $\mathcal{G}^{-\alpha}$.*

2) For all $\alpha \in R$ there exists a unique T-equivariant group morphism:

$$exp_\alpha : W(\mathcal{G}^\alpha) \longrightarrow G,$$

where T acts on $\mathcal{G}^\alpha$ by the adjoint action and on G by conjugation, inducing the natural inclusion Lie algebra morphism $\mathcal{G}^\alpha = \mathrm{Lie}(W(\mathcal{G}^\alpha)) \hookrightarrow \mathcal{G}$.

3) There is a unique group morphism $\alpha^\vee : \mathbb{G}_{mS} \longrightarrow T$ *satisfying for all* $(X, Y) \in \mathcal{G}^\alpha \times \mathcal{G}^{-\alpha}$, *so that* $1+ < X, Y >$ *is an invertible section of* $\mathcal{O}_S$, *the following formula:*

$$exp_\alpha(X) \cdot exp_\alpha(Y) = exp_{-\alpha}\left(\frac{Y}{1+ < X, Y >}\right) \cdot \alpha^\vee\left(1+ < X, Y >\right)$$

$$\cdot exp_\alpha\left(\frac{X}{1+ < X, Y >}\right).$$

One calls $\alpha^\vee$ *the* **co-root** *associated with* α, *and one writes* $R^\vee = \{\alpha^\vee | \alpha \in R\}$ *(cf. Definition 1.5. of loc. cit., Exp. XXII, Proposition 1.10).*

Remark 11.7 *Let* $M = \underline{Hom}_{S-gr}(T, \mathbb{G}_{mS})$ *(resp.* $M^* = \underline{Hom}_{S-gr}(\mathbb{G}_{mS}, T)$*) Remark that*

$$\underline{Hom}_{S-gr}(T, \mathbb{G}_{mS}) \simeq M_S \ (resp. \underline{Hom}_{S-gr}(\mathbb{G}_{mS}, T) \simeq M_S^*).$$

Thus each constant function $\alpha \in R$ *(resp.* $\alpha^\vee \in R^*$*) is given by some element of the finitely generated* $\mathbb{Z}$-*module* M *(resp.* M^**)*

11.3.2   Conjugation automorphisms of a maximal torus defined by the roots

Definition 11.8 *Denote by* s_α *the automorphism of* T *defined by:*

$$s_\alpha(t) = t \cdot \alpha^\vee(\alpha(t))^{-1} .$$

This automorphism acts on $M_S = \underline{Hom}_{S-gr}(T, \mathbb{G}_{mS})$ *by Cartier duality by:*

$$s_\alpha(m) = m - (\alpha^\vee, m)\alpha ,$$

and finally on $M_S^* = \underline{Hom}_{S-gr}(\mathbb{G}_{mS}, T)$ *by:*

$$s_\alpha(u) = u - (\alpha, u)\alpha^\vee .$$

Where m *(resp.* u*) denotes a section of* $\underline{Hom}_{S-gr}(T, \mathbb{G}_{mS})$ *(resp.* $\underline{Hom}_{S-gr}(\mathbb{G}_{mS}, T)$*).*

11.3.3   Splittings and frames of a reductive group scheme

Assume that:

1. The roots (resp. the corresponding co-roots) $\alpha \in \text{Hom}_{S-gr}(T, \mathbb{G}_{mS})$ (resp. $\alpha^\vee \in \text{Hom}_{S-gr}(\mathbb{G}_{mS}, T)$) are given by constant functions of S with values in M (resp. M^*).

2. Each $\mathcal{G}^\alpha \subset \mathcal{G}$ is a free $\mathcal{O}_S$-module.

Definition 11.9 *(cf. loc. cit., Exp. XXII, Definition 1.13.)*
*It is said that the above data (G, T, M, R) defines a **split** (resp. **deployed**)*
S-reductive group G, or that the data $T = \underline{Hom}_{S\text{-}gr}(M_S, \mathbb{G}_{mS})$, $R \subset M$ (M
*being a finitely generated $\mathbb{Z}$-module) is a **splitting** of the S-reductive group*
G.

It results from the general theory of S-reductive groups G that

$$\mathcal{R} = (M, M^*, R, R^\vee)$$

is a $\mathbb{Z}$-root data. It is said that $\mathcal{R}$ is the $\mathbb{Z}$-root data defined by the split
S-reductive group (G, T, M, R). Write

$$\mathcal{R} = \mathcal{R}(G).$$

Definition 11.10 *(cf. loc. cit., Exp. XXIII, Definition 1.1.)*
*A **frame** E of a splitted S-reductive group (G, T, M, R) is given by the follow-*
ing additional data:

1. *A simple root system $R_0 \subset R$.*

2. *The choice for each $\alpha \in R_0$ of a basis (X_α) of $\mathcal{G}^\alpha$.*

Remark 11.11
One renders by "frame" the french word "épinglage".

There is the

Proposition 11.12 *(cf. loc. cit., Exp. XXII, Proposition 2.1.)*
Let $T \subset G$ be a trivial maximal torus of the S-reductive group G. Then
for each $s \in S$ there exists a Zariski neighborhood U_s of s and a splitting
(G_{U_s}, T_{U_s}, M, R) of G, i.e. a splitting of G_{U_s} with maximal torus $T_{U_s} =$
restriction of T to $U_s \subset S$.

The following corollary explains how to proceed for obtaining a splitting
of the S-reductive group G.

Corollary 11.13
Let it be supposed the S-reductive group G is endowed with a maximal torus
$T \subset G$. Let
$$(S_i \to S)_{i \in I}$$
be an etale covering of S trivializing T, i.e. so that the maximal torus T_{S_i} is
trivial. Then after composing this covering with open coverings $(S_{ij} \to S_i)_{j \in L_i}$
of the S_i one obtains an etale covering $(S_{ij} \to S)_{(i,j) \in K}$ of S such that $G_{S_{ij}}$
is split (resp. admits a frame).

Corollary 11.14 *The root scheme $\mathscr{R}$ associated with the maximal torus T is representable by a locally finite etale S-scheme.*

The following proposition resumes the main interest of frames.

Proposition 11.15 *(cf. loc. cit., Exp. XXIV, Lemme 1.5.)*
Let E_1 and E_2 be two frames of the S-reductive group G. Then there exists a unique inner automorphism
$$\alpha : G \to G$$
of G such that $\alpha(E_1) = E_2$. Write $\alpha = \alpha(E_1, E_2)$.

Definition 11.16 *(cf. loc. cit., Exp. XXII, 2.6.)*
Let $\mathcal{R}$ be a $\mathbb{Z}$-root system. It is said that the S-reductive group G is of **type** $\mathcal{R}$ *if for each $s \in S$ there exists an etale neighborhood $\tilde{U}_s$ such that there exists a splitting $\left(G_{\tilde{U}_s}, T, M, R\right)$ with the corresponding Root Data satisfying*
$$\mathcal{R}\left(G_{\tilde{U}_s}, T, M, R\right) = \mathcal{R}.$$

Definition 11.17 *Let*
$$E = \left(G, T, M, R, R_0, (X_\alpha)_{\alpha \in R_0}\right)$$
be a frame of G. It is associated with E a $\mathbb{Z}$-root system $\mathcal{R}(E)$ endowed with a frame of root system, i.e. a simple system of roots R_0, namely
$$\mathcal{R}(E) = (M, M^*, R, R^\vee, R_0)$$
(cf. Exp. XXIII, §1).

Observe that the main goal of [23], is the proof of the following

Theorem 11.18 *(cf. loc. cit., Exp. XXV, Théorème 1.1.)*
The functor
$$(G, E) \to \mathcal{R}(E)$$
from the category of S-reductive groups G endowed with a frame E to that of $\mathbb{Z}$-root data endowed with a simple system of roots is a category equivalence.

Definition 11.19 *(cf. [23], Exp. XXIII, p. 317)*
Given a $\mathbb{Z}$-root system $\mathcal{R}$ there exists a reductive group $\mathbb{Z}$-scheme $\underline{Ep}_{\mathbb{Z}}(\mathcal{R})$ endowed with a canonical frame $E_{\mathcal{R}}$, of type $\mathcal{R}$, called the **Chevalley scheme of type** $\mathcal{R}$.

It will be seen that the Smooth Resolutions of the Schubert Schemes of $\mathbb{Z}$-scheme $\underline{Ep}_{\mathbb{Z}}(\mathcal{R})$ have a Universal property. On the other hand, observe that the data of a frame E of type $\mathcal{R}$ of a reductive group S-scheme G amounts to that of an isomorphism

$$G \simeq \underline{Ep}_{\mathbb{Z}}(\mathcal{R}) \times S = \underline{Ep}_S(\mathcal{R}).$$

More precisely, with the notation of [23], Exp. XXIV, the $\underline{Aut}_{S-gr}(\underline{Ep}_S(\mathcal{R}))$-principal fiber space of isomorphisms

$$\underline{Isom}_{S-gr}(\underline{Ep}_S(\mathcal{R}), G),$$

may be seen as the scheme of frames of G of type $\mathcal{R}$. (cf. [23], Exp. XXIV, Remarque 1.20).

This result suggests that functorial constructions concerning objects naturally associated with G may be described in terms of the combinatorial data $\mathcal{R}(E)$.

11.4   Parabolic subgroups

Definition 11.20 *A* **Parabolic Subgroup**

$$P \subset G$$

is a smooth S-subgroup scheme of G, so that for every $s \in S$, $P_{\bar{s}}$ is a parabolic subgroup of $G_{\bar{s}}$, i.e. $G_{\bar{s}}/P_{\bar{s}}$ is a proper $\overline{\kappa}(s)$-scheme, or which amounts to the same, $P_{\bar{s}}$ contains a Borel subgroup of $G_{\bar{s}}$. A **Borel subgroup** *of G is a Minimal Parabolic Subgroup.*

The **Parabolic Subgroups Functor** $\mathcal{P}ar(G)$ (resp. **Borel Subgroups** S-**sub-functor** $\mathcal{B}or(G)$), is obtained naturally from the above definition (cf. 11.42) and is representable by a smooth and projective S-scheme with integral geometric fibers as will be seen (cf. also loc.cit., *Exp. XXVI*, §3.2).

The main property of a Parabolic Subgroup P is given by:
"P is a closed subgroup, with connected fibers, it is equal to its own normalizer, i.e. $Norm_G(P) = P$, and the quotient sheaf G/P is representable by a projective smooth S-scheme"
(cf. loc. cit., *Exp. XXVI*, *Proposition 1.2*).

Definition 11.21 *According to [23], Exp. XXVI, Definition 1.11, it is said that $E = (T, M, R, R_0, (X_\alpha)_{\alpha \in R_0})$ is a* **frame of G adapted to the parabolic subgroup** *P if the Lie algebra $Lie(P)$ of P may be written as:*

$$Lie(P) = \mathcal{G}^0 \oplus \left(\underset{\alpha \in R_P}{\oplus} \mathcal{G}^\alpha \right)$$

where $R_P \subset R$ is a **parabolic subset** *containing R_0.*

11.5   Standard position couples of Parabolics scheme

Given two facets F and F' of a building I, defined by a reductive group G over an algebraically closed field k, there exists an apartment $\mathcal{A} \subset I$ which contains both. This means that there exists a maximal torus $T \subset P_F \cap P_{F'}$ such that $\mathcal{A} = \mathcal{A}_T$. Let G be an reductive S-group scheme. Given two parabolic subgroups P and Q, it is not always true that there exists a maximal torus T locally contained in both P and Q. This motivates the following developments.

Let (P, Q) be a couple of parabolics of G. The following conditions are equivalent:

1. $P \cap Q$ is smooth.

2. $P \cap Q$ locally contains a maximal torus of G for the fpqc-topology.

3. $P \cap Q$ locally contains a maximal torus of G for the Zariski-topology.

(cf. loc. cit., Exp. XXVI, 4.5.1.)

Definition 11.22 *(cf. loc. cit., 4.5.1.)*
If (P, Q) verifies one of the three equivalent conditions above, it is said that P and Q are in **standard position**, *or that the couple of parabolics (P, Q) is in standard position.*

Let
$$\mathcal{S}tand(G) \subset \mathcal{P}ar(G) \times_S \mathcal{P}ar(G)$$
be the representable sub-functor whose sections are the couples (P, Q) of parabolics in standard position (cf. loc. cit., 4.5.3.).

Remark 11.23 *It will be seen that, in the relative case, $\mathcal{S}tand(G)$ plays the same role as $I \times I$ in the case of a building I of a k-reductive group G.*

PROPOSITION - DEFINITION 11.24 *(cf. loc. cit., Exp. XXVI, 4.5.3.)*
Given two couples (P, Q) and (P', Q') of parabolics in standard position the following assertions are equivalent:

1. *(P, Q) and (P', Q') are locally conjugate for the fpqc-topology, i.e. there exists a locally defined section x of G with $int(x)(P) = xPx^{-1} = P'$ (resp. $int(x)(Q) = xQx^{-1} = Q'$).*

2. *(P, Q) and (P', Q') are locally conjugate for the etale topology.*

3. *For every $s \in S$, $(P_{\bar{s}}, Q_{\bar{s}})$ and $(P'_{\bar{s}}, Q'_{\bar{s}})$ are conjugate, i.e. there exists $x \in G_{\bar{s}}((\overline{\kappa}(s))$ with*

$$(int(x)\,(P_{\bar{s}}),\ int(x)\,(Q_{\bar{s}})) = (P'_{\bar{s}}, Q'_{\bar{s}}).$$

It is said that (P, Q) and (P', Q') define the same **type of relative position**.

11.5.1 Relative position types quotient scheme

The relative position types index naturally the Bruhat (resp. Schubert) Cells. There is a natural action of G on $\mathcal{S}\mathrm{tand}(G)$:

$$G \times \mathcal{S}\mathrm{tand}(G) \to \mathcal{S}\mathrm{tand}(G)$$

defined by

$$(x, (P, Q)) \to (\mathrm{int}(x)(P),\ \mathrm{int}(x)(Q)).$$

The quotient $\mathcal{S}\mathrm{tand}(G)/G$ of $\mathcal{S}\mathrm{tand}(G)$ under the action of G is representable by a twisted locally constant finite S-scheme (denoted by $\underline{T.st}$ in loc. cit., *Exp. XXVI*, §4.5.3). Denote by

$$t_2 : \mathcal{S}\mathrm{tand}(G) \to \mathrm{Relpos}\,_G$$

the quotient morphism.

Proposition 11.25

1. *The functor $\mathcal{S}\mathrm{tand}(G)$ is representable*

2. *The morphism $t_2 : \mathcal{S}\mathrm{tand}(G) \to \mathrm{Relpos}\,_G$ is S-smooth, of finite presentation, with irreducible geometrical fibers.*

There is a natural morphism

$$t_2 \times pr_1' : \mathcal{S}\mathrm{tand}(G) \to \mathrm{Relpos}\,_G \times_S \mathcal{P}\mathrm{ar}(G),$$

where pr_1' denotes the restriction to $\mathcal{S}\mathrm{tand}(G) \subset \mathcal{P}\mathrm{ar}(G) \times_S \mathcal{P}\mathrm{ar}(G)$ of the first projection

$$pr_1 : \mathcal{P}\mathrm{ar}(G) \times_S \mathcal{P}\mathrm{ar}(G) \to \mathcal{P}\mathrm{ar}(G).$$

Let (τ, P) be a section of $\mathrm{Relpos}\,_G \times_S \mathcal{P}\mathrm{ar}(G)$, and write:

$$\mathcal{S}\mathrm{tand}(\tau, P) = (t_2 \times pr_1')^{-1} ((\tau, P))$$

$$\left(\text{resp. } \mathcal{S}\mathrm{tand}(\tau) = t_2^{-1} (\tau),\ \mathcal{S}\mathrm{tand}(P) = (pr_1')^{-1} (P) \right).$$

Remark that it depends on a compatibility condition between the type of relative position τ and the type t of the parabolic P, as it will be later defined, for $\mathcal{S}\mathrm{tand}(\tau, P)$ to be empty or not.

Let

$$\mathcal{S}\mathrm{tand}(G)' = \mathcal{S}\mathrm{tand}(G) \cap (\mathcal{B}\mathrm{or}(G) \times \mathcal{P}\mathrm{ar}(G)).$$

Definition 11.26

Define $\mathrm{Relpos}'_G \subset \mathrm{Relpos}_G$ as the quotient S-subscheme

$$\mathrm{Relpos}'_G = \mathcal{S}\mathrm{tand}(G)'/G.$$

Let $B \subset G$ be a Borel subgroup of G, thus defining a section of $\mathcal{B}or(G)$. The quotient of

$$\mathcal{S}\text{tand}(B) \subset \mathcal{S}\text{tand}(G)',$$

by B is canonically identified with $\text{Relpos}(G)'$:

$$\mathcal{S}\text{tand}(B)/B \simeq \text{Relpos}'_G,$$

where the isomorphism is induced by the above inclusion.

Thus there is a morphism

$$\rho_B : \mathcal{S}\text{tand}(B) \to \mathcal{S}\text{tand}(B)/B \simeq \text{Relpos}'_G,$$

induced by the quotient morphism.

For the sake of briefness one calls, in what follows, a reductive group S-scheme G simply an S-reductive group G.

11.6   Dinkyn scheme and the typical simplex scheme

Each one of the following basic sets associated to a building I:

1) the typical simplex typ I;

2) the set gall $_I$ (resp. gall $_I^m$) of generalized galleries (resp. Minimal Generalized Galleries) of the typical simplex typ I;

3) the set of relative positions Relpos I,

have a corresponding object in the setting of S-reductive groups G. This objects are twisted locally constant S-schemes. One proceeds to define them by descent from the case G split. Let a description of typ I in terms of a frame E of G be given. Denote by

$$I^m \subset I$$

the set of minimal facets of I, i.e. the set of vertices of I, and it is written

$$\text{typ } I^m = I^m/G$$

for the image of I^m by typ $: I^m \to I/G$. One has then a canonical bijection

$$\text{typ } I \,\tilde{\to}\, \mathcal{P}\,(\text{typ } I^m)\,.$$

Every facet F of I may be written as

$$F = \bigcup_{F' \in I^m,\, F' \subset F} F' \,.$$

Thus with the class of F in typ $I = I/G$, is associated the set of classes of the $F' \in \text{typ } I^m$ with $F' \subset F$ which determines F, i.e. F is the upper bound of its vertices (cf. [50]).

The natural building morphism $\mathcal{A}_E \hookrightarrow I$ induces the bijections typ $\mathcal{A}_E \simeq$ typ I (resp. typ $\mathcal{A}_E^m \simeq$ typ I^m), where $\mathcal{A}_E$ denotes the building of parabolic subsets of the root system R_E given by the frame E, and typ $\mathcal{A}_E^m$ denotes the types of the set of maximal parabolics subsets. Given $\alpha \in R_0$, write

$$R^{(\alpha)} = \mathbb{Z}(R_0 - \{\alpha\}) \cap R \cup R_+ \quad \left(\text{resp. } F^{(\alpha)} = F_{R^{(\alpha)}}, \ P^{(\alpha)} = P_{F^{(\alpha)}} \right).$$

$R^{(\alpha)}$ is a maximal parabolic set of R. There is a natural bijection $R_0 \simeq$ typ $\mathcal{A}_E^m$ defined by $\alpha \mapsto R^{(\alpha)}$. One obtains a canonical bijection

$$R_0 = \mathrm{Dyn}(E) \simeq \mathrm{typ}\, I^m,$$

(cf. loc. cit., Exp. XXIV, 3.) defined by $\alpha \mapsto F^{(\alpha)}$.
Denote by $\Delta(E)$ the set $\mathcal{P}(\mathrm{Dyn}(E))$. There is an order preserving bijection

$$\Delta(E) \to \mathrm{typ}\, I,$$

defined by

$$R' \mapsto \mathrm{typ} \left(\bigcup_{\alpha \in R'} F_{R^{(\alpha)}} \right).$$

Clearly, $\bigcup F_{R^{(\alpha)}}$ $(\alpha \in R')$ gives the facet defined by the parabolic set $\bigcap R^{(\alpha)}$ $(\alpha \in R')$.

Identify the simplex $\Delta(E)$ to the subcomplex of I (resp. $\mathcal{A}_E$), given by the set of parabolics P (resp. facets F) so that

$$B_{R_E} \subset P \quad (\text{resp. } C_{R_E} \supset F).$$

On the other hand, under the canonical isomorphism

$$C(W_{\mathcal{A}_E}, S_E) \simeq \mathcal{A}_E$$

the simplex $\Delta(E)$ is identified with the class of subgroups $(W_t)_{t \in \Delta(E)}$ of $W = W_{\mathcal{A}_E}$, where one writes

$$W_t = \mathrm{Stab}_W F_{\underset{\alpha \in S-t}{\bigcap} R^{(\alpha)}},$$

i.e. $t \in \Delta(E)$ may be identified with the set of canonical generators $X_t \subset S$ of the Weyl group of $P_{\underset{\alpha \in S-t}{\bigcap} R^{(\alpha)}}$.

Given two frames E and E' of G, it is denoted by

$$\alpha(E, E')_{\mathrm{Dyn}} : \mathrm{Dyn}(E) \to \mathrm{Dyn}(E'),$$

the Dynkin diagram isomorphism induced by $\alpha(E, E')$. One obtains an inductive (resp. transitive) system of isomorphisms

$$(\mathrm{Dyn}(E), \ \alpha(E, E')_{\mathrm{Dyn}}).$$

It is easy to see that there is a natural isomorphism

$$\varinjlim_{E} \Delta(E) \xrightarrow{\sim} \operatorname{typ} I = I/G,$$

where the inductive limit is given by the inductive system of isomorphisms $(\Delta(E), \alpha(E, E')_\Delta)$ associated with the above system. Let

$$\operatorname{Gall}_{\Delta(E)} \quad \left(\text{resp. } \operatorname{Gall}_{\Delta(E)}^{m}\right)$$

be the set of MGG of $\Delta(E)$ (resp. the set of types of MGG of I, which is defined taking on account the isomorphism $\Delta(E) \xrightarrow{\sim} \operatorname{typ} I$). The above isomorphism induces the bijection

$$\varinjlim_{E} \operatorname{Gall}_{\Delta(E)} \;\rightarrow\; \operatorname{Gall}_{\operatorname{typ} I} = \operatorname{gall}_{I}$$

$$\left(\text{resp. } \quad \varinjlim_{E} \operatorname{Gall}_{\Delta(E)}^{m} \;\rightarrow\; \operatorname{Gall}_{\operatorname{typ} I}^{m} = \operatorname{gall}_{I}^{m}\right).$$

The above preliminaries motivate the following definitions. Let G be an S-reductive group.

Definition 11.27 (The Dynkin S-scheme $\mathcal{D}yn(G)$ of G) *(cf. loc. cit., Exp. XXIV, 3.)*
First suppose G split. In this case the set of frames G is not empty. Given a frame E of G write
$$\mathcal{R}(E) = (M, M^*, R, R^\vee, R_0),$$

for the $\mathbb{Z}$-root system, endowed with a system of simple roots R_0, defined by E. Let $Dyn(E)$ be the Dynkin diagram defined by $\mathcal{R}(E)$ (cf. loc. cit., Exp. XXI, Definition 7.4.2).
Write
$$Dyn(G) = \varinjlim_{E} Dyn(E)_S,$$

where $Dyn(E)_S$ denotes the constant S-scheme defined by $Dyn(E)$. The inductive limit is defined by the inductive system $\left(Dyn(E)_S, (\alpha(E, E')_{Dyn})_S\right)$. In the general case it is considered an etale covering $(S_i \to S)$ so that G_{S_i} is split. Denote by (c_{ij}) the corresponding cocycle (resp. descent data) defining G.
The set $(\mathcal{D}yn(G_{S_i}))$ is endowed with a descent data $((c_{ij})_{Dyn})$ induced by (c_{ij}).
Define
$$\mathcal{D}yn(G)$$

by descent from this data.
This definition is independent of the etale covering $(S_i \to S)$.

The Dynkin scheme as defined in *SGA III* is a more complete data than ours. It consists of a Dynkin diagram structure defined by a morphism $Dyn(G) \longrightarrow \{1, 2, 3\}_S$. Recall that the Dynkin diagram structure defined by a system of simple roots S amounts to that of the Cartan matrix of S. Thus a system of roots R may be obtained from the Dynkin diagram. The root scheme R is defined by etale descent by means of the following proposition.

Proposition 11.28 *Given adjoint root data R and R', a simple root system S of R and a simple root system S' of R', and a bijection $u : S \longrightarrow S'$ transforming the Cartan matrix of S in that of S'. Then there exists a unique isomorphism $R \simeq R'$ induced by u.*

Definition 11.29 (The relative typical simplex S-scheme $\underline{\Delta}(G)$) *(cf. loc. cit., Exp. XXVI, 3.1.)*
Given a frame E of G write

$$\Delta(E) = \Delta(Dyn(E)) = \mathcal{P}(Dyn(E))$$

(the combinatorial simplex given by the vertices of $Dyn(E)$). Define $\underline{\Delta}(G)$ following the same pattern as in the definition of $Dyn(G)$.
If G is split write

$$\underline{\Delta}(G) = \varinjlim_{E} \Delta(E)_S,$$

where E runs on the (non-empty) set of frames of G and the transition isomorphisms are induced by $(\alpha(E, E'))$.
In the general case one defines $\underline{\Delta}(G)$ by descent from an etale covering $(S_i \to S)$ such that G_{S_i} is split as was done for $Dyn(G)$.

By loc. cit., the sections of $\underline{\Delta}(G)$ over an S-scheme S' may be characterized as follows:

$$\underline{\Delta}(G)(S) = \text{the set of open and closed subsets of } Dyn(G)_{S'},$$

From this characterization it follows that the locally trivial S-scheme $\underline{\Delta}(G)$ is endowed with a functorial inclusion relation "$\subset$", naturally allowing the definition of generalized galleries of $\underline{\Delta}(G)$.

Remark 11.30 *According to loc. cit., with a constant "twisted" finite S-scheme X is associated the S-functor $\mathcal{P}(X)$ defined by*

$$\mathcal{P}(X)(S') = \text{the set of open and closed subsets of } X_{S'},$$

which is representable by a constant "twisted" finite S-scheme. More explicitly if $X = A_S$ then $\mathcal{P}(X) = (\mathcal{P}(A))_S$. The assertion is obtained in general by descent of open and closed subschemes. This remark clearly applies to $Dyn(G)$ and allows defining $\underline{\Delta}(G)$ otherwise.

11.7 Parabolic type morphism

According to [23], Exp. XXVI, Definition 1.11., in order to define the
type section $t_1(P)$ for a parabolic $P \subset G$, one may proceed locally for
the etale topology, and may suppose that G is endowed with a frame
$E = (T, M, R, R_0, (X_\alpha)_{\alpha \in R_0})$ adapted to P.

It is recalled that this means that $P \supset T$, and that the Lie algebra $\mathrm{Lie}(P)$
of P may be written as

$$\mathrm{Lie}(P) = \mathcal{G}^0 \oplus (\oplus \mathcal{G}^\alpha) \ (\alpha \in R_P)$$

where $R_P \subset R$ is a parabolic subset containing R_0. In other terms there exists
$F \in \Delta(C_E)$ such that $R_P = R_F$.

There is a canonical isomorphism

$$(R_0)_S \simeq \mathcal{D}\mathrm{yn}(G) \ (R_0 = \mathrm{Dyn}(E)),$$

and the image $t_1(P)$ of P is given by the section $(R_0 - R_0 \cap (-R_P))_S$ of
$\mathcal{P}(R_0)_S$ defined by $R_0 - R_0 \cap (-R_P) \subset R_0$. Remark that in loc. cit., $t_1(P)$
is defined by $(R_0 \cap (-R_P))_S$. The definition of t_1 given here is coherent with
the view point of buildings.

The morphism t_1 allows the identification of the quotient S-scheme
$\mathcal{P}\mathrm{ar}(G)/G$ with $\mathcal{P}(\mathcal{D}\mathrm{yn}(G))$. One may then write

$$t_1 : \mathcal{P}\mathrm{ar}(G) \to \mathcal{P}\mathrm{ar}(G)/G \simeq \mathcal{P}(\mathcal{D}\mathrm{yn}(G)) = \underline{\Delta}(G).$$

11.8 (R)-subgroups

Definition 11.31 *A S-sub-group scheme $H \subset G$ is of type (R) if:*

1) *H is a smooth S-scheme of finite presentation, with connected fibers, i.e.
$H \times_S \kappa(s)$ is connected for all $s \in S$.*

2) *H contains a maximal torus locally for the etale topology of S.*

Notation 11.32 *Let $\mathcal{H} \subset \mathcal{G}$ be a Lie subalgebra of $\mathrm{Lie}(G) = \mathcal{G}$. Denote by
$\mathrm{Norm}_G(\mathcal{H}) \subset G$ the S-subgroup functor whose sections g on the S-scheme
$(S' \to S)$ satisfy: adj $g \ (\mathcal{H}_{S'}) = \mathcal{H}_{S'}$. $\mathrm{Norm}_G(\mathcal{H})$ is representable by a closed
and finite presentation subscheme of G.*

The following Proposition shows that a (R)-group is characterized by its
Lie algebra.

Proposition 11.33 *Let H be a (R)-subgroup of G with Lie algebra $\mathcal{H}$. Then
$\mathrm{Norm}_G(\mathcal{H})$ is smooth along the identity section, and*

$$H = \mathrm{Norm}_G(\mathcal{H})^\circ ,$$

where $\mathrm{Norm}_G(\mathcal{H})^\circ$ denotes the connected component of the identity.

Corollary 11.34 *Let H (resp. H') be a (R)-subgroup of G. Then:*

$$(H = H') \Longleftrightarrow (\mathcal{H} = \mathcal{H}') \ .$$

(cf. loc. cit., Exp. XXII, Corollaire 5.3.5.)

11.8.1 (R)-subgroups and closed subsets of roots

Let G be endowed with a splitting $G = (G, T, M, R)$. The Lie algebra of a (R)-subgroup is of the form:

$$\mathcal{H} = \mathcal{T} \oplus \left(\bigoplus_{\alpha \in R'} \mathcal{G}^\alpha \right) ,$$

where $R' \subset R$ is by definition a subset of type (R).

It is recalled that a subset $R' \subset R$ is a closed subset if

$$(\alpha, \beta \in R' \text{ and } \alpha + \beta \in R) \Rightarrow (\alpha + \beta \in R) \ .$$

Proposition 11.35 *Every closed subset $R' \subset R$ is a subset of type (R).*

Remark 11.36 *This Proposition characterizes all the subsets of type (R) if S satisfies:*

$$\text{for all } s \in S \ ch(\kappa(s)) \neq 2 \text{ or } 3 \ .$$

Corollary 11.37   *1) A parabolic subgroup P of G is a subgroup of type (R).*

   2) Given a couple of parabolics in standard position (P, Q) the intersection subgroup $P \cap Q$ is a subgroup of type (R).

Proof *The property a subgroup of being of type (R) is local in S, one may suppose that G is endowed with a splitting: $G = (G, T, M, R)$. Thus $\mathcal{P} = \mathcal{T} \oplus \left(\bigoplus_{\alpha \in F_P} \mathcal{G}^\alpha \right.$ where F_P denotes a parabolic subset of R.*

It may be supposed that $T \subset P \cap Q$. Thus $\mathcal{P} \cap \mathcal{Q} = \mathcal{T} \oplus \left(\bigoplus_{\alpha \in F_P \cap F_Q} \mathcal{G}^\alpha \right).$

The following Proposition allows adapting to the relative case important building constructions.

Proposition 11.38 *Two maximal tori $T, T' \subset H$ of a type (R) subgroup H of G are locally conjugate for the etale topology.*

11.8.2 Representability of the (R)-subgroups functor

Definition 11.39 *Define the type (R)-subgroups of G functor $\mathscr{H}$ by:*

$$\forall\ (S' \longrightarrow S)\quad \mathscr{H}(S') = \{subgroups\ of\ type\ (R)\ of\ G\}\ .$$

There is a canonical morphism $u : \mathscr{H} \longrightarrow Grass(\mathcal{G})$ associating to a (R)-subgroup H its Lie algebra $\mathcal{H}$. Corollary 11.34 implies that this morphism is a monomorphism. In fact this monomorphism is representable by a finite presentation embedding. It results then the

Theorem 11.40 *$\mathscr{H}$ is representable by a quasi-projective and finite presentation S-scheme.*

(cf. loc. cit., Exp. XXII, Theoreme 5.8.1)

This theorem has an important consequence. Observe that the normalizer $Norm_G(H)$ of a type (R) subgroup H, is a closed and smooth S-subgroup scheme of G, thus

Proposition 11.41   *1) The morphism $v : G \longrightarrow Grass(\mathcal{G})$ defined by:*
$$v : g \mapsto u\ int(g) \cdot H\ \textit{factors as } G \longrightarrow G/Norm_G(H) \xrightarrow{\bar{v}} Grass(\mathcal{G})\ \textit{with}$$
$\bar{v}$ an embedding, identifying $G/Norm_G(H)$ with an open subscheme U of $\mathscr{H}$. Thus $G/Norm_G(H)$ is a **quasi-projective** *S-scheme.*

The following functors play an important role in the relative building constructions. A **Killing couple** of G is a couple (T, B) formed by a maximal torus T and a Borel subgroup $T \subset B$ containing T of G.

Definition 11.42 *The functor $\mathcal{P}ar(G)$ of parabolic subgroups of G is defined by:*
$$\mathcal{P}ar(G)(S') = \{parabolic\ subgroups\ of\ G_{S'}\}\ ;$$
the functor of Borel subgroups by:
$$\mathcal{B}or(G)(S') = \{Borel\ subgroups\ of G_{S'}\ \}\ ;$$
the functor of maximal tori of G by:
$$\mathcal{T}or(G)(S') = \{maximal\ tori\ of\ G_{S'}\}\ ;$$
and the functor of Killing couples of G by:
$$\mathcal{K}il(G)(S') = \{Killing\ couples\ of\ G_{S'}\}\ .$$

Let T (resp. B a Borel subgroup of G, $T \subset B$ a Killing couple of G, and P a type t parabolic subgroup of G). From the conjugation properties of maximal tori, and of parabolic subgroups of the same type it follows:

Proposition 11.43 *There are isomorphisms:*

1) $G/Norm_G(T) \simeq \text{Tor}(G);$

2) $G/B \simeq \mathcal{B}or(G);$

3) $G/T \simeq \mathcal{K}il(G);$

4) $G/P \simeq \mathcal{P}ar_t(G),$

defined by:

1) $g \mapsto int(g) \cdot T;$

2) $g \mapsto int(g) \cdot B;$

3) $g \mapsto int(g) \cdot T \subset int(g) \cdot B;$

4) $g \mapsto int(g) \cdot P.$

From the Proposition 11.41 it follows:

Corollary 11.44 *The functor* $\text{Tor}(G)$ *(resp.* $\mathcal{K}il(G)$*) is representable by a smooth quasi-projective S-scheme. The functor $\mathcal{B}or(G)$ (resp. $\mathcal{P}ar_t(G)$) is representable by a projective, smooth S-scheme with integral geometric fibers.*

11.9   Weyl complex scheme

Given frames E, E', the isomorphism of Proposition clearly transforms parabolic subsets of $A(R_E)$ in parabolic subsets of $A(R_{E'})$, and its unicity allows defining, by etale descent, the following S-schemes. The **root scheme** $\mathcal{R}(G)$, the **apartment scheme** $\mathcal{A}(G)$, and the **Weyl group scheme** $\underline{W}_G$. Given a frame $E = (T, M, R, R_0, (X_\alpha)_{\alpha \in R_0})$ of G, let $\mathcal{A}_E$, be the apartment given by the $\mathbb{Z}$-root system $\mathcal{R}(E)$ defined by E. A facet F of $\mathcal{A}_E$ corresponds to a parabolic set of roots $R_F \subset R$.
Let W_{R_E} be the Weyl group of R_E as defined in *loc. cit.*, *Exp. XXI, Definition* 1.1.8. In fact W_{R_E} is the Weyl group of the apartment $\mathcal{A}_E$, i.e. the group $W_{\mathcal{A}_E}$ of type preserving automorphisms of $\mathcal{A}_E$. Denote by $C_E \in \text{Ch}\mathcal{A}_E$ the chamber defined by the positive root system $R_+ = (\mathbb{N}R_0) \cap R$, i.e. such that $R_{C_E} = R_+$.
If G is split write

$$\mathcal{A}(G) = \varinjlim_E (\mathcal{A}_E)_S \ \left(\text{resp. } \mathcal{R}(G) = \varinjlim_E (R_E)_S\right),$$

where E runs on the set of frames of G, and the transition isomorphisms are induced by the family of automorphisms $(\alpha(E, E'))$, where (E, E') runs on the set of the couples of frames of G. In the case of an S-reductive group G,

$\mathcal{A}(G)$ (resp. $\mathcal{R}(G)$) is defined by descent from an etale covering $(S_i \to S)$ such that G_{S_i} is splitted.

Define the Weyl group S-scheme $\underline{W}_G$ by

$$\underline{W}_G = \varinjlim_E (W_{A_E})_S$$

if G is split, and by descent for an S-reductive group G.

11.10 Weyl complex typical simplex scheme and the Type morphism

The group $\underline{W}_G$ acts as a group of automorphisms of $\mathcal{A}(G)$. One may thus consider the quotient scheme

$$\mathcal{A}(G)/\underline{W}_G$$

(The S-scheme of types of facets of $\mathcal{A}(G)$). If G is split one has

$$\mathcal{A}(G)/\underline{W}_G \simeq \varinjlim_E (A_E/W_{A_E})_S \ .$$

For a frame E of G there is a building morphism

$$\mathcal{P}(\mathrm{Dyn}(E)) \to A_E,$$

induced by the mapping $\mathrm{Dyn}(E) \to A_E$, which correspond to $\alpha \in R_0 = \mathrm{Dyn}(E)$, the maximal parabolic set $R^{(\alpha)}$ defined by:

$$R^{(\alpha)} = (\mathbb{Z}(R_0 - \{\alpha\})) \cap R_E \cup (R_E)_+ \ .$$

Thus to $R_0' \subset R_0 = \mathrm{Dyn}(E)$ corresponds the parabolic subset $\bigcap R^{(\alpha)}$ ($\alpha \in R_0'$) of R_E.

Given a facet $F \in A_E$, satisfying $(R_E)_+ \subset R_F$, i.e. so that $F \in \Delta(C_E) \subset A_E$, there exists a unique subset $R_0(F) \subset R_0$, such that $\mathbb{Z}(R_0 - R_0(F)) = R_F \cap (-R_F)$ which satisfies

$$R_F = \bigcap R^{(\alpha)} \ (\alpha \in R_0(F)).$$

The set $R_0(F)$ is given by $R_0(F) = R_0 - R_0 \cap (-R_F)$, and is in fact the set of vertices of the facet F. Thus it is concluded that $\mathcal{P}(\mathrm{Dyn}(E)) \to A_E$ gives a building isomorphism between $\mathcal{P}(\mathrm{Dyn}(E))$ and the subcomplex $\Delta(C_E)$, given by the set of faces of C_E. Observe that

$$R_0' \subset R_0'' \Rightarrow \bigcap_{\alpha \in R_0'} R^{(\alpha)} \supset \bigcap_{\alpha \in R_0''} R^{(\alpha)}.$$

Thus according to the definition of the ordering of A_E, the mapping $R_0' \mapsto \bigcap R^{(\alpha)}$ ($\alpha \in R_0'$) is order preserving. Since the restriction of the quotient morphism $\mathcal{A}_E \to \mathcal{A}_E/W_{\mathcal{A}_E}$, to $\Delta(C_E)$ induces an isomorphism of buildings $\Delta(C_E) \simeq \mathcal{A}_E/W_{\mathcal{A}_E}$, it is deduced that the composed mapping

$$\mathcal{P}(\mathrm{Dyn}(E)) \to \mathcal{A}_E/W_{\mathcal{A}_E},$$

is a building isomorphism. If G is split one obtains an isomorphism

$$\underline{\Delta}(G) = \varinjlim_{\vec{E}} \mathcal{P}(\mathrm{Dyn}(E))_S \to \varinjlim_{\vec{E}} (\mathcal{A}_E/W_{\mathcal{A}_E})_S = \mathcal{A}(G)/\underline{W}_G.$$

It is deduced that there is a canonical isomorphism defined by etale descent

$$\underline{\Delta}(G) \simeq \mathcal{A}(G)/\underline{W}_G$$

for an S-reductive group G. Denote by

$$t_{1,\mathcal{A}} : \mathcal{A}(G) \to \mathcal{A}(G)/\underline{W}_G \simeq \underline{\Delta}(G)$$

the quotient morphism (**Weyl Complex type morphism**). The isomorphism $\mathcal{A}(G)/\underline{W}_G \simeq \underline{\Delta}(G)$ admits the factorization:

$$\mathcal{A}(G)/\underline{W}_G \to \mathcal{P}ar(G)/G \to \underline{\Delta}(G) \ .$$

11.11   Weyl Complex type of relative position morphism

With a frame E of G there is an associated natural morphism

$$(\mathcal{A}_E)_S \times_S (\mathcal{A}_E)_S \to \mathcal{S}\mathrm{tand}(G),$$

that correspond to (F_S, F'_S) $((F, F') \in \mathcal{A}_E \times \mathcal{A}_E)$ the couple of parabolics in standard position

$$(P_F, P_{F'}),$$

where P_F (resp. $P_{F'}$) is given by the parabolic subset R_F (resp. $R_{F'}$).
    This morphism induces, by composition with $t_2 : \mathcal{S}\mathrm{tand}(G) \longrightarrow \mathrm{Relpos}_G$, a morphism

$$t_{2,E} : (\mathcal{A}_E)_S \times_S (\mathcal{A}_E)_S \to \mathrm{Relpos}_G \ ,$$

which factors as

$$t_{2,E} : (\mathcal{A}_E)_S \times_S (\mathcal{A}_E)_S \to ((\mathcal{A}_E \times \mathcal{A}_E)/W_{\mathcal{A}_E})_S \to \mathrm{Relpos}_G \ ,$$

There are isomorphisms

$$\varinjlim_{\vec{E}}((\mathcal{A}_E \times \mathcal{A}_E)/W_{\mathcal{A}_E})_S \simeq \mathcal{A}(G) \times_S \mathcal{A}(G)/\underline{W}_G \ ,$$

i.e.

$$\varinjlim_{\vec{E}} (\mathrm{Relpos}\, \mathcal{A}_E)_S \stackrel{\sim}{\to} \mathrm{Relpos}\, \mathcal{A}(G) \ ,$$

and
$$\text{Relpos } \mathcal{A}(G) = \mathcal{A}(G) \times_S \mathcal{A}(G)/\underline{W}_G \xrightarrow{\sim} \text{Relpos}_G.$$

Assume that G is split. Define $t_{2,\mathcal{A}}$ as the quotient morphism

$$t_{2,\mathcal{A}} : \mathcal{A}(G) \times_S \mathcal{A}(G) \rightarrow \text{Relpos } \mathcal{A}(G) \xrightarrow{\sim} \text{Relpos}_G.$$

$$\left(\text{resp. } t_{2,\mathcal{A}} = \varinjlim_{E} t_{2,E} : \varinjlim_{E} \left((\mathcal{A}_E)_S \times_S (\mathcal{A}_E)_S \right) \longrightarrow \text{Relpos}_G \right).$$

Clearly for a reductive S-group there exists a morphism

$$t_{2,\mathcal{A}} : \mathcal{A}(G) \times_S \mathcal{A}(G) \rightarrow \text{Relpos } \mathcal{A}(G) \xrightarrow{\sim} \text{Relpos}_G.$$

which corresponds to the above one if G is split, and in fact is the canonical quotient morphism.

The morphism

$$t_1 \times t_1 : \mathcal{P}ar(G) \times_S \mathcal{P}ar(G) \rightarrow \underline{\Delta}(G) \times_S \underline{\Delta}(G)$$

induces by restriction the morphism

$$\mathcal{S}tand(G) \rightarrow \underline{\Delta}(G) \times_S \underline{\Delta}(G),$$

which factors through

$$t_2 : \mathcal{S}tand(G) \rightarrow \text{Relpos}_G.$$

Denote by

$$\epsilon = \epsilon_1 \times \epsilon_2 : \text{Relpos}_G \rightarrow \underline{\Delta}(G) \times_S \underline{\Delta}(G)$$

the induced morphism.

If G is endowed with a frame E, then in view of isomorphism $\mathcal{A}(G) \xrightarrow{\sim} \text{Relpos}_G$, ϵ may be described as follows. Write

$$\text{Relpos } \mathcal{A}_E = \coprod_{(t,s)\in \text{typ } \mathcal{A}_E \times \text{typ } \mathcal{A}_E} \text{Relpos}_{(t,s)} \mathcal{A}_E \ ,$$

with

$$\text{Relpos}_{(t,s)} \mathcal{A}_E = (\mathcal{A}_E)_t \times (\mathcal{A}_E)_s/W(\mathcal{R}(E)).$$

Define:

$$\epsilon_E = (\epsilon_E)_1 \times (\epsilon_E)_2 : \text{Relpos } \mathcal{A}_E \rightarrow \text{typ } \mathcal{A}_E \times \text{typ } \mathcal{A}_E$$

by

$$\epsilon_E : \tau \mapsto (t,s),$$

if $\tau \in \text{Relpos}_{(t,s)} \mathcal{A}_E$. One has $\epsilon = (\epsilon_E)_S$.

11.12 Relative Position Types scheme of a Weyl Star Complex scheme

It is supposed that G is endowed with a frame E of G. For

$$t \in \operatorname{typ} \mathcal{A}_E,$$

one defines

$$\operatorname{Relpos}^t \mathcal{A}_E \subset \operatorname{Relpos} \mathcal{A}_E,$$

as the image of the injective mapping

$$\operatorname{St}_{\mathrm{F}}^{\mathcal{A}_{\mathrm{E}}} \times \operatorname{St}_{\mathrm{F}}^{\mathcal{A}_{\mathrm{E}}}/\mathrm{W}_{\mathrm{F}} \to \mathcal{A}_{\mathrm{E}} \times \mathcal{A}_{\mathrm{E}}/\mathrm{W}(\mathcal{R}(\mathrm{E}))$$

where F denotes a facet of type t of $\mathcal{A}_E$, and one writes

$$W_F = \operatorname{Stab}_{W(\mathcal{R}(E))}.$$

This image is independant of the choice of F.

Let $t \in \operatorname{typ} \mathcal{A}_E$, and $\underline{t} = t_S$. Define:

$$\operatorname{Relpos}_G^{\underline{t}} = \operatorname{Relpos}_G^{t_S} = \left(\operatorname{Relpos}^t \mathcal{A}_E\right)_S.$$

Given an S-reductive group G, and a section $\underline{t}$ of $\underline{\Delta}(G)$, there is a subscheme

$$\operatorname{Relpos}_G^{\underline{t}} \subset \operatorname{Relpos}_G,$$

so that if E is a locally (for the etale topology) defined frame of G, the canonical isomorphism

$$(\operatorname{Relpos} \mathcal{A}_E)_{S'} \simeq (\operatorname{Relpos}_G)_{S'}$$

induces the isomorphism

$$\operatorname{Relpos}_G^{t_{S'}} \simeq \left(\operatorname{Relpos}_G^{\underline{t}}\right)_{S'},$$

if $(\underline{t})_{S'} = t_{S'}$ for $t \in \operatorname{typ} \mathcal{A}_E$.

It is recalled that there is a canonical isomorphism

$$\operatorname{Relpos} \mathcal{A}(G) \simeq \operatorname{Relpos}_G \quad (\text{resp.} \ \mathcal{A}(G)/\underline{W}_G \simeq \underline{\Delta}(G)),$$

allowing identification of both members. Define

$$\operatorname{Relpos}^{\underline{t}} \mathcal{A}(G) = \operatorname{Relpos}_G^{\underline{t}},$$

for a section $\underline{t}$ of $\underline{\Delta}(G)$.

The morphism ϵ induces a morphism

$$\epsilon_{\mathcal{A}(G)} : \operatorname{Relpos} \mathcal{A}(G) \to \underline{\Delta}(G) \times_S \underline{\Delta}(G),$$

by composition with $\mathrm{Relpos}\,\mathcal{A}(G) \simeq \mathrm{Relpos}\,_G$.

Given a section $(\underline{s}, \underline{s}')$ of $\underline{\Delta}(G) \times_S \underline{\Delta}(G)$ one defines:

$$\mathrm{Relpos}\,_{(\underline{s}, \underline{s}')} = \epsilon^{-1}\left((\underline{s}, \underline{s}')\right) \subset \mathrm{Relpos}\,_G$$

$$\left(\text{resp. } \mathrm{Relpos}\,_{(\underline{s}, \underline{s}')}\,\mathcal{A}(G) = \left(\epsilon_{\mathcal{A}(G)}\right)^{-1}\left((\underline{s}, \underline{s}')\right)\right) ,$$

and

$$\mathrm{Relpos}\,^t_{(\underline{s}, \underline{s}')} = \mathrm{Relpos}\,_{(\underline{s}, \underline{s}')} \cap \mathrm{Relpos}\,^t_G$$

$$\left(\text{resp. } \mathrm{Relpos}\,^t_{(\underline{s}, \underline{s}')}\,\mathcal{A}(G) = \mathrm{Relpos}\,_{(\underline{s}, \underline{s}')}\,\mathcal{A}(G) \cap \mathrm{Relpos}\,^t\,\mathcal{A}(G)\right).$$

Chapter 12

Associated Schemes to the Relative Building

Preceding definitions and constructions about buildings are adapted to the schematic and functorial point of view. The main constructions carried out in this chapter are those of the Root functors of the Apartment scheme, the Convex Hull scheme defined by two parabolic subgroups and the canonical affine open covering of a parabolics standard position scheme.

The Data given by:

- the Parabolics scheme $\mathcal{P}ar(G)$;

- the scheme of Couples of Parabolics in Standard Position $\mathcal{S}tand(G) \longrightarrow Relpos_G$ seen as a $Relpos_G$-scheme;

- the **subscheme of Incident Parabolics** of $\mathcal{S}tand(G)$ corresponding to the couples of parabolics in **osculating relative position** of [23];

- and the **Apartments scheme** given by the scheme of maximal tori $\mathcal{T}or(G)$ of G,

play the role of a **Relative Building** for the Reductive S-Group scheme G. The following schemes are naturally defined in terms of these Data: The **Root functors**. The **Universal Schubert Cell scheme** Σ of G is defined as the graph $\mathbf{gr}(t_2)$ of the type of relative position morphism $t_2 : \mathcal{S}tand(G) \longrightarrow Relpos_G$, and the **Universal Schubert scheme** $\overline{\Sigma} \longrightarrow \mathbf{Relpos_G}$ of G is defined as its **Schematic Closure** in $Relpos_G \times_S \mathcal{P}ar(G) \times_S \mathcal{P}ar(G)$, i.e. $\overline{\Sigma} = \Sigma^{\mathrm{sche}}$. This Schematic Closure is well defined as $\overline{\Sigma} \longrightarrow S$ is a quasi-compact morphism and $Relpos_G \times_S \mathcal{P}ar(G) \times_S \mathcal{P}ar(G)$ an S-projective scheme. The Convex Hull scheme $\mathcal{F}ix(P,Q)$ of a couple of parabolics in standard position (P,Q) is defined in this setting. The **Tautological Couple of Parabolics** $(\tilde{\mathbf{P}}, \tilde{\mathbf{Q}})$ on $\mathcal{S}tand(G)$ is the section of $\mathcal{S}tand(G)$ given by the diagonal section $\Delta : \mathcal{S}tand(G) \longrightarrow \mathcal{S}tand(G) \times_S \mathcal{S}tand(G)$. By means of the **Big Cell open covering** of the Universal Schubert Cell it is proven that the Tautological Couple is in Standard Position. Thus, one obtains the finite **Convex Hull scheme** $\mathcal{F}ix(\tilde{P}, \tilde{Q})$ over the Universal Schubert Cell scheme. The fiber of $\mathcal{F}ix(\tilde{P}, \tilde{Q})$ over the couple (P,Q) is given by $\mathcal{F}ix(P,Q)$. It will be seen that this scheme has the following important property: With a section **g** of the "Typical Simplex scheme (resp. Dynkin scheme) minimal generalized galleries of types scheme", defined in the next chapter, is associated a section $\tau_{\mathbf{g}}$ of $Relpos_G$, and a **unique section** $\gamma_{\mathbf{g}}(\tilde{\mathbf{P}}, \tilde{\mathbf{Q}})$ of the **galleries scheme** **associated with** $\mathcal{F}ix(\tilde{P}, \tilde{Q})_{\tau_{\mathbf{g}}} \longrightarrow \Sigma_{\tau_{\mathbf{g}}}$.

12.1   Root functors

Let T be a torus of the reductive S-group G, and α a root defined by T, i.e. a section of the root functor $\mathcal{R}_T$ (cf. [23], *Exp. XIX*, §3). One recalls that if α is a root of G (cf. 11.2) then $-\alpha$ is also a root of G, and that the subfunctor of $W(\mathcal{G})$ defined by

$$W(\mathcal{G})^\alpha(S') = \{x \in W(\mathcal{G})(S') \, / \, \mathrm{ad}(t)x = \alpha(t)x \text{ for every } t \in T(S''), \, S'' \to S'\},$$

satisfies:

There exists a rank 1 direct factor $\mathcal{O}_S$-submodule $\mathcal{G}^\alpha \subset \mathcal{G}$ so that:

$$W(\mathcal{G})^\alpha = W(\mathcal{G}^\alpha) \quad \text{(cf. loc. cit., } Exp. \, XIX, \, \S 4.1).$$

Definition 12.1
Let
$$\mathcal{F}_\alpha \subset \mathcal{P}ar(G)$$
be the S-subfunctor defined by
$$\mathcal{F}_\alpha(S') = \{P \text{ parabolic of } G_{S'} \, / \, \mathcal{G}^\alpha_{S'} = \mathcal{O}_{S'} \otimes_{\mathcal{O}_S} \mathcal{G}^\alpha \subset Lie(P)\}.$$

*$\mathcal{F}_\alpha$ is called the **Root** S-functor defined by the root α of G.*

Definition 12.2
*Define the **Hyperplane** S-functor $\mathcal{H}_\alpha$ defined by the root α of G by*

$$\mathcal{H}_\alpha = \mathcal{F}_\alpha \cap \mathcal{F}_{-\alpha} \subset \mathcal{P}ar(G).$$

It is proved below that $\mathcal{F}_\alpha$ is a representable S-functor, and a fortiori that $\mathcal{H}_\alpha$ is also representable.

12.1.1 The Plücker embedding of the Parabolics scheme and representability of Root functors

It follows from Proposition 11.33 and the fact that a parabolic subgroup is its own normalizer that a parabolic subgroup is characterized by its Lie algebra.

Proposition 12.3
Let $P \subset G$ be a parabolic subgroup of the reductive S-group G. Write $\mathcal{P} = Lie(P)$. We then have:
$$\mathcal{N}orm_G(\mathcal{P}) = P$$

From Proposition 11.40 it is obtained:

Proposition 12.4 *The morphism*
$$u_{par} : \mathcal{P}ar(G) \longrightarrow Grass(\mathcal{G}),$$

defined by: $u_{par} : P \mapsto \mathcal{P} = Lie(P)$, defines a closed embedding.

Let one make explicit this embedding into a grassmannian. Let $n(t)$ be the rank of the Lie algebra of a parabolic group P of type t. Let Plücker's embedding morphism be recalled

$$\omega_t : \text{Grass}_{n(t)}(\mathcal{G}) \to \text{Grass}_1 \left(\bigwedge^{n(t)} \mathcal{G} \right) = \mathbb{P}\left(\bigwedge^{n(t)} \mathcal{G} \right) \quad \text{(cf. [24], Ch. 1, 9.8)}.$$

Let η be a section of $\text{Grass}_{n(t)}(\mathcal{G})$, given by a rank $n(t)$ submodule $\mathcal{N} \subset \mathcal{G}$ with locally free quotient $\mathcal{G}/\mathcal{N}$, then the image $\omega_t(\eta)$ of η is obtained as follows.

The $n(t)$-th exterior product $\bigwedge^{n(t)} \mathcal{N}$ of $\mathcal{N}$ is a rank 1 locally free submodule of $\bigwedge^{n(t)} \mathcal{G}$, with locally free quotient

$$\bigwedge^{n(t)} \mathcal{G} / \bigwedge^{n(t)} \mathcal{N}.$$

Thus the section $\omega_t(\eta)$ of $\mathbb{P}\left(\bigwedge^{n(t)} \mathcal{G} \right)$ corresponds to the rank 1 locally free, direct factor submodule $\bigwedge^{n(t)} \mathcal{N}$ of $\bigwedge^{n(t)} \mathcal{G}$.

To see that ω_t is an embedding it suffices to see that the submodule $\bigwedge^{n(t)} \mathcal{N} \subset \bigwedge^{n(t)} \mathcal{G}$ characterizes $\mathcal{N} \subset \mathcal{G}$. Let $e_1, \cdots, e_{n(t)}$ be a basis of $\mathcal{N}$.

Then $e = e_1 \wedge \cdots \wedge e_{n(t)}$ is a basis of $\bigwedge^{n(t)} \mathcal{N}$.
Let S' be a local S-scheme. One thus has

$$\mathcal{N}(S') = \{x \text{ section of } \mathcal{O}_{S'} \otimes_{\mathcal{O}_S} \mathcal{G} / x \wedge e_{S'} = 0\}.$$

If $S' = \mathrm{Spec}(k)$, k a field, the equality is immediate. The general case follows from Nakayama's lemma (cf. [40]). It results from this that ω_t is an embedding. Write $n = n(t)$.

Definition 12.5
Let
$$\mathcal{G}^\alpha \subset \mathcal{G}$$
be the rank 1 submodule of G defined by the root α. Define the S-subfunctor $Z_\alpha \subset Grass_n(\mathcal{G})$, by:
$$Z_\alpha(S') = \{\mathcal{N} \subset \mathcal{G}_{S'} \mid \mathcal{G}_{S'}/\mathcal{N} \text{ locally free, rank } \mathcal{N} = n, \text{ and } \mathcal{G}_{S'}^\alpha \subset \mathcal{N}\}.$$

Let it be seen that Z_α is a representable S-subfunctor of $\mathrm{Grass}_n(\mathcal{G})$. In fact Z_α is a Schubert cell. It suffices to see that the image $\omega_t(Z_\alpha)$ by Plücker's embedding is a representable subfunctor of $\mathbb{P}\left(\bigwedge^n \mathcal{G}\right)$.

Notation 12.6 *Write*

$$\mathbb{P}\left(\bigwedge^n \mathcal{G}\right) = Sph\left(\mathcal{O}_S[\Delta(i_1, \cdots, i_n)]\right) \ (1 \leqslant i_1 < \cdots < i_n \leqslant rk \ \mathcal{G}).$$

where $(\Delta(i_1, \cdots, i_n))_{1 \leqslant i_1 < \cdots < i_n \leqslant rk \ \mathcal{G}}$ denotes a set of variables. Denote by

$$I_\alpha \subset \mathcal{O}_S[\Delta(i_1, \cdots, i_n)],$$

the homogeneous ideal generated by the set of variables

$$(\Delta(i_1, \cdots, i_n))_{2 \leqslant i_1 < \cdots < i_n \leqslant rk \ \mathcal{G}},$$

and by

$$\mathfrak{I}_\alpha \subset \mathcal{O}_{\mathbb{P}\left(\bigwedge^n \mathcal{G}\right)}$$

the corresponding ideal defined by I_α.

Choose a basis $(e_i)(1 \leqslant i \leqslant rk \ \mathcal{G})$ of $\mathcal{G}$ so that:

$$\mathcal{G}_\alpha = \mathrm{Vect}(e_1).$$

Given a submodule $\mathcal{N} \subset \mathcal{G}$ representing a section of $\mathrm{Grass}_n(\mathcal{G})$, choose a basis $(e'_1, \cdots, e'_n)$ of $\mathcal{N}$ and denote by $N = N(e'_1, \cdots, e'_n)$ the section of $\mathcal{O}_S^{rk \ \mathcal{G} \times n}$ given by the matrix of the coordinates of the vectors $(e'_1, \cdots, e'_n)$ relatively to the basis (e_i) of $\mathcal{G}$. Its j-th column (a_{ij}) $(1 \leqslant i \leqslant rk \ \mathcal{G})$ gives thus the coordinates of e'_j, i.e. $e'_j = \sum_{1 \leqslant i \leqslant rk \ \mathcal{G}} a_{ij}e_i$. By $N(i_1, \cdots, i_n)$, one denotes the $n \times n$-submatrix of N defined by the i_1-th,...,i_n-th rows of N, and one writes

$$\Delta_\mathcal{N}(i_1, \cdots, i_n) = \det N(i_1, \cdots, i_n).$$

The Plücker homogeneous coordinates of $\mathcal{N}$ relatively to (e_i) are thus given by:

$$(\Delta_{\mathcal{N}}(i_1, \cdots, i_n))_{1 \leqslant i_1 < \cdots < i_n \leqslant rk\ \mathcal{G}}.$$

It is supposed that $N(i_1^0, \cdots, i_n^0)$ is an invertible $n \times n$ submatrix of N. This is, locally in S, always the case, i.e. at least one of the homogeneous coordinates of N is given by an invertible section of $\mathcal{O}_S$. The hypothesis $\mathcal{G}/\mathcal{N}$ is locally free which implies that, given $s \in S$, there exists $\Delta(i_1^0, \cdots, i_n^0)$ defining a section of $\mathcal{O}_{S,s}^*$. Write:

$$\overline{N}[i_1^0, \cdots, i_n^0] = N(e_1', \cdots, e_n') \times N(i_1^0, \cdots, i_n^0)^{-1}.$$

Then one has that

$$\overline{N}[i_1^0, \cdots, i_n^0] = N(\overline{e}_1, \cdots, \overline{e}_n),$$

where $(\overline{e}_1, \cdots, \overline{e}_n)$ is the normalized basis of $\mathcal{N}$.

The affine coordinates of the $\overline{e}_j$'s may be calculated in terms of the Plücker homogeneous coordinates $(\Delta_N(i_1, \cdots, i_n))$ as follows. Let

$$\overline{e}_j = \sum_{i=1}^{rk\mathcal{G}} \xi_{ij}\, e_i \quad (1 \leqslant j \leqslant n).$$

Then one has

$$\xi_{ij} = \frac{\Delta_N(i_1^0, \cdots, \overset{\text{j-th place}}{i}, \cdots, i_n^0)}{\Delta_N(i_1^0, \cdots, i_n^0)} \quad (1 \leqslant i \leqslant rk\mathcal{G},\ 1 \leqslant j \leqslant n).$$

Proposition 12.7
Denote by $\mathfrak{I}_\alpha^* \subset \mathcal{O}_{Grass_n(\mathcal{G})}$ the inverse image $\omega_t^*(\mathfrak{I}_\alpha)$ of $\mathfrak{I}_\alpha \subset \mathcal{O}_{\mathbb{P}(\bigwedge^n \mathcal{G})}$ by ω_t.
Thus there is a canonical isomorphism

$$Z_\alpha \simeq Spec\left(\mathcal{O}_{Grass_n(\mathcal{G})}/\mathfrak{I}_\alpha^*\right),$$

which proves that Z_α is representable by a proper subscheme of $Grass_n(\mathcal{G})$.

Proof   *If it is supposed now that for $2 \leqslant i_1 \cdots < i_n \leqslant rk\ \mathcal{G}$ one has*

$$\Delta_N(i_1, \cdots, i_n) = 0.$$

It results immediately from the above expression for ξ_{i1} that

$$\overline{e}_1 = e_1,$$

thus one obtains

$$\mathcal{G}^\alpha = Vect(e_1) \subset \mathcal{N}.$$

On the other hand, if it is supposed that

$$\mathcal{G}^{\alpha} = Vect(e_1) \subset \mathcal{N},$$

it is clear that for $2 \leqslant i_1 \cdots < i_n \leqslant rk\mathcal{G}$ one has:

$$\Delta_{\mathcal{N}}(i_1, \cdots, i_n) = 0.$$

Given $\mathcal{N}$ it is denoted by $\eta_{\mathcal{N}}$ the section of $Grass_n(\mathcal{G})$ defined by $\mathcal{N}$. It has thus been seen:

$\eta_{\mathcal{N}}$ defines a section of $Z_{\alpha} \subset Grass_n(\mathcal{G}) \Leftrightarrow \omega_t(\eta_{\mathcal{N}})$ defines a section of
$V(\mathfrak{I}_{\alpha}) \subset \mathbb{P}\left(\bigwedge^n(\mathcal{G})\right).$

It follows from the definition of $\mathcal{F}_{\alpha}$ (resp. Z_{α}) that:

Corollary 12.8 *There is a canonical isomorphism*

$$\mathcal{F}_{\alpha} \simeq u_{par,t}^*(Z_{\alpha}).$$

Thus the subfunctor $\mathcal{F}_{\alpha} \subset \mathcal{P}ar(G)$ is representable by a projective subscheme.

12.2   Big Cell Open Covering of a Parabolics scheme

Assume that G is endowed with a frame $E = (T, M, R, R_0, (X_{\alpha})_{\alpha \in R_0})$(cf. Proposition 11.12). Let $\mathcal{R} = (M, M^*, R, R^{\vee}, R_0)$ be the corresponding $\mathbb{Z}$-system of roots, and $R_+ = (\mathbb{N} R_0) \cap R$ the system of positive roots defined by R_0. Denote by $\mathcal{A}_E$ the apartment building defined by $\mathcal{R}$. Let $C = C_{R_+}$ be the chamber of $\mathcal{A}_E$ corresponding to the positive roots system R_+. Write $F_t = F_t(C)$ (the facet of type t incident to C). Let

$$B_{R_+} \subset G \quad \left(\text{resp. } B_{R_-} \subset G\right),$$

with $R_- = -R_+$, be the Borel subgroup defined by R_+ (resp. R_-). There is a canonical isomorphism

$$G/B_{R_+} \to \mathcal{B}or(G),$$

induced by the morphism $G \to \mathcal{B}or(G)$, functorially defined by $x \mapsto int(x) \cdot (B_{R_+})$. More generally, denote by P_{t,R_+} the parabolic subgroup of type t, containing B_{R_+}. There is an isomorphism

$$G/P_{t,R_+} \to \mathcal{P}ar_t(G),$$

induced by $x \mapsto int(x)(P_{t,R_+})$ from G to $\mathcal{P}ar_t(G)$. (cf. [23], Exp. XXVI, Th. 3.3.)

Let $P_\alpha \subset G$ $(\alpha \in R)$ be the subgroup defined as the image of the vector group $W(\mathcal{G}^\alpha)$ by $\exp_\alpha : W(\mathcal{G}^\alpha) \to G$. (cf. loc. cit., Exp. XXII., Théorème 1.1). For each $\mathcal{G}^\alpha$ one chooses a basis X_α. Let

$$\exp'_\alpha : \mathrm{Spec}(\mathcal{O}_S[t_\alpha]) \to P_\alpha$$

denote the morphism defined by composing the isomorphism induced by X_α, $\mathrm{Spec}(\mathcal{O}_S[t_\alpha]) \simeq W(\mathcal{G}^\alpha)$, with $\exp_\alpha$.

Definition 12.9

- *The* **Big Open Cell** $\Omega_{R_+} \subset G$ *is the relatively schematically dense open subscheme of G defined as the image of the S-morphism*

$$\prod_{\alpha \in R_-} P_\alpha \times T \times \prod_{\alpha \in R_+} P_\alpha \to G,$$

induced by the product of G, and the $(\exp_\alpha)_{\alpha \in R}$, where the cartesian products have taken over S, and R_+ (resp. R_-) is endowed with some total order. (cf. loc. cit., Exp. XXII, §4.1.)

- *Define the* **Big Open Cell of** G/B_{R_+}

$$\tilde{\Omega}_{R_+} \subset G/B_{R_+}$$

as the image of Ω_{R_+} by the quotient morphism $G \to G/B_{R_+}$. Let

$$\overline{\Omega}_{R_+} \subset \mathcal{B}or(G),$$

denote the open subscheme image of $\tilde{\Omega}_{R_+}$ by $G/B_{R_+} \tilde{\to} \mathcal{B}or(G)$;

- *Define the* **Big Open Cell**

$$\tilde{\Omega}_{t,R_+} \subset G/P_{t,R_+},$$

as the image of the Big Open Cell Ω_{R_+} by the quotient morphism $G \to G/P_{t,R_+}$, and let

$$\overline{\Omega}_{t,R_+} \subset \mathcal{P}ar_t(G)$$

be the image of $\tilde{\Omega}_{t,R_+}$ by the isomorphism

$$G/P_{t,R_+} \tilde{\to} \mathcal{P}ar_t(G),$$

induced by $x \mapsto int(x)(P_{t,R_+})$ from G to $\mathcal{P}ar_t(G)$.

There is an induced isomorphism of S-schemes

$$\prod_{\alpha \in R_-} P_\alpha \tilde{\to} \tilde{\Omega}_{R_+}$$

which by right composition with $\prod \exp'_\alpha (\alpha \in R_-)$ gives rise to an isomorphism

$$\prod_{\alpha \in R_-} \operatorname{Spec}(\mathcal{O}_S[t_\alpha]) = \operatorname{Spec}(\mathcal{O}_S[t_\alpha]_{\alpha \in R_-}) \xrightarrow{\sim} \tilde{\Omega}_{R_+}.$$

It results in an isomorphism

$$\prod_{\alpha \in R_-} \operatorname{Spec}(\mathcal{O}_S[t_\alpha]) \xrightarrow{\sim} \overline{\Omega}_{R_+},$$

which may be paraphrased as follows. For every section of $\left(\overline{\Omega}_{R_+}\right)_{S'}$, given by some Borel subgroup $B \subset G_{S'}$, there exists a unique section (x_α) of $\prod\limits_{\alpha \in R_-} \mathcal{O}_{S'}$, with $B = \operatorname{int}\left(\prod \exp(x_\alpha X_\alpha)\right) \cdot (B_{R_+})_{S'}$. More generally one has: Let $R_t \supset R_+$ be the parabolic set of roots given by P_{t,R_+}, i.e. R_t verifies $P_t(E) = P_{R_t}$. Denote by

$$\mathcal{U}_{t,R_+} \subset B_{R_-} \ (R_- = -R_+),$$

the subgroup of G defined by the closed set of roots $R - R_t$. This subgroup parameterizes the big cell $\overline{\Omega}_{t,R_+}$. Thus there is an isomorphism

$$\mathcal{U}_{t,R_+} \simeq \overline{\Omega}_{t,R_+},$$

defined by $y \mapsto \operatorname{int}(y)(P_t)$. Remark that $\mathcal{U}_{t,R_+}$ is isomorphic as an S-scheme to the product $\prod\limits_{\alpha \in R - R_t} P_\alpha$. The quotient morphism $G \to G/P_{t,R_+}$ being smooth, and the reciprocal image of $\overline{\Omega}_{t,R_+}$ being equal to the relatively schematically open subscheme Ω_{R_+} of G, it results that the big cell $\overline{\Omega}_{t,R_+}$ is an **open relatively schematically dense** subscheme of $\mathcal{P}ar_t(G)$.

From the **Bruhat Cell decomposition** of $\mathcal{P}ar_t(G)$ given by B_{R_+}, the fact that a Cell may be embedded in a Big cell is defined by some Borel subgroup adapted to the splitting of G, one may state the following

Definition 12.10
The family of **relatively affine** *open sets of* $\mathcal{P}ar_t(G)$

$$\left(\overline{\Omega}_{t,R_+}\right),$$

where R_+ runs on the set of systems of positive roots of R, is called the **Big Cell Open Covering** *of* $\mathcal{P}ar_t(G)$. *One has*

$$\mathcal{P}ar_t(G) = \bigcup \overline{\Omega}_{t,R_+}.$$

12.3 The Big cell Open Covering trivializes a Standard Position scheme

Let τ be a type of relative position of $\mathcal{A}_E$, and (t,s) the corresponding couple of types of parabolics. Let $\mathcal{S}\mathrm{tand}(\tau_S)$ be seen as defining a locally trivial fibration

$$\operatorname{pr}_{\mathcal{P},\tau} : \mathcal{S}\mathrm{tand}(\tau_S) \to \mathcal{P}ar_t(G),$$

where pr $_{\mathcal{P},\tau}$ is induced by the first projection $pr_1 : \mathcal{P}ar_t(G) \times_S \mathcal{P}ar_s(G) \to \mathcal{P}ar_t(G)$.

Proposition 12.11 *The Big Cell Open Covering of* $\mathcal{P}ar_t(G)$

$$\mathcal{P}ar_t(G) = \bigcup \overline{\Omega}_{t,R_+} \quad (cf. \text{ Definition 12.10}),$$

where $G = (G, T, M, R)$ *and* R_+ *runs on the positive root system of* R, *trivializes the locally trivial fibration defined by* pr$_{\mathcal{P},\tau}$. *One has:*

$$\mathcal{S}tand(\tau_S) = \bigcup \mathcal{S}tand(\tau_S)_{\overline{\Omega}_{t,R_+}} = \bigcup \sigma_{R_+}^{(\tau)}((\overline{\Omega}_{t,R_+} \times \mathcal{S}tand(\tau_S, P_{t,R_+})) .$$

Proof *Identify* $\mathcal{U}_{t,R_+}$ *with* $\overline{\Omega}_{t,R_+}$ *by means of the isomorphism obtained from the definition of* $\overline{\Omega}_{t,R_+}$, $\mathcal{U}_{t,R_+} \xrightarrow{\sim} \overline{\Omega}_{t,R_+}$. *Define an isomorphism of* $\overline{\Omega}_{t,R_+}$-*schemes:*

$$\sigma_{R_+}^{(\tau)} : (\mathcal{U}_{t,R_+} \simeq) \overline{\Omega}_{t,R_+} \times_S \mathcal{S}tand(\tau_S, P_{t,R_+}) \to \mathcal{S}tand(\tau_S)_{\overline{\Omega}_{t,R_+}} = (pr_{\mathcal{P},\tau})^{-1}(\overline{\Omega}_{t,R_+}) ,$$

by

$$\sigma_{R_+}^{(\tau)} : (y, (P_{t,R_+}, Q)) \mapsto (int(y)(P_{t,R_+}), int(y)(Q)) ,$$

$(y, (P_{t,R_+}, Q))$ *being a section of* $\mathcal{U}_{t,R_+} \times_S \mathcal{S}tand(\tau_S, P_{t,R_+})$. *Clearly there is*

$$\mathcal{S}tand(\tau_S) = \bigcup \mathcal{S}tand(\tau_S)_{\overline{\Omega}_{t,R_+}} .$$

Thus the family of open subschemes $(\overline{\Omega}_{t,R_+})$ *of* $\mathcal{P}ar_t(G)$ *trivializes the locally trivial fibration defined by* pr$_{\mathcal{P},\tau}$.

12.4 Big Cell Open Covering of a Standard Position scheme

Let τ a type of relative position and (t, s) be the corresponding couple of parabolic types. There is a canonical section of $\mathcal{S}tand(\tau)$ over $S' = \mathcal{S}tand(G)$ given by the restriction of the diagonal section $\Delta : \mathcal{P}ar(G) \times_S \mathcal{P}ar(G) \to (\mathcal{P}ar(G) \times_S \mathcal{P}ar(G)) \times_S (\mathcal{P}ar(G) \times_S \mathcal{P}ar(G))$ to $\mathcal{S}tand(\tau)$. This section is given by a couple of parabolics $(\tilde{P}, \tilde{Q})$. The Big Cell Open Covering of $\mathcal{S}tand(\tau)$ allows one to prove that $(\tilde{P}, \tilde{Q})$ is in standard position. Thus the relative convex hull of $(\tilde{P}, \tilde{Q})$ may be defined. This construction is the corner stone needed to define the minimal generalized galleries in the relative case.

Given a parabolic $P = P_F$ ($F \in \mathcal{A}_E$) define the **centers of** $\mathcal{S}tand(\tau_S, P)$ as the facets $F' \in \mathcal{A}_E$ with $\tau(F, F') = \tau$, or equivalently the facets $F' \in \mathcal{A}_E$ such that $P_{F'}$ is a section of $\mathcal{S}tand(\tau_S, P)$. Remark that the stabilizer W_F of F in $W = W_{\mathcal{A}_E}$ acts transitively on the set of centers, as it follows from the fact that the intersection subgroup $P_F \cap P_{F'}$ is an (R)-subgroup containing T, and acts transitively on the set of maximal tori contained in $P_F \cap P_{F'}$. Thus if $P_{F'}$ and $P_{F''}$ are conjugate under the action of P_F, as all the three contain T, it results that they are also conjugate under the normalizer $N(T)$ of T.

Consider a chamber $C' \in \mathrm{Ch}\, \mathrm{St}_{F_t} \subset \mathrm{Ch}\, \mathcal{A}_E$, where $F_t = F_t(C)$, and $C = C_{R_+}$ is the chamber given by the splitting of G. The **Bruhat cell decomposition** of $\mathcal{S}\mathrm{tand}(\tau_S, P_{t,R_+})$ under the action of $B_{C'}$ may be written as

$$\mathcal{S}\mathrm{tand}(\tau_S, P_{t,R_+}) = \bigcup \mathcal{S}\mathrm{tand}(\tau(C', F')_S, B_{C'}),$$

where F' runs over the centers of $\mathcal{S}\mathrm{tand}(\tau_S, P_{t,R_+})$. Recall that $P_{t,R_+} = P_{F_t}$. More precisely one has

Proposition 12.12 *The natural morphism*

$$\coprod_{F' \in \mathcal{A}_E \ and \ \tau(F_t, F') = \tau} \mathcal{S}\mathrm{tand}(\tau(C', F')_S, B_{C'}) \to \mathcal{S}\mathrm{tand}(\tau_S, P_{t,R_+})$$

is a surjective monomorphism.

Remark that the center F' of a $B_{C'}$-Bruhat cell $\mathcal{S}\mathrm{tand}(\tau(C', F')_S, B)$ contained in $\mathcal{S}\mathrm{tand}(\tau_S, P_{t,R_+}))$ is a center of $\mathcal{S}\mathrm{tand}(\tau(C', F')_S, B_{C'})$. Given $C' \in \mathrm{St}_{F_{t,R_+}}$ there is a unique facet

$$F_{\tau,C'} \in \mathcal{A}_E.$$

characterized by

1. $\tau(F_t, F_{\tau,C'}) = \tau$;

2. $d(C', F_{\tau,C'}) = \max \{d(C', F') \mid F' \in \mathcal{A}_E, \tau(F_t, F') = \tau\}$

(cf. 9.22).

The facet $F_{\tau,C'}$ is the center of a $B_{C'}$-Bruhat cell contained in $\mathcal{S}\mathrm{tand}(\tau_S, P_{t,R_+})$ which is **open** in $\mathcal{S}\mathrm{tand}(\tau_S, P_{t,R_+})$. In fact there is one and only one center of $\mathcal{S}\mathrm{tand}(\tau_S, P_{t,R_+})$ satisfying this condition.

Lemma 12.13
The S-subscheme

$$\mathcal{S}\mathrm{tand}((\tau(C', F_{\tau,C'})_S, B_{C'}) \subset \mathcal{S}\mathrm{tand}(\tau_S, P_{t,R_+}),$$

*is **open** and **relatively schematically dense**.* *Consequently* $\mathcal{S}\mathrm{tand}((\tau(C', F_{\tau,C'})_S, B_{C'})^{schc.} = \mathcal{S}\mathrm{tand}(\tau_S, P_{t,R_+})^{schc.}$.

Proof *Let* $\mathbf{g}$ *be a minimal gallery of types of* $\mathcal{A}_E$ *with corresponding type of relative position given by* τ, *i.e. so that* $\pi_{(\mathbf{g}, P_{t,R_+})} : \mathcal{C}onf_G^m(\mathbf{g}_S, P_{t,R_+}) \longrightarrow \mathcal{S}\mathrm{tand}(\tau_S, P_{t,R_+})^{schc}$ *is a smooth resolution of singularities. Thus*

- *There is an open relatively schematically dense subscheme $\mathcal{C}onf_G^m(\mathbf{g}_S, P_{t,R_+})' \subset \mathcal{C}onf_G^m(\mathbf{g}_S, P_{t,R_+})$ so that the restriction of the resolving morphism $\pi_{(\underline{\tau}, P_{t,R_+})} : \mathcal{C}onf_G^m(\mathbf{g}_S, P_{t,R_+}) \longrightarrow \mathcal{C}onf_G^m(\underline{\tau}, P_{t,R_+})$ to $\mathcal{C}onf_G^m(\mathbf{g}_S, P_{t,R_+})'$ is an isomorphism;*

- $\mathcal{C}onf_G^m(\mathbf{g}_S, P_{t,R_+})'$ *is the reciprocal image of $\mathcal{S}tand(\tau_S, P_{t,R_+}) \hookrightarrow \mathcal{S}tand(\tau_S, P_{t,R_+})^{schc}$. In fact $\mathcal{C}onf_G^m(\mathbf{g}_S, P_{t,R_+})'$ is the image of a canonical section on $\mathcal{S}tand(\tau_S, P_{t,R_+})$.*

Let $\overline{s}$ be the type of a chamber, and consider $\overline{\mathbf{g}}$ the gallery of types obtained by composing $\overline{s} \supset t$ with $\mathbf{g}$. By definition of a Minimal Generalized Gallery of types it follows that $\overline{\mathbf{g}}$ is a minimal gallery of types. Consider the Generalized Minimal Gallery $\gamma_{\mathbf{g}}(F_t(R_+), F_{\tau,C'})$; as C' is incident to $F_t(R_+)$ and at maximal distance from $F_{\tau,C'}$, then the gallery $\gamma_{\overline{\mathbf{g}}}(C', F_{\tau,C'})$ obtained by composition of $\gamma_{\mathbf{g}}(F_t(R_+), F_{\tau,C'})$ with $C' \supset F_t(R_+)$ is minimal by definition of Minimal Generalized Gallery, and of type $\overline{\mathbf{g}}$. On the other hand, there are a natural isomorphism

$$\mathcal{C}onf_G^m(\overline{\mathbf{g}}_S, B_{C'}) \simeq \mathcal{C}onf_G^m(\mathbf{g}_S, P_{t,R_+}) ,$$

and an inclusion of open subsets

$$\mathcal{C}onf_G^m(\overline{\mathbf{g}}_S, B_{C'})' \subset \mathcal{C}onf_G^m(\mathbf{g}_S, P_{t,R_+})' ,$$

thus $\mathcal{C}onf_G^m(\overline{\mathbf{g}}_S, B_{C'})'$ is open and schematically dense in $\mathcal{C}onf_G^m(\mathbf{g}_S, P_{t,R_+})'$.

The unicity of $F_{\tau,C'}$ also follows from the above result. If F' satisfies the same conditions as $F_{\tau,C'}$ then $\mathcal{S}tand(\tau(C', F')_S, B_{C'})$ is relatively schematically dense in $\mathcal{S}tand(\tau_S, P_{t,R_+})$. The geometric fibers $\kappa(\overline{s})$, $\mathcal{S}tand(\tau(C', F')_S, B_{C'})_{\kappa(\overline{s})}$ and $\mathcal{S}tand((\tau(C', F_{\tau,C'})_S, B_{C'})_{(\kappa(\overline{s})}$ are open subschemes of the irreducible scheme $\mathcal{S}tand(\tau_S, P_{t,R_+})_{(\kappa(\overline{s})}$ and have a not empty intersection. The centers of these Bruhat cells are respectively F' and $F_{\tau,C'})$ thus it is concluded that $F' = F_{\tau,C'}$. Write

$$\tau_{C'} = \tau(C', F_{\tau,C'}) \in \mathrm{Relpos}\,\mathcal{A}_E.$$

The following proposition results from the transitivity of the action of W_F on the set of chambers incident to F.

Proposition 12.14
Keep the above notation.

- *Given $\tau \in \mathrm{Relpos}\,\mathcal{A}_E$ and $F \in \mathcal{A}_E$ of type t, the type of relative position $\tau_{C'}$ defined as above by F is independant of the choice of C'.*

- *Given a center F of $\mathcal{S}tand(\tau_S, P_{t,R_+})$ there is a chamber $C' \in \mathrm{Ch}\,\mathrm{St}_{F_t(C)}$, so that $F = F_{\tau,C'}$, i.e. every center of $\mathcal{S}tand(\tau_S, P_{t,R_+})$ is also the center of a $B_{C'}$-open cell for some chamber $C' \in \mathrm{Ch}\,\mathrm{St}_{\mathbf{F}_t}$.*

Observe that there may be several chambers C' satisfying the property of the proposition. Denote by $P_{\tau,C'}$ the parabolic subgroup corresponding to $F_{\tau,C'}$.

Definition 12.15
Let $C \in \operatorname{Ch} \mathcal{A}_E$ and $F \in \mathcal{A}_E$,

$$R(C,F) = \{\alpha \in R_C \mid H_\alpha \in \mathcal{H}(C,F)\}\,,$$

where R_C denotes the positive system of roots defined by $C \in \operatorname{Ch} \mathcal{A}_E$, H_α the wall of $\mathcal{A}_E$ defined by the root $\alpha \in R$, and $\mathcal{H}(C,F)$ the set of walls separating C and F. Denote by B_C the Borel subgroup given by the chamber C, i.e. by R_C, and

$$\mathcal{U}(C,F) \subset B_C$$

the subgroup defined by the closed system of roots $R(C,F)$.

The main property of $\mathcal{U}(C,F)$ is given by:

Lemma 12.16
There is an isomorphism of S-schemes

$$\mathcal{U}(C,F) \xrightarrow{\sim} \mathcal{S}tand\left(\tau(C,F)_S, B_C\right),$$

defined by $x \mapsto int(x)(P_F)$.

Proposition 12.17
There is an open covering by affine open sets (if S is an affine scheme)

$$\mathcal{S}tand(\tau_S, P_t) = \bigcup_{C' \in \operatorname{Ch} \operatorname{St}_{F_t}} \mathcal{S}tand((\tau_{C'})_S, B_{C'})$$

(The Big Cell Open covering *of the P_t-cell $\mathcal{S}tand(\tau_S, P_t)$). Moreover there are isomorphisms $\mathcal{U}(C', F_{\tau,C'}) \xrightarrow{\sim} \mathcal{S}tand((\tau_{C'})_S, B_{C'})$.*

It is clear that this covering corresponds to the Big Cell Open Covering

$$\mathcal{P}ar_t(G) = \bigcup \overline{\Omega}_{t,R_+}\,,$$

where R_+ runs over the set of positive systems of roots given by $\mathcal{R}$.

In view of the above cell decomposition (cf. 12.12) it is easy to see that this Proposition follows from the following

Proposition 12.18
Given $F' \in \mathcal{A}_E$ with $\tau(F_t, F') = \tau$, i.e. a center of $\mathcal{S}tand(\tau_S, P_t)$, there exists $C(F') \in \operatorname{Ch} \operatorname{St}_{F_t}$, such that:

1. $\tau(C(F'), F') = \tau(C, F_{\tau,C})$, *i.e. $\mathcal{S}tand(\tau(C(F'), F')_S, B_{C(F')})$ is the $B_{C(F')}$-open cell, with center F', in $\mathcal{S}tand(\tau_S, P_t)$;*

2. $Stand(\tau(C, F')_S, B_C) \subset Stand(\tau(C(F'), F')_S, B_{C(F')})$,

i.e. the B_C-cell $Stand(\tau(C, F')_S, B_C)$ may be embedded in the big cell $Stand(\tau(C(F'), F')_S, B_{C(F')})$ with the same center F'.

Proof Define $C(F') = proj_{c(F_t, F')} C$. Let it be proved that:

a) $\tau(C(F'), F') = \tau(C, F_{\tau, C})$;

b) $R_+(C, F') \subset R_+(C(F'), F')$.

Clearly condition a) corresponds to condition 1. On the other hand, in view of the isomorphism

$$\mathcal{U}(C(F'), F') \xrightarrow{\sim} Stand(\tau(C(F'), F')_S, B_{C(F')}),$$

one deduces that b) $\Rightarrow$ 2.

By definition of $C(F')$ one has $c(F_t, F') \in C(F')$ From Proposition 9.18 it follows that $C(F') \subset$ Ch St$_{F_t}$ is at maximal distance from F' with $\tau(F_t, F') = \tau$. Thus $Stand(\tau(C(F'), F')_S, B_{C(F')})$ is the $B_{C(F')}$-open cell in $Stand(\tau_S, P_t)$, and it is concluded that $\tau(C(F'), F') = \tau(C, F_{\tau, C})$.

To prove b) it suffices to show that there is a minimal gallery $\Gamma(C(F'), F')$ containing C. This gives

$$\mathcal{H}(C, F') \subset \mathcal{H}(C(F'), F'),$$

and consequently

$$R_+(C, F') \subset R_+(C(F'), F') \ .$$

Let

$$\Gamma_1 \subset St_{F_t},$$

be a minimal gallery between $c(F_t, F')$ and C, which may be seen as a gallery between $C(F') = proj_{c(F_t, F')} C$ and C. Choose a minimal gallery

$$\Gamma_2 = \Gamma(C, F') \subset Env(C, F') \ ,$$

and write

$$\Gamma = \Gamma(C(F'), F') = \Gamma_1 \circ \Gamma_2.$$

Remark that if $H \in \mathcal{H}(c(F_t, F'), C) \subset \mathcal{H}_{F_t}$ then:

$$\text{``}H \text{ separates } c(F_t, F') \text{ from } F' \text{ ''}.$$

Thus as H separates also $c(F_t, F')$ from C, it is deduced that

$$H \notin \mathcal{H}(C, F') \ (cf. \ Definition \ 9.17).$$

It follows that the set of walls $\mathcal{H}(\Gamma)$ crossed by Γ is given by:

$$\mathcal{H}(\Gamma) = \mathcal{H}(c(F_t, F'), C) \coprod \mathcal{H}(C, F').$$

Thus Γ crosses each wall it encounters only once.

By the above remark, if $H \in \mathcal{H}(c(F_t, F'), C)$, as $H \in \mathcal{H}_{F_t}$, then H separates $c(F_t, F)$ from F'. If $H \in \mathcal{H}(C, F')$ and $H \notin \mathcal{H}_{F_t}$, then H separates F_t and F' and thus H separates $c(F_t, F')$ and F'.

If $H \in \mathcal{H}(C, F') \cap \mathcal{H}_{F_t}$, then $F' \notin H$, thus $c(F_t, F') \notin H$, and by the above remark H separates $c(F_t, F')$ from F'. (The carrier of $c(F_t, F')$ is the same as that of the convex hull of (F_t, F').) Thus all the hyperplanes that Γ crosses separate $C(F')$ from F', and it crosses each one of them only once. It is concluded that Γ is a minimal gallery. This achieves the proof.

Denote by $\mathcal{F}\mathrm{ram}(\mathcal{R})$ the set of positive root systems of $\mathcal{R}$. Write

$$I_\tau(\mathcal{R}) = \left\{ (R_+, C') \in \mathcal{F}\mathrm{ram}(\mathcal{R}) \times \mathrm{Ch}\, \mathcal{A}_E \mid C' \in \mathrm{St}_{\mathrm{F}_t(\mathrm{C}_{R_+})} \right\},$$

where as usual C_{R_+} denotes the chamber of $\mathcal{A}_E$ given by $R_+ \in \mathcal{F}\mathrm{ram}(\mathcal{R})$.

Let

$$\sigma_{R_+,C'}^{(\tau)} : \mathcal{U}_{t,R_+} \times_S \mathcal{U}(C', F_{\tau,C'}) \to \mathcal{S}\mathrm{tand}(\tau_S)_{\overline{\Omega}_{t,R_+}},$$

be the morphism defined by

$$\sigma_{R_+,C'}^{(\tau)} : (y, x) \mapsto \left(\mathrm{int}(y)(P_{t,R_+}), \mathrm{int}(yx)(P_{\tau,C'}) \right).$$

Observe that $\mathcal{U}_{t,R_+} = \mathcal{U}(C_{R_-}, F_{t(R_+)})$.

From 12.11 immediately follows

Proposition 12.19

*There is a **canonical open covering** of the Universal Bruhat cell $\mathcal{S}\mathrm{tand}(\tau_S)$ of the split group G defined by τ:*

$$\mathcal{S}\mathrm{tand}(\tau_S) = \bigcup_{(R_+,C')\in I_\tau(\mathcal{R})} \sigma_{R_+,C'}^{(\tau)} \left(\overline{\Omega}_{t,R_+} \times_S \mathcal{U}(C', F_{\tau,C'}) \right),$$

*which, if S is affine, is a refinement by **affine** open sets, which are indeed isomorphic to **affine spaces**, of the covering of Proposition 12.11.*

The properties of the morphisms $\left(\sigma_{R_+,C'}^{(\tau)} \right)_{(R_+,C')\in I_\tau(\mathcal{R})}$ are clearly resumed in the following

Proposition 12.20 *The morphism $\sigma_{R_+,C'}^{(\tau)}$ defines an open embedding of $\mathcal{U}_{t,R_+} \times_S \mathcal{U}(C', F_{\tau,C'})$ in $\mathcal{S}\mathrm{tand}(\tau_S)_{\overline{\Omega}_{t,R_+}}$, namely the image*

$$\sigma_{R_+,C'}^{(\tau)} \left(\mathcal{U}_{t,R_+} \times_S \mathcal{U}(C', F_{\tau,C'}) \right) \subset \mathcal{S}\mathrm{tand}(\tau_S)_{\overline{\Omega}_{t,R_+}},$$

*is a **relatively schematically dense open affine** subscheme and the induced morphism by $\sigma_{R_+,C'}^{(\tau)}$*

$$\mathcal{U}_{t,R_+} \times_S \mathcal{U}(C', F_{\tau,C'}) \to \sigma_{R_+,C'}^{(\tau)} \left(\mathcal{U}_{t,R_+} \times_S \mathcal{U}(C', F_{\tau,C'}) \right),$$

is an isomorphism.

12.5   Apartment subschemes

Definition 12.21 *Let T be a maximal torus of G. Define the S-functor $\mathcal{F}ix(T) \subset \mathcal{P}ar(G)$ by:*

$$\mathcal{F}ix(T): \; S' \; \to \; \{P \subset G_{S'} \text{ parabolic subgroup } / \, T_{S'} \subset P\},$$

where S' is an S-scheme. (**The Apartment subscheme defined by the maximal torus T**)

Let $E = (T, M, R, R_0, (X_\alpha)_{\alpha \in R_0})$ be a frame of G given by a trivialization of T. Thus there is an isomorphism $\mathcal{A}_E \simeq \mathcal{F}ix(T)$ so that the root Φ_α of $\mathcal{A}_E$ corresponds to the representable subfunctor $\mathcal{F}_\alpha \cap \mathcal{F}ix(T) \subset \mathcal{F}ix(T)$.

Remark 12.22 *As a parabolic subgroup P of G is its own normalizer it is obtained that:*

$$(T_{S'} \subset P) \Longleftrightarrow ((\forall t \in T_{S'}) \; int(t) \cdot (P) = P).$$

Thus $\mathcal{F}ix(T) \subset \mathcal{P}ar(G)$ is the fixed points functor of the conjugation action of T on $\mathcal{P}ar(G)$. It is concluded that $\mathcal{F}ix(T)$ is representable. However an ad-hoc proof of this fact is given.

Recall the isomorphism $\prod_{\alpha \in R_-} \mathrm{Spec}\,(\mathcal{O}_S[t_\alpha]) \overset{\sim}{\to} \overline{\Omega}_{R_+}$, which may be para-phrased as follows. For every section of $(\overline{\Omega}_{R_+})_{S'}$, given by some Borel subgroup $B \subset G_{S'}$, there exists a unique section (x_α) of $\prod_{\alpha \in R_-} \mathcal{O}_{S'}$, with

$$B = int\left(\prod \exp(x_\alpha X_\alpha)\right) \cdot (B_{R_+})_{S'}.$$

Let the action of T relatively to the (x_α)-coordinates in $\overline{\Omega}_{R_+}$ be described. For every section t of T one has

$$int(t)(B) = B' \Leftrightarrow x'_\alpha = \alpha(t)x_\alpha (\alpha \in R_-),$$

where (x_α) (resp. x'_α)) denotes the coordinates of B (resp. B') in $\tilde{\Omega}_{R_+}$, i.e.

$$B = int\left(\prod \exp(x_\alpha X_\alpha)\right)(B_{R_+}) \; \left(\text{resp. } B' = int\left(\prod \exp(x'_\alpha X_\alpha)\right)(B_{R_+})\right).$$

Then the invariance condition on B, under normalization by T, translates as:

For every section t of T : $\alpha(t)\, x_\alpha = x_\alpha$.

This implies that the sections of $\mathcal{F}ix\,(T) \cap \overline{\Omega}_{R_+}$, are characterized as the zeroes of the ideal $\mathfrak{I}((t_\alpha))$ of $\mathcal{O}[t_\alpha](\alpha \in R_-)$, generated by $(t_\alpha)(\alpha \in R_-)$. With the usual notation one has

$$V((t_\alpha)) = V(\mathfrak{I}((t_\alpha))) \simeq \mathcal{F}ix\,(T) \cap \overline{\Omega}_{R_+}.$$

By considering the big cell open covering of $\mathcal{B}or(G)$

$$\mathcal{B}or(G) = \bigcup \overline{\Omega}_{R_+},$$

where R_+ runs over all possible systems of positive roots of R, we deduce the

Proposition 12.23
The S-functor

$$\mathcal{F}ix(T) \cap \mathcal{B}or(G) \subset \mathcal{B}or(G)$$

is representable by the constant S-scheme

$$(\mathrm{Ch}\, \mathcal{A}_\mathrm{E})_S,$$

i.e. there is a canonical isomorphism

$$(\mathrm{Ch}\, \mathcal{A}_\mathrm{E})_S \simeq \mathcal{F}ix(T) \cap \mathcal{B}or(G),$$

associating to $(R_+)_S$ the parabolic (resp. Borel) subgroup $B_{R_+} = P_{R_+}$, given by "the center of $\overline{\Omega}_{R_+}$", i.e. with (x_α)-coordinates, $x_\alpha = 0$ $(\alpha \in R_-)$.

Let P_{t,R_+} be the parabolic of G of type t, containing B_{R_+}. Proceeding as in the case $\mathcal{P}ar_t(G) = \mathcal{B}or(G)$, one introduces $(t_\alpha)(\alpha \in R - R_t)$-coordinates. Here $R_t \subset R$ denotes the parabolic set defining P_{t,R_+}, i.e. $P_{t,R_+} = P_{R_t}$. It is then proved, following the same reasoning as above, that

$$\mathcal{F}ix(T) \cap \overline{\Omega}_{t,R_+},$$

is given by $V((t_\alpha)_{\alpha \in R - R_t})$. It has thus been proved the following

Proposition 12.24
There are canonical isomorphisms:

1. $((\mathcal{A}_E)_t)_S \simeq \mathcal{F}ix(T) \cap \mathcal{P}ar_t(G)$;

2. $(\mathcal{A}_E)_S = \coprod ((\mathcal{A}_E)_t)_S \, (t \in typ\, \mathcal{A}_E) \simeq \mathcal{F}ix(T)$.

Thus $\mathcal{F}ix(T)$ is representable by a twisted constant S-scheme.

Remark 12.25
$\mathcal{N}orm_G(T)$ *acts as a group of automorphisms on* $\mathcal{F}ix(T)$. *This action factors through the quotient*

$$\mathcal{N}orm_G(T) \to W_T = \mathcal{N}orm_G(T)/T,$$

i.e. the Weyl group S-scheme W_T defined by the maximal torus T acts on the S-scheme $\mathcal{F}ix(T)$.

12.6   The Retraction morphism of the Parabolics scheme on an Apartment scheme

Keep the above hypothesis on G and denote by E its frame. Write $A = \mathcal{A}_E$. Let $C \in \mathrm{Ch}\, A$ be the chamber given by the system of positive roots R_+ defined by E, and B the corresponding Borel subgroup. There is a bijection

$$\mathrm{Relpos}'\, A = \{\tau(C, F) \mid F \in A\} \overset{\sim}{\to} A,$$

associating to $\tau \in \mathrm{Relpos}'\, A$ the facet $F_\tau \in A$ such that $\tau(C, F_\tau) = \tau$. By Proposition 12.24 there is a canonical isomorphism of S-schemes $A_S \overset{\sim}{\to} \mathcal{F}\mathrm{ix}(T)$, defined by $F_S \mapsto P_F$, for $F \in A$. On the other hand, there is a canonical isomorphism $\mathrm{Relpos}'_G \simeq (\mathrm{Relpos}'\, A)_S$. Thus one deduces a canonical isomorphism

$$\mathrm{Relpos}'_G \overset{\sim}{\to} \mathcal{F}\mathrm{ix}(T).$$

Definition 12.26
Define the **retraction morphism**

$$\rho_E : \mathcal{S}\mathrm{tand}(B) \to \mathcal{F}\mathrm{ix}(T),$$

as the composition of the morphism

$$t'_2 : \mathcal{S}\mathrm{tand}(B) \to \mathrm{Relpos}'_G$$

induced by t_2, followed by the isomorphism $\mathrm{Relpos}'_G \overset{\sim}{\to} \mathcal{F}\mathrm{ix}(T)$.

For example given a section (B, P) of $\mathcal{S}\mathrm{tand}(B)$ with $t_2(B, P) = \tau_S$ ($\tau \in \mathrm{Relpos}\, A$), one has $\rho_E : (B, P) \mapsto P_\tau$, where P_τ is the parabolic corresponding to the center of $\mathcal{S}\mathrm{tand}(t_2(B, P), B)$.

Observe that the fibers of the Retraction morphism are precisely the Bruhat cells defined by B_C and the centers F_τ.

12.7   Convex Hull of a couple of parabolics in standard position

Keep the above hypothesis on G. Let (P, Q) be a couple of parabolics of G in standard position. Recall that the intersection subgroup $P \cap Q$ is an (R) subgroup (cf. [23], *Exp. XXII*, §5), i.e. $P \cap Q$ is S-smooth, of finite presentation, with connected geometric fibers, and containing a maximal torus T locally for the etale topology. Assume that $T \subset P \cap Q$ (resp. $T \subset (P, Q)$), where T is the maximal torus given by E.

Let $\mathcal{R}$ be the $\mathbb{Z}$-root system defined by the frame E of G and $\mathcal{A}_E$ the apartment defined by $\mathcal{R}$. Let $\mathcal{D}_{\mathcal{A}_E}$ denote the set of combinatorial roots of $\mathcal{A}_E$. The system of roots $\mathcal{R}$ being reduced, it is recalled that there exists a bijection

$$R \to \mathcal{D}_{\mathcal{A}_E},$$

associating with $\alpha \in R$, the combinatorial root

$$\Phi_\alpha = \{F \in \mathcal{A}_E \mid \alpha \in R_F\} \subset \mathcal{A}_E.$$

Given $(F, F') \in \mathcal{A}_E \times \mathcal{A}_E$ write

$$\mathcal{D}_{\mathcal{A}_E}(F, F') = \{\Phi \in \mathcal{D}_{\mathcal{A}_E} \mid F, F' \in \Phi\}.$$

By definition of the convex hull $\mathrm{Env}(F, F')$ one has

$$\mathrm{Env}(F, F') = \bigcap \mathcal{D}_{\mathcal{A}_E}(F, F').$$

Consider the subgroups corresponding to the facets F and F',

$$P = P_F = P_{R_F} \quad (\text{resp. } Q = Q_{F'} = P_{R_{F'}}).$$

From loc. cit., it follows that the subgroup $P \cap Q$ is characterized by its Lie algebra

$$\mathrm{Lie}(P \cap Q) = \mathcal{G}^0 \oplus (\oplus \mathcal{G}^\alpha(\alpha \in R_F \cap R_{F'})),$$

where $\mathcal{G}^0 = \mathrm{Lie}(T)$. On the other hand, $R_F \cap R_{F'}$ is a closed system of roots, and the bijection $R \overset{\sim}{\to} \mathcal{D}_{\mathcal{A}_E}$ induces a bijection

$$R_F \cap R_{F'} \overset{\sim}{\to} \mathcal{D}_{\mathcal{A}_E}(F, F').$$

Let

$$P_\alpha = \exp(W(\mathcal{G}^\alpha)) \subset G$$

be the image of the vector group $W(\mathcal{G}^\alpha)$ by the morphism

$$\exp : W(\mathcal{G}^\alpha) \to G$$

(cf. loc. cit., Exp. XXII, Théorème 1.1).

The group $P \cap Q$ is generated by the maximal torus T and by the set of subgroups $(P_\alpha)_{\alpha \in R_F \cap R_{F'}}$. Thus one obtains the

Proposition 12.27
The following four statements concerning a parabolic subgroup $\overline{P} \subset G$ are equivalent:

1. $P \cap Q \subset \overline{P}$;

2. $(P_\alpha) \subset \overline{P} \ (\alpha \in R_F \cap R_{F'})$, $T \subset \overline{P}$ ($\overline{P}$ defines a section of $\mathcal{F}_\alpha$);

3. $\mathcal{G}^\alpha \subset \mathrm{Lie}(\overline{P}) \ (\alpha \in R_F \cap R_{F'})$, $T \subset \overline{P}$;

4. there exists $\overline{F} \in \mathrm{Env}^{\mathcal{A}_E}(F, F')$ with $\overline{P} = P_{\overline{F}}$.

Definition 12.28
Define the S-subfunctor

$$\mathcal{F}ix(P,Q) \subset \mathcal{P}ar(G),$$

by

$$\mathcal{F}ix(P,Q)(S') = \left\{ \overline{P} \text{ parabolic of } G_{S'} \mid (P \cap Q)_{S'} \subset \overline{P} \right\}.$$

(The Convex Hull scheme defined by the couple (P,Q))

Let it be proved that $\mathcal{F}ix(P,Q)$ is representable by giving a local description of this functor in terms of a splitting of G. From Proposition 12.27,3, and Definition 12.28 one obtains the

Proposition 12.29
Let $G = (G,T,M,R)$, one has

$$\mathcal{F}ix(P,Q) = \mathcal{F}ix(T) \cap \left(\bigcap \mathcal{F}_\alpha \left(\alpha \in R_F \cap R_{F'} \right) \right).$$

It is immediate that for $\alpha \in R$ there is an isomorphism of S-schemes

$$(\Phi_\alpha)_S \xrightarrow{\sim} \mathcal{F}\text{ix}(T) \cap \mathcal{F}_\alpha,$$

where $\Phi_\alpha \subset \mathcal{A}_E$ denotes the combinatorial root defined by $\alpha \in R$, defined by

$$F_S \longmapsto P_F (= P_{R_F}).$$

It is then deduced that

$$\mathcal{F}\text{ix}(P,Q) = \bigcap_{\alpha \in R_F \cap R_{F'}} (\mathcal{F}\text{ix}(T) \cap \mathcal{F}_\alpha) = \bigcap_{\alpha \in R_F \cap R_{F'}} (\Phi_\alpha)_S = \left(\bigcap_{\alpha \in R_F \cap R_{F'}} \Phi_\alpha \right)_S.$$

One finally obtains the

Proposition 12.30
Let $F, F' \in \mathcal{A}_E$. Write $P = P_F$, $Q = P_{F'}$. There is a canonical isomorphism

$$Env(F,F')_S \xrightarrow{\sim} \mathcal{F}ix(P,Q),$$

given by

$$\overline{F}_S \mapsto P_{\overline{F}},$$
$$\left(\overline{F} \in Env(F,F') \right).$$

Corollary 12.31
The S-subfunctor

$$\mathcal{F}ix(P,Q) \subset \mathcal{P}ar(G)$$

is representable by a twisted constant finite S-scheme.

Given a maximal torus $T \subset P \cap Q$ of G the functor $\mathcal{F}\mathrm{ix}(P,Q)$ may be seen as a subcomplex scheme of $\mathcal{F}\mathrm{ix}\,(T)$. We recall that two maximal tori of G contained in the (R)-subgroup $P \cap Q$ are conjugate locally for the etale topology by an element of $P \cap Q$. It follows that

Proposition 12.32

Given two maximal tori $T, T' \subset P \cap Q$, i.e. satisfying $\mathcal{F}ix(P,Q) \subset \mathcal{F}ix\,(T)$ and $\mathcal{F}ix(P,Q) \subset \mathcal{F}ix\,(T')$ a conjugation isomorphism α satisfying $\alpha(\mathcal{F}ix\,(T)) = \mathcal{F}ix\,(T')$ and $\alpha|_{\mathcal{F}ix(P,Q)} = Id_{\mathcal{F}ix(P,Q)}$ exists locally for the etale topology.

Remark 12.33

From Proposition 12.32 it follows that the convex hull $\mathrm{Env}(F, F')$ of two facets of the building I, of a reductive k-reductive group G over an algebraically closed field k, is independant of the apartment A containing F and F'. In fact the statement of this proposition corresponds to one of the definition axioms of abstract buildings. This justifies the introduction of (R)-subgroups in this work. The main example is thus given by the intersection of two parabolic subgroups and their main properties are:

1) *Two maximal tori in an (R)-subgroup are conjugate locally for the etale topology;*

2) *the R-groups of a reductive S-group G functor is representable by a quasi-projective S-scheme.*

The latter allows one to carry out the building constructions in the relative case. In fact the representability of the apartment functor (resp. convex hull functor) follows from it. It will be seen that another consequence is that minimal generalized galleries configurations depend uniquely on the convex hull defined by its extremities and not on the apartment containing it.

12.8 Convex Hull of the Tautological Couple of parabolics

Keep the above notation and hypothesis on G. Fix a type of relative position τ with corresponding couple of types (t, s).

Definition 12.34

- Let G be a reductive S-group scheme and $\underline{\tau}$ a section of $Relpos_G$. The graph of the natural embedding $j_{\underline{\tau}}: Stand(\underline{\tau}) \longrightarrow \mathcal{P}ar(G) \times_S \mathcal{P}ar(G)$ defines a section

$$\xi_{\underline{\tau}}: Stand(\underline{\tau}) \longrightarrow (\mathcal{P}ar(G) \times_S \mathcal{P}ar(G)) \times_S Stand(\underline{\tau})$$

of the product $\mathcal{P}ar(G) \times_S \mathcal{P}ar(G)$ over $Stand(\underline{\tau})$. It will be seen that $\xi_{\underline{\tau}}$ is in fact given by a couple of parabolics $\left(\tilde{P}, \tilde{Q}\right)_{\underline{\tau}}$ in **standard position**, and thus defines a section of $Stand(\underline{\tau})$ over itself. More precisely, write $S' = Stand(\underline{\tau})$, then $\xi_{\underline{\tau}}$ may be seen as a section of $Stand(\underline{\tau})$ over S'. One calls $\left(\tilde{P}, \tilde{Q}\right)_{\underline{\tau}}$ the **couple of tautological parabolics of type** $\underline{\tau}$.

- The section $\xi_{\underline{\tau}}$ may be extended to a section ξ_Σ of the **Universal Schubert Cell** Σ (cf. Definition 12.44) over Σ itself. This section is characterized by the property: Given a section $\underline{\tau}$ of $Relpos_G$ then:

$$\xi_\Sigma|_{Stand(\underline{\tau})} - \xi_{\underline{\tau}}.$$

Write $\left(\tilde{P}, \tilde{Q}\right)_\Sigma$ for the corresponding couple of parabolics in standard position.

- If G is endowed with a frame E and τ is a type of relative position of $\mathcal{A}_E$ with corresponding couple of types (t, s) write

$$\xi_\tau = \xi_{\tau_S}: Stand(\tau_S) \longrightarrow (\mathcal{P}ar_t(G) \times_S \mathcal{P}ar_s(G)) \times_S Stand(\tau_S)$$

and $\left(\tilde{P}, \tilde{Q}\right)_\tau = \left(\tilde{P}, \tilde{Q}\right)_{\tau_S}$. There are identifications

$$\xi_\Sigma \simeq \coprod_{\tau \in Relpos \ A(\mathcal{R}(E).)} \xi_\tau \ \left(resp. \ \left(\tilde{P}, \tilde{Q}\right)_\Sigma \simeq \coprod_{\tau \in Relpos \ A(\mathcal{R}(E).)} \left(\tilde{P}, \tilde{Q}\right)_\tau\right).$$

Remark 12.35 The section $(\tilde{P}, \tilde{Q})_{\underline{\tau}}$ of $\mathcal{P}ar(G)_t \times_S \mathcal{P}ar(G)_s \times_S Stand(\underline{\tau})$, over $Stand(\underline{\tau})$, is obtained as the restriction to

$$Stand(\underline{\tau}) \subset \mathcal{P}ar(G)_t \times_S \mathcal{P}ar(G)_s,$$

of the diagonal section

$$\Delta_X: X \to X \times_S X,$$

with $X = \mathcal{P}ar(G)_t \times_S \mathcal{P}ar(G)_s$.

Let the above terminology be justified by proving that $\left(\tilde{P}, \tilde{Q}\right)_\tau$ is in **standard position**. Consider the restriction of the section ξ_τ of $Stand(\tau_{S'})$ corresponding to $(\tilde{P}, \tilde{Q})_\tau$ to the open subset of the Big Cell open covering indexed by (R_+, C'). Write $S' = Stand(\tau_S)$, and $U = \mathcal{U}_{t,R_+} \times_S \mathcal{U}(C', F_{\tau,C'})$. The pull-back G_U is endowed with the pull-back frame E_U of E. Denote by T the maximal torus given by E. The section $(P_{t,R_+}, P_{\tau,C'})$ of $Stand(\tau_S)$ gives rise to the section $(P_{t,R_+}, P_{\tau,C'})_U$ of $Stand(\tau_{S'})$ given by $(\xi_\tau)_U$. Observe that $T \subset (P_{t,R_+}, P_{\tau,C'})$ implies $T_U \subset (P_{t,R_+}, P_{\tau,C'})_U$. On the other hand, the section $\left(\tilde{P}, \tilde{Q} \right)_{\tau_U}$ over the section (x,y) of $U = \mathcal{U}_{t,R_+} \times_S \mathcal{U}(C', F_{\tau,C'})$ is given by $(x,y) \mapsto \left(\mathrm{int}(y)(P_{t,R_+}), \mathrm{int}(yx)(P_{\tau,C'}) \right)$. It results that

$$\mathrm{int}(yx)(T) \subset \left(\mathrm{int}(y)(P_{t,R_+}), \mathrm{int}(yx)(P_{\tau,C'}) \right),$$

i.e.

$$\mathrm{int}(yx)(T) \subset \mathrm{int}(y)(P_{t,R_+}) \cap \mathrm{int}(yx)(P_{\tau,C'}),$$

and thus $\tilde{P}$ and $\tilde{Q}$ contain a common maximal torus locally, namely

$$\mathrm{int}(yx)(T) \subset \tilde{P} \cap \tilde{Q}.$$

It has thus been proved the

Proposition 12.36
The couple $(\tilde{P}, \tilde{Q})_\tau$ *of parabolics of* $G_{S'}$ *is in standard position, and one has*

$$t_2((\tilde{P}, \tilde{Q})_\tau) = \tau_{S'}.$$

Definition 12.37
Let G denote a reductive S-group scheme, $\underline{\tau}$ a section of $Relpos_G$.

- *Write $S' = Stand(\underline{\tau})$. The couple of parabolics $(\tilde{P}, \tilde{Q})_{\underline{\tau}}$ of $G_{S'}$, is in standard position, and thus defines a section of $Stand(G_{S'}) = Stand(G) \times_S S'$. Define the finite S'-scheme*

$$\mathcal{F}ix_{\underline{\tau}} = \mathcal{F}ix(\tilde{P}, \tilde{Q})_{\underline{\tau}} \subset \mathcal{P}ar(G_{S'})$$

  (The Convex Hull of the tautological couple $(\tilde{P}, \tilde{Q})_{\underline{\tau}}$**).**

- *Denote by $\mathcal{F}ix_\Sigma \longrightarrow \Sigma$ the finite scheme characterized by: for all sections $\underline{\tau}$ of $Relpos_G$,*

$$\mathcal{F}ix_\Sigma|_{Stand(\underline{\tau})} = \mathcal{F}ix_{\underline{\tau}}$$

  (The Convex Hull of the tautological couple $(\tilde{P}, \tilde{Q})_\Sigma$**).**

Proposition 12.38 *(The Universal Property of $\mathcal{F}ix_{\underline{\tau}}$)*
Let (P,Q) be a section of $Stand(\underline{\tau})$ defined over the S-scheme S', and $S' \to Stand(\underline{\tau})$ the corresponding morphism. Then one has

$$\mathcal{F}ix_{\underline{\tau}} \times_{Stand(\underline{\tau})} S' \simeq \mathcal{F}ix(P,Q).$$

Proof *Retain the above notation. Let $E_{S'}$ be the frame of $G_{S'}$ pull-back of E_U by $S' \to U$. There is a natural building isomorphism $\mathcal{A}_{E_U} \simeq \mathcal{A}_{E_{S'}}$. Let $F, F' \in \mathcal{A}_{E_U}$ so that*
$(\tilde{P}, \tilde{Q}) = (P_F, P_{F'})$ *in* G_U *and* $(P, Q) = (P_F, P_{F'})$ *in* $G_{S'}$. *There is an induced isomorphism of constant S'-schemes* $(Env^{\mathcal{A}_{E_U}}(F, F')_U) \times_U S' \simeq Env^{\mathcal{A}_{E_{S'}}}(F, F')_{S'}$, *which proves that the natural morphism*

$$\mathcal{F}ix(\tilde{P}, \tilde{Q}) \times_U S' \to \mathcal{F}ix(P, Q)$$

is also an isomorphism, on account of the isomorphisms $(Env^{\mathcal{A}_{E_U}}(F, F')_U) \times_U S' \simeq \mathcal{F}ix(\tilde{P}, \tilde{Q}) \times_U S'$ *(resp.* $Env^{\mathcal{A}_{E_{S'}}}(F, F')_{S'} \simeq \mathcal{F}ix(P, Q))$, *and a commutative square with vertices these four schemes and with arrows these four morphisms.*

12.8.1 The Projection Morphism

Let (P, Q) be a couple of parabolics of the reductive S-group scheme G in standard position. Assume that G_S is endowed with a frame. Then there is an isomorphism: $Env(P, Q)_S \simeq \varinjlim_E Env(F_E, F'_E)_S$, where E runs on the set of frames of G_S "adapted" to the couple (P, Q), and (F_E, F'_E) represents the corresponding couple of facets in $\mathcal{A}_E$. Suppose that $(S_i \longrightarrow S)$ is an etale covering such that G_{S_i} is endowed with a frame E_i. Denote by G_i (resp. G_j) the pull-back of G_{S_i} (resp. G_{S_j}) by the projection $S_i \times_S S_j \longrightarrow S_i$ (resp. $S_i \times_S S_j \longrightarrow S_j$), and by $\alpha_{i,j} : G_i \simeq G_j$ the cocyle isomorphism. Denote also E_i (resp. E_j) the frame of G_i (resp. G_j) given by the pull-back of the frame of G_{S_i} (resp. G_{S_j}).
Fix (i, j), write $\alpha = \alpha_{ij}$, $E = E_i$. Let x be a section of G_i. Denote by $int(x)(E)$ the frame obtained from E by conjugation, and by $\alpha(E)$ the image frame by α of G_j. There is a commutative diagram:

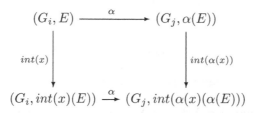

On the other hand given an isomorphism of split groups $\beta : (G_1, E_1) \longrightarrow (G_2, E_2)$, i.e. an isomorphism $\beta : G_1 \longrightarrow G_2$ such that $\beta(E_1) = E_2$, and $F, F' \in \mathcal{A}_E$ then: $\beta(proj_{F'} F) = proj_{\beta(F')} \beta(F)$. On account of these developements the following can be stated

Definition 12.39
Keep the above notation.

- Let (P, Q) be a couple of parabolics in standard position of G. Denote by $proj_{(P,Q)}$ the section of $\mathcal{F}ix(P, Q)$ so that if E is a frame of $G_{S'}$, where S' denotes an etale scheme, then $(proj_{(P,Q)})_{S'}$ corresponds to the constant section $(proj_{F'}\ F)_{S'}$ by the isomorphism $Env(F, F')_{S'} \simeq \mathcal{F}ix(P, Q)$. Define the $\underline{\tau}$-**projection section**

$$proj^{(\underline{\tau})} : Stand(\underline{\tau}) \longrightarrow \mathcal{F}ix_{\underline{\tau}} \subset Stand(\underline{\tau}) \times_S Par(G)$$

  as the section $proj_{(\tilde{P}, \tilde{Q})}$ given by the tautological couple $(\tilde{P}, \tilde{Q})_{\underline{\tau}}$ of type $\underline{\tau}$ of the finite $Stand(\underline{\tau})$-scheme $\mathcal{F}ix_{\underline{\tau}} = \mathcal{F}ix(\tilde{P}, \tilde{Q})$.

- Define the $\underline{\tau}$-**projection morphism**

$$\mathcal{P}roj^{(\underline{\tau})} : Stand(\underline{\tau}) \longrightarrow \mathcal{P}ar(G)$$

  as the composition of $proj^{(\underline{\tau})}$ followed by the projection on $\mathcal{P}ar(G)$.

- Define the Σ-**projection morphism**

$$\mathcal{P}roj^{(\Sigma)} : \Sigma \longrightarrow \mathcal{P}ar(G)$$

  as the morphism characterized by: for all sections $\underline{\tau}$ of $Relpos_G$

$$\mathcal{P}roj^{(\Sigma)}|_{Stand(\underline{\tau})} = \mathcal{P}roj^{(\underline{\tau})} \ .$$

Write $\mathcal{P}roj^{(\Sigma)}_{(P,Q)}$ for the image of (P, Q). Define the section $proj^{(\Sigma)} : \Sigma \longrightarrow \mathcal{F}ix_\Sigma$ following the same pattern.

12.9   Schematic closure of a subscheme

The following definitions are of use in defining Schubert schemes in the next section. For more details see §16.3.

Definition 12.40
Let $f : X \to Y$ be a morphism of schemes. It is said that the smallest closed subscheme $Y' \subset Y$, so that the canonical embedding $j_{Y'} : Y' \to Y$ factors f, is **the schematic image** of f, if it exists.

  If X is a subscheme of Y, and $j_X : X \to Y$ the canonical embedding, **the schematic closure** X^{schc} of X in Y is defined as the schematic image of j_X. (cf. [24], Ch. 1, Definition (6.10.1)).

  If the canonical injection $j_X : X \to Y$ is schematically dominant, it is said that X is **schematically dense** in Y (cf. [24], Ch. 1, Definition (5.4.2))

Definition 12.41
A morphism $f : X \to Y$ is **schematically dominant** if for every open

subset $U \subset Y$ and every closed subscheme $Z \subset U$, so that the restriction $f^{-1}(U) \to U$ of f factors as

$$f^{-1}(U) \xrightarrow{g} Z \xrightarrow{j_Z} U,$$

it is

$$Z = U.$$

The **transitivity of schematic images** is recalled.

Proposition 12.42
Let $f : X \to Y$ and $g : Y \to Z$ two morphisms. It is supposed that the schematic image Y' of f exists, and that the schematic image Z' of Y' by the restriction g' of g to Y' also exists. Then the schematic image of X by $g \circ f$ exists and is equal to Z'. (cf. loc. cit., Proposition (6.10.3))

The following Proposition gives sufficient conditions assuring the existence of the schematic image (resp. schematic closure) of $f : X \to Y$ (resp. $j_X : X \to Y$).

Proposition 12.43
The schematic image Y' of X by the morphism $f : X \to Y$ exists in each of the following two cases:

1. $f_*(\mathcal{O}_X)$ is a **quasi-coherent** $\mathcal{O}_Y$-module (this condition is satisfied if f is **quasi-compact** and **quasi-separated**).

2. X is **reduced**.

The underlying subspace of Y' is given by $\overline{f(X)}$ (= the closure in Y of the image of f), and f factors as

$$X \xrightarrow{g} Y' \xrightarrow{j_{Y'}} Y,$$

where g is schematically dominant. (cf. [24], Ch 1, Proposition (6.10.5))

It is proved in §16.3 that the scheme $\mathcal{S}\mathrm{tand}(\tau_S, P)$ of parabolics of type s in Standard Position with P as a **quasi-compact scheme** if S is affine, so that the scheme of couples of parabolics in standard position $\mathcal{S}\mathrm{tand}(\underline{\tau})$ is also a quasi-compact scheme. Thus the **Schematic Closure** $\mathcal{S}\mathrm{tand}(\tau_S, P)^{schc}$ (**resp.** $\mathcal{S}\mathrm{tand}(\underline{\tau})$) of $\mathcal{S}\mathrm{tand}(\tau_S, P)$ (resp. $\mathcal{S}\mathrm{tand}(\underline{\tau})^{schc}$) in $\mathcal{P}\mathrm{ar}_s(G)$ (resp. $\mathcal{P}\mathrm{ar}_t(G) \times_S \mathcal{P}\mathrm{ar}_S(G)$) **exists**.

12.10   The Universal Schubert Cell and the Universal Schubert scheme

Definition 12.44   *1.* The **Universal Schubert Cell** (resp. **Bruhat Cell**) of G *is by definition the G-orbit scheme of $Stand(G)$, given by the graph*

$$gr(t_2) \subset Stand(G) \times_S Relpos_G,$$

of the type of relative position morphism $t_2 : Stand(G) \to Relpos_G = Stand(G)/G$. *Write* $\Sigma = gr(t_2)$. *There is a natural embedding* $j_\Sigma : \Sigma \to (Par(G) \times_S Par(G)) \times_S Relpos_G$.

2. Define the **Universal Schubert scheme** $\overline{\Sigma}$ *of G as* **the schematic image** *of j_Σ (resp. the* **schematic closure** Σ^{schc} *of the Universal Schubert cell Σ) in*

$$(Par(G) \times_S Par(G)) \times_S Relpos_G .$$

Denote by $j_{\overline{\Sigma}} : \overline{\Sigma} \to Par(G) \times_S Par(G) \times_S Relpos_G$ *the natural embedding.*

Recall that:

1. The functor $Stand(G)$ is representable

2. The morphism $t_2 : Stand(G) \to Relpos_G$ is the quotient morphism $Stand(G) \longrightarrow Stand(G)/G$ and it is S-smooth, of finite presentation, with irreducible geometrical fibers, and thus faithfully flat.

Let

$$\left(\tilde{P}, \tilde{Q}\right)_\Sigma$$

be **the tautological couple** of parabolics in standard position of G_Σ. This couple is characterized locally for the etale topology as follows. Let E be a frame of $G_{S'}$, where $S' \longrightarrow S$ is an etale morphism. One has $\Sigma_{S'} = \coprod_{\tau \in Relpos\,\mathcal{A}_E} Stand(\tau_{S'})$, and $\left(\left(\tilde{P}, \tilde{Q}\right)_\Sigma\right)_{Stand(\tau_{S'})} = \left(\tilde{P}, \tilde{Q}\right)_{\tau_{S'}}$, where the first member denotes the restriction of $\left(\tilde{P}, \tilde{Q}\right)_\Sigma$ to $Stand(\tau_{S'})$, and the second the tautological couple $\left(\tilde{P}, \tilde{Q}\right)_{\tau_{S'}}$.

Define

$$pr_\mathcal{R} : \Sigma \longrightarrow Relpos_G \;\left(\text{resp. } pr_{\mathcal{R},\overline{\Sigma}} : \overline{\Sigma} \longrightarrow Relpos_G\right)$$

by $pr_\mathcal{R} = pr_3 \circ j_\Sigma$ (resp. $pr_{\mathcal{R},\overline{\Sigma}} = pr_3 \circ j_{\overline{\Sigma}}$), where $pr_3 : (Par(G) \times_S Par(G)) \times_S Relpos_G \longrightarrow Relpos_G$ denotes the canonical projection. Given a section $\underline{\tau}$ of $Relpos_G$ write $\Sigma_{\underline{\tau}} = pr_\mathcal{R}^{-1}(\underline{\tau})$. Then one has

$$\Sigma_{\underline{\tau}} = Stand(\underline{\tau}).$$

Let

$$\mathrm{pr}_{\mathcal{P}} = \mathrm{pr}_1 \circ j_{\Sigma} : \Sigma \longrightarrow \mathcal{P}\mathrm{ar}(G)$$

$$\left(\text{resp. } \mathrm{pr}_{\mathcal{P},\overline{\Sigma}} = \mathrm{pr}_1 \circ j_{\overline{\Sigma}} : \overline{\Sigma} \longrightarrow \mathcal{P}\mathrm{ar}(G)\right)$$

be the morphism obtained as the composition of j_{Σ} (resp. $j_{\overline{\Sigma}}$) followed by the first projection $\mathrm{pr}_1 : \mathcal{P}\mathrm{ar}(G) \times_S \mathcal{P}\mathrm{ar}(G) \times_S \mathrm{Relpos}_G \longrightarrow \mathcal{P}\mathrm{ar}(G)$. Given a section (τ, P) of $\mathrm{Relpos}_G \times_S \mathcal{P}\mathrm{ar}(G)$ (resp. a parabolic P of G) write:

$$\Sigma_{(\tau,P)} = (\mathrm{pr}_{\mathcal{R}} \times \mathrm{pr}_{\mathcal{P}})^{-1} \left((\tau, P)\right) \ \left(\text{resp. } \Sigma_P = \mathrm{pr}_{\mathcal{P}}^{-1}(P)\right).$$

Where P denotes the section of $\mathcal{P}\mathrm{ar}(G)$ given by the parabolic subgroup $P \subset G$.

There is a natural embedding

$$j_{\tau} : \Sigma_{\tau} \longrightarrow \mathcal{P}\mathrm{ar}(G) \times_S \mathcal{P}\mathrm{ar}(G)$$

$$\left(\text{resp. } j_{(\tau,P)} : \Sigma_{(\tau,P)} \longrightarrow \mathcal{P}\mathrm{ar}(G), \ j_P : \Sigma_P \longrightarrow \mathcal{P}\mathrm{ar}(G)_{\mathrm{Relpos}_G}\right),$$

as it results from the definition of

$$\Sigma_{\tau} \ \left(\text{resp. } \Sigma_{(\tau,P)}, \Sigma_P\right).$$

Define

$$\overline{\Sigma}_{\tau} \ \left(\text{resp. } \overline{\Sigma}_{(\tau,P)}, \overline{\Sigma}_P\right)$$

as the schematic image of

$$j_{\tau} \ \left(\text{resp. } j_{(\tau,P)}, j_P\right),$$

or equivalently the schematic closure of

$$\Sigma_{\tau} \ \left(\text{resp. } \Sigma_{(\tau,P)}, \Sigma_P\right),$$

in

$$\mathcal{P}\mathrm{ar}(G) \times_S \mathcal{P}\mathrm{ar}(G) \ \left(\text{resp. } \mathcal{P}\mathrm{ar}(G), \mathcal{P}\mathrm{ar}(G)_{\mathrm{Relpos}_G}\right).$$

One calls $\overline{\Sigma}_{\tau}$ (resp. $\overline{\Sigma}_{(\tau,P)}$, $\overline{\Sigma}_P$) the **Universal Schubert scheme of** G **of type** τ (resp. **the Schubert scheme of type** (τ, P)**, the Universal Schubert scheme of** G **defined by the parabolic** $P \subset G$). Observe that $\overline{\Sigma}_{\tau} = \left(\mathrm{pr}_{\mathcal{R},\overline{\Sigma}}\right)^{-1}(\tau)$.

To make evident the dependence of the above defined objects on G write:

$$\Sigma^G = \Sigma \ \left(\text{resp. } \Sigma^G_{\tau} = \Sigma_{\tau}, \ \overline{\Sigma}^G = \overline{\Sigma}, \ \overline{\Sigma}^G_{\tau} = \overline{\Sigma}_{\tau} \cdots \text{etc}\right).$$

Recall that given $S' \longrightarrow S$, there is the following isomorphism

$$\mathcal{P}\mathrm{ar}(G_{S'}) \simeq \mathcal{P}\mathrm{ar}(G)_{S'}$$

$$(\text{resp. } \mathcal{S}\text{tand}(G_{S'}) \simeq \mathcal{S}\text{tand}(G)_{S'}) \,,$$

which results immediately from the definition of the functor $\mathcal{P}\text{ar}(G)$ (resp. $\mathcal{S}\text{tand}(G)$) (cf. [23], Exp. XXVI, 3.2. (resp. 4.5.3)). An isomorphism is deduced

$$\text{Relpos}_{G_{S'}} \to (\text{Relpos}_G)_{S'}.$$

On the other hand, the morphism

$$t_2' : \mathcal{P}\text{ar}(G_{S'}) \to \text{Relpos}_{G_{S'}} \,,$$

is obtained as $t_2' = (t_2)_{S'}$. From the above remark one obtains the identifications

$$\Sigma^{G_{S'}} = \left(\Sigma^G\right)_{S'}$$

$$\left(\text{resp. } \Sigma^{G_{S'}}_{\underline{\tau}_{S'}} = \left(\Sigma^G_{\underline{\tau}}\right)_{S'}\right),$$

where $\underline{\tau}$ denotes a section of Relpos_G.

12.11 The group action on the Universal Schubert scheme

By definition of Σ as the G-orbit scheme $\text{gr}(t_2)$ = graph of t_2, there is a natural action of G on Σ that is induced by the action of G on

$$\mathcal{P}\text{ar}(G) \times_S \mathcal{P}\text{ar}(G) \times_S \text{Relpos}_G,$$

given by

$$(g, ((P, Q), \underline{\tau})) \mapsto ((\text{int}(g)(P), \text{int}(g)(Q)), \underline{\tau}) \,.$$

Clearly, the subscheme

$$\Sigma \hookrightarrow \mathcal{P}\text{ar}(G) \times_S \mathcal{P}\text{ar}(G) \times_S \text{Relpos}_G$$

is invariant under G.

Let $\underline{\tau}$ be a section of Relpos_G. It follows that

$$\Sigma_{\underline{\tau}} \subset \mathcal{P}\text{ar}(G) \times_S \mathcal{P}\text{ar}(G)$$

$$\left(\text{resp. } \Sigma_{(\underline{\tau}, P)} \subset \mathcal{P}\text{ar}(G)\right),$$

is G-invariant (resp. P-invariant).

Proposition 12.45
For every section g of G (resp. h of P) there is

$$\text{int}(g)(\overline{\Sigma}) = \overline{\Sigma}$$

$$\left(\text{resp. } \text{int}(g)(\overline{\Sigma}_{\underline{\tau}}) = \overline{\Sigma}_{\underline{\tau}}, \text{int}(h)(\overline{\Sigma}_{(\underline{\tau}, P)}) = \overline{\Sigma}_{(\underline{\tau}, P)}\right).$$

Proof *For every section g of G $int(g)(\Sigma) = \Sigma$. It follows that*

$$\overline{\Sigma} \subset int(g)(\overline{\Sigma})$$

by definition of the schematic closure $\overline{\Sigma}$, as $int(g)(\overline{\Sigma})$ is a closed subscheme of $\mathcal{P}ar(G) \times_S \mathcal{P}ar(G) \times_S Relpos\,G$. It is deduced that

$$int(g^{-1})(\overline{\Sigma}) \subset \overline{\Sigma} \text{ for every section } g \text{ of } G.$$

It is concluded that

$$int(g^{-1})(\overline{\Sigma}) = \overline{\Sigma} \text{ for every section } g \text{ of } G ,$$

and finally that
$$int(g)(\overline{\Sigma}) = \overline{\Sigma}.$$

This proves the first equality. The second one follows from

$$\overline{\Sigma}_{\underline{\tau}} = pr_{\mathcal{R}}^{-1}(\underline{\tau}) = \text{fiber of } \overline{\Sigma} \text{ over } \underline{\tau},$$

and the G-invariance of $pr_{\mathcal{R}}$. The following one is immediate from the second equality as P is equal to its own normalizer in G.

12.12 The Universal Schubert scheme locally trivial fibration over the Parabolics scheme

Assume that G **is endowed with a frame** E. Let $\tau \in$ Relpos $\mathcal{A}_E$ and let (t, s) be the couple of types determined by τ. Write $\underline{\tau} = \tau_S$. Recall that the Universal Schubert scheme defined by the type of relative position $\underline{\tau} \in Relpos_G$, $\Sigma_{\underline{\tau}} = \mathcal{S}tand(\underline{\tau}) \longrightarrow \mathcal{P}ar_t(G)$ is a locally trivial fibration with typical fiber $\Sigma_{(\underline{\tau}, P_t)} = \mathcal{S}tand(\underline{\tau}, P_t)$, which is trivialized by the big cell open covering $(\overline{\Omega}_{t,R_+})_{R_+ \in \mathcal{F}ram(\mathcal{R}(E))}$ (That is why it is assumed that G is endowed with a frame). With the aim of simplifying notations denote by (U_i) the big cell open covering of $\mathcal{P}ar_t(G)$.

There are isomorphisms $\alpha_i : U_i \times_S \Sigma_{(\underline{\tau}, P_t)} \simeq (\Sigma_{\underline{\tau}})_{U_i}$ where α_i is induced by an automorphism $int(g_i)$ and g_i denotes a section of G_{U_i}. In fact $int(g_i)$ induces an U_i- automorphism of $U_i \times_S \mathcal{P}ar(G)$ and sends $U_i \times_S \Sigma_{(\underline{\tau}, P_t)} \subset U_i \times_S \mathcal{P}ar(G)$ on $(\Sigma_{\underline{\tau}})_{U_i} = U_i \times_{\mathcal{P}ar(G)} \Sigma_{\underline{\tau}} \subset \mathcal{P}ar(G)$. As by definition $\overline{\Sigma}_{(\underline{\tau}, P_t)}$ is the schematic closure of $\Sigma_{(\underline{\tau}, P_t)}$ in $\mathcal{P}ar(G)$ and U_i is smooth it is deduced that $U_i \times_S \overline{\Sigma}_{(\underline{\tau}, P_t)}$ is also the schematical closure of $U_i \times_S \Sigma_{(\underline{\tau}, P_t)}$ in $U_i \times_S \mathcal{P}ar(G)$. On the other hand, the automorphism $int(g_i)$ induces an isomorphism $\overline{\alpha}_i : U_i \times_S \overline{\Sigma}_{(\underline{\tau}, P_t)} \simeq (\overline{\Sigma}_{\underline{\tau}})_{U_i}$ of schematical closures which in fact extends α_i.

With the data $(\alpha_i, U_i \times_S \Sigma_{(\underline{\tau}, P_t)})$ (resp. $(\overline{\alpha}_i, U_i \times_S \overline{\Sigma}_{(\underline{\tau}, P_t)})$) is associated the set of transition isomorphisms $(c_{ij} = \alpha_i^{-1} \circ \alpha_j)$ (resp. $(\overline{c}_{ij} = \overline{\alpha}_i^{-1} \circ \overline{\alpha}_j)$). The set (c_{ij}) satisfies the cocycle condition as it defines the $\mathcal{P}ar(G)$-sub-bundle

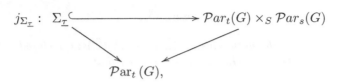

where the couple of types of parabolics (t, s) is defined by $\underline{\tau}$.

From the fact that $\Sigma_{(\underline{\tau}, P_t)}$ is schematically dense in $\overline{\Sigma}_{(\underline{\tau}, P_t)}$ it follows that $(\overline{c}_{ij})$ also satisfies the cocycle condition and thus the data $\left(\overline{\alpha}_i, U_i \times_S \overline{\Sigma}_{(\underline{\tau}, P_t)}\right)$ defines a $\mathcal{P}ar(G)$-sub-bundle $\overline{\Sigma}_{\underline{\tau}}^f$ of $\mathcal{P}ar_t(G) \times_S \mathcal{P}ar_s(G) \to \mathcal{P}ar_t(G)$ which in fact gives an embedding

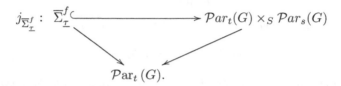

Remark that $\overline{\Sigma}_{\underline{\tau}}^f \to \mathcal{P}ar_t(G)$ is a proper morphism as the fiber $\overline{\Sigma}_{(\underline{\tau}, P_t)} \subset \mathcal{P}ar_t(G)$ is a projective scheme. Thus the morphism $\overline{\Sigma}_{\underline{\tau}}^f \to S$ is proper as $\mathcal{P}ar_t(G) \to S$ is a projective scheme. It is concluded that the $\overline{\Sigma}_{\underline{\tau}}^f \subset \mathcal{P}ar_t(G) \times_S \mathcal{P}ar_s(G)$ is a closed subscheme.

As the embedding $\Sigma_{\underline{\tau}} \hookrightarrow \mathcal{P}ar_t(G) \times_S \mathcal{P}ar_s(G)$ is quasi-compact and separated the schematic closure $\overline{\Sigma}_{\underline{\tau}}$ exists (cf. Proposition 12.43), satisfies $\overline{\Sigma}_{\underline{\tau}} \subset \overline{\Sigma}_{\underline{\tau}}^f$, and commutes with the flat extension $X \to \mathcal{P}ar_t(G) \times_S \mathcal{P}ar_s(G)$, where $X = \coprod U_i \times_S \mathcal{P}ar_t(G)$. Thus there is an isomorphism

$$\coprod \overline{\alpha}_i : U_i \times_S \overline{\Sigma}_{(\underline{\tau}, P_t)} \simeq \coprod \overline{(\Sigma_{\underline{\tau}})}_{U_i} = \overline{(\Sigma_{\underline{\tau}})}_X = \overline{(\Sigma_{\underline{\tau}})}_X \ .$$

Thus it follows that the image of the embedding $j_{\overline{\Sigma}_{\underline{\tau}}^f}$ is equal to $\overline{\Sigma}_{\underline{\tau}}$.

One concludes that

Proposition 12.46

Keep the above notation. Suppose G endowed with a frame. Then $\overline{\Sigma}_{\underline{\tau}} \longrightarrow \mathcal{P}ar_t(\mathbf{G})$ is a locally trivial fiber bundle with typical fiber $\overline{\Sigma}_{(\underline{\tau}, \mathbf{P}_t)}$, and $\Sigma_{\underline{\tau}} \longrightarrow \mathcal{P}ar_t(\mathbf{G})$ is a sub-bundle of it with typical fiber $\Sigma_{(\underline{\tau}, \mathbf{P}_t)}$.

Chapter 13

Incidence Type Schemes of the Relative Building

By G denote a reductive S-group scheme. By a **Building type scheme** associated with G one understands an S-scheme X so that for all $s \in S$ the geometric fiber $X_{\bar{s}}$ "may be written" in terms of the Building $I(G_{\bar{s}})$ of the geometric fiber $G_{\bar{s}}$ of G. The schemes associated with the Relative Building which has been constructed in the preceding chapter are examples of Building type schemes. An **Incidence type scheme** is a subscheme of a finite product of Building type schemes defined in terms of the **Incidence scheme** $\mathcal{I}nc(G)$. With the following Building type schemes

- the Typical Simplex scheme $\underline{\Delta}(G)$;

- the Weyl Complex scheme $\mathcal{A}(G)$;

- the Types of Relative Position scheme $Relpos_G$;

- the Parabolic scheme $\mathcal{P}ar(G)$,

are associated Incidence type schemes, whose geometric fibers over $\bar{s}$ are respectively given by sets of generalized galleries of the Typical Simplex, the Weyl Complex, the Types of Relative Positions of $I(G_{\bar{s}})$, and $I(G_{\bar{s}})$ itself. The smooth resolutions are constructed for the Schubert schemes are particular cases of these objects. More precisely stated one has:

The **Minimal Galleries of Types** scheme and the **Relative Positions Galleries** scheme associated with the reductive S-group scheme G are first defined, along with the morphism δ_2 associating with a minimal gallery of types **g** a type of relative position section $\tau_{\mathbf{g}}$. The **Incidence scheme** plays the role, in the setting of relative buildings, of the incidence relation of a building. The **Galleries Configurations scheme** (resp. **Galleries Configurations over minimal galleries of types scheme**) is defined in terms of the Incidence scheme, as a subscheme of the disjoint union of all finite products of the Parabolics scheme $\mathcal{P}ar(G)$. The **Minimal Galleries Configurations** are defined in terms of the Convex Hull scheme $\mathcal{F}\mathrm{ix}(\tilde{\mathbf{P}}, \tilde{\mathbf{Q}}) \longrightarrow \Sigma$. They are particular sections of the Convex Hull scheme over the Universal Schubert Cell Σ. There is a morphism associating with a gallery configuration a gallery of types. It will be seen in the next chapter that the fiber of the Galleries Configurations scheme over a section $\mathbf{g}$ of the Minimal Galleries of types scheme provides a smooth resolution of singularities of the Schubert scheme $\overline{\Sigma}_{\tau_{\mathbf{g}}}$.

13.1   Typical Simplex Generalized Galleries scheme

Recall that the set of sections of the **relative typical simplex** $\underline{\Delta}(G)(S')$ is given by the **set of open and closed subsets** of $\mathcal{D}\mathrm{yn}(G)_{S'}$ and thus it is naturally endowed with an order relation, namely the inclusion relation "$\subset$". Define the S-functor

$$\Gamma(\underline{\Delta}(G)) \subset \coprod_{N \in \mathbb{N}} \underline{\Delta}(G) \times_S \cdots \times_S \underline{\Delta}(G) \ (N\text{-times})$$

of **Generalized Galleries of types** (resp. **Generalized Galleries** of $\underline{\Delta}(G)$) by: "The set

$$\Gamma(\underline{\Delta}(G))(S') \cap (\underline{\Delta}(G)(S') \times_S \cdots \times_S \underline{\Delta}(G)(S') \ (N\text{-times}))$$

is given by the N-uples of sections of $\underline{\Delta}(G)$ on S' satisfying, relatively to '$\subset$', the defining relations of a Generalized Gallery in a building as given in Definition 5.1". If G is split one has

$$\Gamma(\underline{\Delta}(G)) \simeq \varinjlim_{\vec{E}} \left(\mathrm{Gall}_{\Delta(E)} \right)_S$$

where $\Delta(E) = \mathrm{typ}\, \mathcal{A}_E = \mathcal{A}_E / W(\mathcal{R}(E))$. Thus $\Gamma(\underline{\Delta}(G))$ is locally representable by a disjoint union of constant S-schemes, namely $\left(\varinjlim_{\vec{E}} \mathrm{Gall}_{\Delta(E)} \right)_S$. It is concluded that in general $\Gamma(\underline{\Delta}(G))$ is representable by a locally constant S-scheme.

13.2   Minimal Generalized Galleries of types scheme

Let G be split and E a frame of G. Denote by

$$\mathrm{gall}^{\,m}_{\mathcal{A}_E} \subset \mathrm{gall}_{\mathcal{A}_E}$$

the subset of Generalized Galleries of types, given by the **images** of the minimal Generalized Galleries of the apartment $\mathcal{A}_E$ by the mapping

$$\mathrm{typ}^g : \mathrm{Gall}_{\mathcal{A}_E} \to \mathrm{gall}_{\mathcal{A}_E},$$

associating with $\gamma \in \mathrm{Gall}_{\mathcal{A}_E}$ the Generalized Gallery of types $\mathrm{typ}\,\gamma \in \mathrm{gall}_{\mathcal{A}_E}$ given by the images by $\mathrm{typ} : \mathcal{A}_E \to \mathrm{typ}\,\mathcal{A}_E$ of its facets. Observe that the subset $\mathrm{gall}^m_{\mathcal{A}_E}$ is invariant under the action of the group $Aut(\mathcal{A}_E)$ of automorphisms of the apartment $\mathcal{A}_E$. Given a reductive S-group scheme one may thus define an S-subfunctor

$$\Gamma^m(\underline{\Delta}(G)) \subset \Gamma(\underline{\Delta}(G))$$

characterized by the following property. If G is split then there is a canonical isomorphism

$$\varinjlim_{E} \left(\mathrm{gall}^m_{\mathcal{A}_E} \right)_S \simeq \Gamma^m\left(\underline{\Delta}(G) \right),$$

where E runs on the set of frames of G. The sections of $\Gamma^m(\underline{\Delta}(G))$ are characterized as the sections of $\Gamma(\underline{\Delta}(G))$ which are locally of the form $\mathrm{typ}\gamma(F, F')_{S'}$, where $S' \to S$ is an etale morphism so that $G_{S'}$ is endowed with a frame E and $\gamma(F, F')$ is a Minimal Gallery of $\mathcal{A}_E$.

The canonical isomorphism

$$\underline{\Delta}(G) \simeq \mathcal{A}(G)/\underline{W}_G,$$

allows the identification of the sections of $\Gamma(\underline{\Delta}(G))$ with the gg of the S-scheme of types $\mathcal{A}(G)/\underline{W}_G$ of $\mathcal{A}(G)$. Thus $\Gamma^m(\underline{\Delta}(G))$ appears as the image of $\Gamma^m_{\mathcal{A}(G)}$ by the restriction $\Gamma^m_{\mathcal{A}(G)} \to \Gamma(\mathcal{A}(G)/\underline{W}_G)$ of the morphism $\Gamma_{\mathcal{A}(G)} \to \Gamma(\mathcal{A}(G)/\underline{W}_G)$ induced by the quotient morphism $\mathcal{A}(G) \to \mathcal{A}(G)/\underline{W}_G$.

13.3   Weyl Complex Generalized Galleries scheme

For a split S-group G write

$$\Gamma_{\mathcal{A}(G)} = \varinjlim_{E} \left(\mathrm{Gall}_{\mathcal{A}_E} \right)_S$$

$$\left(\text{resp. } \Gamma^m_{\mathcal{A}(G)} = \varinjlim_{E} \left(\mathrm{Gall}^m_{\mathcal{A}_E} \right)_S \right).$$

Given in general an S-reductive group G one defines the generalized galleries of $\mathcal{A}(S)$, S-scheme $\Gamma_{\mathcal{A}(G)}$ (resp. the MGG, S-scheme $\Gamma^m_{\mathcal{A}(G)}$ of $\mathcal{A}(S)$) by descent, following the pattern of the definition of $\mathcal{A}(G)$ itself.

13.4   The Weyl Complex Gallery type morphism

Let G be endowed with a frame E. Let $\mathrm{typ}^g : \mathrm{Gall}_{\mathcal{A}_E} \to \mathrm{gall}_{\mathcal{A}_E}$, be defined by $\mathrm{typ}^g : \gamma \to \mathrm{typ}\,\gamma$, and

$$t_{1,E}^{\Gamma} = (\mathrm{typ}^g)_S : (\mathrm{Gall}_{\mathcal{A}_E})_S \to (\mathrm{gall}_{\mathcal{A}_E})_S\,.$$

Write for a split reductive S-group G:

$$t_{1,\mathcal{A}}^{\Gamma} = \varinjlim_E t_{1,E}^{\Gamma} : \Gamma_{\mathcal{A}(G)} \to \Gamma(\underline{\Delta}(G))\ ,$$

where E runs on the set of frames of G. Thus there exists a morphism

$$t_{1,\mathcal{A}}^{\Gamma} : \Gamma_{\mathcal{A}(G)} \to \Gamma(\underline{\Delta}(G))$$

locally characterized as above, defined by descent, if it is supposed only G a reductive S-group. Let

$$t_{1,\mathcal{A}}^m : \Gamma_{\mathcal{A}(G)}^m \to \Gamma^m(\underline{\Delta}(G)),$$

denote the induced morphism. It is clear that the Weyl group $\underline{W}_G$ acts naturally on $\Gamma_{\mathcal{A}(G)}^m$. Assume that G is endowed with a frame E, then there is a canonical isomorphism

$$\Gamma_{\mathcal{A}(G)}^m \simeq \left(\mathrm{Gall}_{\mathcal{A}_E}^{\,m}\right)_S$$

$$\left(\text{resp. } \Gamma_{\mathcal{A}(G)}^m/\underline{W}_G \simeq \left(\mathrm{Gall}_{\mathcal{A}_E}^{\,m}/W(\mathcal{R}(E))\right)_S\right).$$

On the other hand, there is an isomorphism

$$\bar{t}_{1,E}^{\,m} : \left(\mathrm{Gall}_{\mathcal{A}_E}^{\,m}/W(\mathcal{R}(E))\right)_S \simeq \left(\mathrm{gall}_{\mathcal{A}_E}^{\,m}\right)_S\,,$$

induced by $t_{1,E}^{\Gamma}$.

It is clear that $t_{1,\mathcal{A}}^m$ factors through the quotient morphism

$$\Gamma_{\mathcal{A}(G)}^m \to \Gamma_{\mathcal{A}(G)}^m/\underline{W}_G,$$

and thus that it induces a morphism

$$\bar{t}_{1,\mathcal{A}}^{\,m} : \Gamma^m{}_{(G)}/\underline{W}_G \to \Gamma^m(\underline{\Delta}(G))\,.$$

As

$$\bar{t}_{1,\mathcal{A}}^{\,m} = \varinjlim_E \bar{t}_{1,E}^{\,m}$$

it is deduced the following

Lemma 13.1
The morphism $\bar{t}_{1,\mathcal{A}}^{\,m}$ is an isomorphism.

13.5 Relative position types galleries scheme

A twisted constant scheme is defined which indexes, as will be seen, a cellular decomposition of the galleries scheme. Let t (resp. $\underline{s}', \underline{s}$) be a section of $\underline{\Delta}(G)$. Suppose that

$$\underline{t} \subset \underline{s}', \underline{s}$$

and that $\underline{s}'$ denotes the type section of a Borel subgroup of G. Write

$$\mathrm{Relpos}_{\underline{s}}^{\underline{t}} = \mathrm{Relpos}_{(\underline{s}', \underline{s})}^{\underline{t}}$$

(cf. §11.12).

Definition 13.2

1. *Suppose that the S-group scheme G is endowed with a frame E.*
   Let $g \in \mathrm{gall}_{\mathcal{A}_E}$, with $l(g) = r + 1$, such that

   $$g = g_1, g_1' \quad (\text{resp. } g = g_2, g_2').$$

   Write

   $$\mathrm{Relpos}_G^{gall}(g_S) = \prod \mathrm{Relpos}_{s_{i-1}(g)_S}^{t_i(g)_S} \quad (r + 1 \geqslant i \geqslant 1)(\text{resp. } r \geqslant i \geqslant 1),$$

   where the products are products of S-schemes.
   Given a section $\underline{g}$ of $\Gamma(\underline{\Delta}(G))$ define

   $$\mathrm{Relpos}_G^{gall}(\underline{g})$$

   for a reductive S-group scheme G by descent from the above definition, by taking into account that locally, there is an isomorphism

   $$(\mathrm{gall}_{\mathcal{A}_E})_{S'} \simeq \Gamma(\underline{\Delta}(G)).$$

   Here $S' \to S$ denotes an étale morphism and E a frame of $G_{S'}$, and $\underline{g}_{S'}$ is supposed to be of the form $\underline{g}_{S'} = g_{S'}$ for some $g \in \mathrm{gall}_{\mathcal{A}_E}$.

2. *If G is endowed with a frame E define*

   $$\mathrm{Relpos}_G^{gall} = \coprod_{g \in \mathrm{gall}\, A(\mathcal{R}(E))} \mathrm{Relpos}_G^{gall}(g_S).$$

   *Define in general Relpos_G^{gall} (**The Relative Position Types galleries scheme**) for a reductive S-group scheme G by descent, following the same pattern as in 1.*

3. *If G is endowed with a frame E, there is a morphism*

$$\epsilon^{gall} : Relpos_G^{gall} = \coprod Relpos_G^{gall}(g_S) \rightarrow \Gamma(\underline{\Delta}(G)),$$

so that given a section $\underline{\tau}_{\mathscr{C}}$ of $Relpos_G^{gall}$, then

$$\epsilon^{gall}(\underline{\tau}_{\mathscr{C}}) = g_S,$$

if $\underline{\tau}_{\mathscr{C}}$ is a section of $Relpos_G^{gall}(g_S)$.

In the general case define ϵ^{gall} by descent.

13.6   The type of relative position associated to a minimal gallery of types

Let

$$\mathcal{E}_A = \mathcal{E}_{1\,A} \times \mathcal{E}_{2\,A} : \Gamma_{A(G)} \rightarrow A(G) \times_S A(G)$$

$$(\text{resp. } e = e_1 \times e_2 : \Gamma(\underline{\Delta}(G)) \rightarrow \underline{\Delta}(G) \times_S \underline{\Delta}(G))$$

be the morphism associating with a section γ (resp. g) of $\Gamma_{A(G)}$ (resp. $\underline{\Delta}(G)$) the couple of its extremities

$$(\mathcal{E}_{1\,A}(\gamma), \mathcal{E}_{2\,A}(\gamma)) \quad (\text{resp. } (e_1(g), e_2(g))) .$$

The restriction of $\mathcal{E}_A$ (resp. e) to

$$\left(\prod^N A(G) \right) \cap \Gamma_{A(G)} \quad \left(\text{resp. } \left(\prod^N \underline{\Delta}(G) \right) \cap \Gamma(\underline{\Delta}(G)) \right)$$

is the morphism induced by

$$p_{1\,A(G)} \times p_{N\,A(G)} \quad \left(\text{resp. } p_{1\,\underline{\Delta}(G)} \times p_{N\,\underline{\Delta}(G)} \right),$$

where

$$p_{1\,A(G)} \text{ (resp. } p_{N\,A(G)}) : \prod^N A(G) \rightarrow A(G)$$

$$\left(\text{resp. } p_{1\,\underline{\Delta}(G)} \text{ (resp. } p_{N\,\underline{\Delta}(G)}) : \prod^N \underline{\Delta}(G) \rightarrow \underline{\Delta}(G) \right)$$

denotes the first projection (resp. the N-th projection).
    Write

$$\mathcal{E}_A^m = \mathcal{E}_{1\,A}^m \times \mathcal{E}_{2\,A}^m \quad (\text{resp. } e^m = e_1^m \times e_2^m)$$

for the restriction of $\mathcal{E}_A$ (resp. e) to the minimal galleries scheme

$$\Gamma_{A(G)}^m \quad (\text{resp. } \Gamma^m(\underline{\Delta}(G))) .$$

The morphism giving the relative position of the extremities of a minimal gallery

$$t_{2,\mathcal{A}} \circ \mathcal{E}_{\mathcal{A}}^m : \Gamma_{\mathcal{A}(G)}^m \to \text{Relpos } \mathcal{A}(G),$$

factors through

$$\Gamma_{\mathcal{A}(G)}^m \to \Gamma_{\mathcal{A}(G)}^m / \underline{W}_G.$$

Let:

$$\overline{\delta_2} : \Gamma_{\mathcal{A}(G)}^m / \underline{W}_G \to \text{Relpos } \mathcal{A}(G),$$

be the induced morphism.

On the other hand, on the basis of Lemma 13.1 one has an isomorphism

$$\overline{t}_{1,\mathcal{A}}^m : \Gamma_{\mathcal{A}(G)}^m / \underline{W}_G \to \Gamma^m \left(\underline{\Delta}(G) \right),$$

induced by $t_{1,\mathcal{A}}^m$. Thus one may define

$$\delta_2 = \overline{\delta}_2 \circ (\overline{t}_{1,\mathcal{A}}^m)^{-1} : \Gamma^m \left(\underline{\Delta}(G) \right) \to \text{Relpos } \mathcal{A}(G)$$

$\Big($**Minimal Generalized Gallery of types Type of Relative Position Morphism**$\Big)$.

Locally δ_2 may be described as follows. Let E be a frame of G. Then there is a canonical isomorphism

$$\left(\text{gall}_{\mathcal{A}_E}^m \right)_S \simeq \Gamma^m \left(\underline{\Delta}(G) \right)$$

$$(\text{resp. } (\text{Relpos } \mathcal{A}_E)_S \simeq \text{Relpos } \mathcal{A}(G)).$$

Thus the morphism induced by δ_2

$$\delta_2' : \left(\text{gall}_{\mathcal{A}_E}^m \right)_S \to (\text{Relpos } \mathcal{A}_E)_S,$$

is given by

$$\delta_2' = (\tau_\bullet)_S,$$

where $\tau_\bullet : \text{gall}_{\mathcal{A}_E}^m \to \text{Relpos } \mathcal{A}_E$, is defined by associating with a minimal gallery of types the relative position of the extremities of a minimal gallery of this type (cf. Definition 9.20).

13.7 The Incidence morphisms in the Weyl Complex

Given a section $(\underline{s}, \underline{t})$ of $\underline{\Delta}(G) \times_S \underline{\Delta}(G)$, satisfying $\underline{s} \supset \underline{t}$ define the **Incidence morphism of type** $(\underline{s}, \underline{t})$

$$\mathcal{F}_{\underline{s},\underline{t}}^{\mathcal{A}} : \mathcal{A}(G)_{\underline{s}} \to \mathcal{A}(G)_{\underline{t}}$$

locally as follows. Suppose that G is endowed with a frame E. Thus there is a canonical isomorphism

$$(\mathcal{A}_E)_S \simeq \mathcal{A}(G).$$

Given $(s,t) \in \operatorname{typ} \mathcal{A}_E \times \operatorname{typ} \mathcal{A}_E$, define the incidence mapping

$$\mathcal{F}^A_{s,t} : (\mathcal{A}_E)_s \to (\mathcal{A}_E)_t,$$

as the mapping associating with a facet $F \in (\mathcal{A}_E)_s$ the unique facet $F' \in (\mathcal{A}_E)_t$ incident to F. Thus one has

$$\mathcal{F}^A_{s,t}(F) = F'.$$

Let

$$(\underline{s}, \underline{t}) = (s_S, t_S).$$

Define

$$\mathcal{F}^{\mathcal{A}}_{\underline{s},\underline{t}} = \left(F^A_{s,t} \right)_S$$

One obtains $\mathcal{F}^{\mathcal{A}}_{\underline{s},\underline{t}}$ in the general case by etale descent.

The corresponding incidence morphism

$$\mathcal{F}_{\underline{s},\underline{t}} : \mathcal{P}ar_{\underline{s}}(G) \to \mathcal{P}ar_{\underline{t}}(G)$$

is easier to define. Let $Q \subset G$ be a parabolic subgroup of type $\underline{s}$.

Then $\mathcal{F}_{\underline{s},\underline{t}}$ associates with Q the unique parabolic subgroup $P \subset G$, so that P is of type $\underline{t}$, and $Q \subset P$ (cf. [23], *Exp. XXVI, Lemme* 3.8).

The product of incidence morphisms:

$$\mathcal{F}^{\mathcal{A}}_{\underline{s},\underline{t}} \times \mathcal{F}^{\mathcal{A}}_{\underline{s}',\underline{t}'} : A(G)_{\underline{s}} \times_S A(G)_{\underline{s}'} \to A(G)_{\underline{t}} \times_S A(G)_{\underline{t}'}$$

$$\left(\text{resp. } \mathcal{F}_{\underline{s},\underline{t}} \times \mathcal{F}_{\underline{s}',\underline{t}'} : \mathcal{P}ar_{\underline{s}}(G) \times_S \mathcal{P}ar_{\underline{s}'}(G) \to \mathcal{P}ar_{\underline{t}}(G) \times_S \mathcal{P}ar_{\underline{t}'}(G) \right)$$

induces a morphism

$$\operatorname{Relpos}_{(\underline{s},\underline{s}')} A(G) \to \operatorname{Relpos}_{(\underline{t},\underline{t}')} A(G)$$

$$\left(\text{resp. } (\operatorname{Relpos} G)_{(\underline{s},\underline{s}')} \to (\operatorname{Relpos} G)_{(\underline{t},\underline{t}')} \right)$$

which is called **the Relpos incidence morphism in** $A(G)$ (resp. **the Relpos incidence morphism**) defined by the couples of types $(\underline{s},\underline{s}')$ and $(\underline{t},\underline{t}')$ satisfying $\underline{s} \supset \underline{t}$, and $\underline{s}' \supset \underline{t}'$. If $\mathcal{F}^{\mathcal{A}}_{\underline{s},\underline{t}} \times \mathcal{F}^{\mathcal{A}}_{\underline{s}',\underline{t}'} : (Q,Q') \mapsto (P,P')$ and (Q,Q') is in standard position then (P,P') is also in standard (resp. transversal) position.

Definition 13.3
It is recalled that $\underline{\Delta}(G)$ is ordered by a functorial order relation (cf. §11.27).

One endows $\prod \underline{\Delta}(G) = \coprod \overset{N}{\prod} \underline{\Delta}(G) \ (1 \leqslant N)$ with the order given by the product order $\leqslant$ and the linear order of the products. Given a couple of sections (σ', σ) of $\overset{N'}{\prod} \underline{\Delta}(G) \times \overset{N}{\prod} \underline{\Delta}(G)$ with $N' \leqslant N$, one writes

$$\sigma' \underset{(\bar{\sigma},p)}{\prec} \sigma$$

if σ' is obtained from σ as follows. There exists a section $\overline{\sigma}$ of $\prod^{N} \underline{\Delta}(G)$ with

$$\overline{\sigma} \leqslant \sigma$$

so that σ' is obtained as the image of $\overline{\sigma}$ by some projection

$$p : \prod^{N} \underline{\Delta}(G) \to \prod^{N'} \underline{\Delta}(G).$$

Given two sections $\underline{g}'$ and $\underline{g}$ of $\Gamma(\underline{\Delta}(G))$, one writes

$$\underline{g}' \underset{(\overline{\sigma},p)}{\prec} \underline{g}$$

if the corresponding sections of $\prod \underline{\Delta}(G)$, verify this relation.
Remark that

$$\Gamma_{\mathcal{A}(G)} \subset \coprod \prod^{N} \mathcal{A}(G).$$

Thus given $\underline{g}' \underset{(\overline{\sigma},p)}{\prec} \underline{g}$ one introduces a natural morphism

$$\overline{\mathcal{F}}^{\Gamma_{\mathcal{A}(G)}}_{\underline{g},(\overline{\sigma},p)} : (t^{\Gamma}_{1,\mathcal{A}})^{-1}(\underline{g}) \to \prod^{N'} \mathcal{A}(G) \cap \Gamma_{\mathcal{A}(G)}.$$

This morphism is induced by the composition of a product of incidence morphisms

$$\mathcal{F}^{\mathcal{A}}_{\sigma,\overline{\sigma}} : \prod^{N} \mathcal{A}(G) \to \prod^{N} \mathcal{A}(G),$$

followed by the projection

$$\prod^{N} \mathcal{A}(G) \to \prod^{N'} \mathcal{A}(G)$$

given by p. These morphisms allow considering a hierarchical relation between galleries. They depend on the choice of some $(\overline{\sigma}, p)$. In practice this choice is natural.

13.8 The Incidence scheme

The incidence relation in buildings plays a very important role in the description of their associated geometry (cf. [50]). In the schematic setting the graph of the incidence relation appears as a subscheme of $\mathcal{S}\text{tand}(G)$. The galleries configurations that are introduced are sections of fiber products of the Incidence scheme with itself. One may say that the types of relative position between couples of parabolics are generated by the "types of incidence" between couples of parabolics (cf. galleries of types of relative position associated with minimal galleries of types). Schubert cells are obtained as fiber products of cells defined in terms of the incidence relation, along galleries of types of relative position. These decompositions give rise to smooth resolutions of their schematic closures;

Let G be a reductive S-group and (P, Q) a couple of parabolics of G.

Definition 13.4
It is said that P and Q are **incident parabolics** *if the following equivalent statements hold:*

1. *$P \cap Q$ is a parabolic subgroup of G*

2. *$P \cap Q$ contains a Borel subgroup locally for the étale topology of S'.*

(cf. loc. cit., Exp. XXVI, Proposition 4.4.1)

Let
$$\mathfrak{Inc}(G) \subset \mathcal{P}\mathrm{ar}(G) \times_S \mathcal{P}\mathrm{ar}(G)$$

be the subfunctor whose sections over $S' \to S$ are the couples (P, Q) of incident parabolics of $G_{S'}$. One calls $\mathfrak{Inc}(G)$ the **Incidence scheme**.

Clearly one has the inclusion

$$\mathfrak{Inc}(G) \subset \mathcal{S}\mathrm{tand}(G) \ ,$$

and the diagonal action of G stabilizes $\mathfrak{Inc}(G)$. From the following proposition it results that the type of relative position of a couple of incident parabolics is characterized by their corresponding couple of types.

Proposition 13.5 *(cf. [23], Exp. XXVI, Corollaire 4.4.3.)*
Let (P, Q) (resp. (P', Q')) be a couple of incident parabolics of type $(\underline{t}, \underline{s})$ (resp. a couple of parabolics of type $(\underline{t}, \underline{s})$).
  Then (P', Q') is a couple of incident parabolics, if and only if, the couple (P', Q') is conjugate, under the diagonal inner action of G, to the couple (P, Q), locally for the etale topology of S'.

There is a morphism

$$\underline{\Delta}(G) \times_S \underline{\Delta}(G) \to \mathrm{Relpos}\,_G,$$

which associates with a couple of types $(\underline{t}, \underline{s})$, the type of relative position $t_2(P, Q)$ of a couple of incident parabolics of type $(\underline{t}, \underline{s})$. By $\tau(\underline{t}, \underline{s})$ denote **the type of relative position** $t_2(P, Q)$ of a couple of **incident parabolics** (P, Q), with $(t_1 \times t_1)((P, Q)) = (\underline{t}, \underline{s})$.

It results

Corollary 13.6
The morphism

$$(t_1 \times t_1)' : \mathfrak{Inc}(G) \to \underline{\Delta}(G) \times_S \underline{\Delta}(G),$$

obtained as the restriction of $t_1 \times t_1$ to $\mathfrak{Inc}(G) \subset \mathcal{P}ar(G) \times_S \mathcal{P}ar(G)$, allows identifying $\underline{\Delta}(G) \times_S \underline{\Delta}(G)$ with the quotient scheme $\mathfrak{Inc}(G)/G \subset \mathcal{R}elpos\ G$, of the subscheme $\mathfrak{Inc}(G) \subset \mathcal{S}tand(G)$ under the diagonal action of G. The restriction of the type of relative position morphism $t_2' : \mathfrak{Inc}(G) \to \mathcal{R}elpos_G$ factors through $(t_1 \times t_1)'$ followed by $(\underline{t}, \underline{s}) \mapsto \tau(\underline{t}, \underline{s})$. The S-functor $\mathfrak{Inc}(G)$ is representable. Assume G endowed with a frame E, then

$$\mathfrak{Inc}(G) = \coprod_{(t,s) \in typ\ \mathcal{A}_E \times typ\ \mathcal{A}_E} \mathcal{S}tand(\tau(t_S, s_S)) \ .$$

Given a section $(\underline{t}, \underline{s})$ of $\underline{\Delta}(G) \times_S \underline{\Delta}(G)$, by definition of $\tau(\underline{t}, \underline{s})$ one has:

$$(t_1 \times t_1)'^{-1}(\underline{t}, \underline{s}) = t_2^{-1}(\tau(\underline{t}, \underline{s})) = \mathcal{S}tand(\tau(\underline{t}, \underline{s})).$$

From the Proposition 11.25 it results that $(t_1 \times t_1)'^{-1}(\underline{t}, \underline{s})$ is an S-scheme smooth, projective, with integral fibers.

Let $(\underline{t}, \underline{t}')$ be a section of $\underline{\Delta}(G) \times_S \underline{\Delta}(G)$ such that $\underline{t}' \subset \underline{t}$. Recall that the **incidence morphism of type** $(\underline{t}, \underline{t}')$, $\mathcal{F}_{\underline{t}, \underline{t}'} : \mathcal{P}ar_{\underline{t}}(G) \to \mathcal{P}ar_{\underline{t}'}(G)$, by definition, associates with a parabolic P of type $\underline{t}$ the parabolic $P' \supset P$ of type $\underline{t}'$. Given a couple of sections $(\underline{t}, \underline{s})$ of $\underline{\Delta}(G)$ denote by $\underline{t} \cup \underline{s}$ the section of $\underline{\Delta}(G)$ given by sup $(\underline{t}, \underline{s})$ defined relatively to the order of $\underline{\Delta}(G)$. It is easy to see that there is a canonical isomorphism

$$\mathcal{F}_{\underline{t} \cup \underline{s}, \underline{t}} \times \mathcal{F}_{\underline{t} \cup \underline{s}, \underline{s}} : \mathcal{P}ar_{\underline{t} \cup \underline{s}}(G) \tilde{\to} \mathcal{S}tand(\tau(\underline{t}, \underline{s})).$$

The reciprocal isomorphism is given by the isomorphism

$$\mathcal{S}tand(\tau(\underline{t}, \underline{s})) \to \mathcal{P}ar_{\underline{t} \cup \underline{s}}(G),$$

defined by
$$(P, Q) \mapsto P \cap Q.$$

It is deduced that the geometrical fibers of

$$(t_1 \times t_1)' : \mathfrak{Inc}(G) \to \underline{\Delta}(G) \times_S \underline{\Delta}(G)$$

(resp. of the restriction $t_2' : \mathfrak{Inc}(G) \to \mathcal{R}elpos_G$ of t_2)

are S-smooth, projective, and irreducible. Observe that the Stein factorization of $\mathfrak{Inc}(G) \to S$ is given by

$$\mathfrak{Inc}(G) \to \underline{\Delta}(G) \times_S \underline{\Delta}(G) \to S.$$

13.9   Galleries of Parabolics Configurations schemes

Definition 13.7
The Typical Simplex generalized galleries scheme $\Gamma(\underline{\Delta}(G))$ *may be seen as a subscheme of* $\prod_N \underline{\Delta}(G) = \coprod_N \prod^N \underline{\Delta}(G)$. *Let*

$$\prod^N {}^{inc} \mathcal{P}ar(G) \subset \prod^N \mathcal{P}ar(G)$$

be the S-subfunctor whose sections $(P_i)(1 \leqslant i \leqslant N)$ *satisfy:*

$$for \ 1 \leqslant i < N, \ (P_i, P_{i+1}) \ is \ a \ couple \ of \ incident \ parabolics$$

$$\left(resp. \ \prod {}^{inc} \mathcal{P}ar(G) = \coprod_N \prod^N {}^{inc} \mathcal{P}ar(G) \ (The \ Scheme \ of \right.$$

$$\left. \textbf{Chains of incident parabolics} \right)$$

Clearly

$$\prod^N {}^{inc} \mathcal{P}ar(G) = \mathcal{I}nc(G) \times_{\mathcal{P}ar(G)} \cdots \times_{\mathcal{P}ar(G)} \mathcal{I}nc(G)$$

$$((N-1) \ times \ product \ of \ \mathcal{I}nc(G) \ over \ \mathcal{P}ar(G)).$$

Write

$$\mathcal{C}onf_G^N = \left(\prod^N {}^{inc} \mathcal{P}ar(G) \right) \cap \left(\left(\prod^N t_1 \right)^{-1} \left(\left(\prod^N \underline{\Delta}(G) \right) \cap \Gamma(\underline{\Delta}(G)) \right) \right).$$

Remark that a couple of incident parabolics (P, Q) of type $(\underline{t}, \underline{s})$ with $\underline{t} \subset \underline{s}$ (resp. $\underline{t} \supset \underline{s}$) verifies $P \supset Q$ (resp. $P \subset Q$). Thus if (P_i) is a section of $\mathcal{C}onf_G^N \mathcal{P}ar(G)$, then the image

$$\left(\prod^N t_1 \right)((P_i)) = (t_1(P_i)) = \underline{g}$$

defines a section of $\Gamma(\underline{\Delta}(G))$, let it be said

$$\underline{g} : \underline{t}_1 \subset \underline{t}_2 \supset \cdots \underline{t}_{N-1} \supset \underline{t}_N.$$

From the above remark it is obtained that (P_i) satisfies

$$P_1 \supset P_2 \subset \cdots P_{N-1} \subset P_N.$$

From Definition 13.7 a natural embedding is obtained

$$\mathcal{C}\mathrm{onf}_G^N \subset \prod^N \mathcal{P}\mathrm{ar}(G).$$

One thus obtains an embedding

$$\mathcal{C}\mathrm{onf}_G = \coprod_N \mathcal{C}\mathrm{onf}_G^N \subset \prod^N \mathcal{P}\mathrm{ar}(G) = \coprod_N \prod^N \mathcal{P}\mathrm{ar}(G).$$

There is a natural morphism

$$t_1^\Gamma : \mathcal{C}\mathrm{onf}_G \;\to\; \Gamma(\underline{\Delta}(G)),$$

induced by $\coprod \prod^N \ell_1$ (compare with $t_{1,\mathcal{A}}^\Gamma$ of §13.4).
Given a section $\underline{g}$ of $\Gamma(\underline{\Delta}(G))$ let it be written

$$\mathcal{C}\mathrm{onf}_G(\underline{g}) = (t_1^\Gamma)^{-1}(\underline{g}).$$

If $\underline{g}$ is of the form $\underline{g} = g_S$, with $g \in \mathrm{gall}_{\mathcal{A}_E}$, for some E frame of G, one denotes by $N(\underline{g})$ the integer N so that

$$\mathcal{C}\mathrm{onf}_G(\underline{g}) \subset \prod^N \mathcal{P}\mathrm{ar}(G).$$

Let the extremities morphism $\mathcal{E} = \mathcal{E}_1 \times \mathcal{E}_2 : \mathcal{C}\mathrm{onf}_G \to \mathcal{P}\mathrm{ar}(G) \times_S \mathcal{P}\mathrm{ar}(G)$ be defined as follows. Write $\mathcal{E} = \coprod_N \mathcal{E}^N$, where $\mathcal{E}^N : \mathcal{C}\mathrm{onf}_G^N \to \mathcal{P}\mathrm{ar}(G) \times_S \mathcal{P}\mathrm{ar}(G)$, is induced by the morphism

$$\pi_1 \times \pi_N : \prod^N \mathcal{P}\mathrm{ar}(G) \;\to\; \mathcal{P}\mathrm{ar}(G) \times \mathcal{P}\mathrm{ar}(G),$$

denoting by π_1 (resp. π_N) the first (resp. last) projection morphism.

Notation 13.8 *For a section P of $\mathcal{P}\mathrm{ar}(G)$ (resp. $(\underline{g}, P)$ of $\Gamma(\underline{\Delta}(G)) \times_S \mathcal{P}\mathrm{ar}(G)$) one writes:*

$$\mathcal{C}\mathit{onf}_G(P) = (\mathcal{E}_1)^{-1}(P)$$

$$\left(\mathit{resp.}\ \mathcal{C}\mathit{onf}_G(\underline{g}, P) = \left(t_1^\Gamma \times \mathcal{E}_1\right)^{-1}(\underline{g}, P)\right).$$

The sections of $\mathcal{C}\mathit{onf}_G$ (resp. $\mathcal{C}\mathit{onf}_G(\underline{g})$, $\mathcal{C}\mathit{onf}_G(\underline{g}, P)$) are called **Generalized Galleries Configurations (GG)** *of G (resp. Generalized Galleries (GG) configurations of* **type** *$\underline{g}$, Generalized Galleries (GG) configurations of* **type** *$\underline{g}$* **issued from** *P).*

13.10   Galleries Configurations schemes fiber decomposition

Suppose that G is endowed with a frame E. Given a section $\underline{g}$ of $\Gamma(\underline{\Delta}(G))$ of the form $\underline{g} = g_S$ with $g \in \mathrm{gall}_{A_E}$, we write $\mathrm{Conf}_G(\underline{g}) = \mathrm{Conf}_G(g_S)$.

Let

$$\overline{p}^{[\alpha,\alpha+1]} : \mathcal{C}onf_G(g^{(\alpha)}) \rightarrow \mathcal{C}onf_G(g^{(\alpha+1)})$$

be the natural morphism associating with a section $\sigma_{\mathscr{C}}$ of $\mathcal{C}onf_G(g^{(\alpha)})$ the $(\alpha+1)-th$ truncated configuration $\sigma_{\mathscr{C}}^{(\alpha+1)}$. Where $\sigma_{\mathscr{C}}^{(\alpha+1)}$ is defined in terms of $\sigma_{\mathscr{C}}^{(\alpha)}$, following the pattern of Proposition 10.9.1 giving $\gamma^{(\alpha)}$ in terms of γ.

Proposition 13.9
The morphism $\overline{p}^{[\alpha,\alpha+1]}$ defines a locally trivial fibration, with typical fiber proper and smooth.

Proof   *Let*

$$q^{[\alpha,\alpha+1]} : \mathcal{P}ar_{s_\alpha(g)}(G) \rightarrow \mathcal{P}ar_{t_{\alpha+1}(g)}(G),$$

be the natural morphism associating with a parabolic Q of type $s_\alpha(g)$ the unique parabolic of type $t_{\alpha+1}(g)$, containing Q, i.e. $q^{[\alpha,\alpha+1]} = \mathcal{F}_{s_\alpha(g),t_{\alpha+1}(g)}$ (cf. §13.7). It is easy to see that $q^{[\alpha,\alpha+1]}$ is a locally trivial fibration with typical fiber being the quotient P'/Q' of a parabolic P' of type $t_{\alpha+1}(g)$ by a parabolic $Q' \subset P'$ of type $s_\alpha(g)$.

On the other hand, there is a natural isomorphism

$$par_{t_{\alpha+1}} : \mathcal{C}onf_G(g^{(\alpha+1)}) \rightarrow \mathcal{P}ar_{t_{\alpha+1}(g)}(G),$$

associating with the configuration $\sigma_{\mathscr{C}}$ of type $g^{(\alpha+1)}$ the parabolic $P_{\alpha+1}$ of type $t_{\alpha+1}(g)$ of $\sigma_{\mathscr{C}}$.

It is clear that the $\mathcal{C}onf_G(g^{(\alpha+1)})$-scheme

$$\mathcal{C}onf_G(g^{(\alpha+1)}) \times_{\mathcal{P}ar_{t_{\alpha+1}(g)}(G)} \mathcal{P}ar_{s_\alpha(g)}(G),$$

is canonically isomorphic to the $\mathcal{C}onf_G(g^{(\alpha+1)})$-scheme $\mathcal{C}onf_G(g^{(\alpha)})$.

It is concluded by remarking that the morphism

$$\mathcal{C}onf_G(g_S^{(\alpha+1)}) \times_{\mathcal{P}ar_{t_{\alpha+1}(g)}(G)} \mathcal{P}ar_{s_\alpha(g)}(G) \rightarrow \mathcal{C}onf_G(g_S^{(\alpha+1)}),$$

obtained from $q^{[\alpha,\alpha+1]}$ (resp.   from the fiber product given by $(par_{t_{\alpha+1}}, q^{[\alpha,\alpha+1]}))$ by base change, defines a locally trivial fibration with typical fiber P'/Q'.

13.11   Minimal Galleries Configurations

In this section the definition of a Minimal Generalized Gallery for a reductive S-group scheme is given.

Definition 13.10
Define the **Minimal Galleries type Configurations scheme** *of G as the pull-back of*
$$\Gamma^m(\underline{\Delta}(G)) \hookrightarrow \Gamma(\underline{\Delta}(G)) \ by \ \mathcal{C}onf_G \to \Gamma(\underline{\Delta}(G)),$$

$$\mathcal{C}onf_G^m = (t_1^\Gamma)^{-1}\left(\Gamma^m(\underline{\Delta}(G))\right) \subset \mathcal{C}onf_G.$$

NOTATION 13.11

- *Write*
$$\mathcal{C}onf_G^m(\underline{\tau}) = \left(t_1^\Gamma\right)^{-1}\left(\Gamma^m(\underline{\Delta}(G))_{\underline{\tau}}\right),$$
where $\underline{\tau}$ denotes a section of $\mathcal{R}elpos_G$, and
$$\Gamma^m(\Delta(G))_{\underline{\tau}} = \delta_2^{-1}(\underline{\tau}),$$
$$\mathcal{C}onf_G^m(\underline{\tau}, (P', Q')) = \mathcal{C}onf_G^m(\underline{\tau}) \cap \mathcal{E}^{-1}((P, Q)),$$
where (P, Q) denotes a couple of parabolics. It will be seen in the next chapter that the Configurations scheme $\mathcal{C}onf_G^m(\mathbf{g})$ defined by a section $\mathbf{g}$ of $\Gamma^m(\underline{\Delta}(G))_{\underline{\tau}}$ contains a relatively schematically dense open subscheme isomorphic to the schematic closure of $\mathcal{S}tand(\underline{\tau})$, Σ_{τ}, where $\underline{\tau}$ denotes the type of relative position associated to $\mathbf{g}$, i.e. $\delta_2(\mathbf{g}) = \underline{\tau}$. Remark that $\mathcal{C}onf_G^m(\mathbf{g}, (P, Q)) = \mathcal{C}onf_G^m(\mathbf{g}) \cap \mathcal{E}^{-1}((P, Q))$ is the fiber over (P, Q) of the restriction of $\mathcal{E}$ to $\mathcal{C}onf_G^m(\mathbf{g})$.

- *Suppose G endowed with a frame E. Given a subcomplex $K \subset A = \mathcal{A}_E$ one writes*

$$\mathcal{G}all_K \subset \mathcal{G}all_A \quad (resp. \ \mathcal{G}all_K^m = \mathcal{G}all_K \cap \mathcal{G}all_A^m)$$

for the set of gg of A (resp. MGG of A) contained in K.

Given $F, F' \in A$, let $P = P_F$ (resp. $Q = P_{F'}$) be written. $\mathcal{F}ix(P, Q) \subset \mathcal{P}ar(G)$ has been defined as the subfunctor whose sections over S' are the parabolics $\overline{P}$ of $G_{S'}$ satisfying $(P \cap Q)_{S'} \subset \overline{P}$.

It is known that
$$\mathrm{Env}(F, F')_S \simeq \mathcal{F}ix(P, Q).$$

Write
$$\mathcal{C}onf_{\mathcal{F}ix(P,Q)} = \mathcal{C}onf_G \cap \left(\prod \mathcal{F}ix(P, Q)\right),$$

where
$$\prod \mathcal{F}ix(P, Q) = \coprod_N \prod^N \mathcal{F}ix(P, Q).$$

From the definition of $\mathcal{C}onf_{\mathcal{F}ix(P,Q)}$ it follows that

$$\left(\mathcal{G}all_{\mathrm{Env}(F,F')}\right)_S \simeq \mathcal{C}onf_{\mathcal{F}ix(P,Q)}.$$

One may thus define, given a section (P, Q) of $\mathcal{S}\mathrm{tand}(G)$,

$$\mathrm{Conf}^{m}_{\mathcal{F}\mathrm{ix}(P,Q)} = \mathrm{Conf}^{m}_{G} \cap \mathrm{Conf}_{\mathcal{F}\mathrm{ix}(P,Q)}$$

(resp. $\mathrm{Conf}^{m}_{\mathcal{F}\mathrm{ix}(P,Q)}\,(\underline{\tau}) = \mathrm{Conf}^{m}_{G}\,(\underline{\tau}) \cap \mathrm{Conf}_{\mathcal{F}\mathrm{ix}(P,Q)}$,

$\mathrm{Conf}^{m}_{\mathcal{F}\mathrm{ix}(P,Q)}\,(\underline{\tau},(P',Q')) = \mathrm{Conf}^{m}_{G}\,(\underline{\tau},(P',Q')) \cap \mathrm{Conf}_{\mathcal{F}\mathrm{ix}(P,Q)}$).

Definition 13.12

A **Minimal Gallery Configuration** $\sigma_{\mathscr{C}}$ *with extremities* (P, Q) *of* G, *is a section* $\sigma_{\mathscr{C}}$ *of* $\mathcal{C}onf_{\mathcal{F}ix(P,Q)}$ *which is locally for the etale topology in* S *of the form* $(\sigma_{\mathscr{C}})_S = \gamma(F, F')_S$, *where*

$$\gamma(F, F') \in Gall^{m}_{Env(F,F')}.$$

In other words, $\sigma_{\mathscr{C}}$ is local in S equal to the image of some minimal generalized gallery $\gamma(F, F')$, by the above isomorphism induced by a splitting of G. Remark that $\mathcal{C}onf_{\mathcal{F}ix(P,Q)}$ has been defined for any reductive S-group scheme G, and a couple of parabolics (P, Q) of G defining a section of $\mathcal{S}\mathrm{tand}(G)$. A section $\sigma_{\mathscr{C}}$ of $\mathcal{C}onf_{\mathcal{F}ix(P,Q)}$ is a minimal configuration given by the couple (P, Q) of $\mathcal{S}\mathrm{tand}(G)$, if there exists an étale covering $(S_i \to S)$, so that G_{S_i} is endowed with a frame E_i, whose maximal torus T is contained in P_{S_i} (resp. P_{Q_i}), and $(\sigma_{\mathscr{C}})_{S_i}$ is a minimal configuration defined by (P_{S_i}, Q_{S_i}) as above.

The following proposition is a criterion for a section $\sigma_{\mathscr{C}}$ of $\mathrm{Conf}^{m}_{\mathcal{F}\mathrm{ix}(P,Q)}$ to be a minimal gallery configuration.

Proposition 13.13

Let (P, Q) be a section of $\mathcal{S}\mathrm{tand}(\underline{\tau})$, thus $t_2(P, Q) = \underline{\tau}$. Then the set of sections over the S-scheme S' of the S-scheme

$$\mathcal{C}onf^{m}_{\mathcal{F}ix(P,Q)}\,(t_2(P, Q), (P, Q)),$$

correspond to the minimal generalized galleries configurations $\sigma_{\mathscr{C}}$ given by (P, Q) (resp. with extremities (P, Q)).

Proof *One may suppose $S' = S$. Let (P, Q) be a section of $\mathcal{S}\mathrm{tand}(\underline{\tau})$ over S and $\sigma_{\mathscr{C}}$ a configuration of G giving a section of $\mathcal{C}onf_{\mathcal{F}ix(P,Q)}$, then $\sigma_{\mathscr{C}}$ is a section of $\mathcal{C}onf^{m}_{\mathcal{F}ix(P,Q)}\,(t_2(P, Q), (P, Q))$, if $\mathcal{E}(\sigma_{\mathscr{C}}) = (P, Q)$, and the image $\underline{g} = t_1^{\Gamma}(\sigma_{\mathscr{C}})$ of $\sigma_{\mathscr{C}}$ gives a section of $\Gamma^{m}(\underline{\Delta}(G))$, satisfying $\delta_2(\underline{g}) = t_2(P, Q)$. The statement of the proposition may be verified locally. With the aim of comparing the former defining conditions of the proposition on $\sigma_{\mathscr{C}}$ with the later, one may thus suppose that there is a frame E of G, so that*

$$P = P_F \quad (resp. \ Q = P_{F'}),$$

with $F, F' \in \mathcal{A}_E$, $\gamma(F, F') \in Gall_{\mathcal{A}_E}$ so that $\gamma(F, F') \subset Env(F, F')$, and $\sigma_\mathscr{C} = \gamma(F, F')_S$. Let it be written $g = typ^g \gamma(F, F') \in gall^m_{\mathcal{A}_E}$. Recall that:

$$\gamma(F, F') \text{ is a } MGG \Leftrightarrow \tau_\bullet(g) = \tau_g = \tau(F, F')$$

(cf. Proposition 9.38, and Definition 9.20).

One has by Definition of $\sigma_\mathscr{C}$:

$$t_2(P, Q) = \tau(F, F')_S \quad \left(resp. \ \left(\delta_2 \circ t_1^\Gamma\right)(\sigma_\mathscr{C}) = (\tau_g)_S\right),$$

and on the other hand that

$$\left(\delta_2 \circ t_1^\Gamma\right)(\sigma_\mathscr{C}) = t_2(P, Q) \text{ if and only if } \tau_g = \tau(F, F').$$

One concludes that a section $\sigma_\mathscr{C}$ of $Conf_{\mathcal{F}ix(P,Q)}$ with

$$\mathcal{E}(\sigma_\mathscr{C}) = (P, Q)$$

is a minimal configuration if and only if $\sigma_\mathscr{C}$ defines a section of $Conf^m_{\mathcal{F}ix(P,Q)}(t_2(P, Q), (P, Q))$.

Chapter 14

Smooth Resolutions of Schubert Schemes

The relation between the Minimal Galleries type Configurations scheme over the Minimal Galleries of Types scheme, $\mathbf{t}_1^{\Gamma^m} : \mathcal{C}onf_{\mathbf{G}}^{\mathbf{m}} \to \mathbf{\Gamma^m}(\mathbf{\underline{\Delta}(G)})$ and the Universal Schubert scheme over the Types of Relative positions scheme $\overline{\Sigma} \longrightarrow \mathrm{Relpos}_{\mathbf{G}}$ is made explicit. Observe that the former scheme is also a Relpos_{G}-scheme through the finite morphism $\delta_2 : \Gamma^m(\underline{\Delta}(G)) \longrightarrow \mathrm{Relpos}_{G}$. It is proven that the natural morphism

$$\mathcal{C}onf_{\mathbf{G}}^{\mathbf{m}} \longrightarrow \mathbf{\Gamma^m}(\mathbf{\underline{\Delta}(G)}) \times_{\mathrm{Relpos}_{\mathbf{G}}} \overline{\Sigma}$$

is a **Smooth Resolution of Singularities** (cf. 14.17).

Given a section $\mathbf{g}$ of $\Gamma^m(\underline{\Delta}(G))$, with associated type of relative position $\delta_2(\mathbf{g}) = \underline{\tau}$, there is a **relatively schematically dense open subscheme** $\mathcal{C}onf_G^m(\mathbf{g})'$ of the fiber $\mathcal{C}onf_G^m(\mathbf{g})$, and an isomorphism $\mathcal{C}onf_{\mathbf{G}}^{\mathbf{m}}(\mathbf{g})' \simeq \mathbf{\Sigma}_{\underline{\tau}} = \mathcal{S}\mathrm{tand}(\underline{\tau})$, induced by the above resolving morphism. One constructs a section

$$\Theta_{\mathbf{g}} : \mathcal{S}\mathrm{tand}(\underline{\tau}) \longrightarrow \mathcal{C}onf_G^m(\mathbf{g})'$$

which is its reciprocal isomorphism (cf. 12.34).

The **Universal Cellular scheme** $\mathcal{C}onf\,'^{\mathrm{std}}_{\mathbf{G}}$ is defined in terms of the **projection morphism**, giving rise, by specialization, to cellular decompositions of Galleries Configurations schemes whose sections begin by a Borel subgroup. Thus the Configurations scheme $\mathcal{C}onf_{\mathbf{G}}^{\mathbf{m}}(\mathbf{g}, \mathbf{P})$ whose sections are the galleries configurations of type $\mathbf{g}$ with left extremity P admits the following cell decomposition. Let $B \subset P$ be a Borel subgroup and suppose that the finite scheme of galleries of types of relative positions $\mathrm{Relpos}\,_G^{\mathrm{gall}}(g_S)$ is trivial a surjective monomorphism is obtained

$$\coprod j_{[\tau_\mathscr{C},E]} : \mathcal{C}onf\,'_G^{std}(\mathbf{g},B) = \coprod \mathcal{C}onf\,'_G^{std}(\tau_\mathscr{C},B) \to \mathcal{C}onf_G^m(\mathbf{g},P)\;,$$

where $\tau_\mathscr{C}$ runs on its set of sections. The cells $\mathcal{C}onf\,'_G^{std}(\tau_\mathscr{C},B)$ are isomorphic to affine spaces, and are parametrized in terms of **Contracted products** defined by root subgroups. There is only one open cell (**The Big Cell**), and it corresponds to a Minimal Generalized Gallery. Both of these results are proven in the next chapter where contracted products are introduced. A **retraction morphism** is defined from the Galleries Configurations scheme of G to the Galleries Configuration scheme of an Apartment scheme, which is in fact an interpretation of the above cellular decomposition.

14.1 Universal Minimal Galleries of Types Configurations Scheme

Recall that with the notation of 13.12 a **Minimal Gallery Configuration** is a section of $\mathcal{C}onf_G$ locally of the form $(\gamma(F,F')_g)_S$. Recall that $\Gamma^m(\underline{\Delta}(G)) \subset \Gamma(\underline{\Delta}(G))$ is the subscheme whose sections are locally given by the images, by the gallery type morphism $t_1^\Gamma : \mathcal{C}onf_G \longrightarrow \Gamma(\underline{\Delta}(G))$, of Minimal Gallery Configurations. A section of $\Gamma^m(\underline{\Delta}(G)) \subset \Gamma(\underline{\Delta}(G))$ is called a **Minimal Gallery of Types Configuration**. Write

$$\mathcal{C}onf_G^m = (t_1^\Gamma)^{-1}\,(\Gamma^m(\underline{\Delta}(G))) \subset \mathcal{C}onf_G \;(resp.\; \mathcal{C}onf_G^m(\underline{\tau}) = t_1^\Gamma)^{-1}((\delta_2)^{-1}(\underline{\tau}))\;,$$

where $\delta_2 : \Gamma^m(\underline{\Delta}(G)) \longrightarrow \mathrm{Relpos}\,_G$ associates with a Minimal Gallery of Types the type of relative position of the extremities of a representative Minimal Gallery Configuration of this gallery. One calls $\mathcal{C}onf_G^m$ the **Universal Minimal Galleries of Types Configurations Scheme**. Thus the sections of $\mathcal{C}onf_G^m$ are the sections γ of $\mathcal{C}onf_G$ so that its image $t_1^\Gamma(\gamma)$ is a section of $\Gamma^m(\underline{\Delta}(G))$.

Define $t_1^{\Gamma^m} : \mathcal{C}onf_G^m \to \Gamma^m(\underline{\Delta}(G))$, as the restriction of t_1^Γ to $\mathcal{C}onf_G^m$. From the definition of $\mathcal{C}onf_G$ in terms of the incidence relation, and the functorial formula $\mathcal{P}ar(G) \times_S S' = \mathcal{P}ar(G_{S'})$, it is deduced

$$\mathrm{Conf}_G \times_S S' = \mathrm{Conf}_{G_{S'}} \left(resp.\; \mathrm{Conf}_G^m \times_S S' = \mathrm{Conf}_{G_{S'}}^m \right).$$

As a particular case with $S' = \mathcal{S}tand(\underline{\tau})$, one obtains

$$\mathrm{Conf}_G \times_S \mathcal{S}tand(\underline{\tau}) = \mathrm{Conf}_{G_{\mathcal{S}tand(\underline{\tau})}}$$

$$\left(resp.\; \mathcal{C}onf_G^m \times_S \mathcal{S}tand(\underline{\tau}) = \mathrm{Conf}_{G_{\mathcal{S}tand(\underline{\tau})}}^m \right).$$

Recall that $\mathcal{F}ix_{\underline{\tau}} = \mathcal{F}ix\left(\tilde{P},\tilde{Q}\right)$, where $(\tilde{P},\tilde{Q})_{\underline{\tau}}$ denotes the tautological couple in standard position of type $\underline{\tau}$ over $\mathcal{S}tand(\underline{\tau})$. Thus $\mathcal{F}ix_{\underline{\tau}}$ is a finite $\mathcal{S}tand(\underline{\tau})$-scheme whose fiber over a section given by a couple of parabolics (P,Q), defined over an algebraically closed field $\overline{k}$, is the convex hull of the couple of facets (F_P, F_Q) of the building of $G_{\overline{k}}$ (cf. §12.7). By definition, the sections of the subscheme $Conf_{\mathcal{F}ix_{\underline{\tau}}} \subset Conf_{G_{Stand(\underline{\tau})}}$ are the configurations contained in $\mathcal{F}ix_{\underline{\tau}}$. Write

$$Conf_{\mathcal{F}ix_{\underline{\tau}}}^{m} = Conf_{\mathcal{F}ix_{\underline{\tau}}} \cap Conf_{G_{Stand(\underline{\tau})}}^{m}$$

$$\left(resp. \; Conf_{\mathcal{F}ix_{\underline{\tau}}}^{m}(\underline{\tau}) = Conf_{\mathcal{F}ix_{\underline{\tau}}} \cap Conf_{G_{Stand(\underline{\tau})}}^{m}(\underline{\tau})\right) \; ,$$

and

$$Conf_{\mathcal{F}ix_{\underline{\tau}}}^{m}\left(\underline{\tau}, (\tilde{P}, \tilde{Q})_{\underline{\tau}}\right) = Conf_{\mathcal{F}ix_{\underline{\tau}}} \cap Conf_{G_{Stand(\underline{\tau})}}^{m}(\underline{\tau}) \cap \mathcal{E}^{-1}((\tilde{P}, \tilde{Q})_{\underline{\tau}}).$$

Where $\mathcal{E} = (\mathcal{E}_1, \mathcal{E}_2)$ denotes the extremities morphism. From this definition a canonical embedding of $Stand(\underline{\tau})$-schemes results immediately

$$Conf_{\mathcal{F}ix_{\underline{\tau}}}^{m}\left(\underline{\tau}, (\tilde{P}, \tilde{Q})_{\underline{\tau}}\right) \subset Conf_{G_{Stand(\underline{\tau})}}^{m}(\underline{\tau}) = Conf_{G}^{m}(\underline{\tau}) \times_{S} Stand(\underline{\tau})$$

Following the pattern of the proof of the isomorphism

$$\mathcal{F}ix_{\underline{\tau}} \times_{Stand(\underline{\tau})} S' \simeq \mathcal{F}ix(P, Q),$$

where (P, Q) is a section of $Stand(\underline{\tau})$ over the S-scheme S', i.e. (P, Q) is a couple of parabolics of $G_{S'}$ in standard position, with

$$t_2(P, Q) = \underline{\tau}_{S'} \; (\text{cf. Proposition 12.38}),$$

one obtains the following

Proposition 14.1
With the above notation there is a natural isomorphism

$$Conf_{\mathcal{F}ix_{\underline{\tau}}}^{m}\left(\underline{\tau}, (\tilde{P}, \tilde{Q})_{\underline{\tau}}\right) \times_{Stand(\underline{\tau})} S' \simeq Conf_{\mathcal{F}ix_{(P,Q)}}^{m}(\underline{\tau}, (P, Q)).$$

Corollary 14.2 *(The universal property of $Conf_{\mathcal{F}ix_{\underline{\tau}}}^{m}\left(\underline{\tau}, (\tilde{P}, \tilde{Q})_{\underline{\tau}}\right)$)*
Let (P, Q) be a section of $Stand(\underline{\tau})$ over S'. The set of minimal configurations $\sigma_{\mathscr{C}}$ over (P, Q) corresponds naturally to the set of sections of $Conf_{\mathcal{F}ix_{\underline{\tau}}}^{m}\left(\underline{\tau}, (\tilde{P}, \tilde{Q})_{\underline{\tau}}\right)$, over the section $\tau(P, Q)$ of $Stand(\underline{\tau})$ defined by (P, Q), i.e. $Conf_{\mathcal{F}ix(P,Q)}^{m}(\underline{\tau}, (P, Q))$ is given by the fiber product defined by the couple $(\mathcal{E}, \sigma_{(P,Q)})$.

14.2 The Canonical section of a Minimal Galleries of Types Configurations Scheme

There are natural morphisms

$$\mathcal{E} \times t_1^{\Gamma^m} : \mathcal{C}onf_G^m \longrightarrow \mathcal{P}ar(G) \times_S \mathcal{P}ar(G) \times_S \Gamma^m(\underline{\Delta}(G))$$

$$\left(\text{resp. } \mathcal{E} \times (\delta_2 \circ t_1^{\Gamma^m}) : \mathcal{C}onf_G^m \longrightarrow \mathcal{P}ar(G) \times_S \mathcal{P}ar(G) \times_S \mathrm{Relpos}_G\right).$$

Definition 14.3
Let $\underline{\tau}$ be a section of $Relpos_G$ over S. Observe that

$$\Gamma^m(\underline{\Delta}(G)_{\underline{\tau}})_{\mathcal{S}tand(\underline{\tau})} = \mathcal{S}tand(\underline{\tau}) \times_S \delta_2^{-1}(\underline{\tau}) \subset \mathcal{P}ar(G) \times_S \mathcal{P}ar(G) \times_S \Gamma^m(\underline{\Delta}(G)) .$$

Write

$$\mathcal{C}onf_G^m(\underline{\tau})' = \left(\mathcal{E} \times t_1^{\Gamma^m}\right)^{-1} \left(\mathcal{S}tand(\underline{\tau}) \times_S \delta_2^{-1}(\underline{\tau})\right) \subset \mathcal{C}onf_G^m(\underline{\tau}),$$

and denote by

$$\left(\mathcal{E} \times t_1^{\Gamma^m}\right)' : \mathcal{C}onf_G^m(\underline{\tau})' \longrightarrow \mathcal{S}tand(\underline{\tau}) \times_S \delta_2^{-1}(\underline{\tau}),$$

the morphism induced by $\mathcal{E} \times t_1^{\Gamma^m}$.

Observe that the sections of $\mathcal{C}onf_G^m(\underline{\tau})'$ over $S' \to S$ are the galleries configurations γ with associated gallery of types $t_1^{\Gamma^m}(\gamma) = g$ satisfying $\delta_2(g) = \underline{\tau}$, and extremities $\mathcal{E}(\gamma) = (P, Q)$ such that (P, Q) is in standard position with type of relative position equal to $\underline{\tau}$. Thus $\mathcal{C}onf_G^m(\underline{\tau})'$ is naturally a $\mathcal{S}tand(\underline{\tau})$-scheme. In fact it will be seen that $\mathcal{C}onf_G^m(\underline{\tau})'$ is a canonical **open subscheme relatively schematically dense** in $\mathcal{C}onf_G^m(\underline{\tau})$ and that the morphism

$$\mathcal{C}onf_G^m(\underline{\tau})' \simeq \mathcal{S}tand(\underline{\tau}) \times_S \delta_2^{-1}(\underline{\tau})$$

is in fact an isomorphism. Thus the above definition is a functorial description of $\mathcal{C}onf_G^m(\underline{\tau})'$, i.e. a characterization of the sections of $\mathcal{C}onf_G^m(\underline{\tau})$ which are in fact sections of $\mathcal{C}onf_G^m(\underline{\tau})'$. Observe that $\mathcal{E}$ and $t_1^{\Gamma^m}$ are G-equivariant morphisms. The next aim is the construction of a canonical section of $\left(\mathcal{E} \times t_1^{\Gamma^m}\right)'$.

One can look at $\mathcal{C}onf_{\mathcal{F}\mathrm{ix}_{\underline{\tau}}}^m(\underline{\tau}, (\tilde{P}, \tilde{Q})_{\underline{\tau}})$ as an S-scheme. A section of $\mathcal{C}onf_{\mathcal{F}\mathrm{ix}_{\underline{\tau}}}^m\left(\underline{\tau}, (\tilde{P}, \tilde{Q})_{\underline{\tau}}\right)$ on the S-scheme S' is given by a couple

$$((P, Q), \sigma_\mathscr{C}),$$

formed by a section (P, Q) of $\mathcal{S}tand(\underline{\tau})$ over S', and a Minimal Gallery Configuration $\sigma_\mathscr{C}$ given by (P, Q) (cf. Proposition 14.1) and contained in the fiber $\left(\mathcal{F}\mathrm{ix}_{\underline{\tau}}\right)_{(P,Q)} = \mathcal{F}\mathrm{ix}(P, Q)$. Thus there is an inclusion of S-functors

$\mathcal{C}\mathrm{onf}^m_{\mathcal{F}\mathrm{ix}_{\underline{\tau}}}\left(\underline{\tau}, (\tilde{P}, \tilde{Q})_{\underline{\tau}}\right) \subset \mathcal{C}\mathrm{onf}^m_G(\underline{\tau})'$, which is an isomorphism.

On the other hand there is a natural morphism of $\mathcal{S}\mathrm{tand}(\underline{\tau})$-schemes induced by the restriction of $\left(\mathcal{E} \times t_1^{\Gamma^m}\right)'$:

$$\mathcal{C}\mathrm{onf}^m_{\mathcal{F}\mathrm{ix}_{\underline{\tau}}}\left(\underline{\tau}, (\tilde{P}, \tilde{Q})_{\underline{\tau}}\right) \longrightarrow (\Gamma^m(\underline{\Delta}(G))_{\underline{\tau}})_{\mathcal{S}\mathrm{tand}(\underline{\tau})} = \mathcal{S}\mathrm{tand}(\underline{\tau}) \times_S \delta_2^{-1}(\underline{\tau})$$

defined by

$$\sigma_{\mathscr{C}} \mapsto \left(t_1^{\Gamma^m}\right)_{\mathcal{S}\mathrm{tand}(\underline{\tau})}(\sigma_{\mathscr{C}}).$$

It may be seen that it is an isomorphism. One proceeds locally in S. Assume that G is endowed with a frame E. With the notations of the proof of 12.36 the group $G_{\mathcal{S}\mathrm{tand}(\underline{\tau})}$ is thus endowed with a frame $\tilde{E}$ over the open subscheme given by the open set of the big cell covering, $\mathcal{U}_{t,R_+} \times_S \mathcal{U}(C', F_{\tau,C'})$, adapted to the tautological couple $(\tilde{P}, \tilde{Q})_{\underline{\tau}}$. The maximal torus $\tilde{T}$ given by $\tilde{E}$ is related to the maximal torus T of E as follows.

Let (y, x) be a section of $U = \mathcal{U}_{t,R_+} \times_S \mathcal{U}(C', F_{\tau,C'})$. By definition of $P_{\tau,C'}$ one has, with the obvious notation,

$$T \subset (P_{t,R_+}, P_{\tau,C'}),$$

where T is the maximal torus defined by E, and thus

$$\tilde{T} = \mathrm{int}(yx)(T) \subset \left(\mathrm{int}(y)(P_{t,R_+}), \mathrm{int}(yx)(P_{\tau,C'})\right),$$

i.e.

$$\tilde{T} = \mathrm{int}(yx)(T) \subset \tilde{P} \cap \tilde{Q}.$$

Write $A = \mathcal{A}_{\tilde{E}}$. Assume that there exists $\tau \in \mathrm{Relpos}\, A$ with $\tau = \tau(F, F')$ and $\underline{\tau} = \tau_U$, where $F \in A$ (resp. $F' \in A$) denote the parabolic set of roots corresponding to P_{t,R_+} (resp. $P_{\tau,C'}$), i.e. $P_F = P_{t,R_+}$ (resp. $P_{F'} = P_{\tau,C'}$).

Let the isomorphism $Env(F, F')_U \simeq (\mathcal{F}\mathrm{ix}_{\underline{\tau}})_U$ be made explicit. Let $F'' \in Env(F, F')$, then

$$F''_U \mapsto \mathrm{int}(yx)(P_{F''}).$$

Similarly the following isomorphisms is established. One has

$$\delta_2 = (\tau_{\bullet})_U,$$

where $\tau_{\bullet} : \mathrm{gall}^m_A \to \mathrm{Relpos}\, A$ is defined as in Definition 9.20. Thus

$$\delta_2^{-1}(\tau_U) = \tau_{\bullet}^{-1}(\tau)_U = \mathrm{gall}^m_A(\tau)_U,$$

and

$$\mathrm{gall}^m_A(\tau)_U \simeq (\Gamma^m(\underline{\Delta}(G))_{\underline{\tau}})_U.$$

Write

$$\mathrm{Gall}^m_A(\tau) = (\tau_{\bullet} \circ \mathrm{typ}^g)^{-1}(\tau) = (\mathrm{typ}^g)^{-1}(\mathrm{gall}^m_A(\tau))$$

$$\left(\text{resp. } \mathrm{Gall}^m_{Env(F,F')}(\tau) = \mathrm{Gall}^m_A(\tau) \cap \mathrm{Gall}_{Env(F,F')}\right).$$

One has:

1. $\text{Gall}^m_{\text{Env}(F,F')}(\tau) \cap E^{-1}((F,F')) = \text{Gall}^m_A(\tau) \cap E^{-1}((F,F'))$, since a Minimal Generalized Gallery γ satisfying $E(\gamma) = (F,F')$ and $\tau = \tau(F,F')$ is contained in $\text{Env}(F,F')$;

2. $\left(\text{Gall}^m_{\text{Env}(F,F')}(\tau) \cap E^{-1}((F,F'))\right)_U \simeq \text{Conf}^m_{\mathcal{F}ix(P,Q)}(\tau_U, (\tilde{P}, \tilde{Q})_{\tau_U})$.

Observe that $\text{Gall}^m_{\text{Env}(F,F')}(\tau) \cap E^{-1}((F,F'))$ is the set of Minimal Generalized Galleries γ of A contained in $\text{Env}(F,F')$, so that $E(\gamma) = (F,F')$, and $g = \text{typ}^g(\gamma)$ verifies $\tau_g = \tau$. The morphism $(t_1^{\Gamma^m})_U$ corresponds to the mapping

$$\text{Gall}^m_{\text{Env}(F,F')}(\tau) \cap E^{-1}((F,F')) \longrightarrow \text{gall}^m_A(\tau),$$

given by $\gamma \mapsto \text{typ}^g \gamma$. By proposition (3.22), given $(F,F') \in (A \times A)_\tau$, so that $\tau(F,F') = \tau$, and $g \in \text{gall}^m_A(\tau)$, a minimal gallery of types, there exists a unique Minimal Generalized Gallery $\gamma_g(F,F')$ so that

$$\text{typ}^g \gamma_g(F,F') = g \quad (\text{resp. } E(\gamma_g(F,F')) = (F,F')).$$

It follows that the above mapping is bijective, and thus that $t_1^{\Gamma^m}$ induces an isomorphism. It has thus been proved the

Proposition 14.4
The natural morphism

$$\mathcal{C}onf^m_{\mathcal{F}ix_\tau}(\underline{\tau}, (\tilde{P}, \tilde{Q})_{\underline{\tau}}) \longrightarrow \mathcal{S}tand(\underline{\tau}) \times_S \delta_2^{-1}(\underline{\tau}) = (\Gamma^m(\underline{\Delta}(G))_{\underline{\tau}})_{\mathcal{S}tand(\underline{\tau})}$$

defined above is an isomorphism of $\mathcal{S}tand(\underline{\tau})$-schemes which is G equivariant when it is supposed that $\delta_2^{-1}(\underline{\tau})$ is endowed with the trivial action of G.

Definition 14.5
There is an S-scheme monomorphism

1) *Denote by*

$$\Theta_{\underline{\tau}} : (\Gamma^m(\underline{\Delta}(G))_{\underline{\tau}})_{\mathcal{S}tand(\underline{\tau})} \longrightarrow \mathcal{C}onf^m_G(\underline{\tau})',$$

the composed morphism given by

$$(\Gamma^m(\underline{\Delta}(G))_{\underline{\tau}})_{\mathcal{S}tand(\underline{\tau})} \longrightarrow \mathcal{C}onf^m_{\mathcal{F}ix_{\underline{\tau}}}\left(\underline{\tau}, (\tilde{P}, \tilde{Q})_{\underline{\tau}}\right) \quad (\text{cf. Corollary 14.4})$$

followed by the morphism

$$\mathcal{C}onf^m_{\mathcal{F}ix_{\underline{\tau}}}\left(\underline{\tau}, (\tilde{P}, \tilde{Q})_{\underline{\tau}}\right) \simeq \mathcal{C}onf^m_G(\underline{\tau})'$$

(The $\underline{\tau}$-Canonical Section).

2) Let $\underline{g}$ be a section of $(\Gamma^m(\underline{\Delta}(G))_{\underline{\tau}})_{Stand(\underline{\tau})}$ over $Stand(\underline{\tau})$. Denote by

$$\xi_{\underline{g}} : Stand(\underline{\tau}) \longrightarrow (\Gamma^m(\underline{\Delta}(G))_{\underline{\tau}})_{Stand(\underline{\tau})}$$

the corresponding morphism. Define

$$\Theta_{\underline{g}} : Stand(\underline{\tau}) \longrightarrow \mathcal{C}onf_G^m(\underline{g}) \; (= \mathcal{C}onf_G(\underline{g})) ,$$

*as the composed morphism $\Theta_{\underline{g}} = \Theta_{\underline{\tau}} \circ \xi_{\underline{g}}$ (**The $\underline{g}$-Canonical Section**).*

Remark that the morphism $\Theta_{\underline{\tau}}$ satisfies

$$(\mathcal{E} \circ \Theta_{\underline{\tau}})\,(((P,Q),\underline{g})) = (P,Q) \quad (\text{resp. } (t_1^{\Gamma} \circ \Theta_{\underline{\tau}})\,(((P,Q),\underline{g})) = \underline{g}) \; ,$$

i.e. $\Theta_{\underline{\tau}}$ is a $Stand(\underline{\tau})$-morphism. Thus $\Theta_{\underline{g}}$ defines a **section** of $\mathcal{C}onf_G^m(\underline{g})'$
over $Stand(\underline{\tau})$.
There is a natural G-invariant S-morphism

$$\mathcal{C}onf_G^m(\underline{\tau}) \longrightarrow \mathcal{P}ar(G) \times_S \mathcal{P}ar(G) \times_S \delta_2^{-1}(\underline{\tau}),$$

induced by $\mathcal{E} \times t_1^{\Gamma^m}$. Denote by $\left(\mathcal{E} \times t_1^{\Gamma^m}\right)'$ its restriction to $\mathcal{C}onf_G^m(\underline{\tau})'$ and
remark that $\mathcal{C}onf_G^m(\underline{\tau})'$ is G-invariant. From the above equalities it follows
the

Proposition 14.6
The morphism

$$\Theta_{\underline{\tau}} : Stand(\underline{\tau}) \times_S \delta_2^{-1}(\underline{\tau}) \longrightarrow \mathcal{C}onf_G^m(\underline{\tau})' ,$$

defines a G-equivariant section of the G-equivariant morphism $\left(\mathcal{E} \times t_1^{\Gamma^m}\right)'$.

The next aim is to prove that:

1. $\Theta_{\underline{\tau}}$ is an **open embedding**;

2. The open subscheme $U_{\Theta_{\underline{\tau}}} \subset \mathcal{C}onf_G^m(\underline{\tau})$ defined by $\Theta_{\underline{\tau}}$, i.e. the image of
   $\Theta_{\underline{\tau}}$, is **relatively schematically dense**;

3. $U_{\Theta_{\underline{\tau}}} = \mathcal{C}onf_G^m(\underline{\tau})'$. Observe that $\mathcal{C}onf_G^m(\underline{\tau})'$ is the **reciprocal image**
   of $Stand(\underline{\tau})$ by the extremities morphism $\mathcal{C}onf_G^m(\underline{\tau}) \longrightarrow \mathcal{P}ar(G) \times_S$
   $\mathcal{P}ar(G)$.

The proof of these facts is founded on an isomorphism of this Configurations scheme with a **Contracted Product** and on the induced **Cell Decomposition**.

14.2.1 Local Description of a Canonical Section

The following local description of the section Θ_τ is given. Let it be supposed that G is endowed with a frame E. Write $A = \mathcal{A}_E$ for the apartment defined by $\mathcal{R}(E)$. Let $\tau \in \mathrm{Relpos}\, A$, with $\underline{\tau} = \tau$, and $(t, s) \in \mathrm{typ}A \times \mathrm{typ}A$ the corresponding couple of types defined by τ. With the notation of Definition 9.20 recall the following canonical isomorphisms:

$$\tau_\bullet^{-1}(\tau)_S = \mathrm{gall}_A^m(\tau)_S \simeq \delta_2^{-1}(\tau_S),$$

and

$$\mathcal{S}\mathrm{tand}(\tau_S) \times_S \delta_2^{-1}(\tau_S)) \simeq \coprod_{g \in \tau_\bullet^{-1}(\tau)} \mathcal{S}\mathrm{tand}(\tau_S)^{(g)},$$

where the second member denotes the disjoint union of set of copies of $\mathcal{S}\mathrm{tand}(\tau_S)$, $\left(\mathcal{S}\mathrm{tand}(\tau_S)^{(g)}\right)_{g \in \tau_\bullet^{-1}(\tau)}$ indexed by $\tau_\bullet^{-1}(\tau)$. On the other hand, there are isomorphisms

$$\mathcal{C}\mathrm{onf}_G^m(\tau_S)' \simeq \coprod_{g \in \tau_\bullet^{-1}(g)} \mathcal{C}\mathrm{onf}_G^m(g_S)'$$

$$\left(\text{resp. } \mathcal{C}\mathrm{onf}_G^m(\tau_S) \simeq \coprod_{g \in \tau_\bullet^{-1}(\tau)} \mathcal{C}\mathrm{onf}_G^m(g_S)\right).$$

Accordingly the morphism Θ_{τ_S} (resp. $\overline{\Theta}_{\tau_S}$) splits as

$$\Theta_{\tau_S} = \coprod_{g \in \tau_\bullet^{-1}(\tau)} \Theta_{g_S} \left(\text{resp. } \overline{\Theta}_{\tau_S} = \coprod_{g \in \tau_\bullet^{-1}(\tau)} \overline{\Theta}_{g_S}\right).$$

Where $\overline{\Theta}_{\tau_S}$ (resp. $\overline{\Theta}_{g_S}$) is obtained by composing Θ_{τ_S} (resp. Θ_{g_S}) with the corresponding inclusion morphism.

Let $t = e_1(g)$ i.e. $t =$ the first type of g. One has that Θ_{g_S} may be seen as a $\mathcal{P}\mathrm{ar}\,_t(G)$-morphism, i.e. there is a commutative diagram

$$\Theta_{g_S}: \mathcal{S}\mathrm{tand}(\tau_S) \longrightarrow \mathcal{C}\mathrm{onf}_G^m(g_S)$$

$$pr_{\mathcal{P},\tau_S} \searrow \qquad \swarrow \mathcal{E}_{1,g_S}$$

$$\mathcal{P}\mathrm{ar}\,_t(G)$$

where $\mathcal{E}_{1,g_S}$ is the restriction of $\mathcal{E}_1$ to $\mathcal{C}\mathrm{onf}_G^m(g_S)$, and $pr_{\mathcal{P},\tau_S}$ the morphism induced by the first projection.

Remark 14.7

It follows from the proof of 14.4 that Θ_{g_S} associates with a couple of parabolics $(P, Q) = (P_F, P_{F'})$, containing a maximal torus T, with type of relative position given by τ the Minimal Gallery Configuration $\sigma_{\mathscr{C},g} = \gamma_g(F, F')_S$ with extremities $\mathcal{E}(\sigma_{\mathscr{C},g}) = (P, Q)$. Where $\gamma_g(F, F')$ is the unique mgg in $\mathcal{R}(E)$ defined by a couple of facets (F, F') with type of relative position τ and $\tau_\bullet(g) = \tau$. On the other hand, the morphism Θ_{g_S} is G-invariant. Thus Θ_{g_S} is completely determined by the image of a couple (P, Q).

14.2.2   Restriction of a Canonical Section to the Big Cell Covering

Assume again that G is endowed with a frame $E = (T, M, R, R_0, (X_\alpha)_{\alpha \in R_0})$ (cf. Proposition 11.12). Let $\mathcal{R} = (M, M^*, R, R^\vee, R_0)$ be the corresponding $\mathbb{Z}$-system of roots, and $R_+ = (\mathbb{N} R_0) \cap R$ the system of positive roots defined by R_0. Denote by $\mathcal{A}_E$ the apartment building defined by $\mathcal{R}$. Let $C = C_{R_+}$ be the chamber of $A = \mathcal{A}_E$ corresponding to the positive roots system R_+. Write $F_t = F_t(C)$ for the facet of type t incident to C. Remark that $P_{F_t} = P_{t,R_+}$. For the sake of briefness write $g, \tau, \tau_{\mathscr{C}}, \cdots$, for the constant sections g_S, τ_S, $(\tau_{\mathscr{C}})_S, \cdots$, if no confusion arises. Let one write τ in this section instead of τ_S and $g \in \text{gall}^m A$, such that $\tau_\bullet(g) = \tau_g = \tau$, instead of g_S. Suppose τ corresponds to $(t, s) \in \text{typ}\,\mathcal{A}_E \times \text{typ}\,\mathcal{A}_E$. By Proposition 12.20, 2., there is the big cell open covering of the Universal Cell $\mathcal{S}\text{tand}(\tau)$

$$\mathcal{S}\text{tand}(\tau) = \bigcup_{R_+ \in \mathcal{F}\text{ram}(\mathcal{R})} \sigma_{R_+}^{(\tau)} \left(\overline{\Omega}_{t,R_+} \times_S \mathcal{S}\text{tand}(\tau, P_{t,R_+}) \right)$$

$$= \bigcup_{(R_+, C') \in I_\tau(\mathcal{R})} \sigma_{R_+, C'}^{(\tau)} \left(\mathcal{U}_{t,R_+} \times_S \mathcal{U}(C', F_{\tau, C'}) \right).$$

The morphism

$$\mathcal{E}_{1,g} : \mathcal{C}\text{onf}_G^m(g) \to \mathcal{P}\text{ar}_t(G),$$

defines a locally trivial fibration, with typical fiber $\mathcal{C}\text{onf}_G^m(g, P_{t,R_+})$. It is trivialized by the big cell open covering $(\overline{\Omega}_{t,R_+})_{R_+ \in \mathcal{F}\text{ram}(\mathcal{R})}$ of $\mathcal{P}\text{ar}_t(G)$. There are isomorphisms of $\overline{\Omega}_{t,R_+}$-schemes:

$$\zeta_{R_+}^{(g)} : \overline{\Omega}_{t,R_+} \times_S \mathcal{C}\text{onf}_G^m(g, P_{t,R_+}) \to \mathcal{C}\text{onf}_G^m(g)_{\overline{\Omega}_{t,R_+}},$$

defined by

$$\zeta_{R_+}^{(g)} : \left(y, \sigma_{\mathscr{C}}(P_{t,R_+}) \right) \mapsto \left(\text{int}(y)(P_{t,R_+}), \text{int}(y) \left(\sigma_{\mathscr{C}}(P_{t,R_+}) \right) \right),$$

where y is a section of $\mathcal{U}_{t,R_+} \simeq \overline{\Omega}_{t,R_+}$ and $\sigma_{\mathscr{C}}(P_{t,R_+})$ denotes a configuration of type g, issued from P_{t,R_+}.

Let

$$\Theta_{g, P_{t,R_+}} : \mathcal{S}\text{tand}(\tau, P_{t,R_+}) \to \mathcal{C}\text{onf}_G^m(g, P_{t,R_+}),$$

denote the fiber of the morphism Θ_g over the section of $\mathcal{P}\text{ar}_t(G)$, defined by the parabolic $P_{t,R_+} \subset G$ (resp. $(\Theta_g)_{\overline{\Omega}_{t,R_+}} : \mathcal{S}\text{tand}(\tau)_{\overline{\Omega}_{t,R_+}} \to \mathcal{C}\text{onf}_G^m(g)_{\overline{\Omega}_{t,R_+}}$, the morphism induced by Θ_g over the open set $\overline{\Omega}_{t,R_+} \subset \mathcal{P}\text{ar}_t(G)$). The G-equivariance (cf. Proposition 14.6) of Θ_g implies the commutativity of the following diagram:

$$1_{\overline{\Omega}_{t,R_+}} \times_S \Theta_{g,P_{t,R_+}} : \qquad \overline{\Omega}_{t,R_+} \times_S \mathcal{S}tand(\tau, P_{t,R_+}) \longrightarrow \overline{\Omega}_{t,R_+} \times_S \mathcal{C}onf^{\,m}_G\,(g,P_{t,R_+}) \ ,$$

$$\Big\downarrow \sigma^{(\tau)}_{R_+} \qquad\qquad\qquad\qquad \Big\downarrow \varsigma^{(\tau)}_{R_+}$$

$$(\Theta_g)_{\overline{\Omega}_{t,R_+}} : \qquad \mathcal{S}tand(\tau)_{\overline{\Omega}_{t,R_+}} \longrightarrow \mathcal{C}onf^{\,m}\,(g)_{\overline{\Omega}_{t,R_+}} .$$

which establishes the "compatibility" of the above trivializations.
Let the composed morphism be made explicit:

$$\Theta_{g,P_{t,R_+}} \circ \sigma^{(\tau,C')} : \mathcal{U}(C', F_{\tau,C'}) \to \mathcal{C}onf^{\,m}_G\,(g,P_{t,R_+}).$$

(There is the following relation between $\sigma^{(\tau,C')}$ and $\sigma^{(\tau)}_{R_+,C'}$:

$$\sigma^{(\tau,C')} = \sigma^{(\tau)}_{R_+,C'} \circ j_{\mathcal{U}(C',F_{\tau,C'})} \ ,$$

where $j_{\mathcal{U}(C',F_{\tau,C'})} : \mathcal{U}(C', F_{\tau,C'}) \to \mathcal{U}_{t,R_+} \times_S \mathcal{U}(C', F_{\tau,C'})$ is the closed embedding defined as $j_{\mathcal{U}(C',F_{\tau,C'})} = 1_{\mathcal{U}_{t,R_+}} \times_S Id_{\mathcal{U}(C',F_{\tau,C'})}.$)

Recall that with the notation of Lemma 12.16 there is an isomorphism of S-schemes:

$$\sigma^{(\tau,C')} : \mathcal{U}(C', F_{\tau,C'}) \xrightarrow{\;\sim\;} \mathcal{S}tand(\tau_{C'}, B_{C'}) \subset \mathcal{S}tand(\tau, P_{t,R_+})$$

which is a parametrization of the big cell $\mathcal{S}tand(\tau_{C'}, B_{C'})$ of $\mathcal{S}tand(\tau, P_{t,R_+})$ defined by $B_{C'}$.

Remark that from Proposition 12.17 it results that the set of restrictions

$$\Big(\Theta_{g,P_{t,R_+}} \circ \sigma^{(\tau,C')}\Big)_{C' \in \mathrm{Ch}\,\mathrm{St}_{F_t(R_+)}}$$

gives a complete description of the section $\Theta_{g,P_{t,R_+}}$. Thus one proceeds to describe $\Theta_{g,P_{t,R_+}} \circ \sigma^{(\tau,C')}$. Let $F, F' \in A$ such that $P_{t,R_+} = P_F$, and $F' = F_{\tau,C'}$. As in Proposition 9.38, denote by $\gamma_g(F, F') \in \mathrm{Gall}^{\,m}_A$ the minimal generalized gallery of type g, with $\tau_g = \tau(F, F')$. With the notation of Definition 9.16 let $\gamma_g(C'; F, F')$ be the Generalized Gallery obtained from $\gamma_g(F, F')$ by composition with $C' \supset F$. In fact $\gamma_g(C'; F, F)$ is minimal between the chamber C' and the facet F', by definition of Minimal Generalized Gallery, as C' is at maximal distance from F'. To the closed set of roots in $R_{C'}$ whose walls separate C' and F': $R(C', F') = R_+(C', F') = \{\alpha \in R_{C'} \mid H_\alpha \in \mathcal{H}(C', F')\}$, corresponds the subgroup $\mathcal{U}(C', F_{\tau,C'}) \subset B_{C'}$, parametrizing $\mathcal{S}tand(\tau_{C'}, B_{C'})$ ($=$ the $B_{C'}$-cell contained and open in $\mathcal{S}tand(\tau, P_{t,R_+})$). Let $\sigma_{\mathscr{C}}(\gamma_g(F, F'))$ denote the section of $\mathcal{C}onf^{\,m}_G\,(g, P_{t,R_+})$ obtained by replacing a facet $\overline{F}$ of $\gamma_g(F, F')$ by the corresponding parabolic $P_{\overline{F}}$, and thus the couple (F, F') by $(P_F, P_{F'})$. Observe that $\sigma_{\mathscr{C}}(\gamma_g(F, F'))$ is a Minimal Gallery Configuration, since $\gamma_g(F, F')$ is a Minimal Gallery, i.e. $\sigma_{\mathscr{C}}(\gamma_g(F, F'))$ is a section of $\mathcal{C}onf^{\,m}_G\,(g, P_{t,R_+})' = \mathcal{C}onf^{\,m}_G\,(g)' \cap \mathcal{C}onf^{\,m}_G\,(g, P_{t,R_+})$.
From the G-equivariance of Θ_g (cf. Remark 14.7) one obtains the following

Proposition 14.8
The morphism

$$\Theta_g \circ \sigma^{(\tau)}_{R_+,C'} : \mathcal{U}_{t,R_+} \times \mathcal{U}(C', F_{\tau,C'}) \to \mathcal{C}onf^m_G (g)_{\overline{\Omega}_{t,R_+}}$$

$$\left(resp. \ \Theta_{g,P_{t,R_+}} \circ \sigma^{(\tau,C')} : \mathcal{U}(C', F_{\tau,C'}) \to \mathcal{C}onf^m_G (g, P_{t,R_+}) \right)$$

is given by

$$\Theta_g \circ \sigma^{(\tau)}_{R_+,C'} : (y,x) \mapsto \left(int(y)(P_{t,R_+}), int(y) \left(int(x)(\sigma_{\mathscr{C}}(\gamma_g(F,F'))) \right) \right)$$

$$\left(resp. \ \Theta_{g,P_{t,R_+}} \circ \sigma^{(\tau,C')} : x \mapsto int(x) \left(\sigma_{\mathscr{C}}(\gamma_g(F,F')) \right) \right).$$

14.3 Block decomposition of a Bruhat Cell parametrizing subgroup by a minimal gallery of types

Keep the above hypothesis on G and notations. For the sake of briefness write g, τ, $\tau_{\mathscr{C}}, \cdots$, for the constant sections g_S, τ_S, $(\tau_{\mathscr{C}})_S, \cdots$, if no confusion arises. In this section and the following one makes the morphism (resp. the section) $\Theta_{g,P_{t,R_+}}$ explicit, over the open subscheme of the Big cell covering, corresponding to $\mathcal{U}_{t,R_+} \times_S \mathcal{U}(C', F_{\tau,C'})$, by means of a product decomposition of the parametrizing subgroup $\mathcal{U}(C', F_{\tau,C'})$ of $\mathcal{S}tand(\tau_{C'}, B_{C'})$, $\prod \mathcal{U}(w_i) \simeq \mathcal{S}tand(\tau_{C'}, B_{C'})$. This decomposition is given by a minimal gallery of types g, namely $m^{(g,C')} : \prod \mathcal{U}(C_i, F_{i-1}) \simeq \mathcal{U}(C', F_{\tau,C'})$, where the $\mathcal{U}(C_i, F_{i-1})$ are subgroups of $B_{C'}$ defined by g (see below), and $m^{(g,C')}$ is given by the multiplication in $B_{C'}$. In fact one obtains two representations of $\Theta_{g,P_{t,R_+}}$, one in terms of the minimal gallery $\gamma_g(C'; F, F')$ of $\mathcal{A}_E$, and the other in terms of

$$\sigma_g(C') : (P_{s_i}(C'))(r \geqslant i \geqslant 0), \ (P_{t_j}(C')) \ (r+1 \geqslant j \geqslant 1),$$

where

$$g : (s_i)(r \geqslant i \geqslant 0), \ (t_j) \ (r+1 \geqslant j \geqslant 1),$$

the **basical configuration defined by** (g, C'). Observe that by definition $\sigma_g(C')$ is contained in the **simplex subcomplex** $\Delta(C')$ defined by C'. With the aim of simplifying notation assume that $g = g_1$, with $l(g) = r+1$. One may thus write:

$$\gamma_g(F, F') : (F_i)(r \geqslant i \geqslant 0), \ (F'_j)(r+1 \geqslant j \geqslant 1).$$

One associates with the Minimal Generalized Gallery $\gamma_g(C'; F, F')$ a decomposition of $R_+(C', F')$ as a disjoint union of closed systems of roots. Namely

$$R_+(C', F_{\tau,C'}) = \coprod R_+(C_i, F_{i+1}) \ (r+1 \geqslant i \geqslant 1),$$

where
$$R_+(C_i, F_{i-1}) = \{\alpha \in R_{C_i} \mid H_\alpha \in \mathcal{H}(C_i, F_{i-1})\}\,.$$

Recall that by Definition 9.6 one has

$$\mathcal{H}(C', F') = \coprod \mathcal{H}(C_i, F_{i-1}),$$

from which the above equality follows. Let

$$\mathcal{U}(C_i, F_{i-1}) \subset B_{C'},$$

be the subgroup defined by the closed set of roots $R_+(C_i, F_{i-1})$.

Definition 14.9
From the above equality one obtains an isomorphism of S-schemes

$$m^{(g,C')} : \prod \mathcal{U}(C_i, F_{i-1}) \rightarrow \mathcal{U}(C', F_{\tau,C'}),$$

induced by the multiplication of G. It is called the **block decomposition of** $\mathcal{U}(C', F_{\tau,C'})$ *defined by the* **Minimal Generalized Gallery of types** g *and the* **chamber** C'.

The isomorphism $m^{(g,C')}$ gives rise to **another isomorphism** which is described now. The sequence of chambers constructed is (C_k) $(r+1 \geqslant k \geqslant 0)$ as in §9.2, b). With this sequence of chambers is associated a $\mathcal{S}(C')$-reduced expression of $w(C', \mathrm{proj}_{F'} C')$ $(F' = F_{\tau,C'})$, namely

$$w(C', \mathrm{proj}_{F'} C') = w_{r+1} \cdot w_r \cdots w_1,$$

where the (w_i) $(r+1 \geqslant i \geqslant 1)$ are defined as follows. Let

$$u_i = w(C_i, C_{i+1}) \quad (r+1 \geqslant i \geqslant 1),$$

and
$$v_i = \prod u_\alpha \quad (r+1 \geqslant \alpha \geqslant 1)$$

(resp. $v_{r+2} = 1$). Then
$$w_i = v_{i+1}^{-1} u_i v_{i+1}.$$

Then one has the equality, as is easy to see (otherwise cf. §9.2, b) being aware that what is called w_i (resp. u_i) here, is called u_i (resp. w_i)) there:

$$w_{r+1} \cdot w_r \cdots w_1 = u_1 \cdots u_r u_{r+1}$$

(resp. $w_{r+1} w_r \cdots w_i = u_i \cdots u_r w_{r+1}$).

Let it be written

$$s_i = s_i(g) \ (r \geqslant i \geqslant 0) \quad (\text{resp. } t_j = t_j(g) \ (r+1 \geqslant j \geqslant 1)).$$

With the notation of §9.2, b). one has

$$w_i = w(t_i, s_{i-1}).$$

Given $w \in W$ is denoted by

$$\mathcal{U}(w) \subset B_{C'},$$

the subgroup defined by the closed system of roots

$$R_+(C', w(C')) = \{\alpha \in C' \mid H_\alpha \in \mathcal{H}(C', w(C'))\}.$$

For each $w_i \in W$ a representative in $N(T)$ may be chosen. Denote this representative also by w_i and define v_{i+1} in terms of these w_i's as above.

From Lemma 10.28 it is deduced

$$\mathcal{U}(C_i, F_{i-1}) = v_{i+1} \mathcal{U}(w_i) v_{i+1}^{-1}.$$

In fact the proof of this equality runs accordingly Lemma 10.28, taking on account that

$$w_i^{tr} = w_i.$$

Thus there is an isomorphism of S-schemes

$$\prod \mathcal{U}(w_i) \;\to\; \prod \mathcal{U}(C_i, F_{i-1}),$$

defined by

$$(x_i) \mapsto (\mathrm{int}(v_{i+1})(x_i)).$$

This isomorphism composed with $m^{(g,C')}$ gives rise to an isomorphism of S-schemes

$$\overline{m}^{(g,C')} : \prod \mathcal{U}(w_i) \to \mathcal{U}(C', F_{\tau,C'}).$$

Let

$$\overline{\sigma}^{(g,C')} : \prod \mathcal{U}(w_i) \to \mathcal{S}\mathrm{tand}(\tau_{C'}, B_{C'}),$$

be the composition

$$\overline{\sigma}^{(g,C')} = \sigma^{(\tau,C')} \circ \overline{m}^{(g,C')}.$$

Clearly $\overline{\sigma}^{(g,C')}$ is an **isomorphism** of S-schemes. By definition of $\overline{\sigma}^{(g,C')}$ one obtains:

Proposition 14.10
There is a commutative diagram

$$\overline{\sigma}^{(g,C')} : \quad \prod \mathcal{U}(w_i) \longrightarrow \mathcal{S}\mathrm{tand}(\tau_{C'}, B_{C'}) \; .$$

$$\overline{m}^{(g,C')} \searrow \qquad \uparrow \sigma^{(\tau,C')}$$

$$\mathcal{U}(C', F_{\tau,C'})$$

Let $\overline{\sigma}^{(g,C')}$ be made explicit. From

$$v_1 = w_{r+1} \cdots w_1 = w(C', \mathrm{proj}_{F_{\tau,C'}} C'),$$

it is deduced that

$$v_1(F_s(C')) = F_{\tau,C'}.$$

Finally one obtains

$$\mathrm{int}(v_1)(P_{F_s(C')}) = P_{\tau,C'}.$$

On the other hand, by an easy calculation one obtains

$$\prod v_{i+1}\, x_i\, v_{i+1}^{-1} \; (r+1 \geqslant i \geqslant 1) = \left(\prod x_i\, w_i\right) v_1^{-1} \; (r+1 \geqslant i \geqslant 1),$$

where $x = (x_i)$ denotes a section of $\prod \mathcal{U}(w_i)$. Thus

$$\mathrm{int}\left(\prod v_{i+1}\, x_i\, v_{i+1}^{-1}\right)(P_{\tau,C'}) = \mathrm{int}\left(\prod x_i\, w_i\right) \mathrm{int}(v_1^{-1})\left(\mathrm{int}(v_1)(P_{F_s(C')})\right)$$
$$= \mathrm{int}(\prod x_i\, w_i)(P_{F_s(C')}),$$

i.e.

$$\overline{\sigma}^{(g,C')}(x) = \mathrm{int}(\prod x_i\, w_i)(P_{F_s(C')}) \,,$$

thus it is known to which section of $\mathcal{S}\mathrm{tand}(\tau_{C'}, B_{C'})$ the section (x_i) of the product $\prod \mathcal{U}(w_i)$ corresponds by the isomorphism $\overline{\sigma}^{(g,C')}$.

Next one looks at the composition of Θ_{g,P_t,R_+} with

$$\sigma^{(g,C')} : \prod \mathcal{U}(C_i, F_{i-1}) \;\to\; \mathcal{S}\mathrm{tand}(\tau_{C'}, B_{C'}) \,,$$

where $\sigma^{(g,C')} = \sigma^{(\tau,C')} \mathrm{om}^{(g,C')}$. The following proposition describes $\Theta_{g,P_t,R_+} \circ \sigma^{(g,C')}$ in terms of the block decomposition of $\mathcal{U}(C', F')$ defined by the minimal gallery of types g. The description of

$$\Theta_{g,P_t,R_+} \circ \sigma^{(\tau,C')} : \; x \mapsto \mathrm{int}(x)\left(\sigma_{\mathscr{C}}(\gamma_g(F, F'))\right) \,,$$

as required in proposition 14.8, follows from that one.

Proposition 14.11
Write

$$\gamma_g(F, F') : (F_i) \; (r \geqslant i \geqslant 0), \; (F'_j) \; (r+1 \geqslant j \geqslant 1).$$

Let (y_i) be a section of $\prod \mathcal{U}(C_i, F_{i-1}) \; (r+1 \geqslant i \geqslant 1)$. Then

$$\mathrm{int}\left(\prod_{r+1 \geqslant i \geqslant 1} y_i\right)(P_{F_\alpha}) = \mathrm{int}\left(\prod_{r+1 \geqslant i \geqslant \alpha+1} y_i\right)(P_{F_\alpha}) \; (r \geqslant \alpha \geqslant 0).$$

Proof *Let*

$$\gamma_g^{(\alpha)}(C'; F, F') : (F_i) \ (r + 1 \geqslant i \geqslant \alpha), \ (F'_j) \ (r + 1 \geqslant j \geqslant \alpha + 1),$$

with $F_{r+1} = C'$ be the α-truncated gallery of $\gamma_g(C'; F, F')$. As $\gamma_g^{(\alpha)}(C'; F, F')$ is a MGG, it is

$$R_+(C', F_\alpha) = \coprod R_+(C_i, F_{i-1}) \ (r + 1 \geqslant i \geqslant \alpha + 1).$$

On the other hand, the following equality holds:

$$R_+(C', F_\alpha) = \coprod R_+(C_i, F_{i-1}) \coprod R_{C'} \cap R_{F_\alpha}.$$

It is deduced that

$$\coprod R_+(C_i, F_{i-1})(\alpha \geqslant i \geqslant 1) \subset R_{C'} \cap R_{F_\alpha}.$$

This gives

$$\alpha \geqslant i \geqslant 1 \ \Rightarrow \ \mathcal{U}(C_i, F_{i-1}) \subset P_{F_\alpha},$$

and the image of

$$\prod \mathcal{U}(C_i, F_{i-1}) \ (\alpha \geqslant i \geqslant 1),$$

in $B_{C'}$ is contained in P_{F_α}. It is concluded that $\prod\limits_{\alpha \geqslant i \geqslant 1} y_i$ is a section of P_{F_α}.
This suffices to prove the assertion.

Write:

$$\tau_{\mathscr{C}}^{tr} = (\tau(C_i, F_{i-1})) \in \prod \text{Relpos}_{s_{i-1}(g)}^{t_i(g)} \ (r + 1 \geqslant i \geqslant 1).$$

The couple of facets (C_i, F_{i-1}) is a couple of transversal facets in $St_{F'_i}$. Denote by

$$\sigma_g(C') : (P_{s_i}(C'))(r \geqslant i \geqslant 0), \ (P_{t_j}(C')) \ (r + 1 \geqslant j \geqslant 1),$$

the **basical configuration defined by** C'. Observe that by definition $\sigma_g(C')$ is contained in the simplex subcomplex $\Delta(C')$ defined by C'.

One associates with the section $x = (x_i)$ of $\prod \mathcal{U}(w_i)$, the following configuration:

$$\sigma_g(x, C') : \left(\text{int}(\prod_{r+1 \geqslant \beta \geqslant i+1} x_\beta \, w_\beta)(P_{s_i}(C')) \right) (r \geqslant i \geqslant 0),$$

$$\left(\text{int}(\prod_{r+1 \geqslant \beta \geqslant j+1} x_\beta \, w_\beta)(P_{t_j}(C')) \right) (r + 1 \geqslant j \geqslant 0),$$

where it is written

$$\prod_{r+1 \geqslant \beta \geqslant j+1} x_\beta \, w_\beta = 1$$

if $j = r + 1$, and

$$P_{s_i}(C') = P_{F_{s_i}(C')} \quad \left(\text{resp. } P_{t_j}(C') = P_{F_{t_j}(C')} \right).$$

The next proposition gives a description of $\Theta_{g,P_{t,R_+}} \circ \sigma^{(\tau,C')}$ in terms of the basical configuration $\sigma_g(C')$ instead of $\gamma_g(C'; F, F')$.

Proposition 14.12
The following diagram

$$\lambda_{[\tau_{\mathscr{C}}^{tr},C']} : \prod \mathcal{U}(w_i) \longrightarrow \mathcal{C}onf_G^m(g, P_{t,R_+})$$

with $\overline{m}^{(g,C')}$ and $\Theta_{g,P_{t,R_+}} \circ \sigma^{(\tau,C')}$ mapping to $\prod \mathcal{U}(C', F_{\tau,C'})$,

where $\lambda_{[\tau_{\mathscr{C}}^{tr},C']}$ is defined by

$$\lambda_{[\tau_{\mathscr{C}}^{tr},C']} : x \mapsto \sigma_g(x, C'),$$

is commutative. **Observe that** $\lambda_{[\tau_{\mathscr{C}}^{tr},C']} = \Theta_{g,P_{t,R_+}} \circ \overline{\sigma}^{(g,C')}$.

Proof *Results easily from the two preceding Propositions.*

14.4 The schematic image of the Canonical Section

Keep the notation of the preceding section. Recall that $\overline{\Omega}_{t,R_+} \subset \mathcal{P}ar_t(G)$ is obtained as the image of the Big Cell $\Omega_{t,R_+} \subset G/P_{t,R_+}$, by the isomorphism $G/P_{t,R_+} \longrightarrow \mathcal{P}ar_t(G)$ defined by $\overline{x} \mapsto int(x)(P_{t,R_+})$. On the other hand there is an isomorphism $\mathcal{U}_{t,R_+} \simeq \overline{\Omega}_{t,R_+}$ defined by $x \mapsto int(x)(P_{t,R_+})$, where $\mathcal{U}_{t,R_+}$ is the subgroup generated by the vector subgroups corresponding to the roots $\alpha \in -(R_+ - R_{F_t})$, where $F_t = F_t(C)$ (thus $P_{F_t} = P_{t,R_+}$), and $C = C_{R_+}$ is the chamber given by $\mathcal{R}$. In fact $\overline{\Omega}_{t,R_+}$ is the Big Cell defined by the Borel subgroup B^{opp} opposed to $B = B_C$, i.e. given by $-R_+$, and there is $\mathcal{U}_{t,R_+} = \mathcal{U}(C^{opp}, F_t) \subset B^{opp}$. On the other hand the morphism $Stand(\tau) \longrightarrow \mathcal{P}ar_t(G)$, induced by the first projection, gives a **locally trivial "family" of Schubert Cells**, $(Stand(\tau, P))$ indexed by the sections of $\mathcal{P}ar_t(G)$. There is an $\overline{\Omega}_{t,R_+}$-isomorphism $\sigma_{R_+}^{(\tau)} : \overline{\Omega}_{t,R_+} \times_S Stand(\tau, P_t) \longrightarrow Stand(\tau)_{\overline{\Omega}_{t,R_+}}$ associating with the couple $(P, (P_t, Q))$, formed by a parabolic P section of $\overline{\Omega}_{t,R_+}$, and a section (P_t, Q) of the fiber of $Stand(\tau)_{\overline{\Omega}_{t,R_+}}$ over the center of $\overline{\Omega}_{t,R_+}$, the couple $(P, int(x)(Q) = (int(x)(P_t), int(x)(Q))$, where x is the section of $\mathcal{U}_{t,R_+}$ corresponding to P.

The open covering

$$\mathcal{S}tand(\tau) = \bigcup_{R_+ \in \mathcal{F}ram(\mathcal{R})} \mathcal{S}tand(\tau)_{\overline{\Omega}_{t,R_+}} = \bigcup_{R_+ \in \mathcal{F}ram(\mathcal{R})} \sigma_{R_+}^{(\tau)} \left(\overline{\Omega}_{t,R_+} \times_S \mathcal{S}tand(\tau, P_{t,R_+}) \right)$$

may be refined as follows. Recall that there is an open covering

$$\mathcal{S}tand(\tau, P_{t,R_+}) = \bigcup_{C' \in \mathrm{Ch}\,\mathrm{St}_{\mathrm{F}_t}} \mathcal{S}tand(\tau_{C'}, B_{C'}) \, ,$$

and isomorphisms $\mathcal{U}(C', F_{\tau,C'}) \,\tilde{\rightarrow}\, \mathcal{S}tand\,(\tau_{C'}, B_{C'})$, defined by $x \mapsto int(x)(P_{\tau,C'})$, where $P_{\tau,C'}$ corresponds to $F_{\tau,C'}$. Thus one has

$$\mathcal{S}tand(\tau) = \bigcup_{(R_+, C') \in I_\tau(\mathcal{R})} \overline{\Omega}_{t,R_+} \times_S \mathcal{S}tand(\tau_{C'}, B_{C'})$$

$$\bigcup_{(R_+, C') \in I_\tau(\mathcal{R})} \sigma_{R_+, C'}^{(\tau)} \left(\mathcal{U}_{t,R_+} \times_S \mathcal{U}(C', F_{\tau,C'}) \right) \, .$$

Remark that $\overline{\Omega}_{t,R_+} \times_S \mathcal{S}tand(\tau_{C'}, B_{C'})$ is affine if S is affine.

There is a commutative diagram

$$\lambda_{[\tau_{\mathscr{C}}^{tr}, C']} = \Theta_{g, P_{t,R_+}} \circ \overline{\sigma}^{(g,C')} : \quad \prod \mathcal{U}(w_i) \longrightarrow \mathcal{C}\mathrm{onf}_G^m (g, P_{t,R_+}) \qquad ,$$

$$\prod \mathcal{U}(w_i) w_i \qquad \Big\uparrow j_{[\tau_{\mathscr{C}}^{tr}, E]} \circ \lambda_{[\tau_{\mathscr{C}}^{tr}, E]}$$

where

$$j_{[\tau_{\mathscr{C}}^{tr}, E]} : \mathcal{C}\mathrm{onf}_G^{std}(\tau_{\mathscr{C}}^{tr}, B_{C'}) \hookrightarrow \mathcal{C}\mathrm{onf}_G(g, P_{t,R_+}),$$

denotes the embedding of the open cell defined by $B_{C'}$, and

$$\lambda_{[\tau_{\mathscr{C}}^{tr}, E]} : \prod \mathcal{U}(w_i^m) \, w_i^m \,\tilde{\rightarrow}\, \mathcal{C}\mathrm{onf}_G^{std}(\tau_{\mathscr{C}}^{tr}, B_{C'})$$

denotes the isomorphism of the $\tau_{\mathscr{C}}^{tr}$-**representatives scheme** with the cell $\mathcal{C}\mathrm{onf}_G^{std}(\tau_{\mathscr{C}}^{tr}, B_{C'})$. The sequence $w_{\tau_{\mathscr{C}}^{tr}}^m = (w_i^m)$ is associated with the gallery of types of relative positions $\tau_{\mathscr{C}}^{tr}$ is determined by $\gamma_g(C; F, F')$.

The proof of the following properties of the canonical section $\Theta_{g, P_{t,R_+}}$ are based on the representation of $\mathcal{C}\mathrm{onf}_G(g, P_{t,R_+})$ as a **contracted product** along the generalized minimal gallery of types g,

$$\Pi_G\,[g, E] \simeq \mathcal{C}\mathrm{onf}_G(g, P_{t,R_+}) \, ,$$

given in the next chapter. There is a commutative diagram

$$
\begin{array}{ccc}
\prod \mathcal{U}(w_i^m)\, w_i^m & \xrightarrow{\ q_{[\tau_{\mathscr{C}}^{tr},E]}\ } & \Pi_G\,[g,E] \\
\Big\downarrow{\scriptstyle \lambda_{[\tau_{\mathscr{C}}^{tr},E]}} & & \Big\downarrow{\scriptstyle \lambda_{[g,E]}} \\
\mathrm{Conf}_G^{std}(\tau_{\mathscr{C}}^{tr},B_{C'}) & \xrightarrow{\ j_{[\tau_{\mathscr{C}}^{tr},E]}\ } & \mathrm{Conf}_G(g,P_{t,R_+})\ .
\end{array}
$$

The morphism $q_{[\tau_{\mathscr{C}}^{tr},E]}$ defines an open embedding and its image is a relatively schematically dense subscheme of $\Pi_G\,[g,E]$ (cf. §15.3). As the vertical arrows are isomorphism the following proposition results easily.

Proposition 14.13 *The image of* $j_{[\tau_{\mathscr{C}}^{tr},E]}$ *is an open relatively schematically dense subscheme of* $\mathrm{Conf}_G(g,P_{t,R_+})$. *Thus* $\Theta_{g,P_{t,R_+}}$ *is an* **open embedding relatively schematically dominant**.

On the other hand, there are, a natural isomorphism $\mathrm{Conf}_G^m(g',B_{C'}) \simeq \mathrm{Conf}_G^m(g,P_{t,R_+})$, a relation between the corresponding canonical sections:

$$
\Theta_{g,P_{t,R_+}}\big|_{Stand(\tau_{C'},B_{C'})} = \Theta_{g',B_{C'}}\ ,
$$

where g' denotes the gallery of types defined by $\gamma_g(C;F,F')$, and

$$
(\pi_{(g,P)})^{-1}(Stand(\tau,P_{t,R_+})) = \bigcup_{C'\in\mathrm{Ch}\,St_{F_t}} (\pi_{(g',B_{C'})})^{-1}(Stand(\tau_{C'},B_{C'}))\ .
$$

Where $\pi_{(g',B_{C'})}$ (resp. $\pi_{(g,P_{t,R_+})}$) is the restriction of $\mathcal{E}_2$ to $\mathrm{Conf}_G^m(g',B_{C'})$ (resp. $\mathrm{Conf}_G^m(g,P_{t,R_+})$).

In the next chapter it will be proved that

$$
\mathrm{Conf}_G^m(g',B_{C'})' = (\pi_{(g',B_{C'})})^{-1}(Stand(\tau_{C'},B_{C'})) = Im\ \Theta_{g',B_{C'}}\ .
$$

(Given $C \in St_{F_t}$ write $B = B_C$). There is a **Cellular Decomposition** associated with a Borel subgroup B contained in P_{t,R_+} (cf. Proposition 14.26):

$$
\mathrm{Conf}_G^m(g,P_{t,R_+}) = \coprod_{\tau_{\mathscr{C}}\in\mathrm{Relpos}^{\,gall}(g)} \mathrm{Conf}_G^{std}(\tau_{\mathscr{C}},B)
$$

The set $\mathrm{Relpos}^{\,gall}(g)$ denotes the galleries of types of relative position of $\mathcal{A}_E$ whose associated gallery of types is g. The **Open Cell**, which is relatively schematically dense, is indexed by the gallery $\tau_{\mathscr{C}}^{tr}$ of types of "transversal" relative positions and satisfies the equalities

$$
\mathrm{Conf}_G^{std}(\tau_{\mathscr{C}}^{tr},B) = \mathrm{Conf}_G^m(g',B)' = Im\ \Theta_{g',B}\ .
$$

Remark that the proof of these equalities, given in the next section, follows from the above Cellular Decomposition.
It is deduced that

$$(\pi_{(g,P)})^{-1}(\mathcal{S}tand(\tau, P_{t,R_+})) = Im\, \Theta_{g,P_{t,R_+}},$$

and that there is an isomorphism $\Theta_{g,P_{t,R_+}}$: $\mathcal{S}tand(\tau_S, P_{t,R_+}) \simeq \mathcal{C}onf_G^m(g, P_{t,R_+})'$ whose reciprocal isomorphism is $\pi_{g,P_{t,R_+}}$. It is called $\mathcal{C}onf_G^m(g, P_{t,R_+})'$ the **Canonical Open Subscheme** of $\mathcal{C}onf_G^m(g, P_{t,R_+})$. The sections of this subscheme are characterized as those galleries configurations $\gamma_{\mathscr{C}}$ with associated gallery of types equal to g_S, whose extremities $\mathcal{E}(\gamma_{\mathscr{C}}) = (P_{t,R_+}, Q)$ are in standard position, and whose type of relative position is given by $t_2(P_{t,R_+}, Q) = \tau_S = (\tau_g)_S$.

Thus one obtains the open covering

$$\mathcal{C}onf_G^m(g, P_{t,R_+})' = \bigcup_{C' \in \text{Ch}\, \text{St}_{F_t}} (\pi_{(g', B_{C'})})^{-1}(\mathcal{S}tand(\tau_{C'}, B_{C'})) =$$

$$\bigcup_{C' \in \text{Ch}\, \text{St}_{F_t}} \mathcal{C}onf(g', B_{C'})',$$

which is in fact the image by $\Theta_{g,P_{t,R_+}}$ of the **Big Cell Open covering** of $\mathcal{S}tand(\tau, P_{t,R_+})$),

$$\mathcal{S}tand(\tau, P_{t,R_+}) = \bigcup_{C' \in \text{Ch}\, \text{St}_{F_t}} \mathcal{S}tand(\tau_{C'}, B_{C'}).$$

Proposition 14.14
*Keep the above hypothesis and notation. The morphism $\Theta_{g,P_{t,R_+}}$ is an **open embedding** whose image is a **relatively schematically dense** subscheme $\mathcal{C}onf_G^m(g, P_{t,R_+})'$ of $\mathcal{C}onf_G^m(g, P_{t,R_+})$. It induces an isomorphism $\Theta_{g,P_{t,R_+}}$: $\mathcal{S}tand(\tau, P_{t,R_+}) \simeq \mathcal{C}onf_G^m(g, P_{t,R_+})'$. The image by $\Theta_{g,P_{t,R_+}}$ of the Big Cell open covering of $\mathcal{S}tand(\tau, P_{t,R_+})$ gives rise to the open covering*

$$\mathcal{C}onf_G^m(g, P_{t,R_+}) = \bigcup_{C' \in \text{Ch}\, \text{St}_{F_t}} \mathcal{C}onf(g', B_{C'})'.$$

Corollary 14.15 *Let G be a reductive S-group scheme, $\underline{g}$ a section of $\Gamma^m(\underline{\Delta}(G))$ with associated type of relative position $\underline{\tau} = \delta_2(\underline{g})$, and P a parabolic subgroup of G. The morphism $\Theta_{\underline{g},P}$: $\mathcal{S}tand(\underline{\tau}, P_{t,R_+}) \longrightarrow \mathcal{C}onf_G^m(\underline{g}, P)'$ is an **open embedding** whose image is a **relatively schematically dense** subscheme $\mathcal{C}onf_G^m(\underline{g}, P)'$ of $\mathcal{C}onf_G^m(\underline{g}, P)$.*

It is clear that the proof of the Corollary may be reduced by etale localization to the case where G is endowed with a frame E. In this case it follows immediately from proposition 14.14.

Corollary 14.16 *Keep the hypothesis of the above Corollary. The image of* Θ_{τ} *is the relatively schematically dense open subscheme* $Im\ \Theta_{\tau} = Conf_G^m(\tau)'$.

This Corollary results from the local description of Θ_{τ}.

14.5 Configurations schemes as Schubert schemes Smooth Resolutions

The following definition characterizes the type of resolution of singularities we consider in this work.

Definition 14.17
Let $Y \subset Z$ *be an S-subscheme of a proper S-scheme* Z, *and* Y^{schc} *its schematic closure. We say that a proper S-morphism*

$$f : X \to Y^{schc}$$

is an S-smooth resolution of Y^{schc}, *if:*

1. *X is an S-smooth scheme.*

2. *There exists a section*

$$\Theta : Y \to X$$

   of f, defining an open embedding of Y in X, such that $Im\ \Theta \subset X$ is a relatively schematically dense open subscheme of X, satisfying $Im\ \Theta = f^{-1}(Y)$.

It follows that f is surjective and its schematic image is equal to Y^{sch}. It is briefly said that (X, f) is a **smooth resolution** *of Y^{schc}, if no confusion arises.*

It is observed that the morphism f is a birational morphism of schemes (cf. [24] Ch. 1, 23.4.). The results of this section may be resumed in the following

Theorem 14.18
Let $g \in gall^m(\mathcal{A}_E)$, *and* $\tau = \tau_g$. *The morphism*

$$\overline{\mathcal{E}}_{g,P_{e_1(g)}} : Conf_G^m(g, P_{e_1(g)}) \to Stand(\tau, P_{e_1(g)})^{schc}$$

$$\left(resp.\ \overline{\mathcal{E}}_g : Conf_G^m(g) \to Stand(\tau)^{schc}\right)$$

is an S-smooth resolution of the schematic closure

$$Stand(\tau, P_{e_1(g)})^{schc}\ \left(resp.\ Stand(\tau)^{schc}\right).$$

Where $\overline{\mathcal{E}}_{g,P_{e_1(g)}}$ *(resp. $\overline{\mathcal{E}}_g$) denotes the morphism induced by* $\mathcal{E}_2$.

On account of the Cellular Decomposition of $\mathcal{C}\mathrm{onf}_G^m\,(g, P_{e_1(g)})$ it results, by §16.3, that schematic image of the morphism

$$\mathcal{E}_{g,P_{e_1(g)}} : \mathcal{C}\mathrm{onf}_G^m\,(g, P_{e_1(g)}) \longrightarrow \mathcal{P}\mathrm{ar}_{e_2(g)}(G)$$

exists, and it is known that the schematic image of the canonical embedding

$$j : \mathcal{S}\mathrm{tand}(\tau, P_{e_1(g)}) \to \mathcal{P}\mathrm{ar}_{e_2(g)}(G),$$

in view of the Cellular Decomposition of $\mathcal{S}\mathrm{tand}(\tau, P_{e_1(g)})$, also exists. From the definition of $\Theta_{g,P}$ (cf. Proposition 14.6, Definition 12.34) one obtains:

$$\mathcal{E}_{g,P_{e_1(g)}} \circ \Theta_{g,P_{e_1(g)}} = \mathrm{Id}_{\mathcal{S}\mathrm{tand}(\tau,P_{e_1(g)})}.$$

From the Proposition 14.12 it results that

$$\mathrm{Im}\,\Theta_{g,P_{e_1(g)}} \subset \mathcal{C}\mathrm{onf}_G^m\,(g, P_{e_1(g)}),$$

is a relatively schematically dense open subscheme. By the principle of **transitity of schematic images** (cf. Proposition 12.42) applied with

$$X = \mathcal{S}\mathrm{tand}(\tau, P_{e_1(g)}),\ Y = \mathcal{C}\mathrm{onf}_G^m\,(g, P_{e_1(g)}),\ Z = \mathcal{P}\mathrm{ar}_s(G)$$

$$(resp.\ X = \mathcal{S}\mathrm{tand}(\tau),\ Y = \mathcal{C}\mathrm{onf}_G^m\,(g),\ Z = \mathcal{P}\mathrm{ar}_t(G) \times_S \mathcal{P}\mathrm{ar}_s(G))\,,$$

and with $\tau = \tau_g$,

$$f = \Theta_{g,P_{e_1(g)}},\ g = \mathcal{E}_{g,P_{e_1(g)}} = \text{restriction of } \mathcal{E}_2 \text{ to } \mathcal{C}\mathrm{onf}_G^m(g, P_{e_1(g)})$$

$$(resp.\ f = \Theta_g,\ g = \mathcal{E}_g = \text{fiber of } \mathcal{E}_2 \text{ over } g)\,,$$

and taking into account that

$$g \circ f = 1_{\mathcal{S}\mathrm{tand}(\tau,P_{e_1(g)})}\ (resp.\ 1_{\mathcal{S}\mathrm{tand}(\tau)}),$$

it is deduced:

Proposition 14.19
Let $\tau = \tau_g$. *The schematic image of $\mathcal{E}_{g,P_{e_1(g)}}$ (resp. $\mathcal{E}_g$) is equal to the* **schematic closure**

$$\mathcal{S}\mathrm{tand}(\tau, P_{e_1(g)})^{schc}\ (resp.\ \mathcal{S}\mathrm{tand}(\tau)^{schc})$$

of $\mathcal{S}\mathrm{tand}(\tau, P_{e_1(g)})$ (resp. $\mathcal{S}\mathrm{tand}(\tau)$) in $\mathcal{P}\mathrm{ar}_s(G)$ (resp. $\mathcal{P}\mathrm{ar}_t(G) \times_S \mathcal{P}\mathrm{ar}_s(G)$ with $t = e_1(g)$, $s = e_2(g)$), and thus a **factorization** *is obtained*

$$\overline{\mathcal{E}}_{g,P_{e_1(g)}} : \mathcal{C}\mathrm{onf}_G^m\,(g, P_{e_1(g)}) \longrightarrow \mathcal{S}\mathrm{tand}(\tau, P_{e_1(g)})^{schc}$$

$$\mathcal{E}_{g,P_{e_1(g)}} \searrow \qquad \swarrow$$

$$\mathcal{P}\mathrm{ar}_s(G)$$

$$\left(\text{resp. } \overline{\mathcal{E}}_g : \; \mathcal{C}onf_G^m\,(g) \xrightarrow{\hspace{5cm}} \mathcal{S}tand(\tau)^{schc} \atop \underset{\mathcal{E}_g}{\searrow} \quad \nearrow \atop \mathcal{P}ar_t(G) \times_S \mathcal{P}ar_s(G) \right).$$

The right arrow is the canonical embedding morphism.

Let it be proved that the couple $(\mathcal{C}onf_G^m\,(g, P_{e_1(g)}), \overline{\mathcal{E}}_{g,P_{e_1(g)}})$ (resp. $(\mathcal{C}onf_G^m\,(g), \overline{\mathcal{E}}_g)$) is a **smooth resolution** of $\mathcal{S}tand(\tau, P_{e_1(g)})^{schc}$ (resp. $\mathcal{S}tand(\tau)^{schc}$). Let one prove:

- $Im\;\Theta_{g,P_{e_1(g)}} \subset \mathcal{C}onf_G^m\,(g, P_{e_1(g)})$ is an open subscheme relatively schematically dense;

- $Im\;\Theta_{g,P_{e_1(g)}} = (\overline{\mathcal{E}}_{g,P_{e_1(g)}})^{-1}(\mathcal{S}tand(\tau, P_{e_1(g)})).$

The first statement results from proposition 14.13. Let one prove the equality $Im\;\Theta_{g,P_{e_1(g)}} = (\overline{\mathcal{E}}_{g,P_{e_1(g)}})^{-1}(\mathcal{S}tand(\tau, P_{e_1(g)}))$. Clearly there is an inclusion of open schemes

$$Im\;\Theta_{g,P_{e_1(g)}} \subset (\overline{\mathcal{E}}_{g,P_{e_1(g)}})^{-1}(\mathcal{S}tand(\tau, P_{e_1(g)}))\ ,$$

thus the equality holds if the underlying open sets are equal.

Given an S-scheme X, as usual denote by $X_{\overline{s}} = X \times_S \mathrm{Spec}(\overline{\kappa(s)})$ the geometric fiber of X over $\overline{s}$, and by $X_{(\overline{s})}$ the set of sections of X over $\mathrm{Spec}(\overline{\kappa(s)})$. It suffices to prove that for all geometric point $\overline{s}$ the equality $\left(Im\;\Theta_{g,P_{e_1(g)}}\right)_{(\overline{s})} = \left((\overline{\mathcal{E}}_{g,P_{e_1(g)}})^{-1}(\mathcal{S}tand(\tau, P_{e_1(g)}))\right)_{(\overline{s})}$ holds. Let

$$\left(\mathcal{E}_{g,P_{e_1(g)}}\right)_{(\overline{s})} : \; \mathrm{Conf}_G^m\,(g, P_{e_1(g)})_{(\overline{s})} \to \mathcal{S}tand(\tau, P_{e_1(g)}),$$

be induced by

$$\left(\mathcal{E}_{g,P_{e_1(g)}}\right)_{\overline{s}} = \text{geometric fiber of } \mathcal{E}_{g,P_{e_1(g)}} \text{ over } \overline{s}.$$

One may interpret the geometric fiber $\left(\mathcal{E}_{g,P_{e_1(g)}}\right)_{(\overline{s})}^{-1} (\mathcal{S}tand(\tau, P_{e_1(g)}))_{(\overline{s})}$ in terms of the building $I(G_{\overline{s}})$ of the geometric fiber $G_{\overline{s}}$ of G over s, $G_{\overline{s}}$ being a reductive group over the algebraically closed field $\overline{\kappa(s)}$, namely

$$\left(\mathcal{E}_{g,P_{e_1(g)}}\right)_{(\overline{s})}^{-1} (\mathcal{S}tand(\tau, P_{e_1(g)}))_{(\overline{s})} = \mathrm{Gall}\,_{I(G_{\overline{s}})}^m\,(g, F_{e_1(g)}(C))'\ ,$$

where $\operatorname{Gall}_{I(G_{\overline{s}})}^{m}(g, F_{e_1(g)}(C))'$ denotes the set of Minimal Generalized Galleries in the building $I(G_{\overline{s}})$ of type g, with $\tau = \tau_g$, issued from $F_{e_1(g)}(C)$. On the other hand, observe that $Im\left(\Theta_{g,P_{e_1(g)}}\right)_{(\overline{s})}$ may be interpreted as the set of Minimal Generalized Galleries $\gamma_g(F_{e_1(g)}(C), F')$, where F' is a facet satisfying $\tau(F_{e_1(g)}(C), F') = \tau_g$.

Let $\gamma \in \operatorname{Gall}_{I(G_{\overline{s}})}^{m}(g, F_{e_1(g)}(C))'$. Then γ is a minimal generalized gallery with extremities $(F_{e_1(g)}(C), F')$ such that $\tau(F_{e_1(g)}(C), F') = \tau_g$ and of type g. Recall that there is only one Minimal Generalized Gallery satisfying these properties, thus $\gamma = \gamma_g(F_{e_1(g)}(C), F')$ and $Im\left(\Theta_{g,P_{e_1(g)}}\right)_{(\overline{s})} = \left((\overline{\mathcal{E}}_{g,P_{e_1(g)}})^{-1}(Stand(\tau, P_{e_1(g)}))\right)_{(\overline{s})}$. It is concluded that the corresponding open sets contain the same points with residual field $\kappa(\overline{s})$ and thus that they are equal $Im\left(\Theta_{g,P_{e_1(g)}}\right)_{\overline{s}} = \left((\overline{\mathcal{E}}_{g,P_{e_1(g)}})^{-1}(Stand(\tau, P_{e_1(g)}))\right)_{\overline{s}}$. It has thus been proved the second statement.

From the above reasoning one obtains the following

Corollary 14.20
The mapping

$$\left(\mathcal{E}_{g,P_{e_1(g)}}\right)_{(\overline{s})} : \left(\mathcal{E}_{g,P_{e_1(g)}}\right)^{-1}(Stand(\tau, P_{e_1(g)}))_{(\overline{s})} \rightarrow Stand(\tau, P_{e_1(g)})_{(\overline{s})},$$

induced by $\mathcal{E}_{g,P_{e_1(g)}}$ is bijective, and its inverse is given by the mapping

$$\left(\Theta_{g,P_{e_1(g)}}\right)_{(\overline{s})} : Stand(\tau, P_{e_1(g)}) \rightarrow \left(\mathcal{E}_{g,P_{e_1(g)}}\right)^{-1}(Stand(\tau, P_{e_1(g)}))_{(\overline{s})},$$

induced by the section $\Theta_{g,P_{e_1(g)}}$.

From the corollary immediately one obtains the following

Proposition 14.21
The open embedding

$$\Theta_{g,P_{e_1(g)}} : Stand(\tau, P_{e_1(g)}) \rightarrow \left(\mathcal{E}_{g,P_{e_1(g)}}\right)^{-1}(Stand(\tau, P_{e_1(g)}))$$

is a $P_{e_1(g)}$-equivariant morphism which is in fact an isomorphism, with the relatively schematically dense open subscheme

$$\left(\mathcal{E}_{g,P_{e_1(g)}}\right)^{-1}(Stand(\tau, P_{e_1(g)})) \subset \mathcal{C}onf_G^m(g, P_{e_1(g)}).$$

The corresponding inverse isomorphism

$$\mathcal{E}'_{g,P_{e_1(g)}} : \left(\mathcal{E}_{g,P_{e_1(g)}}\right)^{-1}(Stand(\tau, P_{e_1(g)})) \rightarrow Stand(\tau, P_{e_1(g)}),$$

is the morphism induced by

$$\mathcal{E}_{g,P_{e_1(g)}} : \mathcal{C}onf_G^m\left(g, P_{e_1(g)}\right) \to \mathcal{P}ar_s(G),$$

with $s = e_2(g)$.

Let one now consider the corresponding statement with respect to the couple of morphisms

$$\Theta_g : \mathcal{S}tand(g) \to \mathcal{E}_g^{-1}(\mathcal{S}tand(\tau)) \subset \mathcal{C}onf_G^m(g)$$

$$\left(\text{resp. } \mathcal{E}_g : \mathcal{C}onf_G^m(g) \to \mathcal{P}ar_{e_1(g)}(G) \times_S \mathcal{P}ar_{e_2(g)}(G)\right),$$

with $\tau = \tau_g$:

- $Im\,\Theta_g \subset \mathcal{C}onf_G^m(g)$ is an open subscheme relatively schematically dense;

- $Im\,\Theta_g = (\overline{\mathcal{E}}_g)^{-1}(\mathcal{S}tand(\tau))$.

There is a commutative diagram:

$$
\begin{array}{ccc}
\overline{\Omega}_{e_1(g),R_+} \times_S \mathcal{C}onf_G^m\left(g, P_{e_1(g)}\right) & \xrightarrow{\;\zeta_{R_+}^{(\tau)}\;} & \mathcal{C}onf_G^m\,(g)_{\overline{\Omega}_{e_1(g),R_+}} \\
\Big\downarrow{\scriptstyle Id_{\overline{\Omega}_{e_1(g),R_+}} \times_S \mathcal{E}_{g,P_{e_1(g)}}} & & \Big\downarrow{\scriptstyle (\mathcal{E}_g)_{\overline{\Omega}_{e_1(g),R_+}}} \\
\overline{\Omega}_{e_1(g),R_+} \times_S \mathcal{S}tand(\tau, P_{e_1(g)})^{schc} & \xrightarrow{\;\overline{\sigma}_{R_+}^{(\tau)}\;} & (\mathcal{S}tand(\tau)^{schc})_{\overline{\Omega}_{e_1(g),R_+}}
\end{array}
$$

Observe that $\overline{\sigma}_{R_+}^{(\tau)}$ is the morphism induced by $\sigma_{R_+}^{(\tau)}$. The section $Id_{\overline{\Omega}_{e_1(g),R_+}} \times_S \Theta_{g,P_{e_1(g)}}$ corresponds to the section $(\Theta_g)_{\overline{\Omega}_{e_1(g),R_+}}$, i.e. the following diagram commutes

$$
\begin{array}{cccc}
Id_{\overline{\Omega}_{R_+}} \times_S \Theta_{g,P_{e_1(g)}}: & \overline{\Omega}_{e_1(g),R_+} \times_S \mathcal{S}tand(\tau, P_{e_1(g)}) & \longrightarrow & \overline{\Omega}_{e_1(g),R_+} \times_S \mathcal{C}onf_G^m\left(g, P_{e_1(g)}\right) \\
 & \Big\downarrow{\scriptstyle \sigma_{R_+}^{(\tau)}} & & \Big\downarrow{\scriptstyle \zeta_{R_+}^{(\tau)}} \\
(\Theta_g)_{\overline{\Omega}_{e_1(g),R_+}}: & \mathcal{S}tand(\tau)_{\overline{\Omega}_{e_1(g),R_+}} & \longrightarrow & \mathcal{C}onf_G^m\,(g)_{\overline{\Omega}_{e_1(g),R_+}}
\end{array}
$$

Thus it is deduced that

- The image of Θ_g is relatively schematically dense if and only if image of $\Theta_{g,P_{e_1(g)}}$ is relatively schematically dense.

- $(\Theta_g)_{\overline{\Omega}_{e_1(g),R_+}}$ is an isomorphism, if and only if $\Theta_{g,P_{e_1(g)}}$ is an isomorphism.

Since $(\overline{\Omega}\,{}_{R_+}^{e_1(g)})_{R_+ \in \mathcal{F}\text{ram}(\mathcal{R}(E))}$ is an open covering of $\mathcal{P}ar_{\,e_1(g)}(G)$, by Proposition 14.21 one obtains the following result.

Proposition 14.22
The open embedding

$$\Theta_g : \mathcal{S}tand(\tau) : \longrightarrow (\mathcal{E}_g)^{-1}(\mathcal{S}tand(\tau))$$

is a G-invariant morphism, which is in fact an isomorphism with the relatively schematically dense open subscheme

$$(\mathcal{E}_g)^{-1}(\mathcal{S}tand(\tau)) \subset \mathcal{C}onf_G^m(g).$$

The corresponding inverse isomorphism

$$\mathcal{E}_g' : (\mathcal{E}_g)^{-1}(\mathcal{S}tand(\tau)) \to \mathcal{S}tand(\tau),$$

being the morphism induced by

$$\mathcal{E}_g : \mathcal{C}onf_G^m(g) \to \mathcal{P}ar_t(G) \times \mathcal{P}ar_s(G),$$

with $s = e_2(g)$.

14.6   The Cellular scheme

Consider a gallery of types section g of $\Gamma(\underline{\Delta}(G))$ with left extremity $e_1(g) =$ the type of a Borel subgroup of G. It will be seen that there is a **canonical functorial cell decomposition** of $\mathcal{C}onf_G(g)$ indexed by the sections $\underline{\tau}_{\mathscr{C}}$ of the scheme $Relpos_G^{gall}(g)$ of galleries of types of relative positions.
Given two sections $\underline{\tau}$ and $\underline{\tau}'$ of the subscheme of relative positions $\text{Relpos}_G' \subset \text{Relpos}_G$, corresponding to the subscheme $\mathcal{S}tand(G)' \subset \mathcal{S}tand(G)$, i.e. whose sections are the types of relative positions of couples of parabolics (B, P), with B a Borel subgroup of G (cf. Definition 11.26), let

$$\mathcal{S}tand(\underline{\tau}) \times_{\mathcal{B}or(G)} \mathcal{S}tand(\underline{\tau}') = \mathcal{S}tand(\underline{\tau}) \times_{\left(\mathcal{P}roj^{(\underline{\tau})},\, \text{pr}_{\mathcal{P},\underline{\tau}'}\right)} \mathcal{S}tand(\underline{\tau}'),$$

denote the fiber product defined by the couple $\left(\mathcal{P}roj^{(\underline{\tau})}, \text{pr}_{\mathcal{P},\underline{\tau}'}\right)$, where $\mathcal{P}roj^{(\underline{\tau})}$ is the **projection morphism** (cf. Definition 12.39), and $\text{pr}_{\mathcal{P},\underline{\tau}'}$ denotes the restriction of $\text{pr}_{\mathcal{P}}$ to $\mathcal{S}tand(\underline{\tau}')$, induced by the first projection. For the sake of simplifying notations assume that $g = ((\underline{s}_i); (\underline{t}_j))$ ($N \geq i \geq 0$, $N \geq j \geq 1$). Recall that with the notation of §11.11 given a section $(\underline{s}, \underline{s}')$ of $\underline{\Delta}(G) \times_S \underline{\Delta}(G)$ one defines:

$$\text{Relpos}_{(\underline{s}, \underline{s}')} = \epsilon^{-1}\left((\underline{s}, \underline{s}')\right) \subset \text{Relpos}_G$$

and

$$\text{Relpos}_{(\underline{s}, \underline{s}')}^{\,t} = \text{Relpos}_{(\underline{s}, \underline{s}')} \cap \text{Relpos}_G^{\,t}$$

If $\underline{s}$ is the type of a Borel subgroup write $\mathrm{Relpos}^t_{\underline{s'}} = \mathrm{Relpos}^t_{(\underline{s},\underline{s'})}$. Define
$$\mathcal{S}tand^t_{\underline{s}}(G) = (t_2)^{-1}\left(\mathcal{R}elpos^t_{\underline{s}}\right).$$

Let $\underline{\tau}_{\mathscr{C}} = (\underline{\tau}_i)$ be a section of $\mathcal{R}elpos^{gall}_G(g) = \mathcal{R}elpos^{t_N}_{\underline{s}_{N-1}} \times_S \cdots \times_S \mathcal{R}elpos^{t_1}_{\underline{s}_0}$, thus $\epsilon(\underline{\tau}_i) = (\underline{s}, \underline{s}_{i-1})$, where $\underline{s}$ is the type of a Borel subgroup. Write

$$\mathcal{C}onf^{std}_G(\underline{\tau}_{\mathscr{C}}) = \mathcal{S}tand^{t_N}_{\underline{s}_{N-1}}(\underline{\tau}_N) \times_{\mathcal{B}or(G)} \cdots \times_{\mathcal{B}or(G)} \mathcal{S}tand^{t_1}_{\underline{s}_0}(\underline{\tau}_1)$$

$\left(\text{The } \mathbf{Cell\ of\ }\mathcal{C}onf_G(g) \text{ defined by } \mathbf{the\ gallery\ of\ relative\ positions\ } \underline{\tau}_{\mathscr{C}}\right)$. The sections of this scheme are in fact chamber galleries.

Recall that the **incidence morphism** $\mathcal{F}_{\underline{s},\underline{t}} : \mathcal{P}ar_{\underline{s}}(G) \longrightarrow \mathcal{P}ar_{\underline{t}}(G)$ associates with a parabolic Q of type $\underline{s}$ the unique parabolic P of type $\underline{t} \subset \underline{s}$ with $Q \subset P$ (cf. [23], *Exp. XXVI, Lemme* 3.8). Write $\mathcal{F}_{\underline{t}} = \mathcal{F}_{\underline{s},\underline{t}}$ if $\underline{s}$ is the type of a Borel subgroup. Consider the family of morphisms

$$\mathcal{F}_{\underline{\tau}_i} = \mathcal{F}_{\underline{t}_i} \times \left(\mathcal{F}_{\underline{s}_{i-1}} \circ \mathcal{P}roj^{(\underline{\tau}_i)}\right) : \mathcal{S}tand^{t_i}_{\underline{s}_{i-1}}(\underline{\tau}_i) \to \mathcal{P}ar_{\underline{t}_i}(G) \times_S \mathcal{P}ar_{\underline{s}_{i-1}}(G).$$

$(N \geqslant i \geqslant 1)$. Observe that $\mathcal{F}_{\underline{s}_{i-1}} \circ \mathcal{P}roj^{(\underline{\tau}_i)}$ is induced by the second projection. The product morphism

$$\mathrm{pr}_{\mathcal{P},\underline{\tau}_N} \times_S \prod \mathcal{F}_{\underline{\tau}_i} : \mathcal{S}tand^{t_N}_{\underline{s}_{N-1}}(\underline{\tau}_N) \times_{\mathcal{B}or(G)} \cdots \times_{\mathcal{B}or(G)} \mathcal{S}tand^{t_1}_{\underline{s}_0}(\underline{\tau}_1) \longrightarrow$$
$$\mathcal{B}or(G) \times_S \prod \left(\mathcal{P}ar_{\underline{t}_i}(G) \times_S \mathcal{P}ar_{\underline{s}_{i-1}}(G)\right)$$

$(N \geqslant i \geqslant 1)$ induces an embedding

$$j_{\underline{\tau}_{\mathscr{C}}} : \mathcal{C}onf^{std}_G(\underline{\tau}_{\mathscr{C}}) \to \mathcal{C}onf_G(g) .$$

A section $((B_i, Q_{i-1}))$ of $\mathcal{C}onf^{std}_G(\underline{\tau}_{\mathscr{C}})$ satisfies $B_{N-1} = \mathcal{P}roj^{\underline{\tau}_N}(B_N), \cdots,$ $B_0 = \mathcal{P}roj^{\underline{\tau}_1}(B_1)$. Write $P_N = \mathcal{F}_{\underline{t}_N}(B_N), \cdots, P_1 = \mathcal{F}_{\underline{t}_1}(B_1)$. Thus

$$j_{\underline{\tau}_{\mathscr{C}}} : ((B_i, Q_{i-1})) \mapsto ((Q_i)_{N \geqslant i \geqslant 0}; (P_j)_{N \geqslant i \geqslant 1}) .$$

where $Q_N = B_N$. A section

$$\gamma_{\mathscr{C}} : ((Q_i)\ (N \geq i \geq 0), (P_j)\ (N \geq j \geq 1)),$$

of $\mathcal{C}onf_G(g)$ belongs to $Im\ j_{\underline{\tau}_{\mathscr{C}}}$ if and only if it verifies the following conditions: Let $B_N = Q_N$, then:

1. (B_N, Q_{N-1}) is a couple in standard position and $t_2(B_N, Q_{N-1}) = \underline{\tau}_N$;

2. for $N > i \geqslant 1$ define $B_i = \mathcal{P}roj_{(B_{i+1}, Q_i)}$, then (B_i, Q_{i-1}) is a couple in standard position and $t_2(B_i, Q_{i-1}) = \underline{\tau}_i$.

The configurations satisfying the above conditions are called **standard galleries configurations**.
Let
$$\mathcal{P}roj^{\Sigma} : \Sigma \longrightarrow \mathcal{P}ar(G)$$

be the morphism defined as follows: for all sections $\underline{\tau}$ of $Relpos_G$ the composed morphism of $\mathcal{S}tand(\underline{\tau}) \hookrightarrow \Sigma$ with $\mathcal{P}roj^{\Sigma}$ is equal to $\mathcal{P}roj^{(\underline{\tau})}$. Denote by

$$\Sigma^{St_G} = (t_2)^{-1}(St_G)$$

the pull-back of the subscheme $St_G \subset Relpos_G$ by the morphism t_2. The subscheme St_G is defined locally as follows. Suppose that G is endowed with a frame over the etale covering $S' \to S$ then $(\underset{s \supset t}{\cup} Relpos_s^t)_{S'} \simeq S' \times_S St_G$ by the natural isomorphism.

Definition 14.23
Let
$$\mathcal{C}onf_G^{std} = \coprod \overset{N,std}{\prod} \Sigma^{St_G} \times_{\mathcal{B}or(G)} \Sigma^{St_G} \cdots \times_{\mathcal{B}or(G)} \Sigma^{St_G}$$

(The Universal Cellular scheme). *The symbol* $\overset{N,std}{\prod}$ *means that only products along galleries of relative positions are considered. By definition of $\mathcal{C}onf_G^{std}$ there is a canonical morphism*

$$t_2^{std} : \mathcal{C}onf_G^{std} \longrightarrow Relpos_G^{gall} .$$

such that its restriction

$$t_2^{std,N} : \overset{N,std}{\prod} \Sigma^{St_G} \times_{\mathcal{B}or(G)} \Sigma^{St_G} \cdots \times_{\mathcal{B}or(G)} \Sigma^{St_G} \longrightarrow Relpos_G^{gall}$$

is given by

$$t_2^{std,N} = \overset{N}{\prod}(t_2 \circ pr_1) \times_S (t_2 \circ pr_2) \cdots \times_S (t_2 \circ pr_N) ,$$

i.e. by $t_2^{std} : ((B_i, Q_{i-1})) \mapsto (t_2(B_i, Q_{i-1}))$.

Remark that the sections of $\mathcal{C}onf_G^{std}$ are **chamber galleries**, *i.e. galleries of the form* $\underline{g} = ((\underline{s}_i); (\underline{t}_j))$ ($N \geq i \geq 0$, $N \geq j \geq 1$), *where the $\underline{s}_i$ are the types of Borel subgroups.*

Observe that the above product morphism $pr_1^{(\underline{\tau}_N)} \times_S \prod \mathcal{F}_{\tau_i S}$ associates with a chamber gallery a gallery of type $\underline{g}$.

If G is endowed with a frame E one has

$$\mathcal{C}onf_G^{std} = \coprod_{(\tau_i) \in Relpos_{\mathcal{A}_E}^{gall}} \mathcal{C}onf_G^{std}((\tau_i)_S) .$$

Notation 14.24

Let $\mathcal{E}^{std} : \mathcal{C}onf_G^{std} \longrightarrow \mathcal{B}or(G)$ be the morphism associating with a configuration its left extremity, i.e. induced by $\mathcal{E}_1$. Write

$$t_1^{std} = \epsilon^{gall} \circ t_2^{std} : \mathcal{C}onf_G^{std} \to \Gamma(\underline{\Delta}(G)).$$

Let $\underline{g}$ (resp. $\underline{\tau}_{\mathscr{C}}$, $B \subset G$) be a section of $\Gamma(\underline{\Delta}(G))$ $\left(\text{resp. } \mathcal{R}elpos_G^{gall}, \text{ a Borel subgroup of } G\right)$, and η_B the section of $\mathcal{B}or(G)$ given by B.

Define:

$$\mathcal{C}onf_G^{std}(\underline{g}) = (t_1^{std})^{-1}(\underline{g})$$
$$(\text{resp. } \mathcal{C}onf_G^{std}(\underline{\tau}_{\mathscr{C}}) = (t_2^{std})^{-1}(\underline{\tau}_{\mathscr{C}}),$$
$$\mathcal{C}onf_G^{std}(B) = (\mathcal{E}^{std})^{-1}(\eta_B),$$
$$\mathcal{C}onf_G^{std}(\underline{g}, B) = (t_1^{std} \times \mathcal{E}^{std})^{-1}((\underline{g}, \eta_B)),$$
$$\mathcal{C}onf_G^{std}(\underline{\tau}_{\mathscr{C}}, B) = (t_2^{std} \times \mathcal{E}^{std})^{-1}((\underline{\tau}_{\mathscr{C}}, \eta_B)).$$

Suppose G endowed with a frame E. Given a section $\tau_{\mathscr{C}} \in \mathcal{R}elpos_{\mathcal{A}_E}^{gall}(\underline{g})$ let

$$j_{[\tau_{\mathscr{C}}, E]} : \mathcal{C}onf_G^{std}(\tau_{\mathscr{C}}, B) \to \mathcal{C}onf_G(\underline{g}, B) ,$$

be the natural embedding induced by $j_{\tau_{\mathscr{C}}}$. Where $B = B_C$, and C denotes the chamber of $\mathcal{A}_E$ given by $\mathcal{R}(E)$. Define

$$j_{[\underline{g}, E]} : \coprod_{\tau_{\mathscr{C}} \in \mathcal{R}elpos^{gall}(\underline{g})} \mathcal{C}onf_G^{std}(\tau_{\mathscr{C}}, B) \to \mathcal{C}onf(\underline{g}, B)$$

by

$$j_{[\underline{g}, E]} = \coprod_{\tau_{\mathscr{C}} \in \mathcal{R}elpos^{gall}(\underline{g})} j_{[\tau_{\mathscr{C}}, E]} .$$

If $g = g_1, g_1'$ (resp. $g = g_2, g_2'$) one may write

$$\mathcal{C}onf_G^{std}(\underline{\tau}_{\mathscr{C}}, B) = \mathcal{S}tand(\underline{\tau}_{r+1}, B) \times_{\mathcal{B}or(G)} \mathcal{S}tand(\underline{\tau}_r) \times \cdots \times_{\mathcal{B}or(G)} \mathcal{S}tand(\underline{\tau}_1)$$

$$\left(\text{resp. } \mathcal{C}onf_G^{std}(\underline{\tau}_{\mathscr{C}}, B) = \mathcal{S}tand(\underline{\tau}_r, B) \times_{\mathcal{B}or(G)} \mathcal{S}tand(\underline{\tau}_{r-1}) \times \cdots \times_{\mathcal{B}or(G)} \mathcal{S}tand(\underline{\tau}_1)\right),$$

where

$$\underline{\tau}_{\mathscr{C}} = (\underline{\tau}_i) \, (r + 1 \geqslant i \geqslant 1) \quad (\text{resp. } \underline{\tau}_{\mathscr{C}} = (\underline{\tau}_i) \, (r \geqslant i \geqslant 1)).$$

The fiber products are defined in terms of the couples of morphisms

$$\left(\mathcal{P}roj^{(\underline{\tau}_i)}, \mathrm{pr}_{\mathcal{P}, \underline{\tau}_{i-1}}\right) (r + 1 \geqslant i \geqslant 1) \text{ (resp. } r \geqslant i \geqslant 1).$$

Remark 14.25
*Assume that G is endowed with a frame E, and $\tau_{\mathscr{C}} \in Relpos_{A_E}^{gall}(g)$. One then
has the set of $\overline{s}$-points*

$$\mathcal{C}onf_G^{std}(\tau_{\mathscr{C}}, B)_{(\overline{s})}$$

of the geometrical fiber $\mathcal{C}onf_G^{std}(\tau_{\mathscr{C}}, B)_{(\overline{s})}$ is equal to the cell $\mathscr{C}_C(g, \tau_{\mathscr{C}}) = (\tau_{I,C}(g))^{-1}(\tau_{\mathscr{C}})$, as defined in 10.14.

14.7 The Bruhat cell decomposition of the Configurations Scheme

One proves the following generalization of the Bruhat cell decomposition for
the standard galleries Configurations Scheme.

Proposition 14.26
The morphism

$$j_{[g,E]} : \coprod_{\tau_{\mathscr{C}} \in Relpos^{gall}(g)} \mathcal{C}onf_G^{std}(\tau_{\mathscr{C}}, B) \to \mathcal{C}onf(g, B)$$

is a surjective monomorphism. One calls $\coprod_{\tau_{\mathscr{C}} \in Relpos^{gall}(g)} \mathcal{C}onf_G^{std}(\tau_{\mathscr{C}}, B)$ *the*
Cellular scheme *associated with $\mathcal{C}onf(g, B)$.*

For the sake of simplifying notations assume that $g = ((s_i); (t_j))$ ($N \geq i \geq 0$, $N \geq j \geq 1$). Write

$$t_j(g) = t_j \ (N \geq j \geq 1) \ (\text{resp. } s_i(g) = s_i \ (N \geq i \geq 0)) \ ,$$

and

$$P_{t_j} = P_{t_j}(C) \ (Q_{s_i} = Q_{s_i}(C)) \ .$$

Clearly there is an inclusion $((Q_{s_i}); (P_{t_j})) \subset \Delta(G)$. This gallery is called the
basic gallery defined by the gallery of types g.

Given $\tau_{\mathscr{C}} \in Relpos^{gall}(g)$, let

$$\tau_{\mathscr{C}}^{(\alpha)} = (\tau_N, \cdots, \tau_\alpha) \ (N \geq \alpha \geq 1) \ .$$

Write $g^{(\alpha)} = ((s_i); (t_j))$ ($N \geq i \geq \alpha$, $N \geq j \geq \alpha + 1$). One then has

$$\tau_{\mathscr{C}}^{(\alpha+1)} \in Relpos^{gall}(g^{(\alpha)}).$$

Let $P \subset G$ be a parabolic subgroup. Denote by $\mathcal{P}ar_s(P)$ the functor of
parabolics subgroups of type s of P. Let $t, s \in typ A_E$, with $t \subset s$, and B be
the Borel subgroup defined by E. From the **Bruhat cell decomposition in
double classes** of a reductive group G (cf. [23], Exp. XXVI, 4.5.4) results
that

$$\coprod_{\tau \in Relpos_s^t A} B \overline{w}_\tau Q_s / Q_s \to \mathcal{P}ar_s(P_t)$$

is a **surjective monomorphism**. Given a section $\sigma_\mathscr{C} : ((Q_{s_i}); (P_{t_j}))$ of $\mathcal{C}onf_G(g, B)$, $g = ((s_i); (t_j))$, write $Q_{s_i}(\sigma_\mathscr{C}) = Q_{s_i}$ and $P_{t_j}(\sigma_\mathscr{C}) = P_{t_j}$. Let it be proved by induction on $N \geq \alpha \geq 1$ the

Proposition 14.27
For all $1 \geq \alpha \geq N$ the morphism

$$j_{[g^{(\alpha)}, E]} : \coprod_{\tau_\mathscr{C} \in Relpos^{gall}(g^{(\alpha)})} \mathcal{C}onf_G^{std}(\tau_\mathscr{C}, B) \to \mathcal{C}onf_G(g^{(\alpha)}, B),$$

where one writes:

$$j_{[g^{(\alpha)}, E]} = \coprod_{\tau_\mathscr{C} \in Relpos^{gall}(g^{(\alpha)})} j_{[\tau_\mathscr{C}, E]},$$

is surjective.

Proof *Clearly for $\alpha = N$, this follows from the Bruhat decomposition quoted above. Let $\tau(t_{\alpha+1}, s_\alpha)$ be the type of relative position defined by the couple of* **incident parabolics** $(P_{t_{\alpha+1}}, Q_{s_\alpha})$. *Recall that the natural morphism*

$$\overline{p}^{[\alpha+1, \alpha]} : \mathcal{C}onf_G(g_S^{(\alpha)}, B) \to \mathcal{C}onf(g_S^{(\alpha+1)}, B)$$

defines a locally trivial fibration, with typical fiber

$$\mathcal{P}ar_{s_\alpha}(P_{t_{\alpha+1}}) = \mathcal{S}tand(\tau(t_{\alpha+1}, s_\alpha), P_{t_{\alpha+1}}),$$

(cf. Proposition 13.9). For all $\tau_\mathscr{C} = (\tau_N, \cdots, \tau_{\alpha+2}) \in Relpos^{gall}(g^{(\alpha+1)})$ there is an isomorphism:

$$k_{[\tau_\mathscr{C}, E]} : (j_{[\tau_\mathscr{C}, E]})^*(\mathcal{C}onf_G(g^{(\alpha)}, B)) \simeq \mathcal{C}onf_G^{std}(\tau_\mathscr{C}, B) \times_{\mathcal{P}ar(G)} \mathcal{S}tand(\tau(t_{\alpha+1}, s_\alpha)).$$

Here the fiber product is defined by the couple of morphisms given by

$$\sigma_\mathscr{C} \mapsto P_{t_{\alpha+1}}(\sigma_\mathscr{C}) \ (resp. \ (P, Q) \mapsto P).$$

Where $P_{t_{\alpha+1}}(\sigma_\mathscr{C})$ denotes the parabolic of type $t_{\alpha+1}$ defined by the configuration $\sigma_\mathscr{C}$, i.e. the unique parabolic P of type $t_{\alpha+1}$ containing $Q_{s_{\alpha+1}}(\sigma_\mathscr{C})$.
Remark that the following two statements are equivalent:

1. *$j_{[g^{(\alpha)}, E]}$ is a surjective morphism*

2. *Every section*

$$\sigma_\mathscr{C}^{(\alpha)} : \ Spec(K) \longrightarrow \mathcal{C}onf_G(g^{(\alpha)}, B) \ ,$$

$$\searrow \qquad \downarrow$$

$$S$$

over an algebraically closed field K, factors through some $j_{[\tau_\mathscr{C}, E]}$ ($\tau_\mathscr{C} \in Relpos^{gall}(g^{(\alpha)})$).

Given $\sigma_{\mathscr{C}}^{(\alpha)}$ let it be supposed as inductive hypothesis that:

"$j_{[g^{(\alpha+1)},E]}$ is a surjective morphism"

and prove that it verifies 2. Thus the composed morphism

$$\sigma_{\mathscr{C}}^{(\alpha+1)} = \overline{p}^{[\alpha+1,\alpha]} \circ \sigma_{\mathscr{C}}^{(\alpha)} : Spec(K) \rightarrow \mathcal{C}onf(g^{(\alpha+1)}, B),$$

factors through some cell

$$j_{[\tau_{\mathscr{C}}^{(\alpha+2)},E]} : \mathcal{C}onf_G^{std}(\tau_{\mathscr{C}}^{(\alpha+2)}, B) \rightarrow \mathcal{C}onf_G(g^{(\alpha+1)}, B),$$

for some gallery of relative positions

$$\tau_{\mathscr{C}}^{(\alpha+2)} \in Relpos^{gall}(g^{(\alpha+1)}).$$

One may then write

$$\sigma_{\mathscr{C}}^{(\alpha+1)} = j_{[\tau_{\mathscr{C}}^{(\alpha+2)},E]} \circ \sigma_{\mathscr{C}}^{(\alpha+1)\,\prime},$$

with $\sigma^{(\alpha+1)\,\prime}$ being a section of $\mathcal{C}onf_G^{std}(\tau_{\mathscr{C}}^{(\alpha+2)}, B)$.

Let $Q_{s_{\alpha+1}}(\sigma_{\mathscr{C}}^{(\alpha+1)})$ be the parabolic of type $s_{\alpha+1}$ defined by $\sigma_{\mathscr{C}}^{(\alpha+1)}$, and let $P_{t_{\alpha+1}}(\sigma_{\mathscr{C}}^{(\alpha+1)})$ be the unique parabolic of type $t_{\alpha+1}$ so that

$$Q_{s_{\alpha+1}}(\sigma_{\mathscr{C}}^{(\alpha+1)}) \subset P_{t_{\alpha+1}}(\sigma_{\mathscr{C}}^{(\alpha+1)}).$$

Let $B_{\alpha+2}$ be the Borel subgroup given by definition of the section $\sigma^{(\alpha+1)\,\prime}$ of $\mathcal{C}onf_G^{std}(\tau_{\mathscr{C}}^{(\alpha+2)}, B)$ (cf. Definitions 12.39 and 14.24) and

$$B_{\alpha+1} \subset Q_{s_{\alpha+1}}(\sigma_{\mathscr{C}}^{(\alpha+1)})$$

be the Borel subgroup obtained as the projection of $B_{\alpha+2}$ on $Q_{s_{\alpha+1}}(\sigma_{\mathscr{C}})$. From the isomorphism $k_{[\tau_{\mathscr{C}},E]}$ one deduces an isomorphism

$$\mathcal{C}onf_G(g^{(\alpha)}, B)_{\sigma_{\mathscr{C}}^{(\alpha+1)}} \simeq \mathcal{S}tand(\tau(t_{\alpha+1}, s_\alpha), P_{t_{\alpha+1}}(\sigma_{\mathscr{C}}^{(\alpha+1)}))$$

$$= \mathcal{P}ar_{s_\alpha}(P_{t_{\alpha+1}}(\sigma_{\mathscr{C}}^{(\alpha+1)})),$$

making correspond to $\sigma_{\mathscr{C}}^{(\alpha)}$ a morphism

$$\overline{\sigma}_{\mathscr{C}}^{(\alpha)} : Spec(K) \rightarrow \mathcal{P}ar_{s_\alpha}(P_{t_{\alpha+1}}(\sigma_{\mathscr{C}}^{(\alpha+1)})).$$

Let

$$Relpos^{gall}(g^{(\alpha)})_{\tau_{\mathscr{C}}^{(\alpha+2)}} =$$

$$\left\{ \overline{\tau}_{\mathscr{C}} \in Relpos^{gall}(g^{(\alpha)}) \mid \overline{\tau}_{\mathscr{C}}^{(\alpha+2)} \,((\alpha+2)\text{-truncation of } \overline{\tau}_{\mathscr{C}}) = \tau_{\mathscr{C}}^{(\alpha+2)} \right\}.$$

Observe that there is a bijection

$$Relpos^{gall}(g^{(\alpha)})_{\overline{\tau}_{\mathscr{C}}^{(\alpha+2)}} \simeq Relpos_{s_\alpha}^{t_{\alpha+1}} A,$$

which is denoted by

$$\overline{\tau}_{\mathscr{C}} \mapsto \tau(\overline{\tau}_{\mathscr{C}}).$$

There is a commutative square

$$
\begin{array}{ccc}
\coprod \mathcal{C}onf_G^{std}((\overline{\tau}_{\mathscr{C}})_S, B)_{\sigma_{\mathscr{C}}^{(\alpha+1)}} & \longrightarrow & \mathcal{C}onf_G(g_S^{(\alpha)}, P_{e_1(g)})_{\sigma_{\mathscr{C}}^{(\alpha+1)}} \\
\downarrow & & \downarrow \\
\coprod \mathcal{S}tand(\tau(\overline{\tau}_{\mathscr{C}})_S, B_{\alpha+1}) & \longrightarrow & \mathcal{P}ar_{s_\alpha}(P_{t_{\alpha+1}}(\sigma_{\mathscr{C}}^{(\alpha+1)})) \ .
\end{array}
$$

The upper horizontal arrow is given by $\coprod j_{[\overline{\tau}_{\mathscr{C}}, E]_{\sigma_{\mathscr{C}}^{(\alpha+1)}}}$, *and the lower horizontal arrow by* $\coprod j_{[\tau(\overline{\tau}_{\mathscr{C}})_S, B_{\alpha+1}]}$, *where* $\overline{\tau}_{\mathscr{C}} \in Relpos^{gall}(g^{(\alpha)})_{\overline{\tau}_{\mathscr{C}}^{(\alpha+2)}}$ *and*

$$j_{[\tau(\overline{\tau}_{\mathscr{C}})_S, B_{\alpha+1}]} : \mathcal{S}tand(\tau(\overline{\tau}_{\mathscr{C}})_S, B_{\alpha+1}) \to \mathcal{P}ar_{s_\alpha}(P_{t_{\alpha+1}}(\sigma_{\mathscr{C}}^{(\alpha+1)})),$$

is defined by

$$j_{[\tau(\overline{\tau}_{\mathscr{C}})_S, B_{\alpha+1}]} : (B_{\alpha+1}, Q) \mapsto Q.$$

The vertical arrows are bijective, and the lower horizontal is surjective (Bruhat lemma). One deduces that the upper horizontal arrow is also bijective.

Thus one obtains a factorization of $\overline{\sigma}_{\mathscr{C}}^{(\alpha)}$:

$$\overline{\sigma}_{\mathscr{C}}^{(\alpha)} = j_{[B_{\alpha+1}]} \circ \overline{\sigma}_{\mathscr{C}}^{(\alpha)}{}',$$

which finally gives rise to a factorization of $\sigma_{\mathscr{C}}^{(\alpha)}$, *namely*

$$\sigma_{\mathscr{C}}^{(\alpha)} = j_{[g^{(\alpha)}, E]} \circ \sigma_{\mathscr{C}}^{(\alpha)}{}'.$$

This achieves the proof of the recursive step from $\alpha+1$ *to* α *and thus the proof of the surjectivity of* $j_{[g^{(\alpha)}, E]}$ *for* $r \geqslant \alpha \geqslant 0$ *(resp.* $r > \alpha \geqslant 0$*).*

14.8 Retraction morphism on an Apartment scheme of Standard Galleries Configurations

Let one interpret the morphism t_2^{std} as a retraction morphism. In §10.5 one has defined for $g \in gall_A$, where $A = \mathcal{A}_E$, a bijection

$$\tau_{g,C}^A : \mathrm{Gall}_A\left(g, F_{e_1(g)}(C)\right) \longrightarrow Relpos^{gall}(g),$$

which is recalled here.

By definition 9.64 one has introduced a mapping
$s_{g^*,C} : \mathrm{Gall}_A(g, F_g(C)) \to \mathrm{Gall}_A(g^*, C)$ by $s_{g^*,C} : \gamma \mapsto \gamma^*$, where γ^* is defined as follows:

1) Let $g = g_1$ (resp. g'_1), and
$$\gamma : \; (F_i)_{r \geqslant i \geqslant 0}, \; (F'_j)_{r+1 \geqslant j \geqslant 1} \; (\text{resp. } (F'_j)_{r+1 \geqslant j \geqslant 0}) \; . \quad \text{Define } \gamma^* \text{ by}$$
$$\gamma^* : \; (C_i)_{r+1 \geqslant i \geqslant 0}, \; (F'_j)_{r+1 \geqslant j \geqslant 1} \; (\text{resp. } (F'_j)_{r+1 \geqslant j \geqslant 0}) \; ,$$
where $C_{r+1} = C$, $C_i = \text{proj}_{F_i} C_{i+1}$ $(r \geqslant i \geqslant 0)$.

2) Let $g = g_2$ (resp. g'_2), and
$$\gamma : \; (F_i)_{r \geqslant i \geqslant 0}, \; (F'_j)_{r \geqslant j \geqslant 1} \; (\text{resp. } (F'_j)_{r \geqslant j \geqslant 0}) \; . \quad \text{Define } \gamma^* \text{ by}$$
$$\gamma^* : \; (C_i)_{r \geqslant i \geqslant 0}, \; (F'_j)_{r \geqslant j \geqslant 1} \; (\text{resp. } (F'_j)_{r \geqslant j \geqslant 0}) \; ,$$
where $C_r = C$, $C_i = \text{proj}_{F_i} C_{i+1}$ $(r > i \geqslant 0)$.

Let it be supposed that $F_g(C) = C$. Thus $s_{g^*, C}$ associates with a Generalized Gallery γ of $\mathcal{A}_E$ with first term (resp. left extremity) a chamber a **chamber gallery** γ^*, i.e. such that the set of (s_i) of the gallery of types g^* of γ^* is given by the type of a chamber C of $\mathcal{A}_E$. Define $\tau_i^* = \tau(C_i, C_{i-1})$ (resp. $\tau_i = \tau(C_i, F_{i-1})$) where i runs on one of the sets above according to the type of g. Clearly $(\tau(C_i, F_{i-1}))$ is a gallery of types of relative positions and $(\tau(C_i, F_{i-1})) \in \prod \text{Relpos}'_{\mathcal{A}_E}$. One thus obtains a mapping $\tau_{A,C}(g) : \text{Gall}_{\mathcal{A}_E}(C) \longrightarrow \text{Relpos}^{\text{gall}}_{\mathcal{A}_E}$ associating with a gallery γ issued from $F \subset C$ the gallery of relative positions $(\tau(C_i, F_{i-1}))$. From the reciprocal bijection
$$(\tau^A_{g,C})^{-1} : \; \text{Relpos}^{\text{gall}}(g) \; \to \; \text{Gall}_A \left(g, F_{e_1(g)}(C) \right),$$
one obtains an S-isomorphism
$$(\tau^A_{g,C})^{-1}_S : \; \text{Relpos}^{\,\text{gall}}_G(g) \; \tilde{\to} \; \mathcal{C}\text{onf}_{\mathcal{F}\text{ix}\,(T)} \left(g, P_{e_1(g)}(E) \right)$$
(cf. Definition 10.15).

Definition 14.28
Define the S-morphism
$$\rho^{\,conf}_{E,g} : \; \mathcal{C}onf^{std}_G \left(g, P_{e_1(g)} \right) \; \to \; \mathcal{C}onf_{\mathcal{F}ix(T)} \left(g, P_{e_1(g)}(E) \right),$$
as the composition of the morphism
$$t^{std\,\prime}_2 : \; \mathcal{C}onf^{std}_G \left(g, P_{e_1(g)}(E) \right) \; \to \; \text{Relpos}^{\,gall}_G(g)$$
induced by t^{std}_2, followed by $(\tau_{A,C}(g))^{-1}_S$.

The sub-building of A isomorphic to the typical building
$$\Delta(C) = \{ F_t(C) \mid t \in \text{typ}\, A \}$$
gives rise to a subscheme $\Delta(C)_S \subset A_S$. Denote by
$$\underline{\Delta}(G, E) \subset \mathcal{P}\text{ar}(G),$$

the S-subscheme of $\mathcal{P}ar(G)$ given by the image of $\Delta(C)_S$ by $A_S \dashrightarrow \mathcal{F}ix(T) \subset \mathcal{P}ar(G)$. The couple $(\underline{\Delta}(G, E), \mathcal{F}ix(T))$ corresponds to an apartment A of the building of a reductive group and a chamber C in this apartment. It is observed that $\underline{\Delta}(G, E)$ is a trivialization of the relative typical simplex $\underline{\Delta}(G)$ (cf. Definition 11.29). It is known that Bruhat decomposition of $\mathcal{P}ar(G)$ defined by B_C may be seen as a retraction of the building on A. One introduces here the retraction of the scheme of configurations issued from B_C on the configurations scheme of $\mathcal{F}ix(T)$.

Definition 14.29
Write
$$\mathcal{C}onf_G(\underline{\Delta}(G, E)) = \mathcal{C}onf_G \times_{\mathcal{P}ar(G)} \underline{\Delta}(G, E),$$
where the fiber product is defined by the couple $(\mathcal{E}_1, j_{\underline{\Delta}(G,E)})$, with
$$j_{\underline{\Delta}(G,E)} : \underline{\Delta}(G, E) \to \mathcal{P}ar(G)$$
the canonical embedding, i.e. $\mathcal{C}onf_G(\underline{\Delta}(G, E)) = (\mathcal{C}onf_G)_{\underline{\Delta}(G,E)}$. The sections of this scheme are the galleries configurations issued from a section given by $\underline{\Delta}(G, E)$.

The following identifications are supposed
$$\mathrm{Conf}_G^{std}(\tau_{\mathscr{C}}, P_{e_1(g)}) = \mathrm{Conf}_G^{std}(\tau_{\mathscr{C}}, B)$$

$$\left(\text{resp. } \mathrm{Conf}_G^{std}(g, P_{e_1(g)}) = \coprod_{\tau_{\mathscr{C}} \in \mathrm{Relpos}^{\,\mathrm{gall}}(g)} \mathrm{Conf}_G^{std}(\tau_{\mathscr{C}}, B) \right).$$

One has
$$\mathrm{Conf}_G(\underline{\Delta}(G, E)) = \coprod_{g \in \mathrm{gall}_A} \mathrm{Conf}_G(g, P_{e_1(g)}).$$

Write
$$\mathrm{Conf}_{\mathcal{F}ix(T)}(\underline{\Delta}(G, E)) = \mathrm{Conf}_G(\underline{\Delta}(G, E)) \cap \mathrm{Conf}_{\mathcal{F}ix(T)}$$

$$\left(\text{resp. } \mathrm{Conf}_G^{std}(\underline{\Delta}(G, E)) = \coprod_{g \in \mathrm{gall}_A} \mathrm{Conf}_G^{std}(g, P_{e_1(g)}(E)) \right).$$

Thus from the definition one obtains
$$\mathrm{Conf}_{\mathcal{F}ix(T)}(\underline{\Delta}(G, E)) = \coprod_{g \in \mathrm{gall}_A} \mathrm{Conf}_{\mathcal{F}ix(T)}(g, P_{e_1(g)}(E)).$$

Definition 14.30
Let
$$\rho_E^{conf} : \mathcal{C}onf_G^{std}(\underline{\Delta}(G, E)) \to \mathcal{C}onf_{\mathcal{F}ix(T)}(\underline{\Delta}(G, E)),$$

be given by

$$\rho_E^{conf} := \coprod_{g \in gall_A} \rho_{E,g}^{conf}$$

(The Retraction morphism *of standard galleries configurations).*

Chapter 15

Contracted Products and Galleries Configurations Schemes

It is shown that the galleries configurations scheme $\mathcal{C}\mathrm{onf}_\mathbf{G}(\mathbf{g})$ of a split reductive S-group scheme G, defined by a fixed gallery of types $\mathbf{g}$, is isomorphic to a **Contracted Product**. This isomorphism gives rise to natural parametrizations of the $\mathcal{C}\mathrm{onf}_\mathbf{G}(\mathbf{g})$-Cells. By means of a parametrization it is proved that there is a Cell of $\mathcal{C}\mathrm{onf}_\mathbf{G}(\mathbf{g})$ which is an open relatively schematically dense subscheme (**The Big Open Cell**). A Contracted Product may be decomposed in a sequence of locally trivial fibrations with typical fiber G/P, where P is a parabolic subgroup of G. The Big Open Cell is isomorphic to a contracted product of big open cells of homogeneous spaces G/P. By means of the Contracted Product the image of a minimal gallery type configuration is calculated by the Retraction on an apartment. This calculation amounts to determining the fibers of the resolving morphism.

15.1 Contracted Products

Assume that G is endowed with a frame $E = (T, M, R, R_0, (X_\alpha)(\alpha \in R_0))$. Let $A = \mathcal{A}_E$ be the apartment defined by the $\mathbb{Z}$-root data $\mathcal{R}(E)$. Given $t \in \mathrm{typ}\, A$ let $P_t(E)$ denote the parabolic of type t, such that E is adapted to P, i.e. $P_t(E) = P_t(C)$ where C denotes the chamber of $\mathcal{R}(E)$ given by E.

Let one state some conventions concerning the notation of a generalized gallery g (resp. γ) of typ A (resp. A). Given $g \in \mathrm{gall}\, A$, let $l(g)$ denote its length. Write:

1. $g = g_1$: $l_1(g) = 1$, $l_2(g) = l(g)$;

2. $g = g_1'$: $l_1(g) = 0$, $l_2(g) = l(g)$;

3. $g = g_2$: $l_1(g) = 1$, $l_2(g) = l(g) - 1$;

4. $g = g_2'$: $l_1(g) = 0$, $l_2(g) = l(g) - 1$.

Thus a GG g of typ A is represented by g : $(s_i)(l(g) > i \geqslant 0)$, $(t_j)(l_2(g) \geqslant j \geqslant l_1(g))$. Write

$$s(g) = (s_i) \, (l(g) > i \geqslant 0), \text{ and } s_i(g) = s_i$$

$$(\text{resp. } t(g) = (t_j) \, (l_2(g) \geqslant j \geqslant l_1(g)), \text{ and } t_j(g) = t_j).$$

Given $\gamma \in \text{Gall}_A$ write $l_1(\gamma) = l_1(\text{typ } \gamma)$ (*resp.* $l_2(\gamma) = l_2(\text{typ } \gamma)$). Let

$$\gamma : (F_i) \, (l(\gamma) > i \geqslant 0), (F_j') \, (l_2(\gamma) \geqslant j \geqslant l_1(\gamma)),$$

write $F_i(\gamma) = F_i$ and $F_j'(\gamma) = F_j'$

Let $e^A = e_1^A \times e_2^A$: gall $A \to$ typ $A \times$ typ A be the **extremities mapping** defined as the restriction to $\left(\prod^N \text{typ } A \right) \cap \text{gall}_A$ of the mapping $p_1 \times p_N$, where p_1 (resp. p_N) : $\prod^N \text{typ } A \to$ typ A is the 1st-projection (resp. the N-th projection). Write $e = e^A$ (resp. $e_1 = e_1^A$, $e_2 = e_2^A$) if no confusion arises with the notation of 13.6.

Definition 15.1

*The **basic configuration (resp. gallery)** $\sigma_g(E)$ of G associated to (E, g), $g \in \text{gall}_A$ is, by definition, given by*

$$\sigma_g(E) : (Q_{s_i(g)}(E)) \, (l(g) > i \geqslant 0), (P_{t_j(g)}(E)) \, (l_2(g) \geqslant j \geqslant l_1(g)).$$

Let

$$P_{t(g)}(E) = \prod P_{t_j(g)}(E) \, (l_2(g) \geqslant j \geqslant 1)$$

$$(\text{resp. } Q_{s(g)}(E) = \prod Q_{s_i(g)}(E) \, (l(g) > i \geqslant 0), \text{ if } e_1(g) = t_{l_2(g)} = t_{l(g)},$$

$$Q_{s(g)}(E) = \prod Q_{s_i(g)}(E) \, (l(g) - 1 > i \geqslant 0), \text{ if } e_1(g) = s_{l(g)-1}(g)).$$

Define a right action of $Q_{s(g)}(E)$ on $P_{t(g)}(E)$ following the pattern explained in Definition 9.45.

The quotient scheme

$$\Pi_G \, [g, E] = P_{t(g)}(E) \, / \, Q_{s(g)}(E)$$

*is called the **contracted product along the basic configuration** $\sigma_g(E)$ (resp. $g \in \text{gall}_A$).*

Let $(g^{(\alpha)})$ $(l(g) > \alpha \geqslant 0)$ be the sequence of truncated galleries defined by g as in Definition 9.57. There are natural morphisms

$$p^{[\alpha,\alpha+1]} : \Pi_G [g^{(\alpha)}, E] \to \Pi_G [g^{(\alpha+1)}, E] \ (l(g) - 1 > \alpha \geqslant 0),$$

defining, as is easy to see, locally trivial fibrations with typical fibers respectively $P_{t_{\alpha+1}(g)}(E)/Q_{s_\alpha(g)}(E)$ **(The canonical fibration of a contracted product.** (See Remark 15.6 below). From this remark results the following

Proposition 15.2
The contracted product $\Pi_G [g, E]$ is a smooth S-scheme.

Definition 15.3
Following the pattern of the definition of the combinatorial mapping

$$i_{g,C} : \Pi_W [g, C] = W_{t(g)}/W_{s(g)} \to Gall_A(g, F_{e_1(g)}(C))$$

(cf. Definition 9.53) define

$$\lambda_{[g,E]} : \Pi_W [g, E] \to Conf_G(g_S, P_{e_1(g)}(E))$$

as follows. Given a section $x = (x_j)$ $(l_2(g) \geqslant j \geqslant 1)$ of $P_{t(g)}(E)$, write

$$z_i = \prod x_\alpha \ (l_2(g) \geqslant \alpha \geqslant i) \ (resp. \ z_{l_2(g)+1} = 1).$$

Let $\sigma_g(E, x)$ be the section of $Conf_G(g_S, P_{e_1(g)})$ defined as follows using the conventions of §15.1

$$\sigma_g(E, x) : \big(int(z_{i+1})(Q_{s_i(g)}(E))\big) \ (l(g) > i \geqslant 0),$$

$$\big(int(z_{j+1})(P_{t_j(g)}(E))\big) \ (l_2(g) > j \geqslant l_1(g)).$$

Let

$$\lambda_{[g,E]} : \Pi_G [g, E] = P_{t(g)}/Q_{s(g)} \to Conf_G(g_S, P_{e_1(g)}),$$

be the morphism induced by $x \mapsto \sigma_g(E, x)$.

It is proved that $\lambda_{[g,E]}$ is an isomorphism following the pattern of the proof of Proposition 9.62.

Proposition 15.4
The morphism

$$\lambda_{[g,E]} : \Pi_G [g, E] \to Conf_G(g_S, P_{e_1(g)})$$

is an isomorphism of $P_{e_1(g)})$-schemes.

Observe that the following statements concerning a gg g are equivalent:

$$(1)\ g = g_1, g_1',\ 2)\ l_2(g) = l(g),\ 3)\ e_1(g) = t_{l(g)}$$

$$\left(\text{resp. } (1')\ g = g_2, g_2',\ 2')\ l_2(g) = l(g) - 1,\ 3')\ e_1(g) = s_{l(g)-1}\right).$$

Let g be such that

$$l_2(g) = l(g)\ (\text{resp. } l_2(g) = l(g) - 1).$$

It is then written:

$$\overline{P}_{t(g)}(E) = G \times_S \prod P_{t_j(g)}(E)\ (l_2(g) - 1 \geqslant j \geqslant l_1(g))$$

$$\overline{P}_{t(g)}(E) = G \times_S \prod P_{t_j(g)}(E)\ (l_2(g) \geqslant j \geqslant l_1(g)),$$

and

$$\overline{Q}_{s(g)}(E) = Q_{s(g)}\left(\text{resp. } \prod Q_{s_i(g)}(E)\ (l(g) > i \geqslant 0)\right).$$

A right action of $\overline{Q}_{s(g)}(E)$ on $\overline{P}_{t(g)}(E)$ is defined following the pattern explained in Definition 9.48.

The quotient

$$\overline{\Pi}_G\,[g, E] = \overline{P}_{t(g)}(E)/\overline{Q}_{s(g)}(E)$$

is representable by a smooth and projective S-scheme, as results from Proposition 15.5 below and from Proposition 13.9.

Following the definition of $\lambda_{[g,E]}$ above, a morphism

$$\overline{\lambda}_{[g,E]} : \overline{\Pi}_G\,[g, E] \to \mathcal{Conf}_G(g_S)\ \text{is obtained.}$$

Proposition 15.5
The morphism $\overline{\lambda}_{[g,E]}$ is an isomorphism of G-schemes.

Proof   *One follows the pattern of the proof of Proposition 15.4.*

Let

$$\pi^{[\alpha,\alpha+1]} : \Pi[g^{(\alpha)}, E] \to \Pi_G[g^{(\alpha+1)}, E]$$

be the natural morphism obtained by the definition of $\Pi\,[g^{(\alpha)}, E]$.

Remark 15.6
Denote by

$$\overline{\pi}^{[\alpha,\alpha+1]} : \mathcal{Conf}_G(g_S^{(\alpha)}) \to \mathcal{Conf}_G(g_S^{(\alpha+1)})$$

the morphism associating with a configuration $\sigma_{\mathscr{C}}$ of type $g_S^{(\alpha)}$ the truncated configuration $\sigma_{\mathscr{C}}^{(\alpha+1)}$.

The following compatibility between $\overline{\pi}^{[\alpha,\alpha+1]}$ (resp. $\overline{p}^{[\alpha,\alpha+1]}$) and $\pi^{[\alpha,\alpha+1]}$ (resp. $p^{[\alpha,\alpha+1]}$) holds

$$\overline{\pi}^{[\alpha,\alpha+1]} \circ \overline{\lambda}_{[g^{(\alpha)},E]} = \overline{\lambda}_{[g^{(\alpha+1)},E]} \circ \pi^{[\alpha,\alpha+1]}$$

$$\left(\text{resp. } \overline{p}^{[\alpha,\alpha+1]} \circ \lambda_{[g^{(\alpha)},E]} = \lambda_{[g^{(\alpha+1)},E]} \circ p^{[\alpha,\alpha+1]}\right).$$

Thus from Proposition 13.9 one obtains:

"The morphism $\pi^{[\alpha,\alpha+1]}$ (resp. $p^{[\alpha,\alpha+1]}$) defines a locally trivial fibration with smooth typical fiber $P_{t_{\alpha+1}(g)}(E)/Q_{s_\alpha(g)}(E)$."

15.2 Contracted product parametrization of cells

For $\tau_\mathscr{C} \in \mathrm{Relpos}^{\mathrm{gall}}(g)$ one gives a parametrization of $\mathcal{C}\mathrm{onf}^{std}_G((\tau_\mathscr{C})_S, B)$ in terms of the isomorphism $\lambda_{[g,E]}$.

The same hypothesis about G is kept as in Section 15.1. Given $\tau_\mathscr{C} = (\tau_i) \in \prod \mathrm{Relpos}^{t_i(g)}_{s_{i-1}(g)}$, let

$$\overline{w}_{\tau_\mathscr{C}} = (\overline{w}_i) \quad (\text{resp. } w^m_{\tau_\mathscr{C}} = (w^m_i)),$$

where

$$\overline{w}_i \in W_{t_i(g)}/W_{s_{i-1}(g)},$$

is defined by τ_i, and $w^m_i \in \overline{w}_i$ is the minimal length $(l_{S(C)}(w^m_i))$ element of $\overline{w}_i$. $w^m_{\tau_\mathscr{C}}$ is an $\mathscr{S}$-word $(C = C_{R_+})$, corresponding to the GG

$$\gamma_{\tau_\mathscr{C}} \in \mathrm{Gall}_A(g, F_{e_1(g)}(C)),$$

(cf. §10.5). As usual let

$$\mathcal{U}(w) \subset B_{R_+}$$

be the subgroup defined by the closed system of roots

$$R_+(C, w(C)) = \{\alpha \in R_+ \mid w(\alpha) < 0\}.$$

Write

$$\Pi_G\,[\tau_\mathscr{C}, E] = \prod \mathcal{U}(w^m_i)\,w^m_i$$

$(l(g) \geqslant i \geqslant 1)$ (resp. $l(g) > i > 1$).

Here $w^m_i \in N(T)$ denotes also a representative of $w^m_i \in W$.

Clearly, there is a natural morphism

$$q_{[\tau_\mathscr{C}, E]} : \Pi_G\,[\tau_\mathscr{C}, E] \;\to\; P_{t(g)}/Q_{s(g)} = \Pi_G\,[g, E].$$

Definition 15.7

Let

$$\lambda'_{[\tau_\mathscr{C}, E]} : \Pi_G\,[\tau_\mathscr{C}, E] \;\to\; \mathcal{C}onf_G(g_S, P_{e_1(g)})$$

be the morphism obtained as the composition of the morphism $q_{[\tau_\mathscr{C}, E]}$, followed by the isomorphism

$$\lambda_{[g,E]} : \Pi_G\,[g, E] \;\to\; \mathcal{C}onf_G(g_S, P_{e_1(g)}),$$

i.e.

$$\lambda'_{[\tau_\mathscr{C}, E]} = \lambda_{[g,E]} \circ q_{[\tau_\mathscr{C}, E]}.$$

PROPOSITION - DEFINITION 15.8

The morphism $\lambda'_{[\tau_{\mathscr{C}},E]}$ factors through the embedding

$$j_{[\tau_{\mathscr{C}},E]} : \mathcal{C}onf_G^{std}((\tau_{\mathscr{C}})_S, B) \to \mathcal{C}onf_G(g_S, P_{e_1(g)}),$$

and the factorization induces an isomorphism of S-schemes

$$\lambda_{[\tau_{\mathscr{C}},E]} : \Pi_G[\tau_{\mathscr{C}}, E] \xrightarrow{\sim} \mathcal{C}onf_G^{std}((\tau_{\mathscr{C}})_S, B).$$

Proof *Consider the case $g = g_1, g_1'$. The same proof works for the case $g = g_2, g_2'$ mutatis mutandis. Thus one may write*

$$\sigma_g(E) : (Q_{s_i(g)})(l(g) > i \geqslant 0),\ (P_{t_j(g)})(l_2(g) \geqslant j \geqslant l_1(g))\ \ (l_2(g) = l(g) = r+1).$$

Recall that there is an isomorphism of S-schemes

$$\mathcal{U}(w_i^m)\, w_i^m \xrightarrow{\sim} \mathcal{S}tand(B, \tau_i),$$

by definition of w_i^m.

The definition of $\lambda_{[\tau_{\mathscr{C}},E]}$ of the assertion results from the following points (cf. §12.3). By definition of τ_i (resp. w_i^m) ($l(g) \geqslant i \geqslant 1$), one obtains:

1. $t_2\left(int\left(\prod\limits_{\alpha=i+1}^{r+1} x_\alpha\, w_\alpha^m\right)(B),\ int\left(\prod\limits_{\alpha=i}^{r+1} x_\alpha\, w_\alpha^m\right)(Q_{s_i(g)})\right) = (\tau_i)_S;$

2. $\mathcal{P}roj\ \ :\ \ \left(int\left(\prod\limits_{\alpha=i+1}^{r+1} x_\alpha\, w_\alpha^m\right)(B),\ int\left(\prod\limits_{\alpha=i}^{r+1} x_\alpha\, w_\alpha^m\right)(Q_{s_i(g)})\right) \ \mapsto$

   $int\left(\prod\limits_{\alpha=i}^{r+1} x_\alpha\, w_\alpha\right)(B)$

3. $\lambda'_{[\tau_{\mathscr{C}},E]} : (x_i\, w_i) \mapsto \left(int\left(\prod\limits_{\alpha=i+1}^{r+1} x_\alpha\, w_\alpha\right)(Q_{s_i(g)})\right)(r \geqslant i \geqslant 0),$

   $\left(int\left(\prod\limits_{\alpha=j+1}^{r+1} x_\alpha\, w_\alpha\right)(P_{t_j(g)})\right)(r \geqslant j \geqslant l_1(g))$

   $\left(resp.\ P_{t_{r+1}} = P_{e_1(g)}\right).$

Remark 15.9

The section of $\mathcal{C}onf_G^{std}((g)_S, B)$ is configurations of type $(g^*)_S$ issued from B.

Proposition 15.10

The morphism

$$q_{[\tau_{\mathscr{C}},E]} : \Pi_G[\tau_{\mathscr{C}}, E] \to \Pi_G[g, E]$$

is an embedding.

Proof *By Proposition 15.8 it is known that*

$$\lambda'_{[\tau\mathscr{C},E]} : \Pi_G\,[\tau\mathscr{C},E] \to \mathcal{C}onf_G(g_S, P_{e_1(g)})$$

is an embedding. As by Definition 15.7 one obtains

$$\lambda'_{[\tau\mathscr{C},E]} = \lambda_{[g,E]} \circ q_{[\tau\mathscr{C},E]},$$

it is then deduced that

$$q_{[\tau\mathscr{C},E]} = \left(\lambda_{[g,E]}\right)^{-1} \circ \lambda'_{[\tau\mathscr{C},E]}$$

defines also an embedding.

Definition 15.11
Denote by

$$\Pi'_G\,[\tau\mathscr{C},E] \subset \Pi_G\,[g,E]$$

the S-subscheme defined by $q_{[\tau\mathscr{C},E]}$.

15.3 The canonical relatively schematically dense open subscheme of a Contracted Product

It is shown that with the cell $\mathcal{C}onf_G^{std}\,((\tau_{\mathscr{C}}^{tr})_S, B)$ of $\mathcal{C}onf_G(g_S, P_{e_1(g)})$ corresponds the cell $\Pi_G\,[\tau_{\mathscr{C}}^{tr}, E]$, which is a relatively schematically dense open subscheme, of the corresponding isomorphic contracted product $\Pi_G\,[g,E] \simeq \mathcal{C}onf_G(g_S, P_{e_1(g)})$. Let

$$\tau_{\mathscr{C}} = \tau_{\mathscr{C}}^{tr} \in \mathrm{Relpos}_A^{gall}(g)$$

$$\left(\text{resp. } \overline{w}_{\tau_{\mathscr{C}}^{tr}} = (\overline{w}_i), \text{ i.e. } (\overline{w}_i) \text{ is the word corresponding to } \tau_{\mathscr{C}}^{tr}\right)$$

(cf. Remark 10.26). Write:

1. $w_{\tau_{\mathscr{C}}^{tr}}^m = (w_i^m)$,
   where w_i^m denotes the minimal length $l_{S(C)}(w_i^m)$ representative of $\overline{w}_i$;

2. $t_j = t_j(g)$ (resp. $s_i = s_i(g)$);

3. $P_{t_j} = P_{t_j}(E)$ (resp. $Q_{s_i} = Q_{s_i}(E)$);

4. $W_{t_j} = \mathrm{Stab}\,F_{t_j}(C)$ (resp. $W_{s_i} = \mathrm{Stab}\,F_{s_i}(C)$), where C denotes the chamber of $\mathcal{A}_E$ given by E.

Let T (resp. B) be the maximal torus (resp. the Borel subgroup) defined by E, and $P = P_{t_i}$ (resp. $Q = Q_{s_{i-1}}$). Write P as the semi-direct product

$$P = L(P) \cdot U(P),$$

where $U(P) \subset P$ (resp. $L(P) \subset P$) denotes the unipotent radical of P (resp. the Levi subgroup of P defined by the maximal torus T) (cf. [23], *Exp. XXVI, Proposition* 1.6).

One has the Bruhat decomposition of P in (B, Q)-double classes

$$P = \coprod B w Q,$$

where w runs over some set of representatives in $N(T)$ of the set

$$W_{t_i}/W_{s_{i-1}} \simeq \operatorname{Relpos}_{s_{i-1}}^{t_i}.$$

One may write

$$\coprod_{w} B w Q = \coprod_{w} [B \cap L(P) \, w \, Q \cap L(P)] \, U(P).$$

Let w^m be the minimal length $l_{S(C)}(w^m)$ representative of $\overline{w} \in W_{t_{i-1}}/W_{s_i}$. Observe that

$$\mathcal{U}(w^m) \subset B \cap L(P),$$

thus one obtains

$$\coprod_{w} B w Q = \coprod \mathcal{U}(w^m) \, w^m \, Q,$$

where as usual w^m also denotes a representative of w^m in $N(T)$.

Let

$$\Omega_{L(P)} = \mathcal{U}(w_i^m) \, w_i^m \, Q \cap L(P)$$

be the big cell of $L(P)$ relatively to $(B, Q \cap L(P))$.

It is known that

$$\Omega_{L(P)} \subset L(P)$$

is a relatively schematically dense open S-subscheme.

Thus one obtains

Proposition 15.12
The image $\tilde{\Omega}_{L(P)} \subset P/Q$ of $\Omega_{L(P)}$, by the quotient morphism $P \to P/Q$, is a relatively schematically dense open S-subscheme.

One proceeds to prove the following assertion by induction on α, with the notation of Definition 15.11. One has:

$$\text{``}(\forall \, l(g) > \alpha \geqslant 1) \ \Pi_G' \, [\tau_{\mathscr{C}}^{(\alpha+1)}, E] \subset \Pi_G \, [g^{(\alpha)}, E],$$

is a **relatively schematically dense open** S-subscheme''. Recall that the image of the canonical morphism

$$\mathcal{U}(w_{\alpha+1}^m) \, w_{\alpha+1}^m \to P_{t_{\alpha+1}}/Q_{s_\alpha},$$

is a relatively schematically dense S-subscheme in $P_{t_{\alpha+1}}/Q_{s_\alpha}$. Thus

$$\Pi'_G\left[\tau_{\mathscr{C}}^{(r+1)}, E\right] \subset \Pi_G\left[g^{(r)}, E\right] = P_{t_{r+1}}/Q_{s_r}$$

$$\left(\text{resp. } \Pi'_G\left[\tau_{\mathscr{C}}^{(r)}, E\right] \subset \Pi_G\left[g^{(r-1)}, E\right] = P_{t_{r+1}}/Q_{s_{r-1}}\right),$$

if $g = g_1, g'_2$ (resp. $g = g_2, g'_2$), is a relatively schematically dense open S-subscheme. As inductive hypothesis for $r - 1 > \alpha$ (resp. $r - 2 > \alpha$) assume that:

$$\text{``}\Pi'_G\left[\tau^{(\alpha+2)}, E\right] \subset \Pi_G\left[g^{(\alpha+1)}, E\right]$$

is a relatively schematically dense (r. sch. d.) open S-subscheme". There is a commutative diagram:

$$
\begin{array}{ccc}
q_{[\tau_{\mathscr{C}}^{(\alpha+1)}, E]}: & \Pi'_G\left[\tau_{\mathscr{C}}^{(\alpha+1)}, E\right] \longrightarrow & \Pi_G\left[g^{(\alpha)}, E\right] \\
& \downarrow{\scriptstyle p_{\tau_{\mathscr{C}}}^{[\alpha,\alpha+1]}} & \downarrow{\scriptstyle p^{[\alpha,\alpha+1]}} \\
q_{[\tau_{\mathscr{C}}^{(\alpha+2)}, E]}: & \Pi'_G\left[\tau_{\mathscr{C}}^{(\alpha+2)}, E\right] \longrightarrow & \Pi_G\left[g^{(\alpha+1)}, E\right],
\end{array}
$$

where $p_{\tau_{\mathscr{C}}}^{[\alpha,\alpha+1]}$ is the morphism induced by $p^{[\alpha,\alpha+1]}$. Remark that $p^{[\alpha,\alpha+1]}$ defines a locally trivial fibration and $p_{\tau_{\mathscr{C}}}^{[\alpha,\alpha+1]}$ a sub-bundle given by the restriction of $\Pi_G\left[g^{(\alpha)}, E\right]$ to $\Pi'_G\left[\tau_{\mathscr{C}}^{(\alpha+2)}, E\right]$. The proof of the recursive step results from the following easy lemma.

Lemma 15.13
Let $X \to Y$ be a locally trivial fibration of S-schemes with typical fiber F, i.e. a "bundle", $Y' \subset Y$ a relatively schematically dense sub-scheme of Y, and $X' \to Y'$ a "sub-bundle" of $X_{Y'} \to Y'$, with typical fiber a relatively schematically dense open sub-scheme $F' \subset F$. Then $X' \subset X$ is a relatively schematically dense sub-scheme.

Thus to prove the assertion about the embedding

$$\Pi'_G\left[\tau_{\mathscr{C}}^{(\alpha+1)}, E\right] \subset \Pi_G\left[g_{\mathscr{C}}^{(\alpha)}, E\right]$$

it suffices to see that the image of the natural morphism

$$\Pi'_G\left[\tau_{\mathscr{C}}^{(\alpha+1)}, E\right] \to \Pi'_G\left[\tau_{\mathscr{C}}^{(\alpha+2)}, E\right] \times_{\Pi_G\left[g^{(\alpha+1)}, E\right]} \Pi_G\left[g^{(\alpha)}, E\right]$$

is a relatively schematically dense open S-subscheme. Let

$$G \wedge_{Q_{s_{\alpha+1}}} \left(P_{t_{\alpha+1}}/Q_{s_\alpha}\right) \to G/Q_{s_{\alpha+1}}$$

be the contracted product defined by the right $Q_{s_{\alpha+1}}$ principal space

$$G \to G/Q_{s_{\alpha+1}},$$

and the a left action of $Q_{s_{\alpha+1}}$ on $P_{t_{\alpha+1}}/Q_{s_\alpha}$. This contracted product is isomorphic to the pull-back of the locally trivial fibration $\mathcal{P}ar_{s_\alpha}(G) \to \mathcal{P}ar_{t_{\alpha+1}}(G)$ by the morphism $\mathcal{P}ar_{s_{\alpha+1}}(G) \to \mathcal{P}ar_{t_{\alpha+1}}(G)$ associating with a parabolic Q of type $s_{\alpha+1}$ the parabolic $Q \subset P$ of type $t_{\alpha+1}$. It follows that $G \wedge_{Q_{s_{\alpha+1}}} (P_{t_{\alpha+1}}/Q_{s_\alpha}) \to G/Q_{s_{\alpha+1}}$ is a locally trivial fibration with typical fiber $P_{t_{\alpha+1}}/Q_{s_\alpha}$. On the other hand the image of the schematically dominant morphism $\mathcal{U}(w_{\alpha+1}^m) \, w_{\alpha+1}^m \hookrightarrow P_{t_{\alpha+1}}/Q_{s_\alpha}$ is contained in a big cell $Q_{s_{\alpha+1}}$-cell in $P_{t_{\alpha+1}}$. This big cell is invariant under $Q_{s_{\alpha+1}}$. Observe that $\Pi_G \, [g^{(\alpha+1)}, E]$ may thus be written

$$\Pi_G \, [g^{(\alpha)}, E] = \Pi_G \, [g^{(\alpha+1)}, E] \times_{G/Q_{s_{\alpha+1}}} (G \wedge_{Q_{s_{\alpha+1}}} (P_{t_{\alpha+1}}/Q_{s_\alpha})),$$

where the fiber product is defined by the morphism

$$m^{(\alpha+1)} : \Pi_G \, [g^{(\alpha+1)}, E] \to G/Q_{s_{\alpha+1}},$$

induced by the multiplication in G, and $p^{[\alpha,\alpha+1]}$ corresponds to the first projection $\Pi_G \, [g^{(\alpha+1)}, E]$.

Thus an application of the lemma proves the following

Proposition 15.14
One has that

$$\Pi'_G \, [\tau_{\mathscr{C}}^{(\alpha+1)}, E] \to \Pi_G \, [g^{(\alpha)}, E]$$

is a relatively schematically dense open S-subscheme.

From the Remark 15.6 follows the

Corollary 15.15
The S-subscheme $\mathcal{C}onf_G^{std}((\tau_{\mathscr{C}}^{tr})_S, B) \subset \mathcal{C}onf_G(g, P_{e_1(g)})$ *is open and relatively schematically dense.*

15.4   The Contracted Product and Galleries Configurations retraction morphism

One may resume the results of Chapter 14 as follows. There is a functorial cellular decomposition of $\mathcal{C}onf_G(\underline{\Delta}(G, E))$ given by the morphism $\mathcal{C}onf_G^{std}(\underline{\Delta}(G, E)) \longrightarrow \mathcal{C}onf_G(\underline{\Delta}(G, E))$ induced by $\coprod_{g \in \text{gall}_A} j_{[g,E]}$.
The cells of this decomposition are indexed in terms of the morphism $t_2^{std} : \mathcal{C}onf_G^{std} \longrightarrow \text{Relpos}_G^{gall}$, or more precisely to its restriction to $\mathcal{C}onf_G^{std}(\underline{\Delta}(G, E)) \subset \mathcal{C}onf_G^{std}$. This morphism may be turned into a retraction $\rho_E^{conf} = \coprod_{g \in \text{gall}_A} \rho_{E,g}^{conf} : \mathcal{C}onf_G^{std}(\underline{\Delta}(G, E)) \longrightarrow \mathcal{C}onf_{\mathcal{F}\text{ix}(T)}(\underline{\Delta}(G, E))$ (cf.
14.30). Where $\rho_{E,g}^{conf} = (\tau_{g,C}^A)_S^{-1} \circ t_2^{std}$, $\tau_{g,C}^A : Gall_A(g, F_{e_1(g)}) \to Relpos_A^{gall}(g)$, and $\mathcal{C}onf_{\mathcal{F}\text{ix}(T)}(\underline{\Delta}(G, E)) = \coprod_{g \in \text{gall}_A} Gall_A(g, F_{e_1(g)})_S$.

One has the following description of ρ_E^{conf} (cf. 14.30) in terms of the isomorphisms $\lambda_{[g,E]}$'s.

Proposition 15.16

There is a commutative diagram

$$
\coprod_{\substack{g \in gall_A}} \left(\coprod_{\substack{\tau_{\mathscr{C}} \in Relpos^{gall}(g)}} \Pi_G \left[\tau_{\mathscr{C}}, E \right] \right) \longrightarrow Gall_A(\Delta(C))_S
$$

$$
\lambda_{[E]}^{std} \Big\downarrow \qquad\qquad\qquad\qquad \Big\downarrow
$$

$$
\rho_E^{conf} : \mathcal{C}onf_G^{std}(\Delta(G,E)) \longrightarrow \mathcal{C}onf_{\mathcal{F}ix(T)}(\Delta(G,E)),
$$

where the upper horizontal arrow is given by the morphism:

$$
\coprod_{\substack{g \in gall_A}} \left(\coprod_{\substack{\tau_{\mathscr{C}} \in Relpos^{gall}(g)}} (\tau_{g,C}^A)_S^{-1} \circ I_{\tau_{\mathscr{C}}} \right),
$$

where $I_{\tau_{\mathscr{C}}} : \Pi_G[\tau_{\mathscr{C}}, E] \to Relpos^{gall}(g_S)$ is the constant morphism defined by the section $(\tau_{\mathscr{C}})_S$, and one writes

$$
\lambda_{[E]}^{std} = \coprod_{\substack{g \in gall_A}} \left(\coprod_{\substack{\tau_{\mathscr{C}} \in Relpos^{gall}(g)}} \lambda_{[g,E]} \circ q_{[\tau_{\mathscr{C}}, E]} \right).
$$

The right vertical arrow is induced by the canonical isomorphism $A_S \simeq \mathcal{F}ix(T)$.

Remark 15.17

The mapping

$$
(\tau_{g,C}^A)^{-1} : Relpos^{gall}(g) \to Gall_A(g, F_{e_1(g)}(C))
$$

is obtained as follows. Write $w_{\tau_{\mathscr{C}}}^m = (w_i^m)$. Then

$$
(\tau_{g,C}^A)^{-1} : \tau_{\mathscr{C}} \mapsto \gamma_g((w_i^m)).
$$

Assume that $g \in gall_A^m$. Let $\sigma_{\mathscr{C}}$ be a section of

$$
\mathrm{Conf}_G^{std}((\tau_{\mathscr{C}}^{tr})_S, P_{e_1(g)}) \subset \mathrm{Conf}_G^{std}(g_S, P_{e_1(g)}) ,
$$

i.e. a section of the big cell of $\mathcal{C}onf_G^{std}(g_S, P_{e_1(g)})$. By section §14.3 there exists a section $x = (x_i)$ of $\prod \mathcal{U}(w_i)$, so that, with the notation of Proposition 14.12,

$$
\lambda_{[\tau_{\mathscr{C}}^{tr}, C]}(x) = \sigma_{\mathscr{C}}(x, C) = \sigma_{\mathscr{C}}.
$$

From section §14.3 it follows, that if one writes

$$\sigma_{\mathscr{C}} : (Q_i)(l(g) > i \geqslant 0), (P_j)(l_2(g) \geqslant j \geqslant l_1(g)),$$

then the couple of parabolics

$$(B, Q_i) \text{ (resp. } (B, P_j))$$

is in standard position, and thus defines a section of $\mathcal{S}\text{tand}(B) \subset \mathcal{S}\text{tand}(G)$. One may then define a section $\rho_E(\sigma_{\mathscr{C}})$ of $\mathcal{C}\text{onf}_{\mathcal{F}\text{ix}(T)}$ by:

$$\rho_E(\sigma_{\mathscr{C}}) : (\rho_E(Q_i)), (\rho_E(P_j)).$$

It results easily from definitions the following

Proposition 15.18
With the above notation one has

$$\rho_{E,g}^{conf}(\sigma_{\mathscr{C}}) = \rho_E(\sigma_{\mathscr{C}}).$$

Remark 15.19
Given $\tau_{\mathscr{C}} \in Relpos\,{}^{gall}(g)$, let one write

$$\gamma_{\tau_{\mathscr{C}}} = \left(\tau_{g,C}^A\right)^{-1}(\tau_{\mathscr{C}}) \in Gall_A(g, F_{e_1(g)}(C)).$$

From the definition of $\rho_{E,g}^{conf}$, it results:

$$\mathcal{C}onf_G^{std}((\tau_{\mathscr{C}})_S, P_{e_1(g)}) = \left(\rho_{E,g}^{conf}\right)^{-1}((\gamma_{\tau_{\mathscr{C}}})_S).$$

15.5   The image of a Gallery Configuration by the Building Retraction morphism on an apartment

Let it be assumed that S is the spectre of an algebraically closed fields. Given the closure of a Bruhat cell and a corresponding smooth resolution of singularities by a galleries configurations scheme one gives an algorithm calculating the image by the extremity morphism $\mathcal{E}_2$ of a section of this scheme, and simultaneously the retraction of this section on an apartment. This latter calculation amounts deciding to which Bruhat cell in $\mathcal{P}ar(G)$ this image belongs, and the former allows determining the fibers of the resolution morphism $\mathcal{E}_2$. More precisely, the algorithm allows the explicit calculation of the coordinates of the image of a section by $\mathcal{E}_2$ in its Schubert cell, and that of the corresponding gallery in an apartment. Recall that one denotes by $X_{(\bar{s})}$ the set of $\overline{\kappa(s)}$-points of the S-scheme X, where $\bar{s} = \text{Spec}(\overline{\kappa(s)})$ and $\overline{\kappa(s)}$ is the algebraic closure of the residual field of $s \in X$.

One writes

$$\rho_{\bar{s}}^{std} = \left(\rho_{E,g}^{std}\right)_{\bar{s}},$$

and

$$\rho^{std}_{(\overline{s})} = \left(\rho^{std}_{E,g}\right)_{(\overline{s})} : \mathcal{C}\mathrm{onf}^{std}_G(g_S, P_{e_1(g)})_{(\overline{s})} \to \mathcal{F}\mathrm{ix}(T)_{(\overline{s})}$$

for the induced mapping.

Let

$$\sigma_\mathscr{C} \in \mathcal{C}\mathrm{onf}^{std}_G(g_S, P_{e_1(g)})_{(\overline{s})} = \mathcal{C}\mathrm{onf}^m_G(g_S, P_{e_1(g)})_{(\overline{s})}.$$

The above identification holds from the fact that any couple of parabolics (P, Q) over $\mathrm{Spec}(\overline{\kappa(s)})$ is in standard position (cf. [10], Exp. XXVI, Lemme 4.1.1).

The next aim is the calculation of $\mathcal{E}_2(\sigma_\mathscr{C})$ and "the retraction of $\sigma_\mathscr{C}$" on $\mathcal{F}\mathrm{ix}(T)_{(\overline{s})}$, i.e. the retraction term by term of the gallery $\sigma_\mathscr{C}$: $\rho_E(\sigma_\mathscr{C}) = (\rho_E)_{(\overline{s})}(\sigma_\mathscr{C})$. Recall that $\mathcal{F}\mathrm{ix}(T)_{(\overline{s})}$ is the same thing as an apartment of $G_{(\overline{s})}$ and that the retraction preserves the incidence relation and thus carries a gallery onto a gallery. Remark that in general

$$\rho^{std}_{\overline{s}}(\sigma_\mathscr{C}) \neq (\rho_E)_{(\overline{s})}(\sigma_\mathscr{C}).$$

Let it be written

$$Q' = \mathcal{E}_2(\sigma_\mathscr{C}) = (\mathcal{E}_2)_{(\overline{s})}(\sigma_\mathscr{C}).$$

It is clear that the determination of $\rho_E(\sigma_\mathscr{C})$ may be obtained if one knows how to calculate $\rho_E(Q') = \rho_E(\mathcal{E}_2(\sigma_\mathscr{C}))$, as $\rho_E(\sigma_\mathscr{C})$ is determined by the set of facets

$$\rho_E(\mathcal{E}_2(\sigma_\mathscr{C}^{(\alpha)})),$$

where $\left(\sigma_\mathscr{C}^{(\alpha)}\right)$ denotes the set of α-truncations of $\sigma_\mathscr{C}$. Thus the determination of $(\rho_E)_{(\overline{s})}(\sigma_\mathscr{C})$ follows from that of $(\rho_E)_{(\overline{s})}(\mathcal{E}_2(\sigma_\mathscr{C}))$.

The cellular decomposition (cf. Proposition 14.26)

$$\mathcal{C}\mathrm{onf}^m_G(g_S, P_{e_1(g)})_{(\overline{s})} = \coprod_{\tau_\mathscr{C} \in \mathrm{Relpos}^{\,\mathrm{gall}}(g)} \mathcal{C}\mathrm{onf}^{std}_G((\tau_\mathscr{C})_S, B)_{(\overline{s})}$$

implies that there exists $\tau_\mathscr{C} \in \mathrm{Relpos}^{\,\mathrm{gall}}(g)$ so that

$$\sigma_\mathscr{C} \in \mathcal{C}\mathrm{onf}^{std}_G((\tau_\mathscr{C})_S, B)_{(\overline{s})}.$$

Given $\tau_\mathscr{C} \in \mathrm{Relpos}^{\,\mathrm{gall}}(g)$, denote by $\overline{w}_{\tau_\mathscr{C}} = (\overline{w}_i)$ the uple of classes of W determined by $\tau_\mathscr{C} = (\tau_i)$. Let $w^m_{\tau_\mathscr{C}} = (w^m_i)$, be the uple of minimal length representatives defined by $\overline{w}_{\tau_\mathscr{C}}$. The above cellular decomposition corresponds by the isomorphism

$$\lambda_{[g,E]} : \prod_G [g, E] \xrightarrow{\sim} \mathcal{C}\mathrm{onf}^{std}_G(g_S, P_{e_1(g)}),$$

of $\mathcal{C}\mathrm{onf}^{std}_G(g_S, B)$ with the contracted product $\prod_G [g, E]$, with the disjoint union

$$\prod_G [g, E]_{(\overline{s})} = \coprod_{\tau_\mathscr{C} \in \mathrm{Relpos}^{\,\mathrm{gall}}(g)} \prod_G' [\tau_\mathscr{C}, E]_{(\overline{s})}$$

(cf. Proposition 14.26, and Proposition 15.8). Where $\prod_G'[\tau_{\mathscr{C}}, E]$ denotes the image of

$$\prod_G [\tau_{\mathscr{C}}, E] = \prod \mathcal{U}(w_i^m)\, w_i^m$$

$(l(g) \geqslant i \geqslant 1)$ (resp. $l(g) > i > 1$) by the embedding $q_{[\tau_{\mathscr{C}}, E]} : \prod_G [\tau_{\mathscr{C}}, E] \longrightarrow P_{t(g)}/Q_{s(g)} = \prod_G [g, E]$. By the above isomorphism, to the surjective monomorphism (cf. Proposition 15.8)

$$j_{[g,E]} = \coprod_{\tau_{\mathscr{C}} \in \mathrm{Relpos}^{\,\mathrm{gall}}(g)} j_{[\tau_{\mathscr{C}}, E]},$$

corresponds

$$\coprod_{\tau_{\mathscr{C}} \in \mathrm{Relpos}^{\,\mathrm{gall}}(g)} q_{[\tau_{\mathscr{C}}, E]} : \coprod_{\tau_{\mathscr{C}} \in \mathrm{Relpos}^{\,\mathrm{gall}}(g)} \prod_G [\tau_{\mathscr{C}}, E] \longrightarrow \prod_G [g, E] .$$

There are commutative diagrams:

1)

$$
\begin{array}{ccc}
\coprod \prod_G [\tau_{\mathscr{C}}, E]_{(\overline{s})} & \xrightarrow{\;q_{[g,E](\overline{s})}\;} & \prod_G [g, E]_{(\overline{s})} \\
\Big\downarrow{\scriptstyle \amalg \lambda_{[\tau_{\mathscr{C}}, E]_{(\overline{s})}}} & & \Big\downarrow{\scriptstyle \lambda_{[g,E]_{(\overline{s})}}} \\
\coprod \mathcal{C}\mathrm{onf}_G^{\,std}((\tau_{\mathscr{C}})_S, B)_{(\overline{s})} & \xrightarrow{\;j_{[g,E](\overline{s})}\;} & \mathcal{C}\mathrm{onf}_G(g_S, P_{e_1(g)})_{(\overline{s})} .
\end{array}
$$

Where the $\tau_{\mathscr{C}}$ indexing the terms of the disjoint unions run on $\mathrm{Relpos}^{\,\mathrm{gall}}(g)$, and

$$q_{[g,E]} = \coprod_{\tau_{\mathscr{C}} \in \mathrm{Relpos}^{\,\mathrm{gall}}(g)} q_{[\tau_{\mathscr{C}}, E]} \quad (\text{resp. } j_{[g,E]} = \coprod_{\tau_{\mathscr{C}} \in \mathrm{Relpos}^{\,\mathrm{gall}}(g)} j_{[\tau_{\mathscr{C}}, E]}).$$

2)

$$
\begin{array}{ccc}
\lambda'_{[\tau_{\mathscr{C}}, E]} : \prod_G [\tau_{\mathscr{C}}, E] & \longrightarrow & \mathcal{C}\mathrm{onf}_G(g_S, P_{e_1(g)}) \\
\Big\downarrow & & \Big\downarrow{\scriptstyle \varepsilon_2} \\
G/Q & \xrightarrow{\;\sim\;} & \mathcal{P}ar_{e_2(g)}(G) ,
\end{array}
$$

where the vertical arrow $\prod_G [\tau_{\mathscr{C}}, E] = \prod \mathcal{U}(w_i^m)\, w_i^m \longrightarrow G/Q$ is defined by

$$(x_i\, w_i^m)\,(l(g) \geqslant i \geqslant 1) \longrightarrow \left(\prod x_i\, w_i^m\right) \cdot Q .$$

The horizontal arrow is defined as in 15.7. Thus one has $\mathrm{int}\,(\prod x_i\, w_i^m)\,(Q) = \mathcal{E}_2(\sigma_{\mathscr{C}})$.

Let the notations be simplified by writing, for a fixed point $s \in S$:

$$P = P_{e_1(g)}(E)_{(\overline{s})}$$

$$\left(\text{resp. } Q = Q_{e_2(g)}(E)_{(\overline{s})}, \ G = G_{(\overline{s})}, \ B = B_{(\overline{s})}, \ T = T_{(\overline{s})}, \ N(T) = N(T)_{(\overline{s})}\right).$$

Denote by $S_E \subset W = N(T)/T$, the set of reflexions defined by the system of simple roots R_0 given by E. Thus S_E is a system of generators of W.

We know that (G, B, N, S_E) is a Tits system (cf. [4]), i.e. the following properties hold:

(T1) The set $B \cup N$ generates G, and $B \cap N$ is an invariant subgroup of N.

(T2) The set S_E generates $W = N/B \cap N$, and the elements w of S_E verify $w^2 = 1$.

(T3) $r\,B\,w \subset B\,w\,C \cup B\,r\,w\,B$ for every $r \in S_E$, and $w \in W$.

(T4) For all $r \in S_E$, $\operatorname{int}(r)(B) \neq B$. (cf. loc. cit.)

The building associated to this Tits system is in fact the building of the geometric fiber $G_{(\overline{s})}$ of G.

Let $G = \coprod\limits_{\overline{w} \in W/W_F} B\,\overline{w}\,Q$, with $P_F = Q$, i.e. $F = F_{e_2(g)}(C)$, be the (B, Q)-double class decomposition of G. Given $y \in B\,\overline{w}\,Q$, denote by $\overline{y} \in G/Q$ its class modulo Q, and by $w^m \in \overline{w}$ the minimal length element of the class $\overline{w} \in W/W_F$.

Definition 15.20

It is said that $x\,w^m \in B\,w^m$ is the canonical representative of $\overline{y} \in G/Q$ if w^m is the minimal length element in its class $\overline{w}^m \in W/W_F$,

$$x \in \mathcal{U}(w^m), \text{ and } x\,w^m \in \overline{y}.$$

It follows from the general properties of the Bruhat cell decomposition of G in (B, Q)-double classes that the canonical representative $x\,w^m$ of $\overline{y}$ is uniquely determined and thus well defined.

Given a parabolic $Q' \subset G$ of type $e_2(g)$ we determine $\rho_E(Q')$ as follows. Let $y \in G$ such that $Q' = \operatorname{int}(y)(Q)$, and $x\,w^m$ the canonical representative of $\overline{y} \in G/Q$. One has then

$$Q' = \operatorname{int}(x\,w^m)(Q) = \operatorname{int}(x)\operatorname{int}(w^m)(Q).$$

As $\mathcal{U}(w^m) \subset B$, it is deduced that the couple $(B, \operatorname{int}(w^m)(Q))$, defines the same type of relative position as (B, Q'), and both B and $\operatorname{int}(w^m)(Q)$ contain the maximal torus T. Thus one has that

$$\rho_E(Q') = \operatorname{int}(w^m)(Q).$$

For the sake of briefness, let it be supposed $g = g_1$, so that $P_{e_1(g)} = P_{t_{l(g)}(g)}$. The other cases $g = g_1', g_2, g_2'$ are treated similarly.

From the commutativity of diagram 2) above and by definition of

$$\lambda_{[g,E]} : \Pi_G\,[g,E] = P_{t(g)}/Q_{s(g)} \longrightarrow \mathit{Conf}_G(g_S, P_{e_1(g)}),$$

as $\lambda_{[g,E]} : x \mapsto \sigma_g(E,x)$, where $\sigma_g(E,x)$ denotes the deformation of the basical configuration $\sigma_g(E)$ (cf. Definition 15.3), and the relation

$$\lambda'_{[\tau_\mathscr{C},E]} = \lambda_{[g,E]} \circ q_{[\tau_\mathscr{C},E]}\,,$$

(cf. Definition 15.7, and Proposition 15.10) one obtains:

Proposition 15.21 *Let*

$$\sigma_\mathscr{C}^\natural = \left(\lambda_{[g,E]}\right)_{(\overline{s})}^{-1}(\sigma_\mathscr{C}) \in \Pi'_G\,[\tau_\mathscr{C}, E]_{(\overline{s})}.$$

One has $\sigma_\mathscr{C}^\natural = (\lambda_{[\tau_\mathscr{C},E]})_{(\overline{s})}^{-1}(\sigma_\mathscr{C}) \in \Pi_G\,[\tau_\mathscr{C}, E]_{(\overline{s})}$, *and*

$$\sigma_\mathscr{C}^\natural = (x_i\,w_i^m)\,(l(g) \geqslant i \geqslant 1)$$

with $x_i \in \mathcal{U}(w_i^m)_{(\overline{s})}$, *i.e.* $\sigma_g(E,x) = \sigma_\mathscr{C}$ *with* $x = (x_i)$, *and* $int(\prod x_i\,w_i^m)\,(Q) = \mathcal{E}_2(\sigma_\mathscr{C})$. *Thus the class* $(\prod x_i\,w_i^m) \cdot Q$ *corresponds to* $\mathcal{E}_2(\sigma_\mathscr{C})$.

Let it be explained how to calculate the canonical representative of the class of $\prod x_i\,w_i^m$ in G/Q according to Definition 15.20. Clearly this calculation amounts to that of $\mathcal{E}_2(\sigma_\mathscr{C})$ and the image of the retraction $\rho_E(\mathcal{E}_2(\sigma_\mathscr{C}))$. Write each element w_i^m as a reduced word in S_E:

$$w_i^m = r_1^{(i)}\,r_2^{(i)} \cdots r_{l^{(i)}}^{(i)} \quad (l(g) \geqslant i \geqslant 1),$$

where $l^{(i)} = l_{S_E}(w_i^m)$.

Let it first be determined $w \in W$, so that w is the minimal length element of the class $\overline{w} \in W/W_F$ it determines, and $y \in \mathcal{U}(w)$ so that

$$y\,w\,Q = w_2^m x_1 w_1^m Q = r_1^{(2)}\,r_2^{(2)} \cdots r_{l^{(2)}}^{(2)}\,x_1 w_1^m\,Q.$$

Proceeding by induction on $1 \geqslant k \geqslant l^{(2)}$, let one determine $y^{(l^{(2)})} \in B$, and $w^{(l^{(2)})}$, so that

$$r_{l^{(2)}}^{(2)}\,x_1\,w_1^m\,Q = y^{(l^{(2)})}\,w^{(l^{(2)})}\,Q.$$

By (T3) one has:

$$r_{l^{(2)}}^{(2)}\,x_1\,w_1^m \in B\,w_1^m\,B \cup B\,r_{l^{(2)}}^{(2)}\,w_1^m\,B,$$

and one deduces that there exists some

$$y^{(l^{(2)})} \in B,$$

so that

$$r^{(2)}_{l^{(2)}} x_1 w_1^m Q = y^{(l^{(2)})} w^{(l^{(2)})} Q,$$

with

$$w^{(l^{(2)})} = w_1^m \text{ , or } r^{(2)}_{l^{(2)}} w_1^m.$$

Let t be the type of Q, i.e. $t = e_2(g)$. One chooses $w^{(l^{(2)})}$ as **the minimal length element** w^m in its class in W/W_t. In other terms $w^m = w(C, proj_{F_t(C)})$. It is easy to see that $y^{(l^{(2)})}$ may be chosen in $\mathcal{U}(w^{(l^{(2)})})$. Let $\mathcal{U}'' = B \cap int(w^{(l^{(2)})})(B)$. This subgroup is generated by the root subgroups indexed by the closed set of roots $R_+ \cap w^{(l^{(2)})}(R_+)$ and T.

Since $w^{(l^{(2)})}(R_+ \cap w^{(l^{(2)})}(R_+)) \subset R_+$, where R_+ is the positive root system defined by E, one obtains $\mathcal{U}'' w^{(l^{(2)})} Q = w^{(l^{(2)})} Q$. [1] The subgroup $\mathcal{U}' = \mathcal{U}(w^{(l^{(2)})}) \subset B$ is generated by the root subgroups indexed by the closed set of roots $R_+ - R_+ \cap w^{(l^{(2)})}(R_+) = \{\alpha \in R_+ | w^{(l^{(2)})}(\alpha) \in -R_+\}$ thus one has $B = \mathcal{U}' \cdot \mathcal{U}''$. Finally one obtains $B w^{(l^{(2)})} Q = \mathcal{U}' \cdot \mathcal{U}'' w^{(l^{(2)})} Q = \mathcal{U}' w^{(l^{(2)})} Q$. Thus $y^{(l^{(2)})} w^{(l^{(2)})} Q = y^{(l^{(2)})'} w^{(l^{(2)})} Q$ if $y^{(l^{(2)})'}$ denotes the $\mathcal{U}'$-component of $y^{(l^{(2)})}$. Consequently **one may suppose** $y^{(l^{(2)})} = y^{(l^{(2)})'}$.

Let $k < l^{(2)}$. One supposes that:

$$r^{(2)}_{k+1} \cdots r^{(2)}_{l^{(2)}} x_1 w_1^m Q = y^{(k+1)} w^{(k+1)} Q.$$

Proceeding as above it is deduced that

$$r^{(2)}_k y^{(k+1)} w^{(k+1)} Q = y^{(k)} w^{(k)} Q,$$

with $y^{(k)} \in B$, and

$$w^{(k)} = w^{(k+1)}, \text{ or } r^{(2)}_k w^{(k+1)}.$$

Thus by induction on $1 \leqslant k \leqslant l^{(2)}$, one finally obtains:

"There exists $z_2 \in B$, and $w_2 \in W$ so that $w_2^m x_1 w_1^m Q = z_2 w_2 Q$".

More precisely one may suppose w_2 of minimal length in its class and $z_2 \in \mathcal{U}(w_2)$. Let it be supposed for $2 < j \leqslant l(g)$ that:

$$w_{j-1}^m x_{j-2} w_{j-2}^m \cdots x_1 w_1^m Q = z_{j-1} w_{j-1} Q.$$

By reducing the expression

$$w_j^m x_{j-1} z_{j-1} w_{j-1},$$

---

[1] Let $w \in W$. The roots in $R_+ \cap w(R_+)$ are those roots $\alpha \in R_+$ whose associated hyperplane H_α **does not separate** the chambers C_{R_+} and $C_{w(R_+)}$. On the other hand, the set of roots $\alpha \in R_+$ such that H_α separates C_{R_+} and $C_{w(R_+)}$ is given by $\{\alpha \in R_+ | w(\alpha) < 0\}$. One concludes that

$$\alpha \in R_+ \cap w(R_+) \Longrightarrow 0 < w(\alpha).$$

as above one obtains

$$w_j^m \, x_{j-1} \, w_{j-2}^m \cdots x_1 \, w_1^m \, Q = z_j \, w_j \, Q,$$

with $z_j \in B$, and $w_j \in W$. (Remark that $x_{j-1} \, z_{j-1} \in B$.)

Thus by induction on j, it has been proved that there exists $w = w_{l(g)} \in W$ of minimal length in its class $\overline{w}$ and $z = z_{l(g)} \in \mathcal{U}(w_{l(g)})$, so that

$$\prod_{j=1}^{l(g)} w_j^m \, x_j \, Q = z \, w \, Q,$$

and $\mathcal{E}_2(\sigma_{\mathscr{C}}) = int(z \, w \,)(Q)$. One obtains the retraction of $\mathcal{E}_2(\sigma_{\mathscr{C}})$ on $\mathcal{F}ix(T)_{(\overline{s})}$ by:

$$\rho_E(\mathcal{E}_2(\sigma_{\mathscr{C}})) = int(w^m)(Q).$$

Remark 15.22

The preceding algorithm allows the explicit determination of the fibers of $(\rho_E)_{\overline{s}}$. In fact this calculation may be further developed if combined with the defining relations of G (cf. [23], Exp. XXIII, 3.5).

Chapter 16

Functoriality of Schubert Schemes Smooth Resolutions and Base Changes

The following two questions are considered in this chapter.

- Under which conditions on the base scheme S one has a natural identification

  i.e. when the fiber

  $$\overline{\Sigma}_{(\underline{\tau},P)} \simeq \left(\overline{\Sigma}\right)_{(\underline{\tau},P)} ,$$

  over $(\underline{\tau}, P)$ of the Universal Schubert scheme

  $$\overline{\Sigma} \longrightarrow Relpos_G \times_S \mathcal{P}\mathrm{ar}(G) \times_S \mathcal{P}\mathrm{ar}(G)$$

  gives the schematic closure of the fiber $\Sigma_{(\underline{\tau},P)}$ in $\mathcal{P}\mathrm{ar}(G)$? This question amounts to determine when the formation of the schematic closure of a Schubert cell commutes with base changes $S' \longrightarrow S$. A convenient answer to it is obtained by considering the Chevalley reductive $\mathbb{Z}$-group scheme $\underline{Ep_{\mathbb{Z}}}(\mathcal{R})$ associated with a given root data type $\mathcal{R}$ and observing that given a reductive S-group scheme G of type $\mathcal{R}$, there is an isomorphism $S \times \underline{Ep_{\mathbb{Z}}}(\mathcal{R}) \simeq G$ locally for the etale topology (cf. [23], Exp. XXII, Def. 5.11.). It is obtained that if S is an scheme with residues fields of characteristics distinct from a finite set of primes depending on $\mathcal{R}$ then the above identification holds for all sections $(\underline{\tau}, P)$ of $Relpos_G \times_S \mathcal{P}\mathrm{ar}(G)$. More precisely, for each root data of type $\mathcal{R}$ a not-empty open sub-scheme $U^{\mathcal{R}} \subset Spec(\mathbb{Z})$ exists so that:

"The formation of the schematic closure of the Universal Schubert Cell of the Chevalley scheme $\underline{Ep}_{\mathbb{Z}}(\mathcal{R})$ commutes with base extensions
$$S \to U^{\mathcal{R}} \text{ "}.$$

Moreover the open sub-scheme $U^{\mathcal{R}}$ is maximal with this property. In fact this set is characterized as the set of those primes so that their corresponding geometrical fiber of the Universal Schubert scheme is not integral. It follows that given a reductive group scheme G over a scheme $S \to U^{\mathcal{R}}$ [1], the formation of the schematic closure of the Universal Schubert Cell of G commutes with base changes. Thus the formation of the Smooth Resolution of the Universal Schubert Scheme of G and its cellular decomposition also commutes with base changes $S' \to S$.

- Under which conditions on S the fiber over $(\underline{g}, P)$, where $\delta_2(\underline{g}) = \underline{\tau}$, of the Universal Smooth Resolution $\pi : \widehat{\Sigma} \longrightarrow \overline{\Sigma}$ gives the corresponding resolutions $\pi_{(\underline{\tau},P)} : \widehat{\Sigma}_{(\underline{\tau},P)} \longrightarrow \overline{\Sigma}_{(\underline{\tau},P)}$ of the Schubert scheme $\overline{\Sigma}_{(\underline{\tau},P)}$. The answer to this question relies on the preceding one.

In §16.3 all the results about the schematic closure of a sub-scheme that we need in this work are collected.

16.1   The main theorem

It is recalled that **the Universal Schubert scheme** $\overline{\Sigma}$ **of** G is the schematic image of the embedding $j_{\Sigma} : \Sigma \to (\mathcal{P}\mathrm{ar}(G) \times_S \mathcal{P}\mathrm{ar}(G)) \times_S \mathrm{Relpos}_G$ (resp. the schematic closure of the **Universal Bruhat cell**

$$\Sigma = gr(t_2) \subset (\mathcal{P}\mathrm{ar}(G) \times_S \mathcal{P}\mathrm{ar}(G)) \times_S \mathrm{Relpos}_G).$$

Let $j_{\overline{\Sigma}} : \overline{\Sigma} \to \mathcal{P}\mathrm{ar}(G) \times_S \mathcal{P}\mathrm{ar}(G) \times_S \mathrm{Relpos}_G$ the natural embedding. Denote by $\mathrm{pr}_{\mathcal{R}} : \Sigma \longrightarrow \mathrm{Relpos}_G$ $\left(resp. \ \mathrm{pr}_{\mathcal{R},\overline{\Sigma}} : \overline{\Sigma} \longrightarrow \mathrm{Relpos}_G\right)$ the morphism induced by the third projection, and by $\delta_2 : \Gamma^m(\underline{\Delta}(G)) \longrightarrow \mathrm{Relpos}_G$ the morphism whose geometric fibers associate with a minimal gallery of types g the unique type τ_g of relative position so that the configurations variety defined by g is a resolution of singularities of the Schubert variety defined by τ_g. One has the following important remark: "A generalized gallery $\gamma(F, F')$ of type g, in an apartment, is minimal if and only if $Conf(g, P_F)$ is a smooth resolution of $\overline{\Sigma}(\tau(F, F'), P_F)$".

A more adapted notation is introduced aiming at obtaining a simple formulation of our main result. Let

$$\Sigma_{\Gamma^m} = \Sigma \times_{\mathrm{Relpos}_G} \Gamma^m(\underline{\Delta}(G))$$

---

[1]This condition amounts to: "The residual characteristics of S are not in the finite subset of primes complementary to $U^{\mathcal{R}}$".

$$\left(\text{resp. } \overline{\Sigma}_{\Gamma^m} = \Sigma \times_{\text{Relpos}\, G} \Gamma^m(\underline{\Delta}(G))\right),$$

where the fiber product is defined by the couple of morphisms

$$(\text{pr}_{\mathcal{R}}, \delta_2) \quad (\text{resp. } (\text{pr}_{\mathcal{R}, \overline{\Sigma}}, \delta_2)).$$

Thus Σ_{Γ^m} (resp. $\overline{\Sigma}_{\Gamma^m}$) is obtained from the Relpos G-scheme Σ (resp. $\overline{\Sigma}$) by the finite etale base change $\Gamma^m(\underline{\Delta}(G)) \longrightarrow$ Relpos G. One proves in this chapter that there is a canonical resolution of singularities $(\widehat{\Sigma}, \pi)$ of $\overline{\Sigma}$ of G after this finite etale extension. The Universal Schubert scheme has the following universal property. If the characteristics of the base scheme S are not among a fixed finite set of primes depending on the type of G, then the Schubert scheme $\overline{\Sigma}(\underline{\tau}, P)$ defined by a couple $(\underline{\tau}, P)$, of a type of relative position and a parabolic subgroup, is obtained as the fiber of $\overline{\Sigma}$ over $(\underline{\tau}, P)$. If S satisfies this condition, the resolution of singularities $\widehat{\Sigma}(\underline{g}, P) \longrightarrow \overline{\Sigma}(\underline{\tau}, P)$ given by a configurations scheme defined by a section $\underline{g}$ of $\overline{\Gamma}^m(\underline{\Delta}(G))$ is obtained as the fiber $\pi_{(\underline{\tau}, P)}$ of the morphism $\widehat{\Sigma} \xrightarrow{\pi} \overline{\Sigma}$.

One has

$$\overline{\Sigma}_{\Gamma^m} = \text{schematic closure of } \Sigma_{\Gamma^m} \text{ in } \mathcal{P}ar(G) \times_S \mathcal{P}ar(G) \times_S \Gamma^m,$$

as the formation of the schematic closure commutes with flat quasi-compact and separated base changes $S' \to S$ (cf. [27], Théorème (11.10.5)). Here $S' = \Gamma^m(\underline{\Delta}(G)) \xrightarrow{\delta_2} S = \text{Relpos}\, G$. Let G be an S-reductive group scheme, and $\widehat{\Sigma} = \mathcal{C}onf_G^m = (t_1^{\Gamma})^{-1}(\Gamma^m(\underline{\Delta}(G)))$.

Suppose that G be endowed with a frame E. Thus there are isomorphism $\Gamma^m(\underline{\Delta}(G)) \simeq (\text{gall}_{\mathcal{A}_E}^m)_S$, and

$$\overline{\Sigma}_{\Gamma^m} \simeq \coprod_{g \in \text{gall}_{\mathcal{A}_E}^m} \overline{\Sigma}_{(\tau_g)_S}$$

$$\left(\text{resp. } \widehat{\Sigma} \simeq \coprod_{g \in \text{gall}_{\mathcal{A}_E}^m} \widehat{\Sigma}_{g_S}\right),$$

where one writes, given a section $\underline{g}$ of $\Gamma^m(\underline{\Delta}(G))$ (resp. a section $\underline{\tau}$ of Relpos G):

$$\widehat{\Sigma}_{\underline{g}} = \left(t_1^{\Gamma}\right)^{-1}(\underline{g}) = \mathcal{C}onf_G^m(\underline{g}) \ \left(\text{resp. } \overline{\Sigma}_{(\tau_g)_S} = (t_2)^{-1}(\underline{\tau})\right)$$

(cf. §13.8).
Define $\pi = \overline{\mathcal{E}} : \widehat{\Sigma} \to \overline{\Sigma}_{\Gamma^m}$ by $\pi = \overline{\mathcal{E}} = \coprod_{g \in \text{gall}_{\mathcal{A}_E}^m} \overline{\mathcal{E}}_{g_S}$, where $\overline{\mathcal{E}}_{g_S} :$
$\mathcal{C}onf_G^m(g_S) \longrightarrow \overline{\Sigma}$ associates with a configuration its extremities.

Recall that the morphism Θ_{g_S} (cf. Definition 12.34) gives rise to a section of $\widehat{\Sigma}_{g_S} = \mathcal{C}onf_G^m(g_S)$ over $\Sigma_{(\tau_g)_S} = \mathcal{S}tand((\tau_g)_S)$. Let

$$\Theta^E : \Sigma_{\Gamma^m} \rightarrow \pi^{-1}(\Sigma_{\Gamma^m}) \subset \widehat{\Sigma},$$

be the morphism defined by

$$\Theta^E : \Sigma_{\Gamma^m} = \coprod_{g \in \text{gall } _{\mathcal{A}_E}^m} \mathcal{S}tand((\tau_g)_S) \longrightarrow \coprod_{g \in \text{gall } _{\mathcal{A}_E}^m} \widehat{\Sigma}_{g_S}$$

where $\Theta^E = \coprod_{g \in \text{gall } _{\mathcal{A}_E}^m} \Theta_{g_S}$ and $\Theta_{g_S} : \mathcal{S}tand((\tau_g)_S) \longrightarrow \widehat{\Sigma}_{g_S}$ factors through the embedding $\mathcal{S}tand((\tau_g)_S) \simeq \overline{\mathcal{E}}_{g_S}^{-1}(\mathcal{S}tand((\tau_g)_S)) \subset \widehat{\Sigma}_{g_S}$. It is known that $(\overline{\mathcal{E}}_{g_S})^{-1}(\Sigma_{(\tau_g)_S}) \simeq \overline{\mathcal{E}}_{g_S}^{-1}(\mathcal{S}tand((\tau_g)_S))$ is a relatively schematically dense open subscheme of $\widehat{\Sigma}_{g_S}$ (cf. Proposition 14.22). Thus

$$\text{Im } \Theta^E \subset \widehat{\Sigma}$$

is a relatively schematically dense open subscheme of $\widehat{\Sigma}$.

Let the section Θ^E be described more precisely. Suppose that a section σ of Σ_{Γ^m} over S is of the form

$$\sigma = ((P, Q), g_S),$$

where $g \in \text{gall } _{\mathcal{A}_E}^m$, (P, Q) a section of $\Sigma = \mathcal{S}tand(G)$ so that

$$\tau_S = t_2((P, Q)),$$

with $\tau \in \text{Relpos } \mathcal{A}_E$, and

$$\tau = \tau_g.$$

This last equality means

$$\tau_S = \delta_2(g_S).$$

Denote by $\sigma_{\mathscr{C}}((P, Q), g_S)$ the section of $\mathcal{C}onf_{\mathcal{F}\text{ix}(P,Q)}^m(g_S)$ characterized by

$$\mathcal{E}(\sigma_{\mathscr{C}}((P, Q), g_S)) = (P, Q)$$

$$(\text{resp. } t_1^{\Gamma}(\sigma_{\mathscr{C}}((P, Q), g_S)) = g_S).$$

By definition of $\mathcal{C}onf_{\mathcal{F}\text{ix}(P,Q)}^m(g_S)$ there is an inclusion

$$\mathcal{C}onf_{\mathcal{F}\text{ix}(P,Q)}^m(g_S) \subset \mathcal{C}onf_G^m(g_S),$$

thus $\sigma_{\mathscr{C}}((P, Q), g_S)$ may be seen as a section of $\mathcal{C}onf_G^m(g_S)$. Then

$$\Theta^E(\sigma) = \sigma_{\mathscr{C}}((P, Q), g_S).$$

It is clear that a unique morphism Θ may defined in general by etale descent locally giving the above morphism.

Remark 16.1

*Denote by $(\tilde{P}, \tilde{Q})_\Sigma$ the section of $\mathcal{S}tand(G)$ over Σ so that, given a section $\underline{\tau}$ of $Relpos\,_G$, the restriction to $\Sigma_{\underline{\tau}} = \mathcal{S}tand(\underline{\tau}) \subset \Sigma$ gives $(\tilde{P}, \tilde{Q})_{\underline{\tau}}$ (cf. Definition 12.37) (**The tautological section of $\mathcal{S}tand(G)$ over Σ**). Let $\left(\tilde{P}, \tilde{Q}\right)_{\Sigma_{\Gamma^m}}$ be the section of $\mathcal{S}tand(G)$ over Σ_{Γ^m}, given by the pull-back of $\left(\tilde{P}, \tilde{Q}\right)_\Sigma$ by the canonical morphism*

$$\Sigma_{\Gamma^m} = \Sigma \times_{Relpos\,_G} \Gamma^m(\underline{\Delta}(G)) \to \Sigma.$$

Denote by $\sigma : S \to \Sigma_{\Gamma^m}$ the section of $\Sigma_{\Gamma^m} \to S$ given by $\sigma = ((P, Q), \underline{g})$ with $t_2(P, Q) = \delta_2(\underline{g})$. Then

$$\mathcal{F}ix(P, Q) = \left(\mathcal{F}ix\left(\tilde{P}, \tilde{Q}\right)_{\Sigma_{\Gamma^m}}\right)_\sigma = \left(\mathcal{F}ix\left(\tilde{P}, \tilde{Q}\right)_{\delta_2(\underline{g})}\right)_{(P,Q)},$$

(cf. §12.8).

Theorem 16.2

Keep the above notation. Let G be a reductive S-group scheme.

1) *There exists a unique morphism $\Theta : \Sigma_{\Gamma^m} \longrightarrow \hat{\Sigma}$, so that if $S' \longrightarrow S$ is an etale covering and E a frame of $G_{S'}$ then $\Theta_{S'} = \Theta^E$. Thus Θ is a section of the morphism $\hat{\Sigma} \xrightarrow{\pi} \Sigma_{\Gamma^m}$, and establishes an isomorphism $\Theta : \Sigma_{\Gamma^m} \simeq \pi^{-1}(\Sigma_{\Gamma^m})$. Moreover $Im\ \Theta = \pi^{-1}(\Sigma_{\Gamma^m})$ is a relatively schematically dense subscheme of $\hat{\Sigma}$.*

2) *The quadruple $\left(\hat{\Sigma}, \overline{\Sigma}, \pi, \Theta\right)$ defines a **canonical Smooth Resolution** (cf. Definition 14.17) of the pull-back $\overline{\Sigma}_{\Gamma^m} = \Gamma^m \times_{Relpos\,_G} \overline{\Sigma}$ of the Universal Schubert scheme $\overline{\Sigma} \to Relpos\,_G$ by*

$$\delta_2 : \Gamma^m(\underline{\Delta}(G)) \to Relpos\,_G.$$

The Smooth Resolution $\left(\hat{\Sigma}, \overline{\Sigma}, \pi, \Theta\right)$ is represented by the following commutative cube.

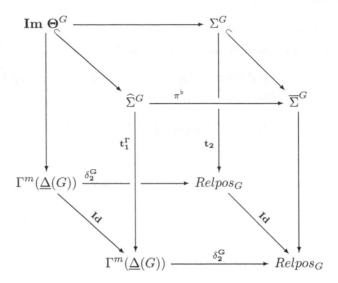

Where $\pi^{\flat}$ is the composition of π followed by $Id_{\mathcal{P}\mathrm{ar}(G)\times_S\mathcal{P}\mathrm{ar}(G)}\times_S\delta_2$. It induces a finite morphism $Im\ \Theta^G \longrightarrow \Sigma^G$. Recall that $\Theta^G : \Sigma^G_{\Gamma^m} \longrightarrow Im\ \Theta^G \subset \widehat{\Sigma}^G$ is a dominant open embedding, and that $\pi : \widehat{\Sigma}^G \longrightarrow \overline{\Sigma}^G_{\Gamma^m}$ is a smooth resolution. The above commutative cube shows that:

"The **Universal Schubert scheme** $\overline{\Sigma}^G \longrightarrow Relpos_G$ admits a **canonical Smooth Resolution** after the twisted finite extension
$$\delta_2 : \Gamma^m(\underline{\Delta}(G)) \longrightarrow Relpos_G."$$

Remark 16.3 *The super index G is naturally introduced with the aim of studying the behaviour of Smooth Resolutions under base changes $S' \longrightarrow S$ and their dependence on G. It will be omitted if no confusion arises.*

Remark that the horizontal arrows of the cube upper face factor as follows:

Where π' denotes the restriction of π to $Im\ \Theta^G \subset \widehat{\Sigma}^G$ and it is in fact an isomorphism. There is a commutative square of $\Gamma^m(\underline{\Delta}(G))$-morphisms which shows that the image $Im\ \Theta^G$ connects $\widehat{\Sigma}^G$ and $\overline{\Sigma}^G_{\Gamma^m}$, where π' is an isomorphism and π a smooth resolution.

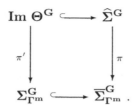

One may go into the details of the above theorem by looking at the fibers of $t^{\Gamma^m} : \widehat{\Sigma} \longrightarrow \Gamma^m(\underline{\Delta}(G))$ as follows. Let $\underline{g} : S \rightarrow \Gamma^m(\underline{\Delta}(G))$ be a section of $\Gamma^m(\underline{\Delta}(G))$ over S, and $(\underline{g})^*(\overline{\Sigma}_{\Gamma^m})$ the pull-back by $\underline{g}$ of the $\Gamma^m(\underline{\Delta}(G))$-scheme $\overline{\Sigma}_{\Gamma^m}$. Write $\underline{\tau} = \delta_2(\underline{g})$, and $\overline{\Sigma}_{\underline{g}} = (\underline{g})^*(\overline{\Sigma}_{\Gamma^m})$. By definition of $\overline{\Sigma}_{\Gamma^m}$ there are canonical identifications $\overline{\Sigma}_{\underline{g}} = \overline{\Sigma}_{\underline{\tau}}$ (resp. $\Sigma_{\underline{g}} = \Sigma_{\underline{\tau}}$). Let

$$\pi_{\underline{g}} : \widehat{\Sigma}_{\underline{g}} \longrightarrow \overline{\Sigma}_{\underline{g}}$$

be the restriction $\pi \circ j_{\widehat{\Sigma}_{\underline{g}}}$ of π to $\widehat{\Sigma}_{\underline{g}}$, where $j_{\widehat{\Sigma}_{\underline{g}}} : \widehat{\Sigma}_{\underline{g}} = \mathcal{C}onf_G^m(\underline{g}) \rightarrow \widehat{\Sigma} = \mathcal{C}onf_G^m$ denotes the natural embedding. Thus $\pi_{\underline{g}}^{-1}(\Sigma_{\underline{g}}) = \mathcal{E}_{\underline{g}}^{-1}(\mathcal{S}tand(\underline{\tau})) \subset \widehat{\Sigma}_{\underline{g}}$. Recall that one has

$$\Theta_{\underline{g}} : \Sigma_{\underline{g}} \rightarrow \pi_{\underline{g}}^{-1}(\Sigma_{\underline{g}}) \subset \widehat{\Sigma}_{\underline{g}}$$

where $\Theta_{\underline{g}} = \Theta \circ j_{\Sigma_{\underline{g}}}$, and

$$j_{\Sigma_{\underline{g}}} : \Sigma_{\underline{g}} = \mathcal{S}tand(\underline{\tau}) \rightarrow \Sigma_{\Gamma^m}$$

denotes the natural embedding. As $\Sigma_{\underline{g}} = \Sigma_{\underline{\tau}}$, the definition of $\Theta_{\underline{g}}$ is coherent with that of $\Theta_{\underline{g}s}$ in Proposition 14.22. Recall that the Universal Schubert scheme of type $\underline{\tau}$, $\overline{\Sigma}_{\underline{\tau}}$ may be seen as the fiber $(\overline{\Sigma})_{\underline{\tau}}$ of the Universal Schubert scheme $\overline{\Sigma} \longrightarrow \mathrm{Relpos}_G$ over the section $\underline{\tau}$.

The connection between the above morphisms is represented by the following commutative cubic diagram of $\Gamma^m(\underline{\Delta}(G))$-morphisms, where all the arrows, with the exception of the two descending vertical ones, are open embeddings. This diagram explicits the connection between $Im\ \Theta$ and $Im\ \Theta_{\underline{g}}$. The later being the fiber of $Im\ \Theta$ over the section $\underline{g}$ of $\Gamma^m(G)$. $\overline{\Sigma}_{\underline{\tau}}$ is identified with the fiber of $\overline{\Sigma}_{\Gamma^m}$ over $(\underline{\tau}, \underline{g})$. Super indices are omitted.

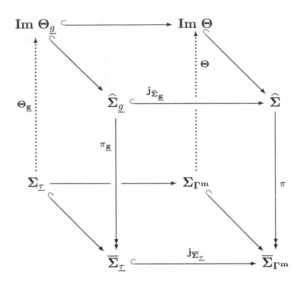

Observe that the right lateral face of this cube is precisely the above commutative square and that the lateral left face is obtained as the fiber over the section g of $\Gamma^m(\underline{\Delta}(G))$ of the right one.

Theorem 16.4
Keep the above notation. Let G be a reductive S-group scheme, and $\underline{g}$ a section of $\Gamma^m(\underline{\Delta}(G))$ with $\delta_2(\underline{g}) = \underline{\tau}$ over S.

  *1) There exists a unique morphism $\Theta_{\underline{g}} : \Sigma_{\underline{\tau}} \longrightarrow \widehat{\Sigma}_{\underline{g}}$, such that if $S' \longrightarrow S$
      is an etale covering and E a frame of $G_{S'}$ then $(\Theta_{\underline{g}})_{S'}$ is locally of the
      form $\Theta_{\underline{g}_{S'}}$. Thus $\Theta_{\underline{g}}$ is a section of the morphism $\widehat{\Sigma}_{\underline{g}} \xrightarrow{\pi_{\underline{g}}} \Sigma_{\underline{\tau}}$ over
      $\Sigma_{\underline{\tau}}$, and establishes an isomorphism $\Theta_{\underline{g}} : \Sigma_{\underline{\tau}} \simeq \pi_{\underline{g}}^{-1}(\Sigma_{\underline{\tau}})$. Moreover
      $Im\, \Theta_{\underline{g}} = \pi_{\underline{g}}^{-1}(\Sigma_{\underline{\tau}})$ is a relatively schematically dense subscheme of $\widehat{\Sigma}_{\underline{g}}$.*

  *2) The quadruple $\left(\widehat{\Sigma}_{\underline{g}}, \overline{\Sigma}_{\delta_2(\underline{g})}, \pi_{\underline{g}}, \Theta_{\underline{g}}\right)$ is a **canonical Smooth Resolu-**
      tion of the schematical closure $\overline{\Sigma}_{\underline{\tau}} = Stand(\underline{\tau})^{schc}$ of $Stand(\underline{\tau})$ in
      $Par(G) \times_S Par(G)$.*

The link between Theorem 16.2 and Theorem 16.4 is resumed by the following diagram which makes it evident that the resolution of Theorem 16.4 is obtained as a fiber over $\underline{g}$ of the resolution of Theorem 16.2 for a section $\underline{g}$ of Γ^m over S. Thus the fiber of $\pi^\flat$ given by $\pi_{\underline{g}}$ defines a smooth resolution of the Universal Schubert scheme $\overline{\Sigma}_{\underline{\tau}}$ defined by the type of relative position $\underline{\tau}$.

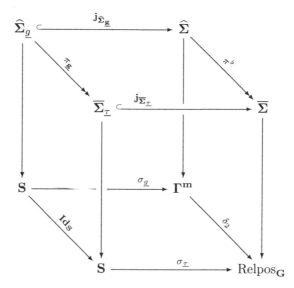

Remark that the front face and the back face of this commutative cube are cartesian diagrams, and that the right lateral face of this cube corresponds to the front face of the first cube. The formation of the schematic closure commutes with flat base changes. Thus, given a section g of Γ^m over a flat scheme $S' \to S$, a similar commutative cube is obtained with S' instead of S, where π_g defines a smooth resolution of $\overline{\Sigma}_{\underline{\tau}}$. In the next sections one investigates under which conditions $S' \to S$ the corresponding morphism π_g defines a Smooth Resolution.

16.2   Smooth Resolutions and base changes

Recall that

Proposition 16.5 *Keep the above notation. Suppose G is endowed with a frame. Then $\overline{\Sigma}_{\underline{\tau}} \longrightarrow \mathcal{P}ar_{\mathbf{t}}(\mathbf{G})$ is a locally trivial fiber bundle with typical fiber $\overline{\Sigma}_{(\underline{\tau}, \mathbf{P_t})}$, and $\Sigma_{\underline{\tau}} \longrightarrow \mathcal{P}ar_{\mathbf{t}}(\mathbf{G})$ is a sub-bundle of it with typical fiber $\Sigma_{(\underline{\tau}, \mathbf{P_t})}$.*

Let G be a reductive S-group scheme and (g, P) be a section of $\Gamma^m(\underline{\Delta}(G)) \times_S \mathcal{P}ar(G)$. Write

$$\widehat{\Sigma}_{(g,P)} = \left(t_1^\Gamma \times \mathrm{pr}_{\widehat{\Sigma}}\right)^{-1}((g, P)),$$

where $\mathrm{pr}_{\widehat{\Sigma}} : \widehat{\Sigma} = \mathcal{C}onf_G^m \to \mathcal{P}ar(G)$ denotes the restriction of the left extremity morphism $\mathcal{E}_1 : \mathcal{C}onf_G \to \mathcal{P}ar(G)$ to $\mathcal{C}onf_G^m$ (cf. Definition 6.10.5). There are cartesian squares:

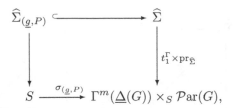

and

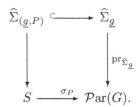

Where $\sigma_{(g,P)}$ denotes the morphism defined by the section (g,P) of $\Gamma^m(\underline{\Delta}(G)) \times_S \mathcal{P}\mathrm{ar}(G)$ and σ_P the morphism defined by the section of $\mathcal{P}\mathrm{ar}(G)$ given by P. It is clear that the second square may be intercalated in the first one.

Theorem 16.6
Keep the above notation. Write $\underline{\tau} = \delta_2(\underline{g})$.

1) *There exists a unique morphism $\Theta_{(g,P)} : \Sigma_{(\underline{\tau},P)} \longrightarrow \widehat{\Sigma}_{(g,P)}$, so that if $S' \longrightarrow S$ is an etale covering and E a frame of $G_{S'}$ then $(\Theta_{(g,P)})_{S'}$ is locally of the form $\Theta_{(g_{S'},P_{S'})}$. Thus $\Theta_{(g,P)}$ is a section of the morphism $\widehat{\Sigma}_g \xrightarrow{\pi_{(g,P)}} \overline{\Sigma}_{(\underline{\tau},P)}$ over $\Sigma_{(\underline{\tau},P)}$, and establishes an isomorphism $\Theta_{(g,P)} : \Sigma_{(\underline{\tau},P)} \simeq \pi_{(g,P)}^{-1}(\Sigma_{(\underline{\tau},P)})$. Moreover $\mathrm{Im}\ \Theta_{(g,P)} = \pi_{(g,P)}^{-1}(\Sigma_{(\underline{\tau},P)})$ is a relatively schematically dense subscheme of $\widehat{\Sigma}_{(g,P)}$.*

2) *The quadruple $\left(\widehat{\Sigma}_{(g,P)}, \overline{\Sigma}_{(\underline{\tau},P)}, \pi_{(g,P)}, \Theta_{(g,P)} \right)$ is a **canonical Smooth Resolution** of the schematical closure $\overline{\Sigma}_{(\underline{\tau},P)} = Stand(\underline{\tau},P)^{schc}$ of $Stand(\underline{\tau},P)$ in $\mathcal{P}\mathrm{ar}(G) \times_S \mathcal{P}\mathrm{ar}(G)$.*

The two following commutative cubes show in detail the connection between $\mathrm{Im}\ \Theta$ and $\mathrm{Im}\ \Theta_{(g,P_t)}$, and that of π and $\pi_{(g,P_t)}$. $\widehat{\Sigma}_{(g,P_t)}$ is the fiber of $\widehat{\Sigma}$ over (g, P_t) however $\overline{\Sigma}_{(\underline{\tau},P_t)}$ **may not be equal** to the fiber of $\overline{\Sigma}_{\Gamma^m}$ over $(\underline{\tau}, P_t)$, and thus $\pi_{(g,P_t)}$ is not in general the fiber of π. To the first cube of the previous section corresponds the following one where all the schemes are considered as $\mathcal{P}\mathrm{ar}(G)$-schemes.

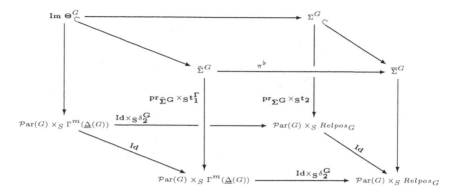

Where $pr_{\widehat{\Sigma}^G} : \widehat{\Sigma}^G \to \mathcal{P}\mathrm{ar}(G)$ associates the left extremity with a gallery configuration, and $pr_{\Sigma^G} : \Sigma^G \to \mathcal{P}\mathrm{ar}(G)$ is given by $(P,Q) \mapsto P$.

The following commutative cubes correspond respectively to the second and third ones of the previous section.

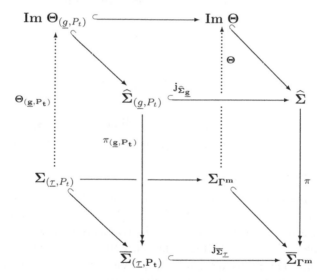

(resp.

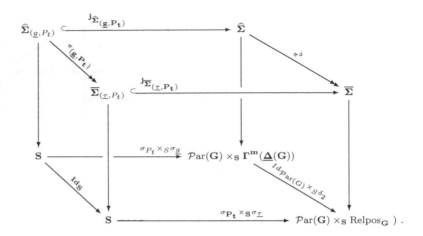

On the other hand, there is a commutative diagram

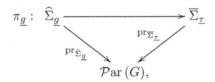

where $\underline{\tau} = \delta_2(\underline{g})$, $\mathrm{pr}_{\widehat{\Sigma}_{\underline{g}}}$ (resp. $\mathrm{pr}_{\overline{\Sigma}_{\underline{\tau}}}$) is the fiber of $\mathrm{pr}_{\widehat{\Sigma}}$ (resp. $\mathrm{pr}_{\overline{\Sigma}} = \mathrm{pr}_{\mathcal{P},\overline{\Sigma}}$) over $\underline{g}$ (resp. $\underline{\tau}$)(cf. Proposition 14.19), and $\pi_{\underline{g}}$ is seen as $\mathcal{P}ar\,(G)$-morphism.

The following question are treated:

Under which conditions on the base scheme S the diagram obtained from the above one by the base change $S \xrightarrow{\sigma_{\mathcal{P}}} \mathcal{P}ar\,(G)$ is a smooth resolution?
If the answer is affirmative it establishes the connection between the Universal Schubert scheme $\overline{\Sigma}_{\underline{\tau}}$ Smooth Resolution and that one of the usual Schubert scheme $\overline{\Sigma}_{(\underline{\tau},P)}$. The following definition provides the terminology needed to state the more general problem about the commutation of the Smooth Resolution of singularities of Schubert schemes with base changes. For the S-reductive group schemes G of a fixed type such that S satisfies the condition stated at the introduction of this chapter, it will be seen that the answers to this question and the preceding one are affirmative. In fact such a condition is easily obtained for a $\mathbb{Z}$-Chevalley group scheme of a given type. From its Universal Property with respect to S-reductive group schemes of the same type the general condition on S results. More generally given a base extension $S' \to S$ such that S' satisfies this condition the Smooth Resolution of the Universal Schubert scheme of $G_{S'}$ is obtained by base change from that of

G. Let $U^{\mathcal{R}}$ be as in the introduction. One has that for all reductive S-group schemes with S in $Sch|_{U^{\mathcal{R}}}$ the formation of the Universal Smooth Resolution commutes with base changes $S' \to S$ in $Sch|_{U^{\mathcal{R}}}$.

Definition 16.7 *The smooth resolution quadruple* $\left(\widehat{\Sigma}^G, \overline{\Sigma}^G, \pi^G, \Theta^G\right)$ $\left(resp.\ \left(\widehat{\Sigma}^G_{\underline{g}}, \overline{\Sigma}^G_{\delta_2(\underline{g})} \pi^G_{\underline{g}}, \Theta^G_{\underline{g}}\right)\right)$ *of the Schubert scheme* $\overline{\Sigma}^G$ $\left(resp.\ \overline{\Sigma}^G_{\delta_2(\underline{g})}\right)$ *of the reductive group* G, *is a* S**-Universal Smooth Resolution**, *if for all* $S' \to S$,

$$\left(S' \times_S \widehat{\Sigma}^G, S' \times_S \overline{\Sigma}^G, Id_{S'} \times_S \pi^G, Id_{S'} \times_S \Theta^G\right) = \left(\widehat{\Sigma}^{G_{S'}}, \overline{\Sigma}^{G_{S'}}, \pi^{G_{S'}}, \Theta^{G_{S'}}\right)$$

$$\left(resp.\ \left(S' \times_S \widehat{\Sigma}^G_{\underline{g}}, S' \times_S \overline{\Sigma}^G_{\delta_2(\underline{g})}, Id_{S'} \times_S \pi^G_{\underline{g}}, Id_{S'} \times_S \Theta^G_{\underline{g}}\right) = \right.$$

$$\left. \left(\widehat{\Sigma}^{G_{S'}}_{\underline{g}_{S'}}, \overline{\Sigma}^{G_{S'}}_{\delta_2(\underline{g}_{S'})}, \pi^{G_{S'}}_{\underline{g}_{S'}}, \Theta^{G_{S'}}_{\underline{g}_{S'}}\right)\right).$$

Let one look at $\widehat{\Sigma}^G_{\underline{g}}$ *(resp.* $\overline{\Sigma}^G_{\delta_2(\underline{g})}$*) as* $\mathcal{P}ar_t(G)$*-schemes and* $\pi^G_{\underline{g}}$, $\Theta^G_{\underline{g}}$ *as* $\mathcal{P}ar_t(G)$*-morphisms. The Smooth Resolution* $\left(\widehat{\Sigma}^G_{\underline{g}}, \overline{\Sigma}^G_{\delta_2(\underline{g})}, \pi^G_{\underline{g}}, \Theta^G_{\underline{g}}\right)$ *of* $\overline{\Sigma}^G_{\delta_2(\underline{g})}$ *is a* $\mathcal{P}ar_t(G)$**-Universal Smooth Resolution** *if for all*

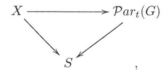

one has

$$\left(X \times_{\mathcal{P}ar_t(G)} \widehat{\Sigma}^G_{\underline{g}}, X \times_{\mathcal{P}ar_t(G)} \overline{\Sigma}^G_{\delta_2(\underline{g})}, Id_X \times_{\mathcal{P}ar_t(G)} \pi^G_{\underline{g}}, Id_X \times_{\mathcal{P}ar_t(G)} \Theta^G_{\underline{g}}\right) =$$

$$\left(\widehat{\Sigma}^G_{(\underline{g}_X, P_X)}, \overline{\Sigma}^G_{(\delta_2(\underline{g}_X), P_X)}, \pi^G_{(\underline{g}_X, P_X)}, \Theta^G_{(\underline{g}_X, P_X)}\right).$$

Where the fiber product $X \times_{\mathcal{P}ar_t(G)} \overline{\Sigma}^G_{\delta_2(\underline{g})}$ *(resp.* $X \times_{\mathcal{P}ar_t(G)} \widehat{\Sigma}^G_{\underline{g}}$*) is defined by the couple of morphisms* $\left(\sigma_X, \overline{\Sigma}_{\delta_2(\underline{g})} \longrightarrow \mathcal{P}ar_t(G)\right)$ *(resp.* $\left(\sigma_X, \mathcal{E}_1 : \widehat{\Sigma}_{\underline{g}} \longrightarrow \mathcal{P}ar_t(G)\right)$*), and* P_X *denotes the parabolic subgroup of* G *over* X *given by the section* σ_X *defined by* $X \to \mathcal{P}ar_t(G)$. *Denote the first members of the above equalities respectively by*

$$\left(\widehat{\Sigma}^G, \overline{\Sigma}^G, \pi^G, \Theta^G\right)_{S'}, \ \left(\widehat{\Sigma}^G_{\underline{g}}, \overline{\Sigma}^G_{\delta_2(\underline{g})}, \pi^G_{\underline{g}}, \Theta^G_{\underline{g}}\right)_{S'}, \ and \ \left(\widehat{\Sigma}^G_{\underline{g}}, \overline{\Sigma}^G_{\delta_2(\underline{g})}, \pi^G_{\underline{g}}, \Theta^G_{\underline{g}}\right)_X.$$

One may now state that the triangle $(\widehat{\Sigma}_{(\underline{g},P)}, \overline{\Sigma}_{\mathcal{T}}, \mathcal{P}ar(G))$ gives rise to a smooth resolution of $\overline{\Sigma}^G_{(\delta_2(\underline{g}_X), P_X)}$ by the base change $X \to \mathcal{P}ar_t(G)$ if and only

if $\left(\widehat{\Sigma}_{\underline{g}}^G, \overline{\Sigma}_{\delta_2(\underline{g})}^G, \pi_{\underline{g}}^G, \Theta_{\underline{g}}^G\right)$ of $\overline{\Sigma}_{\delta_2(\underline{g})}^G$ is a $\mathcal{P}ar_t(G)$-**Universal Smooth Resolu-**
tion. To give an answer to the above stated question one needs the following
definition.

Definition 16.8   • *Consider* $\mathcal{P}ar(G) \times_S \mathcal{P}ar(G)$ *as a* $\mathcal{P}ar(G)$-*scheme by
the first projection. It is said that the* **formation of the Schematic
closure** *(cf. Definition 16.34) of the Schubert cell*

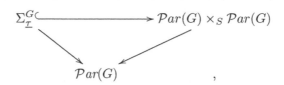

(resp.

$$\Sigma_{\underline{\tau}}^G \subset \mathcal{P}ar(G) \times_S \mathcal{P}ar(G), \quad \Sigma^G \subset \mathcal{R}elpos_G \times_S \mathcal{P}ar(G) \times_S \mathcal{P}ar(G))$$

commutes *with* **the base change** $X \longrightarrow \mathcal{P}ar(G)$ *(resp.* $S' \longrightarrow S)$ *if
the natural morphism*

$$\left(X \times_S \Sigma_{\underline{\tau}}^G\right)^{schc} \longrightarrow X \times_S \overline{\Sigma}_{\underline{\tau}}^G$$

(resp.

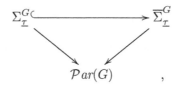

*is an isomorphism, where the superscript "schc" denotes the schematic
closure in* $X \times_S \mathcal{P}ar(G)$ *(resp.* $S' \times_S \mathcal{P}ar(G) \times_S \mathcal{P}ar(G)$, $S' \times_S$
$\mathcal{R}elpos_G \times_S \mathcal{P}ar(G) \times_S \mathcal{P}ar(G))$.

• *Recall that the Schubert cell*

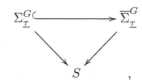

(resp.

$$\Sigma_{\underline{\tau}}^G \hookrightarrow \overline{\Sigma}_{\underline{\tau}}^G$$
$$S$$
,

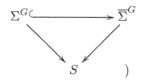

is **universally relatively schematically dense** *(cf. loc. cit.) in the Schubert scheme if for all base change* $X \longrightarrow \mathcal{P}ar(G)$ *(resp.* $S' \longrightarrow S$) *the natural morphism* $\left(X \times_S \Sigma_{\underline{\tau}}^G\right)^{schc} \longrightarrow X \times_S \overline{\Sigma}_{\underline{\tau}}^G$, *(resp.* $\left(S' \times_S \Sigma_{\underline{\tau}}^G\right)^{schc} \longrightarrow S' \times_S \overline{\Sigma}_{\underline{\tau}}^G$, $\left(S' \times_S \Sigma^G\right)^{schc} \longrightarrow S' \times_S \overline{\Sigma}^G$), *is an isomorphism, where the superscript "schc" denotes the schematic closure in* $X \times_S \overline{\Sigma}_{\underline{\tau}}^G$ *(resp.* $S' \times_S \overline{\Sigma}_{\underline{\tau}}^G$, $S' \times_S \overline{\Sigma}^G$).

For the sake of simplifying notation suppose $X = S$. Write P_t instead of P_X, write $\widehat{\Sigma}_{(\underline{g},P_t)}$ instead of $\widehat{\Sigma}_{(\underline{g},P_t)}^G$...etc., and denote by $(\pi_{\underline{g}})_{P_t} : \widehat{\Sigma}_{(\underline{g},P_t)} \to \left(\overline{\Sigma}_{\underline{\tau}}\right)_{P_t}$ the fiber of $\pi_{\underline{g}}$ over the section σ_{P_t}.

The relation between $(\pi_{\underline{g}})_{P_t}$, $\pi_{\underline{g}}$, and the base change $\sigma_{P_t} : S \to \mathcal{P}ar_t(G)$ is represented by the following commutative diagram.

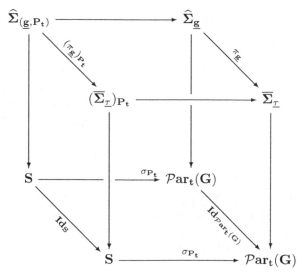

The front face and the back face of the cube are cartesian diagrams. Remark that the morphism $\pi_{\underline{g}}$ as a morphism in $\mathcal{P}ar(G)$ satisfies $\left(\pi_{\underline{g}}\right)_{P_t} = \pi_{(\underline{g},P_t)}$, and one has that its **schematic image** (cf. Definition 16.34) is equal to $\overline{\Sigma}_{(\underline{\tau},P_t)}$. By definition of this schematic image in $\mathcal{P}ar(G)$ the morphism $\left(\pi_{\underline{g}}\right)_{P_t}$ factors as

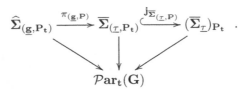

Thus the fiber $(\pi_{\underline{g}})_{P_t} : \widehat{\Sigma}_{(\underline{g},P_t)} \rightarrow (\overline{\Sigma}_{\underline{\tau}})_{P_t}$ coincides with $\pi_{(\underline{g},P_t)} : \widehat{\Sigma}_{(\underline{g},P_t)} \rightarrow \overline{\Sigma}_{(\underline{\tau},P_t)}$ if and only if

$$\overline{\Sigma}_{(\underline{\tau},\mathbf{P_t})} = \left(\overline{\Sigma}_{\underline{\tau}}\right)_{\mathbf{P_t}} = \mathbf{S} \times_{\mathcal{P}\mathrm{ar}(\mathbf{G})} \overline{\Sigma}_{\underline{\tau}} \ .$$

Otherwise stated, as by definition $\overline{\Sigma}_{(\underline{\tau},P_t)}$ is the schematic closure of $\Sigma_{(\underline{\tau},P_t)}$ in $\mathcal{P}\mathrm{ar}_t(G)$, $(\pi_{\underline{g}})_{P_t} = \pi_{(\underline{g},P_t)}$ if and only if "the formation of the schematic closure of $\Sigma_{\underline{\tau}}$ in $\mathcal{P}\mathrm{ar}_t(G)$ commutes with the base change defined by σ_{P_t}".
From Proposition 16.46 results the equivalence of the following statements:

1) The **formation of the schematic closure** of $\Sigma_{\underline{\tau}} \xrightarrow{\mathrm{pr}_{\Sigma_{\underline{\tau}}}} \mathcal{P}\mathrm{ar}_t(G)$ in $\mathcal{P}\mathrm{ar}_t(G) \times_S \mathcal{P}\mathrm{ar}(G) \xrightarrow{\mathrm{pr}_1} \mathcal{P}\mathrm{ar}_t(G)$, **commutes** with all the base changes $X \longrightarrow \mathcal{P}\mathrm{ar}_t(G)$.

2) $\Sigma_{\underline{\tau}} \xrightarrow{\mathrm{pr}_{\Sigma_{\underline{\tau}}}} \mathcal{P}\mathrm{ar}_t(G)$ is **universally relatively schematically dense** in $\overline{\Sigma}_{\underline{\tau}} \xrightarrow{\mathrm{pr}_{\overline{\Sigma}_{\underline{\tau}}}} \mathcal{P}\mathrm{ar}_t(G)$

Assume 1). As the schematic image of the natural embedding $X \times_{\mathcal{P}\mathrm{ar}_t(G)} \Sigma_{\underline{\tau}} \hookrightarrow X \times_{\mathcal{P}\mathrm{ar}_t(G)} \overline{\Sigma}_{\underline{\tau}}$ always exists (cf. Proposition 16.46), it follows from the hypothesis and the transitivity of schematic images that $\left(X \times_{\mathcal{P}\mathrm{ar}_t(G)} \Sigma_{\underline{\tau}}\right)^{schc} = X \times_{\mathcal{P}\mathrm{ar}_t(G)} \overline{\Sigma}_{\underline{\tau}}$. Thus that $\Sigma_{\underline{\tau}}$ is **Universally relatively schematically dense** in $\overline{\Sigma}_{\underline{\tau}}$ as $\mathcal{P}\mathrm{ar}_t(G)$-schemes.
Assume 2). Then $\left(X \times_{\mathcal{P}\mathrm{ar}_t(G)} \Sigma_{\underline{\tau}}\right)^{schc} (in \ X \times_{\mathcal{P}\mathrm{ar}_t(G)} \overline{\Sigma}_{\underline{\tau}}) = X \times_{\mathcal{P}\mathrm{ar}_t(G)} \overline{\Sigma}_{\underline{\tau}}$. Thus by the transitivity of schematic images one has $\left(X \times_{\mathcal{P}\mathrm{ar}_t(G)} \Sigma_{\underline{\tau}}\right)^{schc} (in \ X \times_{\mathcal{P}\mathrm{ar}_t(G)} \mathcal{P}\mathrm{ar}_t(G)) = X \times_{\mathcal{P}\mathrm{ar}_t(G)} \overline{\Sigma}_{\underline{\tau}}$, i.e. the **formation of the schematic closure** commutes with all base changes.
From the equalities $(\pi_{\underline{g}})_{P_t} = \pi_{(\underline{g},P_t)}$ as morphisms in $\mathcal{P}\mathrm{ar}_t(G)$, $\left(\overline{\Sigma}_{\underline{\tau}}\right)_{P_t} = \overline{\Sigma}_{(\underline{\tau},P_t)}$ as subschemes of $\mathcal{P}\mathrm{ar}_t(G)$, and the fact that there is a unique section

$$\Theta_{(\underline{g},P_t)} : \Sigma_{(\underline{\tau},P_t)} \longrightarrow \left(\pi_{(\underline{g},P_t)}\right)^{-1} \left(\Sigma_{(\underline{\tau},P_t)}\right) \subset \widehat{\Sigma}_{(\underline{\tau},P_t)}$$

of $\pi_{(\underline{g},P_t)}$ over $\Sigma_{(\underline{\tau},P_t)}$, it follows that $\left(\Theta_{\underline{g}}\right)_{P_t} = \Theta_{(\underline{g},P_t)}$. It is clear that from this reasoning results the

Proposition 16.9

The smooth resolution $\left(\widehat{\Sigma}_{\underline{g}}^{G}, \overline{\Sigma}_{\delta_2(\underline{g})}^{G}, \pi_{\underline{g}}^{G}, \Theta_{\underline{g}}^{G}\right)$ *of* $\overline{\Sigma}_{\delta_2(\underline{g})}^{G}$ *is a* $\mathcal{P}ar_t(G)$- **Universal Smooth Resolution** *if and only if the Schubert cell* $\Sigma_{\delta_2(\underline{g})}^{G}$ *is* **Universally relatively schematically dense** *in the Schubert scheme* $\overline{\Sigma}_{\delta_2(\underline{g})}^{G}$ *as a* $\mathcal{P}ar_t(G)$-*scheme.*

Observe that the statement 2) is shown by the following diagram:

$$
\begin{array}{ccc}
(\mathbf{X} \times_{\mathcal{P}ar(\mathbf{G})} \Sigma_{\underline{\tau}})^{\mathbf{schc}} = \mathbf{X} \times_{\mathcal{P}ar(\mathbf{G})} \overline{\Sigma}_{\underline{\tau}} & \longrightarrow & \overline{\Sigma}_{\underline{\tau}} \\
\downarrow {\scriptstyle (\mathrm{pr}_{\overline{\Sigma}_{\underline{\tau}}})\mathbf{x}} & & \downarrow {\scriptstyle \mathrm{pr}_{\overline{\Sigma}_{\underline{\tau}}}} \\
\mathbf{X} & \longrightarrow & \mathcal{P}ar(\mathbf{G}),
\end{array}
$$

where $X \longrightarrow \mathcal{P}ar(G)$ denotes a $\mathcal{P}ar(G)$-scheme, and $(X \times_{\mathcal{P}ar(G)} \Sigma_{\underline{\tau}})^{schc}$ the schematic closure of $X \times_{\mathcal{P}ar(G)} \Sigma_{\underline{\tau}}$ in $X \times_{\mathcal{P}ar(G)} \overline{\Sigma}_{\underline{\tau}}$.

Proposition 16.10
Universality *is a* **hereditary property** *of quadruples representing Smooth Resolutions of Schubert schemes. Write* $\delta_2(\underline{g}) = \underline{\tau}$. *More precisely stated:*

1) *If* $\left(\widehat{\Sigma}^{G}, \overline{\Sigma}^{G}, \pi^{G}, \Theta^{G}\right)$ *is an* S-*Universal Smooth Resolution then for all* $S' \to S$,

$$
\left(\widehat{\Sigma}^{G}, \overline{\Sigma}^{G}, \pi^{G}, \Theta^{G}\right)_{S'} = \left(\widehat{\Sigma}^{G_{S'}}, \overline{\Sigma}^{G_{S'}}, \pi^{G_{S'}}, \Theta^{G_{S'}}\right)
$$

is a S'-*Universal Smooth Resolution.*

2) *If* $\left(\widehat{\Sigma}_{\underline{g}}^{G}, \overline{\Sigma}_{\underline{\tau}}^{G}, \pi_{\underline{g}}^{G}, \Theta_{\underline{g}}^{G}\right)$ *is an* S-*Universal Smooth Resolution then for all* $S' \to S$,

$$
\left(\widehat{\Sigma}_{\underline{g}}^{G}, \overline{\Sigma}_{\underline{\tau}}^{G}, \pi_{\underline{g}}^{G}, \Theta_{\underline{g}}^{G}\right)_{S'} = \left(\widehat{\Sigma}_{\underline{g}_{S'}}^{G_{S'}}, \overline{\Sigma}_{\underline{\tau}_{S'}}^{G_{S'}}, \pi_{\underline{g}_{S'}}^{G_{S'}}, \Theta_{\underline{g}_{S'}}^{G_{S'}}\right)
$$

is a S'-*Universal Smooth Resolution.*

3) *If* $\left(\widehat{\Sigma}_{\underline{g}}^{G}, \overline{\Sigma}_{\underline{\tau}}^{G}, \pi_{\underline{g}}^{G}, \Theta_{\underline{g}}^{G}\right)$ *is a* $\mathcal{P}ar_t(G)$-*Universal Smooth Resolution, then for all* $S' \to S$

$$
\left(\widehat{\Sigma}_{\underline{g}}^{G}, \overline{\Sigma}_{\underline{\tau}}^{G}, \pi_{\underline{g}}^{G}, \Theta_{\underline{g}}^{G}\right)_{S'} = \left(\widehat{\Sigma}_{\underline{g}_{S'}}^{G_{S'}}, \overline{\Sigma}_{\underline{\tau}_{S'}}^{G_{S'}}, \pi_{\underline{g}_{S'}}^{G_{S'}}, \Theta_{\underline{g}_{S'}}^{G_{S'}}\right)
$$

is a S'-*Smooth Resolution (resp.* $\mathcal{P}ar_t(G_{S'})$-*Universal Smooth Resolution).*

Proof *The assertions* 1) *and* 2) *are immediate let it be seen* 3). *One proves the first assertion of* 3). *Write* $X = S' \times_S \mathcal{P}ar_t(G)$. *Clearly* $X \times_{\mathcal{P}ar_t(G)}$ $\Sigma_{\underline{\tau}}^G = S' \times_S \Sigma_{\underline{\tau}}^G$ (*resp.* $X \times_{\mathcal{P}ar_t(G)} \overline{\Sigma}_{\underline{\tau}}^G = S' \times_S \overline{\Sigma}_{\underline{\tau}}^G$). *From the hypothesis it follows that the subscheme* $S' \times_S \Sigma_{\underline{\tau}}^G \subset S' \times_S \overline{\Sigma}_{\underline{\tau}}^G$ *is schematically dense. Thus* $S' \times_S \overline{\Sigma}_{\underline{\tau}}^G = \overline{\Sigma}_{\underline{\tau}_{S'}}^{G_{S'}}$. *It results that* $\left(\widehat{\Sigma}_{\underline{g}}^G, \overline{\Sigma}_{\underline{\tau}}^G, \pi_{\underline{g}}^G, \Theta_{\underline{g}}^G\right)$ *is* S-*Universal, and that* $\left(\widehat{\Sigma}_{\underline{g}}^G, \overline{\Sigma}_{\underline{\tau}}^G, \pi_{\underline{g}}^G, \Theta_{\underline{g}}^G\right)_{S'}$ *is* S'-*Universal.*
To prove the respective assertion of 3) *it suffices to see that the schematic closure of* $\Sigma_{\underline{\tau}_{S'}}^{G_{S'}}$ *in* $\mathcal{P}ar_t(G_{S'}) \times_{S'} \mathcal{P}ar_t(G_{S'})$ *commutes with the base changes* $X \to$ $\mathcal{P}ar_t(G_{S'})$. *After replacing in* $X \times_{\mathcal{P}ar_t(G_{S'})} \overline{\Sigma}_{\underline{\tau}_{S'}}^{G_{S'}}$ (*resp.* $X \times_{\mathcal{P}ar_t(G_{S'})} \Sigma_{\underline{\tau}_{S'}}^{G_{S'}}$), $\mathcal{P}ar_t(G_{S'})$ *by* $S' \times_S \mathcal{P}ar_t(G)$, *and* $\overline{\Sigma}_{\underline{\tau}_{S'}}^{G_{S'}}$ *by* $S' \times_S \overline{\Sigma}_{\underline{\tau}}^G$ *one obtains:* $X \times_{\mathcal{P}ar_t(G_{S'})}$ $\overline{\Sigma}_{\underline{\tau}_{S'}}^{G_{S'}} = X \times_{\mathcal{P}ar_t(G)} \overline{\Sigma}_{\underline{\tau}}^G$ (*resp.* $X \times_{\mathcal{P}ar_t(G_{S'})} \Sigma_{\underline{\tau}_{S'}}^{G_{S'}} = X \times_{\mathcal{P}ar_t(G)} \Sigma_{\underline{\tau}}^G$). *From the hypothesis it follows that* $\Sigma_{\underline{\tau}_{S'}}^{G_{S'}} \subset \overline{\Sigma}_{\underline{\tau}_{S'}}^{G_{S'}}$ *is Universally schematically dense relatively to* $\mathcal{P}ar_t(G_{S'})$. *It results that the formation of the schematic closure commutes with base changes* $X \to \mathcal{P}ar_t(G_{S'})$, *so that* $\left(\widehat{\Sigma}_{\underline{g}_{S'}}^{G_{S'}}, \overline{\Sigma}_{\underline{\tau}_{S'}}^{G_{S'}}, \pi_{\underline{g}_{S'}}^{G_{S'}}, \Theta_{\underline{g}_{S'}}^{G_{S'}}\right)$ *is* $\mathcal{P}ar_t(G_{S'})$-*Universal.*

In the next section a sufficient condition on the base scheme S implying that $\Sigma_{\underline{\tau}} \xrightarrow{\mathrm{pr}_{\Sigma_{\underline{\tau}}}} \mathcal{P}ar(G)$ is universally relatively schematically dense in $\overline{\Sigma}_{\underline{\tau}} \xrightarrow{\mathrm{pr}_{\overline{\Sigma}_{\underline{\tau}}}} \mathcal{P}ar(G)$ is given.

16.2.1 Universality of Smooth Resolutions

Let $\mathcal{R}$ be a $\mathbb{Z}$-root system. It is said that a reductive S-group scheme G endowed with a frame E is of type $\mathcal{R}$, if the root data it defines satisfies

$$\mathcal{R}(E) = \mathcal{R}.$$

It is recalled that a reductive S-group scheme G is of type $\mathcal{R}$, if there exists an étale covering $(S_i \to S)$ of S such that G_{S_i} is of endowed with a frame E_i of type $\mathcal{R}$. One recalls the following

Definition 16.11 *(cf. [23], Exp. XXIII, p. 317)*
*Given a $\mathbb{Z}$-root system $\mathcal{R}$ there exists a reductive group $\mathbb{Z}$-scheme $\underline{Ep_{\mathbb{Z}}}(\mathcal{R})$ endowed with a **canonical frame** $E^{\mathcal{R}}$, of type $\mathcal{R}$, called the **Chevalley group scheme of type** $\mathcal{R}$.*

The construction of the group $\underline{Ep_{\mathbb{Z}}}(\mathcal{R})$ is carried out in loc. cit., Exp. XXV. More precisely for all reduced $\mathbb{Z}$-root data $\mathcal{R}$ endowed with a root data frame is associated a $\mathbb{Z}$-group G endowed with a canonical frame $E^{\mathcal{R}}$ so that the root data defined by $E^{\mathcal{R}}$ is isomorphic to $\mathcal{R}$. This construction follows from that of Chevalley's Tohoku memoir [10] later improved by Chevalley himself [9].

Considering the group of automorphisms of a Lie algebra over the complex numbers associated to a Cartan matrix, a germ of group with root data given by the adjoint root data $ad(\mathcal{R})$ is obtained, giving rise to a reductive group scheme over $\mathbb{Z}$ following a procedure by Weil. The reductive group $\underline{Ep}_{\mathbb{Z}}(\mathcal{R})$ is obtained as a covering group of this one. The other achievement of loc. cit., is the functorial presentation of this group in terms of generators and relations, which is recalled in the last chapter, where its correspondence is shown with that of the automorphisms group of a building.

In this section the Universal Schubert scheme and its smooth resolution of singularities of $\underline{Ep}_{\mathbb{Z}}(\mathcal{R})$ are compared with the corresponding objects associated with an S-reductive group scheme G of type $\mathcal{R}$.

The data of a frame E of type $\mathcal{R}$ of a reductive group S-scheme G amounts to that of an isomorphism

$$G \simeq \underline{Ep}_{\mathbb{Z}}(\mathcal{R}) \times S = \underline{Ep}_S(\mathcal{R}).$$

More precisely, with the notation of loc. cit., Exp. XXIV, the $\underline{Aut}_{S-gr}$ $(\underline{Ep}_S(\mathcal{R}))$-principal fiber space of isomorphisms

$$\underline{Isom}_{S-gr}(\underline{Ep}_S(\mathcal{R}), G)$$

may be seen as the scheme of frames of G of type $\mathcal{R}$, i.e. the sections of $\underline{Isom}_{S-gr}(\underline{Ep}_S(\mathcal{R}), G)$ over $S' \longrightarrow S$ are the frames of $G_{S'}$(cf. loc. cit., Exp. XXIV, Remarque 1.20). From the above isomorphism it follows that all the schemes associated to the reductive S-group G endowed with a frame E of type $\mathcal{R}$ are obtained in general by pull-back of the corresponding scheme of $\underline{Ep}_{\mathbb{Z}}(\mathcal{R})$, with the exception of the Schubert scheme. This later one deserves special attention. Assume there is an isomorphism

$$\underline{Ep}_S(\mathcal{R}) \simeq G,$$

where $\underline{Ep}_S(\mathcal{R}) = \underline{Ep}_{\mathbb{Z}}(\mathcal{R}) \times_{\mathrm{Spec}(\mathbb{Z})} S$ (cf. loc. cit., Exp. XXIV). One investigates the link between the schematic closure $\overline{\Sigma}^G$ of

$$\left(\Sigma^{\underline{Ep}_{\mathbb{Z}}(\mathcal{R})}\right)_S = \Sigma^G$$

in $\mathcal{P}ar(G) \times_S \mathcal{P}ar(G) \times_S \mathrm{Relpos}_G$, and the pull-back $\left(\overline{\Sigma}^{\underline{Ep}_{\mathbb{Z}}(\mathcal{R})}\right)_S$ of the corresponding schematic closure $\overline{\Sigma}^{\underline{Ep}_{\mathbb{Z}}(\mathcal{R})}$ of $\Sigma^{\underline{Ep}_{\mathbb{Z}}(\mathcal{R})}$ in $\mathcal{P}ar(\underline{Ep}_{\mathbb{Z}}(\mathcal{R})) \times \mathcal{P}ar(\underline{Ep}_{\mathbb{Z}}(\mathcal{R})) \times \mathrm{Relpos}_{\underline{Ep}_{\mathbb{Z}}(\mathcal{R})}$, by the canonical morphism $S \to \mathrm{Spec}(\mathbb{Z})$. More precisely we establish the existence of an open subscheme $U^{\mathcal{R}}$ is established so that $\left(\Sigma^{\underline{Ep}_{\mathbb{Z}}(\mathcal{R})}\right)_{U^{\mathcal{R}}} \subset \left(\overline{\Sigma}^{\underline{Ep}_{\mathbb{Z}}(\mathcal{R})}\right)_{U^{\mathcal{R}}}$ is an open relatively schematically dense subscheme. Implying that for all $S \to U^{\mathcal{R}}$ and all S-reductive group scheme G, the smooth resolution quadruples are S-universal.

The notation is simplified as follows.

Notation 16.12

Let A be the apartment given by the $\mathbb{Z}$-root system $\mathcal{R} = \mathcal{R}(E^{\mathcal{R}})$, i.e. $A = \mathcal{A}_E$. Given $\tau \in Relpos\, A$, write τ instead of $\tau_{Spec(\mathbb{Z})}$, thus $\Sigma_\tau^{\mathcal{R}}$ instead of $\Sigma_{\tau_{Spec(\mathbb{Z})}}^{\frac{Ep_\mathbb{Z}(\mathcal{R})}{}} \cdots$ etc. For $g \in gall_A^m$ write g instead $g_{Spec(\mathbb{Z})}$, and denote by τ the type of relative position in A given by g. Write $\Sigma^{\mathcal{R}} = \Sigma^{\underline{Ep}_\mathbb{Z}(\mathcal{R})}$, $\widehat{\Sigma}^{\mathcal{R}} = \widehat{\Sigma}^{Ep_\mathbb{Z}(\mathcal{R})}$, $\overline{\Sigma}^{\mathcal{R}} = \overline{\Sigma}^{\underline{Ep}_\mathbb{Z}(\mathcal{R})}$ (**The Absolute Universal Schubert Scheme of root type $\mathcal{R}$**). *Let $\Sigma^{\mathcal{R}} = \coprod\limits_{\tau \in Relpos\, A} \Sigma_\tau^{\mathcal{R}}$*

$$\left(resp.\ \widehat{\Sigma}^{\mathcal{R}} = \coprod\limits_{g \in gall^m\, A} \widehat{\Sigma}_g^{\mathcal{R}},\ \overline{\Sigma}^{\mathcal{R}} = \coprod\limits_{\tau \in Relpos\, A} \overline{\Sigma}_\tau^{\mathcal{R}} \right).$$

be the canonical decompositions. Write

$$\pi^{\mathcal{R}} = \pi^{\underline{Ep}_\mathbb{Z}(\mathcal{R})},\ \Theta^{\mathcal{R}} = \Theta^{\underline{Ep}_\mathbb{Z}(\mathcal{R})} \cdots etc,$$

for π, $\Theta \cdots$ etc, defined as in the previous section with $G = \underline{Ep}_\mathbb{Z}(\mathcal{R})$.

One investigates:

- Under which condition on S the quadruple

$$\left(S \times_{Spec(\mathbb{Z})} \widehat{\Sigma}_g^R, S \times_{Spec(\mathbb{Z})} \overline{\Sigma}_\tau^R, Id_S \times_{Spec(\mathbb{Z})} \pi_g^R, Id_S \times_{Spec(\mathbb{Z})} \Theta_g^R \right)$$

obtained by the base change $S \longrightarrow Spec(\mathbb{Z})$ from the quadruple

$$\left(\widehat{\Sigma}_g^R, \overline{\Sigma}_\tau^R, \pi_g^R, \Theta_g^R \right)$$

giving a resolution of singularities for a Schubert Scheme of the Chevalley group scheme $\underline{Ep}_\mathbb{Z}(\mathcal{R})$, coincides with the resolution of singularities $\left(\widehat{\Sigma}_{\underline{g}}^G, \overline{\Sigma}_{\delta_2(\underline{g})}^G, \pi_{\underline{g}}^G, \Theta_{\underline{g}}^G \right)$ for a Schubert Scheme of $G = \underline{Ep}_S(\mathcal{R})$. A class of schemes forming a full subcategory of that of schemes such that the quadruple

$$\left(\widehat{\Sigma}_g^{\underline{Ep}_\mathbb{Z}(\mathcal{R})}, \overline{\Sigma}_\tau^{\underline{Ep}_\mathbb{Z}(\mathcal{R})}, \pi_g^{Ep_\mathbb{Z}(\mathcal{R})}, \Theta_g^{\underline{Ep}_\mathbb{Z}(\mathcal{R})} \right)$$

is Universal relatively to this subcategory is obtained.

- Under which condition on $\left(\widehat{\Sigma}_{\underline{g}}^G, \overline{\Sigma}_{\delta_2(\underline{g})}^G, \pi_{\underline{g}}^G, \Theta_{\underline{g}}^G \right)$, the quadruple

$$\left(S' \times_S \widehat{\Sigma}_{\underline{g}}^G, S \times_S \overline{\Sigma}_{\delta_2(\underline{g})}^G, Id_{S'} \times_S \pi_{\underline{g}}^G, Id_{S'} \times_S \Theta_{\underline{g}}^G \right)$$

obtained by the base change $S' \longrightarrow S$ from the quadruple $\left(\widehat{\Sigma}_{\underline{g}}^G, \overline{\Sigma}_{\delta_2(\underline{g})}^G, \pi_{\underline{g}}^G, \Theta_{\underline{g}}^G \right)$, coincides with the resolution of singularities

$\left(\widehat{\Sigma}_{\underline{g}_{S'}}^{G_{S'}}, \overline{\Sigma}_{\delta_2(\underline{g}_{S'})}^{G_{S'}}, \pi_{\underline{g}_{S'}}^{G_{S'}}, \Theta_{\underline{g}_{S'}}^{G_{S'}} \right)$ for a Schubert scheme of $G_{S'}$. In other words one looks under which conditions the quadruple associated to G is S-Universal.

The following proposition states the flatness of the Absolute Universal Schubert Scheme of root type $\mathcal{R}$. This allows one to apply the general theorems of schematic closure and base change of which are resumed in §16.3.

Proposition 16.13
Keep the above notation. The morphisms

$$\overline{\Sigma}^{\mathcal{R}} \to Spec(\mathbb{Z}) \left(resp. \, \overline{\Sigma}_{\tau}^{\mathcal{R}} \to Spec(\mathbb{Z}), \, \overline{\Sigma}_{(\tau, P_t(E^{\mathcal{R}}))}^{\mathcal{R}} \to Spec(\mathbb{Z}) \right)$$

*are **proper**, **flat**, and of **finite presentation**; the corresponding schemes are* **integral**.

Proof *It is clear that by definition $\overline{\Sigma} = \overline{\Sigma}^{\mathcal{R}}$ is a closed subscheme of the proper $\mathbb{Z}$-scheme*

$$\mathcal{P}ar(\underline{Ep}_{\mathbb{Z}}(\mathcal{R})) \times_{Spec(\mathbb{Z})} \mathcal{P}ar(\underline{Ep}_{\mathbb{Z}}(\mathcal{R})) \times_{Spec(\mathbb{Z})} Relpos_{\underline{Ep}_{\mathbb{Z}}(\mathcal{R})},$$

and thus proper. Let it be seen that $\overline{\Sigma}$ is flat over $Spec(\mathbb{Z})$. Write

$$\Sigma = \coprod_{\tau \in Relpos \, \mathcal{A}_E} \Sigma_{\tau} = \coprod_{\tau \in Relpos \, \mathcal{A}_E} \mathcal{S}tand(\tau_{Spec(\mathbb{Z})}).$$

Thus

$$\overline{\Sigma} = \coprod \overline{\Sigma}_{\tau},$$

where $\overline{\Sigma}_{\tau}$ is the projective scheme given by the schematic closure of $\mathcal{S}tand(\tau)$ in $\mathcal{P}ar(\underline{Ep}_{\mathbb{Z}}(\mathcal{R})) \times_{Spec(\mathbb{Z})} \mathcal{P}ar(\underline{Ep}_{\mathbb{Z}}(\mathcal{R}))$. It follows that $\overline{\Sigma}_{\tau}$ is a scheme of finite presentation over $\mathbb{Z}$.
To prove that $\overline{\Sigma}_{\tau}$ is flat over $Spec(\mathbb{Z})$, it suffices to see that $\overline{\Sigma}_{\tau}$ is integral. From this it follows that the sheaf of rings $\mathcal{O}_{\overline{\Sigma}_{\tau}}$ has no torsion and is thus flat over $\mathbb{Z}$. Observe that the transitive action of $\underline{Ep}_{\mathbb{Z}}(\mathcal{R})$ on Σ_{τ} implies that the Schubert cell Σ_{τ} is irreducible and thus that $\overline{\Sigma}_{\tau}$ is also irreducible. In Proposition 12.20 there is a covering

$$\Sigma_{\tau} = \mathcal{S}tand(\tau_{Spec(\mathbb{Z})}) = \bigcup_{(R_+, C') \in I_{\tau}(\mathcal{R})} \sigma_{R_+, C'}^{(\tau)}(\mathcal{U}_{t, R_+} \times_{Spec(\mathbb{Z})} U(\tau_{C'}))$$

by open sets. Each of them being isomorphic to the spectrum of a polynomial ring over $\mathbb{Z}$. Given

$$g \in gall_A^m \text{ with } \tau_g = \tau \quad (resp. \, (R_+, C') \in I_{\tau}(\mathcal{R})),$$

it results from Corollary 15.15 and the relation that exists between "the big open cells of Σ_τ and $\widehat{\Sigma}_g$ ", that

$$\overline{\Sigma}_\tau = \sigma_{R_+,C'}^{(\tau)} \left(\mathcal{U}_{t,R_+} \times_{Spec(\mathbb{Z})} U(\tau_{C'}) \right)^{schc}.$$

Thus one has that the schematic closure $\overline{\Sigma}_\tau = \Sigma_\tau^{schc}$ is an integral scheme. (Recall that the schematic closure of an integral subscheme is integral.) It results that the local rings $\mathcal{O}_{\overline{\Sigma}_\tau,x}$ of $\overline{\Sigma}_\tau$ are without $\mathbb{Z}$-torsion, and thus flat over $\mathbb{Z}$.
The integrality of $\overline{\Sigma}^{\mathcal{R}}_{(\tau,P_t(E^{\mathcal{R}}))} \to Spec(\mathbb{Z})$ results from the fact that $\Sigma^{\mathcal{R}}_{(\tau,P_t(E^{\mathcal{R}}))}$ may be covered by open sets which are spectra of polynomial rings over $\mathbb{Z}$ as it is irreducible.

It follows from Proposition 16.13 that the couple $\Sigma_\tau^{\mathcal{R}} \subset \overline{\Sigma}_\tau^{\mathcal{R}}$ satisfies the hypothesis of Proposition 16.44 of the last section.

Proposition 16.14
Assume that $G = \underline{Ep}_S(\mathcal{R})$ so that $\Sigma^G \simeq \left(\Sigma^{\mathcal{R}}\right)_S$. Let $Z \subset Spec(\mathbb{Z})$ be a not-empty subscheme so that the open subscheme

$$\left(\Sigma_\tau^{\mathcal{R}}\right)_Z \subset \left(\overline{\Sigma}_\tau^{\mathcal{R}}\right)_Z \quad (\tau \in Relpos\, \mathcal{A}_E),$$

*is **universally schematically dense relatively to** Z. Let S be a Z-scheme. Then there is the Universal Schubert scheme of G of type τ is obtained as the pull-back of the **Absolute** Universal Schubert scheme of type τ, i.e.*

$$\overline{\Sigma}_{\tau S}^G = \left(\overline{\Sigma}_\tau^{\mathcal{R}}\right)_S = \overline{\Sigma}_\tau^{\mathcal{R}} \times_{Spec(\mathbb{Z})} S.$$

*This means that **the formation of the schematic closure** of $\Sigma_\tau^{\mathcal{R}}$ in $\mathcal{P}ar(\underline{Ep}_{\mathbb{Z}}(\mathcal{R})) \times \mathcal{P}ar(\underline{Ep}_{\mathbb{Z}}(\mathcal{R}))$ **commutes** with base changes of the form*

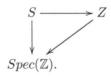

$$Spec(\mathbb{Z}).$$

Proof *The statement of the Proposition follows immediately from Proposition 16.44. Put $S = Z$, $S' = S$, $V = \left(\Sigma_\tau^{\mathcal{R}}\right)_Z$, $X = \left(\overline{\Sigma}_\tau^{\mathcal{R}}\right)_Z$, and*

$$Y = \mathcal{P}ar(\underline{Ep}_Z(\mathcal{R})) \times_Z \mathcal{P}ar(\underline{Ep}_Z(\mathcal{R})) = \left(\mathcal{P}ar(\underline{Ep}_{\mathbb{Z}}(\mathcal{R})) \times_{Spec(\mathbb{Z})} \mathcal{P}ar(\underline{Ep}_{\mathbb{Z}}(\mathcal{R}))\right)_Z.$$

On the other hand, recall that $\overline{\Sigma}_{\tau S}^G$ is by definition the schematic closure of

$$\Sigma_{\tau S}^G = \left(\Sigma_\tau^{\mathcal{R}}\right)_S = \left(\Sigma_\tau^{\mathcal{R}}\right)_Z \times_Z S = V_S$$

in

$$Par(G) \times_S Par(G) = \left(Par(\underline{Ep}_Z(\mathcal{R})) \times_Z Par(\underline{Ep}_Z(\mathcal{R}))\right)_S = Y_S.$$

Thus $\overline{\Sigma}_{\tau_S}^G = \left(\overline{\Sigma}_\tau^{\mathcal{R}}\right)_S = \left(\overline{\Sigma}_\tau^{\mathcal{R}}\right)_Z \times_Z S$ *as it follows from Propositions 16.13 and 16.45.*

The following proposition characterizes the subschemes $Z \subset Spec(\mathbb{Z})$ satisfying the condition of the above proposition.

Proposition 16.15 *The notation of the above proposition is retained.*

- *The open subscheme* $(\Sigma_\tau)_y \subset \left(\overline{\Sigma}_\tau\right)_y$ *is schematically dense if and only if* $\left(\overline{\Sigma}_\tau\right)_y$ *is integral.*

- *A subscheme* $Z \subset Spec(\mathbb{Z})$ *satisfies the hypothesis of Proposition 16.14 if and only if for every* $y \in Z$ *the fiber* $\left(\overline{\Sigma}_\tau\right)_y$ *is an integral scheme.*

Proof *If* $\left(\overline{\Sigma}_\tau\right)_y$ *is integral then the not-empty open subscheme* $(\Sigma_\tau)_y \subset \left(\overline{\Sigma}_\tau\right)_y$ *is dense in* $\left(\overline{\Sigma}_\tau\right)_y$, *and thus schematically dense. Reciprocally if the (integral) open subscheme* $(\Sigma_\tau)_y \subset \left(\overline{\Sigma}_\tau\right)_y$ *is schematically dense, then* $\left(\overline{\Sigma}_\tau\right)_y$ *is integral as the schematic closure of an integral subscheme. The second assertion follows from the first one and Proposition 16.44.*

Remark 16.16 *In fact there is a subscheme* $U \subset Spec(\mathbb{Z})$ *which is open and maximal with respect to the property defining* Z.

This remark justifies the following

Definition 16.17
Given $\tau \in Relpos\ A$ *one associates with* τ *the reduced subscheme* $U_\tau^{\mathcal{R}} \subset Spec(\mathbb{Z})$, *defined by*

$$U_\tau^{\mathcal{R}} = \left\{y \in Spec(\mathbb{Z}) \mid \text{the fiber } \left(\overline{\Sigma}_\tau^{\mathcal{R}}\right)_y \text{ is integral}\right\}.$$

Let

$$U_{(\tau, P_t)} = \left\{y \in Spec(\mathbb{Z}) \mid \text{the fiber } \left(\overline{\Sigma}_{(\tau, P_t)}^{\mathcal{R}}\right)_y \text{ is integral}\right\},$$

where one writes $P_t = P_t(E^{\mathcal{R}})$. *Write*

$$\mathcal{U}_\tau^{\mathcal{R}} = U_\tau^{\mathcal{R}} \times_{Spec(\mathbb{Z})} Par(\underline{Ep}_{\mathbb{Z}}(\mathcal{R})) \subset Par(\underline{Ep}_{\mathbb{Z}}(\mathcal{R})).$$

Proposition 16.18 *The subschemes $U_\tau^\mathcal{R} \subset Spec(\mathbb{Z})$, and $U_{(\tau,P_t)} \subset Spec(\mathbb{Z})$ satisfy:*

- $U_\tau^\mathcal{R} \neq \emptyset$ *(resp. $U_{(\tau,P_t)} \neq \emptyset$).*

- $U_\tau^\mathcal{R} = U_{(\tau,P_t)} = \left\{ y \in Spec(\mathbb{Z}) \mid \text{the geometric fiber } \left(\overline{\Sigma}_\tau^\mathcal{R}\right)_{\overline{y}} \text{ is integral} \right\}.$

- $U_\tau^\mathcal{R}$ *is an open not-empty subscheme of $Spec(\mathbb{Z})$.*

- $\mathcal{U}_\tau^\mathcal{R}$ *is an open not-empty subscheme of $\mathcal{P}ar(\underline{Ep}_\mathbb{Z}(\mathcal{R}))$.*

Proof *Observe that $U_\tau^\mathcal{R} \neq \emptyset$ (resp. $U_{(\tau,P_t)} \neq \emptyset$) as the generic point ξ of $Spec(\mathbb{Z})$ belongs to $U_\tau^\mathcal{R}$ (resp. $U_{(\tau,P_t)}$). Remark that $\left(\overline{\Sigma}_\tau\right) \times_{Spec(\mathbb{Z})} Spec(\mathbb{Q})$ is clearly integral as the schematic closure $\overline{\Sigma}_\tau$ of the integral scheme Σ_τ is also integral (cf. proof of Proposition 16.51). The second assertion is proved in Proposition 16.22. As $\overline{\Sigma}_\tau^\mathcal{R}$ is a proper and faithfully flat $Spec(\mathbb{Z})$-scheme, it is known by loc. cit., that*

$$\left\{ y \in Spec(\mathbb{Z}) \mid \text{the geometric fiber } \left(\overline{\Sigma}_\tau^\mathcal{R}\right)_{\overline{y}} \text{ is integral} \right\}$$

is an open non-empty subscheme of $Spec(\mathbb{Z})$. Thus by definition $\mathcal{U}_\tau^\mathcal{R}$ is an open subscheme of $\mathcal{P}ar(\underline{Ep}_\mathbb{Z}(\mathcal{R}))$.

Apply now the Proposition 16.44 and Proposition 16.45 to show that $U_\tau^\mathcal{R}$ satisfies the condition of the subscheme Z of Proposition 16.14 and is maximal with this property. It follows from Proposition 16.13 that the couple $\Sigma_\tau^\mathcal{R} \subset \overline{\Sigma}_\tau^\mathcal{R}$ satisfies the hypothesis of Proposition 16.44 of the next section. Observe that the geometrical fibers of $\Sigma_\tau^\mathcal{R}$ are integral (cf. loc. cit.).

Theorem 16.19
The open subscheme

$$\left(\Sigma_\tau^\mathcal{R}\right)_{U_\tau^\mathcal{R}} \subset \left(\overline{\Sigma}_\tau^\mathcal{R}\right)_{U_\tau^\mathcal{R}}$$

*is **universally schematically dense** relatively to the open subscheme $U_\tau^\mathcal{R} \subset \mathbf{Spec}(\mathbb{Z})$. This subscheme is maximal for this property, i.e. if $Z \subset Spec(\mathbb{Z})$ is a subscheme so that the open subscheme $(\Sigma_\tau)_Z \subset (\overline{\Sigma}_\tau)_Z$ is universally schematically dense relatively to Z, then*

$$Z \subset U_\tau^\mathcal{R}.$$

The Absolute Universal Schubert scheme of type τ, $\left(\overline{\Sigma}_\tau^\mathcal{R}\right)_{U_\tau^\mathcal{R}}$ satisfies the following property. Let G be a reductive S-group scheme of type $\mathcal{R}$ endowed with *a frame E, and assume that S is a $U_\tau^{\mathcal{R}}$-scheme. There is an isomorphism*

$$\overline{\Sigma}_{TS}^{G} \simeq \left(\overline{\Sigma}_\tau^{\mathcal{R}}\right) \times_{Spec(\mathbb{Z})} S = \left(\overline{\Sigma}_{TU_\tau^{\mathcal{R}}}^{\underline{Ep}_{U_\tau^{\mathcal{R}}}(\mathcal{R})}\right) \times_{U_\tau^{\mathcal{R}}} S$$

(cf. Proposition 16.14).

Remark 16.20

The scheme $\left(\overline{\Sigma}_\tau^{\mathcal{R}}\right)_{U_\tau^{\mathcal{R}}}$ is in fact the Schubert scheme of type τ of the $U_\tau^{\mathcal{R}}$-group $\underline{Ep}_{U_\tau^{\mathcal{R}}}(\mathcal{R}) = U_\tau^{\mathcal{R}} \times \underline{Ep}_{\mathbb{Z}}(\mathcal{R})$, i.e. one has

$$\left(\overline{\Sigma}_\tau^{\mathcal{R}}\right)_{U_\tau^{\mathcal{R}}} = \overline{\Sigma}_{\tau U_\tau}^{\underline{Ep}_{U_\tau}(\mathcal{R})}.$$

When one looks at $\overline{\Sigma}_\tau^{\mathcal{R}}$ as a $\mathcal{P}ar_t(\underline{Ep}_{\mathbb{Z}}(\mathcal{R}))$-scheme one obtains the following result corresponding to $\mathcal{U}_\tau^{\mathcal{R}}$ instead of $U_\tau^{\mathcal{R}}$.

Theorem 16.21
The notation of the above theorem are kept.

- *The open subscheme*

$$\left(\Sigma_\tau^{\mathcal{R}}\right)_{\mathcal{U}_\tau^{\mathcal{R}}} \subset \left(\overline{\Sigma}_\tau^{\mathcal{R}}\right)_{\mathcal{U}_\tau^{\mathcal{R}}}$$

*is **universally schematically dense** relatively to $\mathcal{U}_\tau^{\mathcal{R}} \subset \mathcal{P}ar_t(\underline{Ep}_{\mathbb{Z}}(\mathcal{R}))$. Observe that*

$$\left(\Sigma_\tau^{\mathcal{R}}\right)_{\mathcal{U}_\tau^{\mathcal{R}}} = \left(\Sigma_\tau^{\mathcal{R}}\right)_{U_\tau^{\mathcal{R}}} \quad \left(resp. \ \left(\overline{\Sigma}_\tau^{\mathcal{R}}\right)_{\mathcal{U}_\tau^{\mathcal{R}}} = \left(\overline{\Sigma}_\tau^{\mathcal{R}}\right)_{U_\tau^{\mathcal{R}}}\right).$$

- *$\mathcal{U}_\tau^{\mathcal{R}}$ is **maximal** with respect to this property, i.e. if the subscheme $Z^{\mathcal{P}} \subset \mathcal{P}ar_t(\underline{Ep}_{\mathbb{Z}}(\mathcal{R}))$ satisfies*

$$\left(\Sigma_\tau^{\mathcal{R}}\right)_{Z^{\mathcal{P}}} \subset \left(\overline{\Sigma}_\tau^{\mathcal{R}}\right)_{Z^{\mathcal{P}}}$$

*is **universally schematically dense** relatively to $Z^{\mathcal{P}}$ then $Z^{\mathcal{P}} \subset \mathcal{U}_\tau^{\mathcal{R}}$.*

- *Let S be a $U_\tau^{\mathcal{R}}$-scheme. The embedding of the Schubert cell*

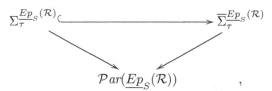

is a dominant morphism, i.e. it defines a universally relatively schematically dense open subscheme of the Schubert scheme.

- Let $S \xrightarrow{\sigma_P} \mathcal{P}ar(\underline{Ep}_S(\mathcal{R}))$ be the section defined by the parabolic subgroup P of $\underline{Ep}_S(\mathcal{R})$ then the following diagram is **cartesian**:

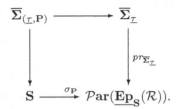

This means that there is a canonical isomorphism

$$\overline{\Sigma}_{(\tau_S,P)}^{\underline{Ep}_S(\mathcal{R}))} \simeq \left(\overline{\Sigma}_{\tau_S}^{\underline{Ep}_S(\mathcal{R}))}\right)_{\sigma_P} = S \times_{\mathcal{P}ar_t(\underline{Ep}_S(\mathcal{R}))} \left(\overline{\Sigma}_{\tau_S}^{\underline{Ep}_S(\mathcal{R}))}\right) .$$

Otherwise stated the Schubert schemes $\overline{\Sigma}_{(\tau_S,P)}^{\underline{Ep}_S(\mathcal{R}))}$ are obtained as the fibers of the Universal Schubert scheme $\overline{\Sigma}_{\tau_S}^{\underline{Ep}_S(\mathcal{R}))} \longrightarrow \mathcal{P}ar_t(\underline{Ep}_S(\mathcal{R}))$.

Proof *Following the pattern of the proof of Proposition 16.13 one obtains that $\Sigma_{(\tau,P_t(E))}$ is an integral scheme. Thus its schematic closure $\overline{\Sigma}_{(\tau,P_t(E))}$ in $\mathcal{P}ar_s(\underline{Ep}_\mathbb{Z}(\mathcal{R}))$ is also an integral scheme. It results that $\overline{\Sigma}_{(\tau,P_t(E))}$ is $Spec(\mathbb{Z})$-flat and finally the locally trivial morphism with $Spec(\mathbb{Z})$-flat typical fiber $\overline{\Sigma}_{(\tau,P_t(E))}$:*

$$pr_{P,\overline{\Sigma}_\tau} : \overline{\Sigma}_\tau^{\mathcal{R}} \to \mathcal{P}ar_t(\underline{Ep}_\mathbb{Z}(\mathcal{R}))$$

is flat, and of finite presentation, as $\overline{\Sigma}_\tau^{\mathcal{R}}$ is a noetherian scheme. One may thus apply Proposition 16.47 or Proposition 16.44. Let it be seen that for

$$z \in \mathcal{U}_\tau^{\mathcal{R}}$$

the open subscheme

$$\left(\Sigma_\tau^{\mathcal{R}}\right)_z \subset \left(\overline{\Sigma}_\tau^{\mathcal{R}}\right)_z$$

is schematically dense or what amounts to the same thing that $\left(\overline{\Sigma}_\tau^{\mathcal{R}}\right)_z$ is an integral scheme.
Suppose that $y \in U_\tau^{\mathcal{R}}$. By definition of $U_\tau^{\mathcal{R}}$ it is known that the fiber

$$\left(\overline{\Sigma}_\tau^{\mathcal{R}}\right)_y = Spec(\kappa(y)) \times \overline{\Sigma}_\tau^{\mathcal{R}}$$

is integral. On the other hand, $pr_{P,\overline{\Sigma}_\tau}$ defines a locally trivial fibration trivialized by the big cell open covering of $\mathcal{P}ar_t(\underline{Ep}_\mathbb{Z}(\mathcal{R}))$. Hence there exists an open sub-scheme $U \subset \mathcal{P}ar_t(\underline{Ep}_\mathbb{Z}(\mathcal{R}))$ isomorphic to a polynomial ring over $\mathbb{Z}$ spectrum and an isomorphism $(\Sigma_\tau^{\mathcal{R}})_U \simeq U \times \overline{\Sigma}_{(\tau,P_t(E))}$. Thus the *open subscheme* $\left(\left(\overline{\Sigma}_\tau^{\mathcal{R}}\right)_U\right)_y$ *of* $\left(\overline{\Sigma}_\tau^{\mathcal{R}}\right)_y$ *is integral. From the isomorphism*

$\left(\left(\overline{\Sigma}_\tau^{\mathcal{R}}\right)_U\right)_y \simeq U_y \times_{Spec(\kappa(y))} \left(\overline{\Sigma}_{(\tau,P_t(E))}\right)_y$, *the fact that*

$$\mathcal{P}ar_t(\underline{Ep}_{\mathbb{Z}}(\mathcal{R}))_y \longrightarrow Spec(\kappa(y)) ,$$

is faithfully flat, and $U_y \subset \mathcal{P}ar_t(\underline{Ep}_{\mathbb{Z}}(\mathcal{R}))_y$ *is an open set it is deduced that* $\left(\overline{\Sigma}_{(\tau,P_t(E))}\right)_y$ *is integral. Thus it results in the inclusion* $U_\tau^{\mathcal{R}} \subset U_{(\tau,P_t)}$.

If $\left(\overline{\Sigma}_{(\tau,P_t)}^{\mathcal{R}}\right)_y$ *is integral then for every set* U *of the open big cell cover-*

ing $\left(\left(\overline{\Sigma}_\tau^{\mathcal{R}}\right)_U\right)_y \simeq U_y \times_{Spec(\kappa(y))} \left(\overline{\Sigma}_{(\tau,P_t(E))}\right)_y$ *is also integral and* $\left(\overline{\Sigma}_\tau^{\mathcal{R}}\right)_y =$

$\bigcup\left(\left(\overline{\Sigma}_\tau^{\mathcal{R}}\right)_U\right)_y$, *where* U *runs over the big cell covering. It is recalled that for a locally noetherian scheme to be integral is equivalent to being a connected scheme that is covered by the spectra of integral domains. It is concluded that* $\left(\Sigma_\tau^{\mathcal{R}}\right)_y$ *is integral and thus the inclusion* $U_{(\tau,P_t)} \subset U_\tau^{\mathcal{R}}$ *and finally that* $U_\tau^{\mathcal{R}} = U_{(\tau,P_t)}$.

Let $z \in \mathcal{U}_\tau^{\mathcal{R}}$. *Its image* y *by the canonical morphism*

$$\mathcal{U}_\tau^{\mathcal{R}} = U_\tau^{\mathcal{R}} \times_{Spec(\mathbb{Z})} \mathcal{P}ar_t(\underline{Ep}_{\mathbb{Z}}(\mathcal{R})) \to U_\tau^{\mathcal{R}}.$$

belongs $U_{(\tau,P_t)}$ *and thus one has that* $\left(\overline{\Sigma}_{(\tau,P_t(E))}\right)_y$ *is integral. The absolutely integral open subscheme* $\left(\Sigma_{(\tau,P_t(E))}\right)_y \subset \left(\overline{\Sigma}_{(\tau,P_t(E))}\right)_y$ *is thus schematically dense and thus relatively schematically dense by Proposition 16.44. Hence from the isomorphism*

$$\left(\overline{\Sigma}_\tau^{\mathcal{R}}\right)_z \simeq Spec(\kappa(z)) \times_{Spec(\kappa(y))} \left(\overline{\Sigma}_{(\tau,P_t(E))}\right)_y$$

by transitivity of fibers (cf. [24], Corollaire 3.4.9), it results that $\left(\overline{\Sigma}_\tau^{\mathcal{R}}\right)_z$ *is integral as the schematic closure of the integral subscheme* $Spec(\kappa(z)) \times_{Spec(\kappa(y))}$ $\left(\Sigma_{(\tau,P_t(E))}\right)_y$. *Consequently on the basis of Proposition 16.44* $\Sigma_\tau^{\mathcal{R}} \subset \overline{\Sigma}_\tau^{\mathcal{R}}$ *is universally relatively schematically dense with respect to* $\mathcal{P}ar_t(\underline{Ep}_{\mathbb{Z}}(\mathcal{R}))$. *This achieves the proof of the first statement. The second statement is proved in the following proposition. The third one is a particular case of the fourth assertion of Proposition 16.44. The last one results from Proposition 16.14 taking on account the third statement.*

Proposition 16.22 *With the notation of the above theorem one has:*

- $U_{(\tau,P_t)} = U_\tau^{\mathcal{R}}$.

- *The open subscheme* $\mathcal{U}_\tau^{\mathcal{R}} \subset \mathcal{P}ar(\underline{Ep}_{\mathbb{Z}}(\mathcal{R}))$ *is maximal with the property* $\left(\Sigma_\tau^{\mathcal{R}}\right)_{\mathcal{U}_\tau^{\mathcal{R}}} \subset \left(\overline{\Sigma}_\tau^{\mathcal{R}}\right)_{\mathcal{U}_\tau^{\mathcal{R}}}$ *is a* **universally relatively schematically dense** *open subscheme.*

Proof *The first assertion results from the proof of the above proposition. Let it be proved the second one. By Propositions 16.44 it suffices to see that the open subscheme*

$$V = \left\{ z \in \mathcal{P}ar(\underline{Ep_{\mathbb{Z}}}(\mathcal{R})) \mid \text{the fiber } \left(\overline{\Sigma}_{\tau}^{\mathcal{R}} \right)_z \text{ is integral} \right\}$$

is equal to $\mathcal{U}_{\tau}^{\mathcal{R}}$. Let y be a point in the image of V by the natural projection. It follows by the proof of the above proposition that $y \in U_{(\tau, P_t)}$ and thus that $y \in U_{\tau}^{\mathcal{R}}$. It is concluded that $V \subset \mathcal{U}_{\tau}^{\mathcal{R}} = \mathcal{P}ar(\underline{Ep_{\mathbb{Z}}}(\mathcal{R})) \times_{Spec(\mathbb{Z})} U_{\tau}^{\mathcal{R}}$. By the above proposition $\mathcal{U}_{\tau}^{\mathcal{R}} = \mathcal{P}ar(\underline{Ep_{\mathbb{Z}}}(\mathcal{R})) \times_{Spec(\mathbb{Z})} \overline{U_{\tau}^{\mathcal{R}}} \subset V$ thus $V = \mathcal{U}_{\tau}^{\mathcal{R}}$.

Proposition 16.23

*For all scheme $S \to U_{\tau}^{\mathcal{R}} \subset Spec(\mathbb{Z})$, and all S-reductive group scheme G of type $\mathcal{R}$, $\Sigma_{\underline{\tau}}^{G} \subset \overline{\Sigma}_{\underline{\tau}}^{G}$ is **Universally schematically dense** relatively to S, where $\underline{\tau}$ denotes a type of relative position such that $\underline{\tau} = \tau_{S}'$ for some etale covering $S' \to S$.*

Proof *Suppose that G is endowed with a frame of type $\mathcal{R}$. Thus given a $U_{\tau}^{\mathcal{R}}$-scheme S one has $G \simeq S \times \underline{Ep_{\mathbb{Z}}}(\mathcal{R}) = S \times_{U_{\tau}^{\mathcal{R}}} \underline{Ep_{U_{\tau}^{\mathcal{R}}}}(\mathcal{R})$. From the definition of $U_{\tau}^{\mathcal{R}}$ it results that the subscheme*

$$\Sigma_{\tau S}^{G} = S \times_{U_{\tau}^{\mathcal{R}}} \Sigma_{\tau}^{\mathcal{R}} \subset S \times_{U_{\tau}^{\mathcal{R}}} \overline{\Sigma}_{\tau}^{\mathcal{R}}$$

is schematically dense. Thus for all $U_{\tau}^{\mathcal{R}}$-scheme S the schematic closure $\overline{\Sigma}_{\tau S}^{G}$ of $\Sigma_{\tau S}^{G}$ in $\mathcal{P}ar_t(G) \times_S \mathcal{P}ar_t(G)$ is equal to $S \times_{U_{\tau}^{\mathcal{R}}} \overline{\Sigma}_{\tau}^{\mathcal{R}}$. Let $S' \to S$. One has $S' \times_S \overline{\Sigma}_{\tau S}^{G} = S' \times_S \left(S \times \overline{\Sigma}_{\tau}^{\mathcal{R}} \right) = S' \times \overline{\Sigma}_{\tau}^{\mathcal{R}}$. Thus from the previous argument with S' instead of S one obtains $\overline{\Sigma}_{\tau S'}^{G_{S'}} = S' \times_S \overline{\Sigma}_{\tau S}^{G}$, i.e. $\Sigma_{\tau S}^{G} \subset \overline{\Sigma}_{\tau S}^{G}$ is Universally schematically dense relatively to S.

Let G be a reductive group scheme of type $\mathcal{R}$ and $S' \to S$. By definition there exists an etale covering $(S_i \to S)$, where S_i may be supposed finite etale over S, and isomorphisms $G_{S_i} \simeq S_i \times_{U_{\tau}^{\mathcal{R}}} \underline{Ep_{U_{\tau}^{\mathcal{R}}}}(\mathcal{R})$. Write $S_i' = S_i \times_S S'$. One has, for all i, $S_i \times_S \left(S' \times_S \overline{\Sigma}_{\tau S}^{G} \right) = S_i' \times_{S_i} \left(S_i \times_S \overline{\Sigma}_{\tau S}^{G} \right) = S_i' \times_{S_i} \overline{\Sigma}_{\tau S_i}^{G_{S_i}}$. The former equality is tautological and the later follows from the commutation of the formation of the schematic closure with finite etale base changes. By the first part of the proof the natural morphisms $\overline{\Sigma}_{\tau S_i'}^{G_{S_i'}} \to S_i' \times_{S_i} \overline{\Sigma}_{\tau S_i}^{G_{S_i}}$ are isomorphisms, as G_{S_i} is endowed with a frame of type $\mathcal{R}$. It is concluded that $\coprod_i S_i \times_S \left(S' \times_S \Sigma_{\tau S}^{G} \right) = \coprod_i \Sigma_{\tau S_i'}^{G_{S_i'}}$ is schematically dense in $\coprod_i S_i \times_S \left(S' \times_S \overline{\Sigma}_{\tau S}^{G} \right) = \coprod_i S_i' \times_{S_i} \overline{\Sigma}_{\tau S_i}^{G_{S_i}}$ and consequently that $S' \times_S \Sigma_{\tau S}^{G}$ is schematically dense in $S' \times_S \overline{\Sigma}_{\tau S}^{G}$. This achieves the proof of the proposition.

The following proposition is an immediate corollary of Theorem 16.21.

Proposition 16.24

Write $\Sigma^{\mathcal{R}} = \coprod\limits_{\tau \in \textbf{Relpos } \mathbf{A}(\mathbf{E}^{\mathcal{R}})} \Sigma_{\tau}^{\mathcal{R}}$ *(resp.* $\overline{\Sigma}^{\mathcal{R}} = \coprod\limits_{\tau \in \textbf{Relpos } \mathbf{A}(\mathbf{E}^{\mathcal{R}})} \overline{\Sigma}_{\tau}^{\mathcal{R}}$ *) and de-fine open subsets by* $\mathcal{U}^{\mathcal{R}} = \bigcap\limits_{\tau \in Relpos \ A} \mathcal{U}_{\tau}^{\mathcal{R}} \subset \mathcal{P}ar(\underline{Ep}_{\mathbb{Z}}(\mathcal{R}))$ *(resp.* $U^{\mathcal{R}} = \bigcap\limits_{\tau \in Relpos \ A} U_{\tau}^{\mathcal{R}} \subset S$*). There is a commutative diagram of* $\mathcal{P}ar(\underline{Ep}_{\mathbb{Z}}(\mathcal{R}))$*-morphisms*

$$\Sigma_{\mathcal{T}_{U}\mathcal{R}}^{\underline{Ep}_{U}\mathcal{R}(\mathcal{R})} = \left(\Sigma^{\mathcal{R}}\right)_{\mathcal{U}^{\mathcal{R}}} \lhook\joinrel\longrightarrow U^{\mathcal{R}} \times_{Spec(\mathbb{Z})} \overline{\Sigma}^{\mathcal{R}} = \left(\overline{\Sigma}^{\mathcal{R}}\right)_{\mathcal{U}^{\mathcal{R}}}$$
$$\mathcal{U}^{\mathcal{R}} ,$$

and a commutative diagram of S*-morphisms*

$$\Sigma_{\mathbf{T}_{U}\mathbf{R}}^{\underline{Ep}_{U}\mathbf{R}(\mathcal{R})} = \left(\Sigma^{\mathcal{R}}\right)_{\mathbf{U}^{\mathcal{R}}} \lhook\joinrel\longrightarrow U^{\mathcal{R}} \times_{Spec(\mathbb{Z})} \overline{\Sigma}^{\mathcal{R}} = \left(\overline{\Sigma}^{\mathcal{R}}\right)_{\mathbf{U}^{\mathcal{R}}}$$
$$U^{\mathcal{R}} ,$$

where the horizontal arrow is respectively a **universally dominant** *embedding relatively to* $\mathcal{U}^{\mathcal{R}}$ *and* $U^{\mathcal{R}}$*. Thus the open subscheme* $\left(\Sigma^{\mathcal{R}}\right)_{\mathcal{U}^{\mathcal{R}}} \subset \left(\overline{\Sigma}^{\mathcal{R}}\right)_{\mathcal{U}^{\mathcal{R}}}$ *is* **universally schematically dense** *relatively to* $\mathcal{U}^{\mathcal{R}}$ *and to* $U^{\mathcal{R}}$*.*

Let S be a $U_{\mathcal{R}}$-scheme, and G a reductive S-group scheme of type $\mathcal{R}$. The Universal property of the above diagram gives rise to the $\mathcal{P}\mathbf{ar}(G)$-**universally dominant** embedding

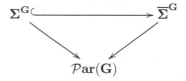

(resp. the S-**universally dominant** embedding

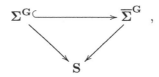

the $\Gamma^m(\underline{\Delta}(G))$-**universally dominant** embedding

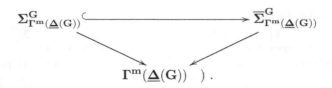

$$\Gamma^m(\underline{\Delta}(G)) \quad) \; .$$

Remark that this last assertion is equivalent to the following: for all section $\underline{g}$ of $\Gamma^m(\underline{\Delta}(G))$ the embedding $\Sigma_{\underline{g}} \hookrightarrow \overline{\Sigma}_{\underline{g}}$ is a dominant S-embedding. Thus

Theorem 16.25
Let S be a $U^{\mathcal{R}}$-scheme, and G a reductive S-group scheme of type $\mathcal{R}$.

1) $\left(\widehat{\Sigma}^G, \overline{\Sigma}^G, \pi^G, \Theta^G \right)$ *is an S-Universal smooth resolution.*

2) $\left(\widehat{\Sigma}^G_{\Gamma^m(\underline{\Delta}(G))}, \overline{\Sigma}^G_{\Gamma^m(\underline{\Delta}(G))}, \pi^G_{\Gamma^m(\underline{\Delta}(G))}, \Theta^G_{\Gamma^m(\underline{\Delta}(G))} \right)$ *is a $\Gamma^m(\underline{\Delta}(G))$-Universal smooth resolution, and also a $\Gamma^m(\underline{\Delta}(G)) \times_S \mathcal{P}ar(G)$-Universal smooth resolution. In fact the notation $\left(\widehat{\Sigma}^G_{\Gamma^m(\underline{\Delta}(G))}, \overline{\Sigma}^G_{\Gamma^m(\underline{\Delta}(G))}, \pi^G_{\Gamma^m(\underline{\Delta}(G))}, \Theta^G_{\Gamma^m(\underline{\Delta}(G))} \right)$ amounts to seeing $\left(\widehat{\Sigma}^G, \overline{\Sigma}^G, \pi^G, \Theta^G \right)$ as formed by $\Gamma^m(\underline{\Delta}(G))$-schemes and morphisms. Recall that $\pi^G_{\Gamma^m}$ denotes the canonical morphism*
$$\pi : \overline{\Sigma}^G \longrightarrow \Gamma^m(\underline{\Delta}(G)) \times_{Relpos_G} \overline{\Sigma}^G = \overline{\Sigma}^G_{\Gamma^m(\underline{\Delta}(G))}.$$

3) *For all section $\underline{g}$ of $\Gamma^m(\underline{\Delta}(G))$ the smooth resolution $\left(\widehat{\Sigma}^G_{\underline{g}}, \overline{\Sigma}^G_{\delta_2(\underline{g})}, \pi^G_{\underline{g}}, \Theta^G_{\underline{g}} \right)$ is S-Universal.*

4) *For all section $(\underline{g}, P_t)$ of $\Gamma^m(\underline{\Delta}(G)) \times_S \mathcal{P}ar_t(G)$ the smooth resolution*

$$\left(\widehat{\Sigma}^G_{(\underline{g}, P_t)}, \overline{\Sigma}^G_{(\delta_2(\underline{g}), P_t)}, \pi^G_{(\underline{g}, P_t)}, \Theta^G_{(\underline{g}, P_t)} \right)$$

is S-Universal.

Assume that G is splitted, i.e. endowed with a frame. Then the third assertion is represented by the following diagram whose front and back faces are cartesian squares.

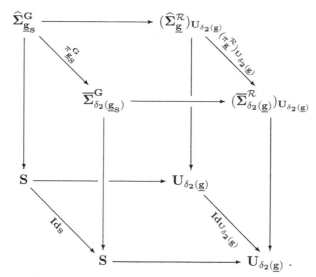

Let $N^{\mathcal{R}}$ be the smallest natural integer N such that $U^{\mathcal{R}} = Spec(\mathbb{Z}[1/N^{\mathcal{R}}])$. Let S be a scheme given in terms of the disjoint sum $\coprod Spec(A_i)$ of affine schemes and the transition cocycle (Φ_{ij}), and an integer $1 \leq N$. Define $Spec(\mathcal{O}_S[1/N])$ as the scheme defined by $\coprod Spec(A_i[1/N])$ and the cocycle $(\Phi_{ij}[1/N])$ obtained by restriction to $\coprod Spec(A_i[1/N]) \cap Spec(A_i[1/N])$ of (Φ_{ij}). The underlying set of points of $Spec(\mathcal{O}_S[1/N])$ is the open subset of S formed by all the points s such that ch $\kappa(s)$ do not divide N. Write $S_{U^{\mathcal{R}}} = Spec(\mathcal{O}_S[1/N^{\mathcal{R}}])$. Remark that the restriction of a smooth resolution to an open subscheme $U \subset S$ is a smooth resolution.

Theorem 16.26

Let G be a reductive S-group scheme of type $\mathcal{R}$. Suppose that there exists at least a residual characteristic of S not dividing $N^{\mathcal{R}}$, i.e. $S_{U^{\mathcal{R}}} \neq \emptyset$.

1) $\left(\widehat{\Sigma}^G, \overline{\Sigma}^G, \pi^G, \Theta^G\right)_{S_{U^{\mathcal{R}}}}$ *is a $S_{U^{\mathcal{R}}}$-Universal smooth resolution.*

2) $\left(\widehat{\Sigma}^G_{\Gamma^m(\underline{\Delta}(G))}, \overline{\Sigma}^G_{\Gamma^m(\underline{\Delta}(G))}, \pi^G_{\Gamma^m(\underline{\Delta}(G))}, \Theta^G_{\Gamma^m(\underline{\Delta}(G))}\right)_{S_{U^{\mathcal{R}}} \times_S \Gamma^m(\underline{\Delta}(G))}$ *is a $S_{U^{\mathcal{R}}} \times_S \Gamma^m(\underline{\Delta}(G))$-Universal smooth resolution, and also a $S_{U^{\mathcal{R}}} \times_S \Gamma^m(\underline{\Delta}(G)) \times_S \mathcal{P}ar(G)$-Universal smooth resolution.*

3) *For all section $\underline{g}$ of $S_{U^{\mathcal{R}}} \times_S \Gamma^m(\underline{\Delta}(G))$ the smooth resolution $\left(\widehat{\Sigma}^G_{\underline{g}}, \overline{\Sigma}^G_{\delta_2(\underline{g})}, \pi^G_{\underline{g}}, \Theta^G_{\underline{g}}\right)_{S_{U^{\mathcal{R}}}}$ is $S_{U^{\mathcal{R}}}$-Universal.*

4) *For all section $(\underline{g}, P_t)$ of $S_{U^{\mathcal{R}}} \times_S \Gamma^m(\underline{\Delta}(G)) \times_S \mathcal{P}ar_t(G)$ the smooth resolution quadruple*

$$\left(\widehat{\Sigma}^G_{(\underline{g}, P_t)}, \overline{\Sigma}^G_{(\delta_2(\underline{g}), P_t)}, \pi^G_{(\underline{g}, P_t)}, \Theta^G_{(\underline{g}, P_t)}\right)_{S_{U^{\mathcal{R}}}}$$

is S_{U^R}-Universal.

16.2.2   A general condition of Universality of smooth resolutions

Proposition 16.27 *Let G be a reductive S-group scheme endowed with a frame of type $\mathcal{R}$. Let $\underline{\tau} = \tau_S$ for some relative position type τ of $\mathcal{A}_E$, and t be the type of a facet defined by τ. Then the locally trivial fibration $\overline{\Sigma}_{\underline{\tau}} \longrightarrow \mathcal{P}ar_t(G)$ is of finite presentation if and only if its typical fiber $\overline{\Sigma}_{(\underline{\tau},P_t)}$ is of finite presentation over S.*

Proof *It is known that $\overline{\Sigma}_{\underline{\tau}} \longrightarrow \mathcal{P}ar_t(G)$ is trivialized by the big cell open covering. Let U be a big cell subscheme of $\mathcal{P}ar_t(G)$. There is an isomorphism*

$$\left(\overline{\Sigma}_{\underline{\tau}}\right)_U \simeq U \times_S \overline{\Sigma}_{(\underline{\tau},P_t)}.$$

On the other hand, the ideal $\mathcal{I} \subset \mathcal{O}_{\overline{\Sigma}_{\underline{\tau}}}$ defining the embedding $j : \overline{\Sigma}_{(\underline{\tau},P_t)} \hookrightarrow \overline{\Sigma}_{\underline{\tau}}$ given by the fiber of $\overline{\Sigma}_{\underline{\tau}}$ over the center of U is finitely generated. Observe that $\overline{\Sigma}_{\underline{\tau}} \longrightarrow S$ is a finite presentation morphism as the composition of finite presentation morphisms, namely $\overline{\Sigma}_{\underline{\tau}} \longrightarrow \mathcal{P}ar_t(G)$ followed by $\mathcal{P}ar_t(G) \longrightarrow S$. It is concluded that the morphism $\overline{\Sigma}_{(\underline{\tau},P_t)} \longrightarrow S$ is locally of finite presentation. On the other hand, it is known that it is projective and thus proper. It is deduced that $\overline{\Sigma}_{(\underline{\tau},P_t)}$ is of finite presentation over S.

Suppose $\overline{\Sigma}_{(\underline{\tau},P_t)}$ of finite presentation over S. From the above isomorphism it results that for all open set U of the big cell open covering $\left(\overline{\Sigma}_{\underline{\tau}}\right)_U \to U$ is of finite presentation. It results that $\overline{\Sigma}_{\underline{\tau}} \longrightarrow \mathcal{P}ar_t(G)$ is of finite presentation.

Proposition 16.28 *Let G be a reductive S-group scheme endowed with a frame of type $\mathcal{R}$. Then the morphism $\overline{\Sigma}_{\underline{\tau}} \longrightarrow \mathcal{P}ar_t(G)$ is flat if and only if $\overline{\Sigma}_{(\underline{\tau},P_t)} \longrightarrow S$ is a flat morphism.*

Proof *Suppose that $\overline{\Sigma}_{(\underline{\tau},P_t)} \longrightarrow S$ is a flat morphism. Thus for all big cell U,*

$$\left(\overline{\Sigma}_{\underline{\tau}}\right)_U \simeq U \times_S \overline{\Sigma}_{(\underline{\tau},P_t)} \longrightarrow U$$

is a flat morphism. It results immediately that $\overline{\Sigma}_{\underline{\tau}} \longrightarrow \mathcal{P}ar_t(G)$ is a flat morphism. Suppose that $\overline{\Sigma}_{\underline{\tau}} \longrightarrow \mathcal{P}ar_t(G)$ is a flat morphism. Let U be a big cell. Then $\left(\overline{\Sigma}_{\underline{\tau}}\right)_U \simeq U \times_S \overline{\Sigma}_{(\underline{\tau},P_t)} \longrightarrow U$ is a flat morphism. By descent of flatness by faithfully flat morphisms it results that $\overline{\Sigma}_{(\underline{\tau},P_t)} \longrightarrow S$ is a flat morphism.

Let k be a field and X a k-scheme of finite type, $V \subset X$ be an integral open subscheme. Then V is schematically dense in X if and only if X is integral. Suppose that V is absolute integral, i.e. for all field extension $k \to K$, $Spec(K) \times_k V$ is an integral scheme, and schematically dense in X. Then for all

K, $Spec(K) \times_k V$ is schematically dense in $Spec(K) \times_k X$ (resp. $Spec(K) \times_k X$ is integral). Reciprocally it is clear by descent that if for some extension K, $Spec(K) \times_k V$ is schematically dense in $Spec(K) \times_k X$ (resp. $Spec(K) \times_k X$ is an integral scheme), then $V \subset X$ is schematically dense (resp. X is integral). Denote by $\overline{k}$ an algebraic closure of k. It follows that $V \subset X$ is schematically dense in X (resp. X is integral) if and only if $Spec(\overline{k}) \times_k V$ is schematically dense in $Spec(\overline{k}) \times_k X$ (resp. $Spec(\overline{k}) \times_k X$ is integral). Let S be a scheme and $V \subset X$ an open subscheme of an S-scheme X. Where X is of finite presentation and flat, and V with absolute integral fibers. Define the subsets of S by

$$U_{(V,X)} = \{\, s \in S \mid V_s \subset X_s \text{ schematically dense}\,\}$$

$$\left(\text{resp. } \tilde{U}_{(V,X)} = \{\, s \in S \mid V_{\overline{s}} \subset X_{\overline{s}} \text{ schematically dense}\,\} \right).$$

By the previous argument $U_{(V,X)} = \tilde{U}_{(V,X)}$, and

$$\tilde{U}_{(V,X)} = \{\, s \in S \mid X_{\overline{s}} \text{ integral}, V_s \text{ not empty}\,\}.$$

Suppose that $X \longrightarrow S$ is **proper**, **flat**, and of **finite presentation**. Under these hypotheses the set of points $s \in S$ such that $X_{\overline{s}}$ is integral is an **open subset** of S. From the above equality we conclude that $U_{(V,X)}$ is an **open set** such that for all $s \in U_{(V,X)}$ the subscheme $V_s \subset X_s$ is schematically dense. As $X \longrightarrow S$ is in particular flat and of finite presentation it results that the subscheme $V_{U_{(V,X)}} \subset X_{U_{(V,X)}}$ is schematically dense, and that $U_{(V,X)}$ is the maximal subscheme of S with this property.

Let G be a reductive S-group scheme of type $\mathcal{R}$. Let $V = \Sigma_{\underline{\tau}}^G$ and $X = \overline{\Sigma}_{\underline{\tau}}^G$. Write

$$U_{\underline{\tau}}^G = U_{(V,X)} \left(\text{resp. } \mathcal{U}_{\underline{\tau}}^G = U_{(V,X)} \text{ if } \Sigma_{\underline{\tau}}^G \text{ and } \overline{\Sigma}_{\underline{\tau}}^G \text{ are seen as } \mathcal{P}ar_t(G) - \text{schemes} \right).$$

Given $z \mapsto y$ there is an isomorphism

$$\left(\overline{\Sigma}_{\underline{\tau}}^G \right)_z \simeq Spec(\kappa(z)) \times_{Spec(\kappa(y))} \left(\overline{\Sigma}_{(\underline{\tau}, P_t)}^G \right)_y$$

(transitivity of fibers). Thus the fiber $\left(\overline{\Sigma}_{\underline{\tau}}^G \right)_z$ is integral if and only if the fiber $\left(\overline{\Sigma}_{(\underline{\tau}, P_t)}^G \right)_y$ is integral. It follows the equality

$$\mathcal{U}_{\underline{\tau}}^G = U_{\underline{\tau}}^G \times_S \mathcal{P}ar_t(G) ,$$

as both members are open subschemes of $\mathcal{P}ar_t(G)$ with the same underlying set of points.

Let $G \simeq S \times \underline{Ep}_{\mathbb{Z}}(\mathcal{R})$ for the etale topology locally in S. Suppose that $\overline{\Sigma}_{\underline{\tau}}^{G} \to$ $\mathcal{P}ar_t(G)$ (t denotes a section of types of G) is for the etale topology locally in S isomorphic to $\overline{\Sigma}_{\tau_S}^{\underline{Ep}_S(\mathcal{R})} \to \mathcal{P}ar_{t'_S}(\underline{Ep}_S(\mathcal{R}))$, where t' (resp. τ) is a type (resp. a type of relative position) of $\mathcal{A}_E$ depending on τ. It is said that $\overline{\Sigma}_{(\tau_S, P_{t'})}^{\underline{Ep}_S(\mathcal{R})}$ is **the typical fiber** of $\overline{\Sigma}_{\underline{\tau}}^{G} \to \mathcal{P}ar_t(G)$. Let $\widehat{\Sigma}_{\underline{g}}^{G}$ define a smooth resolution of $\overline{\Sigma}_{\underline{\tau}}^{G}$, i.e. $\underline{g}$ is a "relative minimal gallery of types" of G with $\delta_2(\underline{g}) = \underline{\tau}$. Suppose there is a minimal gallery of types g of $\mathcal{A}_E$ such that $\widehat{\Sigma}_{g_S}^{\underline{Ep}_S(\mathcal{R})}$ defines a smooth resolution of $\overline{\Sigma}_{\tau_S}^{\underline{Ep}_S(\mathcal{R})}$, and $\widehat{\Sigma}_{g_S}^{\underline{Ep}_S(\mathcal{R})} \simeq \widehat{\Sigma}_{\underline{g}}^{G}$ for the etale topology locally in S. It is said that $\widehat{\Sigma}_{(g_S, P_{t'})}^{\underline{Ep}_S(\mathcal{R})}$ is the typical fiber of $\widehat{\Sigma}_{\underline{g}}^{G}$.

Proposition 16.29
Keep the above notation and hypothesis. The following three assertions are equivalent.

1) *The scheme $\overline{\Sigma}_{\underline{\tau}}^{G}$ is of finite presentation, flat, and with integral fibers relatively to S.*

2) *The scheme $\overline{\Sigma}_{\underline{\tau}}^{G}$ is of finite presentation, flat, and with integral fibers relatively to $\mathcal{P}ar_t(G)$.*

3) *The typical fiber $\overline{\Sigma}_{(\tau_S, P_t)}^{\underline{Ep}_S(\mathcal{R})}$ is of finite presentation, flat, and with integral fibers relatively to S.*

Proposition 16.30
Keep the notation and hypothesis of the preceding proposition. If one of the three assertions of the above theorem is satisfied then:

1) $\left(\widehat{\Sigma}^{G}, \overline{\Sigma}^{G}, \pi^{G}, \Theta^{G} \right)$ *is an S-Universal smooth resolution.*

2) $\left(\widehat{\Sigma}_{\Gamma^m(\underline{\Delta}(G))}^{G}, \overline{\Sigma}_{\Gamma^m(\underline{\Delta}(G))}^{G}, \pi_{\Gamma^m(\underline{\Delta}(G))}^{G}, \Theta_{\Gamma^m(\underline{\Delta}(G))}^{G} \right)$ *is a $\Gamma^m(\underline{\Delta}(G))$-Universal smooth resolution, and also a $\Gamma^m(\underline{\Delta}(G)) \times_S \mathcal{P}ar(G)$-Universal smooth resolution.*

3) *For all section $\underline{g}$ of $\Gamma^m(\underline{\Delta}(G))$ the smooth resolution $\left(\widehat{\Sigma}_{\underline{g}}^{G}, \overline{\Sigma}_{\delta_2(\underline{g})}^{G}, \pi_{\underline{g}}^{G}, \Theta_{\underline{g}}^{G} \right)$ is S-Universal.*

4) *For all section $(\underline{g}, P_t)$ of $\Gamma^m(\underline{\Delta}(G)) \times_S \mathcal{P}ar_t(G)$ the smooth resolution*

$$\left(\widehat{\Sigma}_{(\underline{g}, P_t)}^{G}, \overline{\Sigma}_{(\delta_2(\underline{g}), P_t)}^{G}, \pi_{(\underline{g}, P_t)}^{G}, \Theta_{(\underline{g}, P_t)}^{G} \right)$$

is S-Universal.

Let one recall the Theorem of generical flatness (cf. Theoreme(6.9.1), [26]).

Theorem 16.31 *Let S be an integral scheme locally noetherian and $f :$
$X \longrightarrow S$ a morphism of finite type. There exists a not-empty open subset
U of S so that $\mathcal{O}_X|_U$ is flat over U.*

Theorem 16.32
*Let G be a reductive group scheme over a reduced locally noetherian scheme
S. Then there exists a dense open subset U^G of S satisfying the following
property. The morphism $\left(\overline{\Sigma}^G\right)_{U^G} \longrightarrow U^G$ is a flat morphism of locally finite
presentation so that its fibers are integral. Thus $U^G \times_S \Sigma^G \hookrightarrow \left(\overline{\Sigma}^G\right)_{U^G}$ is
dominant, otherwise stated $U^G \times_S \Sigma^G$ is universally schematically dense in
$\left(\overline{\Sigma}^G\right)_{U^G}$ relatively to U^G. Moreover U^G may be chosen containing S_{U^R}.*

Proof *As density is a local property one may clearly suppose that S is affine
reduced and noetherian. Write $S = Spec(A)$, and denote by $S^{(1)}, \cdots, S^{(N)}$ the
irreducible components of S. There are N prime ideals of A, $\mathfrak{P}^{(1)}, \cdots, \mathfrak{P}^{(N)}$,
with $S^{(1)} = Spec(A/\mathfrak{P}^{(1)}), \cdots, S^{(N)} = Spec(A/\mathfrak{P}^{(N)})$. By the theorem of
generical flatness applied to $\left(\overline{\Sigma}^G\right)_{Spec(A/\mathfrak{P}^{(i)})} \longrightarrow Spec(A/\mathfrak{P}^{(i)})$ there is a
not-empty open subset $U^{(i)}$ of $Spec(A/\mathfrak{P}^{(i)})$ such that $\left(\overline{\Sigma}^G\right)_{U^{(i)}} \longrightarrow U^{(i)}$ is a
flat morphism. Observe that the generic point $\xi^{(i)}$ of $Spec(A/\mathfrak{P}^{(i)})$ belongs to
$U^{(i)}$, and that $Spec(\kappa(\xi^{(i)})) \longrightarrow Spec(A/\mathfrak{P}^{(i)})$ is a quasi-compact flat extension. Thus*

$$Spec(\kappa(\xi^{(i)})) \times_S \Sigma^G \hookrightarrow \left(\overline{\Sigma}^G\right)_{Spec(\kappa(\xi^{(i)}))}$$

*is a dominant morphism. Otherwise stated $\left(\Sigma^G\right)_{\xi^{(i)}}$ is schematically dense in
$\left(\overline{\Sigma}^G\right)_{\xi^{(i)}}$. It follows that $\left(\Sigma^G\right)_{\overline{\xi^{(i)}}}$ is schematically dense in $\left(\overline{\Sigma}^G\right)_{\overline{\xi^{(i)}}}$. Let
V be the set of points s of $Spec(A)$ so that the geometrical fiber $\left(\overline{\Sigma}^G\right)_{\overline{s}}$ of
$\overline{\Sigma}^G$ is integral. It is known that $V \cap U^{(i)}$ is in fact open. On the other hand,
$V \cap U^{(i)}$ coincides with the set of points $s \in U^{(i)}$ so that $\left(\Sigma^G\right)_s$ is schematically
dense in $\left(\overline{\Sigma}^G\right)_s$. Thus $V \cap U^{(i)}$ is an open subscheme of $U^{(i)}$ so that Σ^G is
universally schematically dense in $\overline{\Sigma}^G$ relatively to $V \cap U^{(i)}$ (and is maximal
with this property). The open subscheme $U^G = (\bigcup V \cap U^{(i)}) \cup S_{U^R}$ satisfies
the property of the theorem, as $V \cap U^{(i)}$ is dense in $Spec(A/\mathfrak{P}^{(i)})$, and thus
$\bigcup V \cap U^{(i)}$ is dense in $Spec(A)$.*

Corollary 16.33 *One keeps the same hypothesis and notation of the above
theorem. Then the four statements of Theorem 16.30 hold for G over U^G
instead of S*

16.3   Schematic closure and base change

16.3.1   Generalities about the schematic closure

The following definitions are recalled:

Definition 16.34
Let $f : X \to Y$ be a morphism of schemes. The smallest closed subscheme $Y' \subset Y$, so that the canonical embedding $j_{Y'} : Y' \to Y$ factors f, is **the schematic image** *of f, if it exists.*

- *If X is a subscheme of Y, and $j_X : X \to Y$ the canonical embedding, by definition* **the schematic closure** X^{schc} *of X in Y is the schematic image of j_X. (cf. [24], Ch. 1, Definition (6.10.1)).*

- *X is* **schematically dense** *in Y (cf. [24], Ch. 1, Definition (5.4.2)) if the canonical embedding $j_X : X \to Y$ is* **schematically dominant**. *Otherwise stated: the schematic closure X^{schc_Y} of X in Y is equal to Y.*

- *The S-subscheme $X \subset Y$ is* **universally relatively schematically dense** *if for all base change $S' \to S$ the subscheme $S' \times_S X \subset S' \times_S Y$ is schematically dense. Otherwise stated the canonical embedding j_X is* **universally schematically dominant**, *i.e. for all base change $S' \to S$ the morphism $Id_{S'} \times_S j_X : S' \times_S X \longrightarrow S' \times_S Y$ is schematically dominant.*

Definition 16.35
A morphism $f : X \to Y$ is **schematically dominant** *if for every open subset $U \subset Y$ and every closed subscheme $Z \subset U$, such that the restriction $f^{-1}(U) \to U$ of f factors as*

$$f^{-1}(U) \xrightarrow{g} Z \xrightarrow{j_Z} U,$$

it is
$$Z = U.$$

One recalls **the transitivity of schematic images** which plays an important role in the proof of the main theorem.

Proposition 16.36
Let $f : X \to Y$ and $g : Y \to Z$ be the two morphisms. Suppose that the schematic image Y' of f exists, and that the schematic image Z' of Y' by the **restriction** *g' of g to Y' exists also. Then* **the schematic image** *of X by the composed morphism $g \circ f$* **exists** *and is equal to Z'. (cf. loc. cit., Proposition (6.10.3))*

Let $S' \to S$. Suppose that the schematic closure $(S' \times_S X)^{sch_{Y'}}$ exists (resp. X^{sch_Y} exists). It is said that **the formation of the schematic closure** of the S-subscheme $X \subset Y$ **commutes** with the base change $S' \to S$ if the canonical morphism $(S' \times_S X)^{sch_{Y'}} \to S' \times_S X^{sch_Y}$ is an isomorphism.

The following Proposition gives a sufficient conditions allowing the existence of the schematic image (resp. schematic closure) of a morphism $f : X \to Y$ (resp. an embedding $j_X : X \to Y$).

Proposition 16.37
The schematic image Y' of X by the morphism $f : X \to Y$ exists in each of the following two cases:

1. $f_*(\mathcal{O}_X)$ *is a* **quasi-coherent** $\mathcal{O}_Y$*-module (this condition is satisfied if f is* **quasi-compact** *and* **separated***).*

2. X *is* **reduced** *and as a particular case if X is an* **integral** *scheme.*

The underlying subspace of Y' is given by the closure $\overline{f(X)}$ in Y of the image $f(X)$ of f, and f factors as

$$X \xrightarrow{g} Y' \xrightarrow{j_{Y'}} Y,$$

where g is schematically dominant (cf. [24], Ch. 1, Proposition (6.10.5)). As an embedding is a separated morphism it results the following

Corollary 16.38 *Let $j : X \hookrightarrow Y$ be a quasi-compact embedding, i.e. for all affine open set $U \subset Y$, $j^{-1}(U)$ is quasicompact. Then the schematic closure of X in Y exists.*

As the S-schemes being considered in this work are quasi-projective or projective, it follows from the above proposition that if we suppose one of the following conditions

1. S is a **reduced** scheme.

2. S is a **noetherian** scheme.

the hypothesis for the existence of the schematic image are satisfied. Anyhow in this context one may state results in all generality concerning the base scheme S.

The following proposition allows one to give the schematic closure of a subscheme $j_X : X \hookrightarrow Y$ in terms of descent data once one knows it exists.

Proposition 16.39   *1) Suppose that $f : X \longrightarrow Y$ satisfies one of the conditions of Proposition 16.37. Denote by $f_V : f^{-1}(V) \longrightarrow V$ the restriction of f to an open set V of Y. Then the schematic image of $f^{-1}(V)$ in V exists and is given by the subscheme $inf(V, Y')$ induced by Y' on the open set $V \cap Y'$ of Y'.*

2) More generally if $f : X \longrightarrow Y$ satisfies one of the conditions of Proposition 16.37 the formation of the schematic closure commutes with a flat morphism $Z \to Y$, i.e. $(Z \times_Y X)^{schc_Y z} = Z \times_Y X^{schc_Y}$.

Proposition 16.40
Let $V \subset X$ be a subscheme and $g : X' \to X$ a faithfully flat morphism.

1) Suppose $X' \times_X V \subset X'$ is schematically dense in X'. Then V is schematically dense in X.

2) Let V and X be S-schemes and $S' \to S$ a faithfully flat morphism. Suppose that the subscheme $S' \times_S V \subset S' \times_S X$ is schematically dense. Then V is schematically dense in X.

Proof   *The first assertion is an immediate corollary of Théorème* **(11.10.5)** *of [27] 1966. The following remark proves* 2). *It results from the commutative diagram*

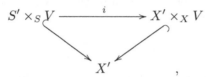

where i is the isomorphism of schemes defined by $(s', v) \mapsto ((s', x_v), v)$, with x_v denoting the image of v in X, that if $S' \times_S V \hookrightarrow X' = S' \times_S X$ is schematically dense then $X' \times_X V \hookrightarrow X'$ is also schematically dense.

Let some general results concerning **separated schemes** be recalled.

1) An affine scheme X is separated, i.e. the diagonal Δ_X is a closed subscheme of the product $X \times X = X \times_{\mathbb{Z}} X$.

2) A projective scheme $X \longrightarrow S$ is separated over S. Thus if one supposes S separated the composed morphism

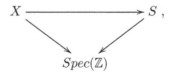

is also separated. It results that a projective scheme is separated. Thus any subscheme of a projective scheme is also separated.

3) Let Y be a separated scheme and $f : X \longrightarrow Y$ a morphism. For all affine open subset U of X and all affine open subset V of Y, $U \cap f^{-1}(V)$ is affine.

Recall that a morphism $f : X \to Y$ is **quasi-compact** if for all quasi-compact open set $V \subset Y$, $f^{-1}(V)$ is quasi-compact. Let S be a scheme and $S = \bigcup_i U_i$ an open covering by affine open subsets. Observe that an S morphism $f : X \to Y$ is quasi-compact if and only if $f^{-1}(V)$ is quasi-compact for all open affine subset V contained in some X_{U_i}

Proposition 16.41 *A morphism* $f : X \longrightarrow Y$ *between a quasi-compact scheme* X *and a separated scheme* Y *is quasi-compact.*

Proof *Write* X *as a finite union of affine open sets,* $X = \bigcup_{1 \le i \le N} U_i$. *Let* $V \subset Y$ *be an affine open set. Then for all* $1 \le i \le N$ *the intersection* $U_i \cap f^{-1}(V)$ *is an affine open set in* X *and* $f^{-1}(V) = \bigcup_{1 \le i \le N} U_i \cap f^{-1}(V)$. *It follows that* $f^{-1}(V)$ *is a quasi-compact open set as a finite union of affine open sets.*

Proposition 16.42
Let S *be an affine scheme,* G *a reductive* S-*scheme endowed with a frame* E, $P \subset G$ *(resp.* $B \subset P$*) a parabolic subgroup of type* t *(resp. a Borel subgroup) adapted to the frame* E, $\tau \in Relpos\, A(R(E))$.

- *The Schubert cell* $\Sigma_{(\tau,\mathbf{P})} \subset \mathcal{P}ar(G)$ *is a quasi-compact subscheme.*

- *The Universal Schubert cell* $\Sigma \subset Relpos_G \times_S \mathcal{P}ar(G) \times_S \mathcal{P}ar(G)$ *is a quasi-compact subscheme.*

- *The schematic closure* $\overline{\Sigma}_{(\tau,\mathbf{P})}$ *(resp.* $\overline{\Sigma}$, $\overline{\Sigma}_\tau$*) of* $\Sigma_{(\tau,\mathbf{P})}$ *(resp.* Σ_τ, Σ*) in* $\mathcal{P}ar(G)$ *(resp.* $\mathcal{P}ar(G) \times_S \mathcal{P}ar(G)$, $Relpos_G \times_S \mathcal{P}ar(G) \times_S \mathcal{P}ar(G)$*) exists.*

Proof *Observe first that as* S *is a separated scheme then* $\mathcal{P}ar(G)$ *is separated as* $\mathcal{P}ar(G) \longrightarrow S$ *is a proper morphism and thus separated and* S *is separated as it is affine. One has* $\Sigma_{(\tau,\mathbf{P})} = \bigcup_{\tau'} \Sigma_{(\tau',\mathbf{B})}$, *where* $\tau' \in Relpos\, A(R(E))$ *runs on the* **finite set** *of types of relative positions indexing the* B-*Schubert cells* Σ' *contained in* $\Sigma_{(\tau,\mathbf{P})}$. *On the other hand, the* B-*Schubert cells* $\Sigma_{(\tau',\mathbf{B})}$ *are affine schemes and thus quasi-compact. It results that* $\Sigma_{(\tau,\mathbf{P})}$ *is a quasi-compact scheme.*
Observe that the Universal Schubert Cell $\Sigma_\tau \longrightarrow \mathcal{P}ar_t(G)$ *of type* τ *is a locally trivial fibration trivialized by the big cell open covering* (U_j) *of* $\mathcal{P}ar(G)$ *and that:*

- $(\Sigma_\tau)_{U_j} \simeq U_j \times_S \Sigma_{(\tau,\mathbf{P})}$;

- $\Sigma_\tau = \bigcup_j (\Sigma_\tau)_{U_j}$.

As the U_j are affine schemes and j runs on a finite set, it is concluded that Σ_τ is a finite union of quasi-compact schemes, and thus quasi-compact. As $\Sigma = \coprod_{\tau \in Relpos\ A(R(E))} \Sigma_\tau$ it results that Σ is quasi-compact.

The last assertion follows immediately from Proposition 16.42 and Proposition 16.37.

Corollary 16.43

1) Let S be an affine scheme, and G a reductive S-group scheme, $\underline{\tau}$ a type of relative position section, and $P \subset G$ a parabolic subgroup. Then the schematic closure $\overline{\Sigma}_{(\tau,\mathbf{P})}$ (resp. $\overline{\Sigma}$, $\overline{\Sigma}_\tau$) of $\Sigma_{(\tau,\mathbf{P})}$ (resp. Σ_τ, Σ) in $\mathcal{P}ar(G)$ (resp. $\mathcal{P}ar(G) \times_S \mathcal{P}ar(G)$, $Relpos_G \times_S \mathcal{P}ar(G) \times_S \mathcal{P}ar(G)$) exists.

2) Same statement as in 1) for S a general scheme.

Proof From Proposition 16.37 it suffices to verify that the canonical embedding of $\Sigma_{(\tau,\mathbf{P})}$ (resp. Σ_τ, Σ) in $\mathcal{P}ar(G)$ (resp. $\mathcal{P}ar(G) \times_S \mathcal{P}ar(G)$, $Relpos_G \times_S \mathcal{P}ar(G) \times_S \mathcal{P}ar(G)$) is quasi-compact.

The verification of the quasi-compacity condition of an S-embedding is local in S for the etale topology. Let $S' \longrightarrow S$ be an affine etale covering so that $G_{S'}$ is endowed with a frame. One has that $\left(\Sigma_{(\underline{\tau},\mathbf{P})}\right)_{\mathbf{S}'} = \Sigma^{\mathbf{G}_{\mathbf{S}'}}_{(\underline{\tau}_{\mathbf{S}'},\mathbf{P}_{\mathbf{S}'})}$ (resp. $\left(\Sigma_{\underline{\tau}}\right)_{\mathbf{S}'} = \Sigma^{\mathbf{G}_{\mathbf{S}'}}_{\underline{\tau}_{\mathbf{S}'}}$, $\left(\Sigma\right)_{\mathbf{S}'} = \Sigma^{\mathbf{G}_{\mathbf{S}'}}$). Thus it suffices to show the quasi-compacity of this embedding for $G_{S'}$, where $S' \to S$ is an affine etale covering such that $G_{S'}$ is endowed with a frame. That has been proved in the proof of Proposition 16.42.

The following remark implies that the proof of the quasi-compacity of the canonical embedding when S is a general base scheme, results from the case where S is affine and thus 2) follows from 1). Let S be a scheme and $S = \bigcup_i S_i$ an open covering by affine open subsets. A S-morphism $f : X \to Y$ is quasi-compact if and only for all open affine subset V contained in some Y_{S_i}, $f^{-1}(V) \subset X_{S_i}$ is quasi-compact, i.e. if an only if f_{S_i} is quasi-compact for all S_i.

16.3.2   Criterion of Universal relative schematical density

Proposition 16.44
Let X be an S-scheme proper and flat of finite presentation, and $V \subset X$ an open subscheme with not empty fibers which are absolute integral, i.e. given $s \in S$ and K a field extension of the residual field $\kappa(s)$ of s then $V_{\tilde{s}}$ is integral, where $\tilde{s} = Spec(K)$.

1) Given a fixed field extension K of $\kappa(s)$ write $\tilde{s} = Spec(K)$. Then $X_{\tilde{s}}$ is integral if and only if X_s is integral. As a particular case one has that X_s is integral if and only if the geometric fiber $X_{\overline{s}}$ is integral.

2) V_s is schematically dense in X_s if and only if X_s is integral.

3) Given a fixed field extension K of $\kappa(s)$ one has that $V_{\bar{s}}$ is schematically dense in $X_{\bar{s}}$ if and only if V_s is schematically dense in X_s. As a particular case V_s is schematically dense in X_s if and only if the geometrical fiber $V_{\bar{s}}$ is schematically dense in the geometrical fiber $X_{\bar{s}}$.

4) The open subscheme V of X is **universally relatively schematically dense** if an only if the fibers V_s (resp. the geometrical fibers $V_{\bar{s}}$) are schematically dense in the fibers X_s (resp. the geometrical fibers $X_{\bar{s}}$).

5) The open subscheme V is **universally relatively schematically dense** in X if and only if the fibers X_s (resp. the geometrical fibers $X_{\bar{s}}$) are integral.

6) There is an open subscheme $U \subset S$ so that if $T \subset S$ satisfies "$T \times_S V \subset T \times_S X$ is a **universally relatively schematically dense**" then $T \subset U$.

The above proposition is complemented by the following corollary which compares the schematic images $V^{sch c_X}{}_{S'}$ and $V^{sch c_Y}{}_{S'}$ for $S' \to S$ and results from the transitivity of schematic images.

Proposition 16.45
Let $V \subset X$ be S-schemes and $X \hookrightarrow Y$ a closed embedding of S-schemes.

1) Suppose that V is universally relatively schematically dense in X. Then for all $S' \to S$ the schematic closure $V^{sch c_Y}{}_{S'}$ of $S' \times_S V$ in $S' \times_S Y$ exists and is equal to $S' \times_S X$, i.e. the formation of the schematic closure of V in Y commutes with base changes $S' \longrightarrow S$.

2) Suppose that for all $S' \to S$ the schematic closure $V^{sch c_Y}{}_{S'}$ exists and is equal to $X_{S'}$, i.e. the formation of the schematic closure of V in Y commutes with base changes $S' \longrightarrow S$. If $V^{sch c_X}{}_{S'}$ exists for all $S' \to S$ then V is universally relatively schematically dense in X.

(resp. $V^{sch c_X}{}_{S'}$)

Proof To see 1) apply the transitivity of schematic images to the sequence of embeddings $V_{S'} \overset{i}{\hookrightarrow} X_{S'} \overset{j}{\hookrightarrow} Y_{S'}$, taking on account that by hypothesis $V_{S'}^{sch c_X}{}_{S'} = X_{S'}$, and that $X_{S'}^{sch c_Y}{}_{S'} = X_{S'}$, as $X_{S'}$ is closed in $Y_{S'}$. This gives that the schematic closure $V_{S'}^{sch c_Y}{}_{S'}$ exists and is equal to $X_{S'}$.

Observe that by the transitivity of schematic images, the schematic image of $V^{sch c_X}{}_{S'} \subset X_{S'}$ by the embedding $X_{S'} \hookrightarrow Y_{S'}$ is equal to $V^{sch c_Y}{}_{S'}$ and thus by hypothesis to $X_{S'}$, thus one obtains 2).

If the embedding $V \hookrightarrow X$ of S-schemes is quasi-compact and X is a projective S-scheme then for all base change $S' \to S$, $S' \times_S V \hookrightarrow S' \times_S X$ is a quasi-compact embedding and $S' \times_S X$ is a projective S'-scheme. Thus for all $S' \to S$ the existence of the schematical closure $V^{schc_{X_{S'}}}$ is granted. Proposition 16.45 may be then re-stated as follows.

Proposition 16.46
Let $V \hookrightarrow X$ be a quasi-compact embedding into a projective scheme X, and $X \hookrightarrow Y$ an embedding. Then V is universally relatively schematically dense in X if and only if the formation of the schematic closure of V in Y commutes with base changes.

The proof of Proposition 16.44 is an immediate consequence of the following ones. The following Proposition states a **Criterion** for an open set V of an S-scheme X to be universally schematically dense in X, relatively to S.

Proposition 16.47
Let
$$f : X \to S$$
be a flat morphism, locally of finite presentation, and
$$V \subset X$$
an "open set". Then V is universally schematically dense in X relatively to S if and only if
$$\forall s \in S, \ V_s = V \cap X_s$$
is schematically dense in X_s. (cf. loc. cit., Proposition 11.10.9).

The following proposition states that the property for a morphism to be schematically dominant, is local for the faithfully flat topology.

Proposition 16.48
Let $g : X' \to X$ be a faithfully flat morphism and
$$f : Z \to X$$
a quasi-compact morphism. Then
$$f_{X'} : Z \times_X X' \to X'$$
is a schematically dominant morphism $\Leftrightarrow f$ is a schematically dominant morphism. (cf. [27], Théorème (11.10.5)).

The following corollary is an immediate consequence of the proposition.

Corollary 16.49 *One keeps the notation and hypothesis of Proposition 16.44. The open sub-scheme V_s is schematically dense in X_s if and only if $V_{\tilde{s}}$ is schematically dense in $X_{\tilde{s}}$.*

Corollary 16.50 *With the notation and hypothesis of Proposition 16.44 one has that V_s (resp. $V_{\tilde{s}}$) is schematically dense in X_s (resp. $X_{\tilde{s}}$) if and only if X_s (resp. $X_{\tilde{s}}$) is integral. Thus X_s is integral if and only if $X_{\tilde{s}}$ is integral.*

Proof *If V_s (resp. $V_{\tilde{s}}$) is schematically dense in X_s (resp. $X_{\tilde{s}}$) then X_s (resp. $X_{\tilde{s}}$ is integral, as the schematic closure of an integral scheme is integral. On the other hand, if X_s (resp. $X_{\tilde{s}}$) is integral then the not void open set V_s (resp. $V_{\tilde{s}}$) is schematically dense.*

One recalls the following result.

Proposition 16.51
Let

$$f : X \to Y$$

be a morphism proper, flat, and of finite presentation.
Then the set of $y \in Y$ such that the geometric fiber

$$X_{\overline{y}} \text{ is } \textbf{integral}$$

*is an **open set** of Y. (cf. loc. cit., Théorème (12.2.4)).*

Corollary 16.52 *One keeps the notation and hypothesis of Proposition 16.44. The set U of points s of S such that V_s is schematically dense in X_s is an open set of S. If U is not empty then $V_U \longrightarrow U$ is universally relatively schematically dense in $X_U \longrightarrow U$.*

Appendix A

About the Coxeter Complex

A.1 Adjacency

The building $I(G)$ of a k-reductive group G corresponds to the Cayley complex $\mathcal{D}(H)$ (cf. [18]) of a discrete group H. It is observed that the contracted product defined by a usual minimal gallery may be seen as a set of paths of the compact real form of G (cf. [7]). The Cayley complex of the Weyl group W of G is obtained in terms of the Weyl complex $C(W, S)$ as follows.

The geometrical representation $\mathcal{C}(\mathcal{H})$ of $C(W, S)$, as a decomposition in simplicial cones of the euclidian space $\mathbb{R}^{(S)}$, gives rise to a decomposition into spherical simplices $\{\sigma_C\}$ $(C \in \mathrm{Ch}\, C(W, S))$ of the unit sphere $S_1^{l-1} \subset \mathbb{R}^{(S)}$. The set of barycenters $\{b_\sigma\}$ $(\sigma \in \mathrm{Ch}\, C(W, S))$ give the vertices of the Cayley diagram $\mathcal{D}(G)$ of W (we recall that $\mathrm{Ch}\, C(W, S)$ is principal homogeneous under the action of W). The edges of $\mathcal{D}(G)$ join the barycenters in pairs as follows. Given a simplex σ denote by $b_{\sigma_1}, \ldots, b_{\sigma_l}$ the barycenters corresponding respectively to the l chambers $C_1, \ldots, C_l$ having a common bounding hyperplane with $C = C_\sigma$. Thus there are l edges issued from b_σ, namely:

$$[b_\sigma, b_{\sigma_1}], \ldots, [b_\sigma, b_{\sigma_l}].$$

If $(n_{ij})_{i,j \in S}$ is the Coxeter matrix of (W, S) then the two dimensional faces of $\mathcal{D}(G)$ are n_{ij}-gons $(i \neq j)$ (cf. [18]).

Let $t \in \mathrm{typ}\, C(W, S)$ be the type of a facet given by some subset $t \subset S$ and $W_t \subset W$ the subgroup generated by t. There is a bijection

$$W/W_t \simeq \mathcal{F}_t$$

between the set of classes W/W_t and the set of facets $\mathcal{F}_t$ of $C(W, S)$ of type t, defined by

$\overline{w} \mapsto w(F_t(C))$. One associates with t a 1-complex $\mathcal{D}_t(W)$ generalizing the Cayley complex. The set of vertices of $\mathcal{D}_t(W)$ is given by the set of classes W/W_t (resp. of facets $\mathcal{F}_t$ of type t of W). The edges issued from some facet F are given by the couples (C, H), where $C \in \mathrm{St}_F$ and H is a bounding hyperplane of C. The couple (C, H) defines an edge between F and the image $s_H(F)$ of F by the reflexion s_H defined by H. Thus the set of edges of $\mathcal{D}_t(W)$ corresponds to the set of generalized galleries of the form

$$F \subset C \supset F' \subset C' \supset F'' ,$$

with typ $F = $ typ F'' and cod $F' = 1$.

Remark that some edges join F to itself, namely those satisfying $F \subset F'$. The group W acts as a group of automorphisms on $\mathcal{D}_t(W)$, as follows easily from its definition.

The geometric realization $\mathcal{C}(\mathcal{H})$ of $C(W, S)$ gives rise to a description of $\mathcal{D}_t(W)$ as in the case of the Cayley complex. The point is that this 1-complex allows a description of the action of W on the set of $\mathcal{F}_t$ in terms of the generators S of W and the relations defined by the Coxeter matrix.

The group W may be obtained as the group of automorphisms of the Cayley complex. Similarly, W may be obtained as a group of automorphisms of $C(W, S)$ preserving the type of facets. In this case the group defining relations are replaced by the incidence relations of $C(W, S)$, i.e. the group of automorphisms of $C(W, S)$ is obtained in term of incidence relations.

The adjacency relation between two chambers C and C' of a building I (cf. [50], 1.2) may be defined in terms of a gg. In fact, C and C' are adjacent chambers if there exists a gg

$$C \supset F \subset C',$$

where cod $F = 1$, and $C \cap C' = F$. The edge relation as defined above between two facets F and F' of the same type t generalises this relation. This gg is minimal if and only if $C \neq C'$.

More generally, given a chamber C of I and an integer $i > 0$, let $E_i(C)$, as in loc. cit., be the set of chambers C' in I such that there exists a gg

$$C \supset F \subset C',$$

with cod $F = i$. Let $\mathcal{A}$ be an apartment of I containing C. It is easy to see that $E_i(C)$ may be written as

$$E_i(C) = \coprod_{C' \in E_i(C) \cap \mathrm{Ch}\, \mathcal{A}} \Sigma_{(B_C, \tau(C,C'))},$$

where $\tau(C, C')$ denotes the type of relative position defined by (C, C'). Observe that $C \supset F \subset C'$ is a MGG if and only if C and C' are transversal in $E\, t_F$.

A.2 Braid relations

Let $\Gamma^{(1)} = \Gamma^{(1)}(C, C')$ (resp. $\Gamma^{(2)} = \Gamma^{(2)}(C, C')$) be the minimal gallery defining the big cell coordinates $(X_\alpha^{(1)})_{\alpha \in R(C, C')}$ (resp. $(X_\alpha^{(1)})_{\alpha \in R(C, C')}$) of $\widehat{\Sigma}(g^{(1)}, B)$ (resp. $\widehat{\Sigma}(g^{(2)}, B)$), i.e. there are isomorphisms

$$\operatorname{Spec}\left(k[X_\alpha^{(1)}]\right) \xrightarrow{\sim} \widehat{\Sigma}'(g^{(1)}, B)$$

(resp.

$$\operatorname{Spec}\left(k[X_\alpha^{(2)}]\right) \xrightarrow{\sim} \widehat{\Sigma}'(g^{(2)}, B)).$$

Where $g^{(1)}$ (resp. $g^{(2)}$) denotes the Minimal Generalized Gallery of types defined by $\Gamma^{(1)}$ (resp. $\Gamma^{(2)}$), $\tau_{g^{(1)}}$ (resp. $\tau_{g^{(2)}}$) its corresponding type of relative position, and $\widehat{\Sigma}'(g^{(1)}, B) \simeq \Sigma(\tau_g^{(1)}, B)$) (resp. $\widehat{\Sigma}'(g^{(2)}, B) \simeq \Sigma(\tau_g^{(2)}, B)$) the big cell of $\widehat{\Sigma}(g^{(1)}, B)$ (resp. $\widehat{\Sigma}(g^{(2)}, B)$). Denote by $[R(C, F')]^{\mathbb{N}}$ the commutative monoid generated by $R(C, F') \subset R_+$, and by $\mathcal{H}^{(i)} = k[[R(C, F')]^{\mathbb{N}}]^{(i)} = k[X_\alpha^{(i)}]_{\alpha \in R(C, F')}$ $(i = 1, 2)$ the corresponding k-algebra. Given $\rho \in \mathbb{N}(R(C, F'))$ write:

$$\mathcal{H}_\rho^{(i)} \subset k[X_\alpha^{(i)}]_{\alpha \in R(C, F')}$$

for the k-subspace generated by the monomials $X_{\alpha_1}^{n_1} X_{\alpha_2}^{n_2} \ldots X_{\alpha_l}^{n_l}$ satisfying $n_1 \alpha_1 + n_2 \alpha_2 + \ldots + n_l \alpha_l = \rho$. Clearly one has $\mathcal{H}_\rho^{(i)} \cdot \mathcal{H}_{\rho'}^{(i)} \subset \mathcal{H}_{\rho + \rho'}^{(i)}$. One may thus graduate the polynomial ring $k[X_\alpha^{(i)}]_{\alpha \in R(C, F')}$ by the k-subspaces $(\mathcal{H}_\rho^{(i)})_{\rho \in \mathbb{N}[R(C, F')]}$. Write

$$\mathcal{H}^{(i)} = k \oplus \left(\bigoplus_{\rho \in \mathbb{N}[R(C, F')]} \mathcal{H}_\rho^{(i)} \right),$$

thus $\mathcal{H}^{(i)}$ is a graded k-algebra obtained from the polynomial algebra $k[X_\alpha]$ by changing its natural graduation. Let $(X_\alpha^{(1)})_{\alpha \in R(C, F')}$ (resp. $(X_\alpha^{(2)})_{\alpha \in R(C, F')}$) be the coordinates of $\Sigma(\tau_g, B)$ given by the minimal gallery $\Gamma^{(1)}$ (resp. $\Gamma^{(2)}$), and

$$\varphi_{(\Gamma^{(1)}, \Gamma^{(2)})} : k[X_\alpha^{(2)}] \longrightarrow k[X_\alpha^{(1)}]$$

the isomorphism given by the coordinates change cocycle

$$X_\alpha^{(2)} = X_\alpha^{(2)}((X_{\alpha'}^{(1)})_{\alpha' \in R(C, F')}) \quad (\alpha \in R(C, F')),$$

defined by the vertical arrow of the diagram:

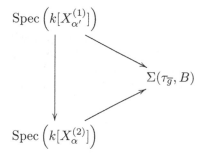

One has:

$$(\forall \rho \in \mathbb{N}[R(C, F')]) \quad \varphi_{(\Gamma^{(1)}, \Gamma^{(2)})}(\mathcal{H}_\rho^{(2)}) = \mathcal{H}_\rho^{(1)},$$

i.e. **the isomorphism** $\varphi_{(\Gamma^{(1)}, \Gamma^{(2)})}$ **is an isomorphism of graded algebras.**
This property results immediately from the commutation defining relations
(Com) of G:

$$\exp(X_\gamma \mathfrak{X}_\gamma)\exp(X_\delta \mathfrak{X}_\delta)\exp(X_\gamma \mathfrak{X}_\gamma)^{-1} = \exp(X_\delta \mathfrak{X}_\delta)\prod \exp_{i\gamma + j\delta}\left(C_{ij\gamma\delta}\, X_\alpha^i\, X_\delta^j\, \mathfrak{X}_{i\gamma + j\delta}\right)$$

with $C_{ij\gamma\delta} \in \mathbb{Z}$, and $(\mathfrak{X}_\gamma)_{\gamma \in R}$ a **Chevalley system**, if it is observed that

$$C_{ij\gamma\delta}\, X_\alpha^i\, X_\delta^j \in \mathcal{H}_{i\gamma + j\delta}^k,$$

with $i = 1, 2$. The factors of the product are indexed by the set of roots of
the form $i\alpha + j\gamma$ with $i, j \in \mathbb{N}^*$. In fact one has the following relation:

$$\mathrm{int}\left(\prod_{\substack{\alpha \in R(C, F') \\ (\text{ordered by } \Gamma^{(1)})}} \exp\left(X_\alpha^{(1)}\, \mathcal{Y}_\alpha\right)\right)(P') = \mathrm{int}\left(\prod_{\substack{\alpha \in R(C, F') \\ (\text{ordered by } \Gamma^{(2)})}} \exp\left(X_\alpha^{(2)}\, \mathcal{Y}_\alpha\right)\right)(P').$$

Thus the second member is obtained by reordering the factors of the first
member according to the order of $R(C, F')$ defined by $\Gamma^{(2)}$ by repeated use
of the commutation relations (Com). As a section of $\Sigma(\tau_g, B)$ given by P, is
uniquely written as $int(\underline{g})(B_{C'}) = B$, with $\underline{g}$ a section of U_B, it is deduced that
the polynomials $X_\alpha^{(2)}((X_{\alpha'}^{(1)})_{\alpha' \in R(C,C')})$ "are generated" by the commutation
relations (Com) and thus that $X_\alpha^{(2)}((X_{\alpha'}^{(1)})_{\alpha' \in R(C,C')}) \in \mathcal{H}^{(1)}$. This proves the
assertion. In fact this operation may be carried out more systematically as
follows.

The change of coordinates cocycle $X_\alpha^{(2)}((X_{\alpha'}^{(1)})_{\alpha' \in R(C,C')}) \in \mathcal{H}^{(1)}$ defined
by the coordinates systems $(X_\alpha^{(2)})$ and $(X_{\alpha'}^{(1)})$ given respectively by $\Gamma^{(2)}$ and

$\Gamma^{(1)}$ may be obtained as the composition of a sequence of change of coordinates cocycles

$$(X_{\alpha,j+1}(X_{\alpha',j})) \quad (j = 0, \ldots, l-1),$$

defined by a sequence of minimal galleries $(\Gamma_j)_{0 \leqslant j \leqslant l}$, with $\Gamma_0 = \Gamma^{(1)}$ and $\Gamma_l = \Gamma^{(2)}$ between the chambers C and $\mathrm{proj}_{F'} C$, i.e. $(X_{\alpha,j+1}(X_{\alpha',j}))$ is the cocycle given by the expression of the coordinates $(X_{\alpha,j+1})$ defined by Γ_{j+1} in terms of the coordinates $(X_{\alpha,j})$ defined by Γ_j. One then writes

$$\left(X_\alpha^{(2)}(X_{\alpha'}^{(1)})\right) = (X_{\alpha,l}(X_{\alpha',l-1})) \circ (X_{\alpha,l-1}(X_{\alpha',l-2})) \circ \ldots \circ (X_{\alpha,1}(X_{\alpha',0})).$$

On the other hand, it is recalled that the apartment associated with a rank 2 reductive group G is a polygon Π (cf. [15], 3.34). Thus given a couple (C, C^{opp}) of antipodal chambers in Π there are exactly 2 minimal galleries namely $\Gamma^{(1)}(C, C^i)$ and $\Gamma^{(2)}(C, C^{opp})$ between C and C^{opp} in Π. The set of cocycles $\left(X_\alpha^{(2)}(X_{\alpha'}^{(1)})\right)$ arising from polygonal apartments may be thus calculated from the tables giving rank 2 reductive groups in terms of defining relations (cf. [10], Exp. XXIII). Given an apartment $\mathcal{A}$ defined by a reductive group G, and two minimal galleries $\Gamma^{(1)} = \Gamma^{(1)}(C, C')$ (resp. $\Gamma^{(2)} = \Gamma^{(2)}(C, C')$) in $\mathcal{A}$ between the couple (C, C') of chambers of $\mathcal{A}$, a sequence of minimal galleries $(\Gamma_j)_{0 \leqslant j \leqslant l}$ between C and C' may be obtained with $\Gamma_0 = \Gamma^{(1)}$ (resp. $\Gamma_l = \Gamma^{(2)}$) such that Γ_{j+1} is obtained from Γ_j by a rank 2 deformation. More precisely there is a chamber $\overline{C}$ of Γ_j and a codimension 2 facet $F \subset \overline{C}$, such that if it is denoted by $\overline{C}^{opp}$ the chamber in St_F opposed to C, and by $\Gamma' = \Gamma'(\overline{C}, C^{opp})$ (resp. $\Gamma'' = \Gamma''(\overline{C}, C^{opp})$) the two minimal galleries in St_F with extremities $\overline{C}$ and C^{opp}, then one may write

$$\Gamma_j = \Gamma_{j,a} * \Gamma' * \Gamma_{j,b}$$

$$(\text{resp. } \Gamma_{j,a} = \Gamma_{j,a} * \Gamma'' * \Gamma_{j,b}),$$

with $\Gamma_{j,a} = \Gamma_{j,a}(C, \overline{C})$ (resp. $\Gamma_{j,b} = \Gamma_{j,b}(\overline{C}^{opp}, C')$). With the terminology of [20] two minimal galleries with the same extremities are **equivalent**.
From this it is deduced that the cocycle $(X_{\alpha,j+1}(X_{\alpha',j}))$ is given by a rank 2 cocycle, and consequently that the cocycle $\left(X_\alpha^{(2)}(X_{\alpha'}^{(1)})\right)$ may be obtained as **the composition of rank 2 coordinates change cocycles.**

From the isomorphisms $\widehat{\Sigma}'(g^{(1)}, B) \simeq \Sigma(\tau_g^{(1)}, B)$ and $\widehat{\Sigma}'(g^{(2)}, B) \simeq \Sigma(\tau_g^{(2)}, B))$ it follows that $\widehat{\Sigma}'(g^{(1)}, B) \times_{\mathcal{P}ar(G)} \widehat{\Sigma}'(g^{(2)}, B) \simeq \Sigma(\tau_g^{(1)}, B) = \Sigma(\tau_g^{(2)}, B)$ is "the big cell" of the Correspondance $\widehat{\Sigma}(g^{(1)}, B) \times_{\mathcal{P}ar(G)} \widehat{\Sigma}(g^{(2)}, B) \longrightarrow \Sigma(\tau, B)$ where $\tau = \tau_g^{(1)} = \tau_g^{(2)}$. This cell may be parametrized by $\Gamma^{(1)}$-coordinates or by $\Gamma^{(2)}$-coordinates, by means of the graph of $\varphi_{(\Gamma^{(1)}, \Gamma^{(2)})}$ (resp. $\varphi_{(\Gamma^{(2)}, \Gamma^{(1)})}$).

Remark that the above assertion results from the relation between the apartment $\mathcal{A}$ and the Cayley diagram $\mathcal{D}(W)$ of (W, S). A geometric interpretation of an algebraic transformation of a path π_1 into a path π_2, of the

Cayley diagram $\mathcal{D}(W)$ by means of the relations $s_i^2 = 1$, $(s_i s_j)^{n_{ij}} = 1$ is given in [18], 4.3 (see also [19])

A.3 Schubert geometry

In the case of $Gl(r+1)$ the Schubert cells are described geometrically by the relative position matrices which in fact resume their classical indexation in terms of the dimensions of the intersection of subspaces of k^{r+1} with a fixed flag. With the canonical basis of k^{r+1} is associated a simplex whose vertices correspond to the basis vectors. The barycenters of its faces correspond to the maximal parabolic subgroups (resp. subspaces of k^{r+1}) "adapted to the canonical basis". The barycentrycal subdivision of this simplex is a geometrical representation of the Coxeter complex associated with the symmetric group $\mathfrak{S}_{r+1}$. The vertices of the simplex thus play an important role. In fact they generate the representation of the Coxeter complex by definition of the barycentic subdivision. It has been seen that a relative position matrix is an abbreviated description of a Minimal Generalized Gallery of types, thus furnishing a smooth resolution of the corresponding Schubert variety. The correspondence between the vertices of the barycentic subdivision of the simplex and the maximal parabolic subgroups of $Gl(r+1)$ is obtained by associating with such a vertex the stabilizer of the subspace of k^{r+1} generated by the corresponding subset of the canonical basis. Remark that the galleries giving smooth resolutions may be obtained solely in terms of subspaces, i.e. without flags of length greater than 1 playing a role. The Coxeter complex and its geometric realization play a central role in the building, and all the galleries giving smooth resolutions of Schubert varieties may be seen as sets of generalized galleries of the building $I(Gl(r+1))$ of $Gl(r+1)$.

The building $I(Gl(r+1))$ is the same as the building $I(PGl(r+1))$ of the projective group $PGl(r+1)$. This is considered its geometric realization in terms of the flags of the projective space $\mathbb{P}(k^{r+1})$, i.e. **the associated geometry** of $Gl(r+1)$, or of its adjoint form $PGl(r+1)$, according to Tits (cf. [50]). On the other hand, one knows that Tits associates an incidence geometry with a reductive group G and thus characterizes the adjoint form of G as the automorphism group of this geometry (cf. loc. cit., see also Freudenthal H., de Vries H., Rosenfeld B., and [38]) thus obtaining a generalization of the fundamental theorem of projective geometry. The following question naturally arises: **How our constructions associated with Schubert cells or varieties are represented in this geometry?** It has been seen that the smooth resolutions of Schubert varieties are calculated in the abstract Coxeter complex. The first step to be taken to answer the above question is to investigate the geometrical realization of the Coxeter complex.

The study of the Galois group of the characteristic Killing equation, i.e. the characteristic equation defined by the adjoint action of a generic element of the Lie algebra, associated with a Lie algebra (cf. [4]) plays an important role in E. Cartan's classification of simple Lie algebras over the complex field $\mathbb{C}$ (cf. [1]). This group is precisely the automorphism group of the corresponding system of roots of the Lie algebra and it has the Weyl group as a distinguished subgroup. In a subsequent paper E. Cartan studies this Galois group and gives Galois resolvents (cf. [2]), i.e. the corresponding Galois extension, for the characteristic Killing equations of each type of Lie algebra. These groups are isomorphic to Galois groups associated to classical algebraic geometrical problems as E. Cartan himself proves it (cf. [3], and [2]). He considers in particular the Galois groups of the classical algebraic geometric problems which are isomorphic to those of the Killing equation of E_6 and E_7, i.e. the Galois group of the 27 lines on a cubic surface in $\mathbb{P}^3$, and that of the 28 bitangents of a quartic curve of $\mathbb{P}^2$. Remark that these Galois groups may be considered as monodromy groups (cf. [32]) and are thus calculated following the pattern of Jordan classical treatise.

E. Cartan establishes this correspondence using the representation of roots given in [1], by associating with "roots configurations" classical "geometric configurations", thus making evident how the respective actions of the Galois groups correspond to each other. On the other hand, these configurations have been extensively studied by classical geometers, and are related to geometric and combinatorial objects. For example, those corresponding to E_6 are associated with the Twenty-seven Lines Polytope studied by Coxeter in [18]. This equivariant correspondence, with respect to the Galois group, between classes of sets of roots and classes of configurations is stated in modern terms by Dolgachev in [28]. As the corresponding Galois principal homogeneous spaces are thus isomorphic, the result is that the set of simple roots correspond to a well known class of configurations. On the other hand, one knows that the Coxeter complex may be described in terms of minimal parabolic sets, i.e. in terms of positive systems of roots (resp. simple systems of roots), thus giving rise to a representation of Schubert cells and minimal generalized galleries in terms of classical configurations and thus generalizing the simplex representation of the Coxeter complex, in the case of the classical Group $Gl(r + 1)$, to a E_6 exceptional group. Thus obtaining a geometrical interpretation of abstract galleries configurations in this cases. A description of the Bruhat decomposition may be obtained in terms of "this geometry" for these groups.

On the other hand, it is known (cf. [50]) that there exists a very simple algorithm by which, from the mere knowledge of the Dynkin diagrams of the groups, one can deduce basic properties of the associated geometries (for instance, the axioms of projective geometry in the case of SL_n) (cf. loc. cit., Théorème 6.3) and relations between geometries associated with different groups.

For each particular simple group G a description of its adapted geometry may thus be obtained. For instance for each of the groups corresponding to Dynkin diagrams without ramification, i.e. groups of types A_n, B_n, C_n, F_4, G_2 the Coxeter complex may be realized as the barycentric subdivision respectively of an n-simplex, an n-cube, an n-octahedron (cf. loc. cit., Ch. 6–Ch. 10), (see also the description of these objects given in [18] and compare it with that of the indexations of the basis of classical algebras as given in [5], §13) a 24-cell polytope, an hexagon (cf. [8]). It has been seen that in the case of a group of type A_n simple descriptions of MGG in terms of the incidence relation of the combinatorial simplex $\Delta^{(r)}$ (or of its barycentric subdivision $\Delta^{(r)\,\prime}$) may be obtained, namely $\mathcal{C}onf(\wedge(M)) \simeq \mathrm{Gall}(g(M))$.

It seems easy to arrive at corresponding simplifications in the description of Minimal Generalized Galleries for the groups of type B_n (resp. C_n, F_4, G_2) by means of configurations varieties defined in terms of the geometry of the polytopes associated with them.

For the groups corresponding to the other Dynkin diagrams, more complicated geometries may serve for the same purposes, for instance, for E_6 the combinatorial geometry of the configuration of the 27 lines on a cubic surface of $\mathbb{P}^3$.

The parabolic sets $A(R)$ of a root system R are determined by the set of minimal parabolics sets. On the other hand there is a natural correspondence between this latter set and certain "configurations" of the 27 lines polytope (cf. [28], [2], [3]) a natural correspondence between $A(R)$ and a set of "configurations" of the 27 lines polytope follows. Consequently geometric representations of $A(R)$ of E_6, of its Schubert varieties, and their smooth resolutions (cf. [38], [37], [47] are obtained.

These descriptions are useful to analyze the fibers of $\widehat{\Sigma} \longrightarrow \overline{\Sigma}$ over $\overline{\Sigma} - \Sigma$ in terms of the canonical cell decomposition of $\widehat{\Sigma}$, and the relations between them and the local rings of $\overline{\Sigma}$. Remark that the relations between geometries of different groups induce relations between their Schubert cells. On the other hand it is well known that the intersection relations between classes of Schubert cycles are described in terms of the Weyl group W and thus these calculations may be re-interpreted, by means of the associated geometry with the Coxeter complex. Thus Ehresmann geometric setting of Schubert calculus (cf. [27]) may be generalized to all groups.

The incidence geometry is particularly interesting in the case of algebraic symmetric spaces, i.e. the hermitian compact symmetric spaces corresponding to the classical groups and to groups of type E_6 and E_7. These spaces are varieties of parabolics, i.e. are isomorphic to the quotient of a reductive group by a parabolic subgroup, and may also be realized as generalized grasmannians (cf. [37], and [39]) by means of the **Magic Square**, and thus obtain the realization of configurations giving resolutions of Schubert varieties in a way similar to that of the linear group. These spaces play a particular role with respect to buildings.

Remark A.1 *The following general proposition plays an important role in [50] as the key to the above mentioned algorithm.*

Proposition A.2 *Let $(\Delta, \mathscr{A})$ be a building, where Δ (resp. $\mathscr{A}$) denotes the corresponding simplicial complex (resp. the set of apartments), and $F \in \Delta$ a facet. Let $\mathscr{A}'$ be the set of all intersections with St_F of all apartments containing F. Then $(St_F, \mathscr{A}')$ is a building whose diagram is obtained by removing from diagr Δ all vertices which belong to typ F.*

Remark that the building blocks of generalized galleries are precisely a couple of facets in these buildings.

Observe that the algorithm which gives the geometry of the building $I(G)$ of a k-reductive group corresponds to "Théorème 1.1." of [23], Exp. XXIII, Théorème 3.5.1, as it results from the discussion in the next section. It asserts that the functor $G \longrightarrow \mathcal{R}(G)$ associating with an S-reductive group G endowed with a frame the root data is an equivalence of categories.

Appendix B

Generators and Relations and the Building of a Reductive Group

The following theorem is proved in [50] and shows the interest of the incidence relation.

Theorem B.1 *A building isomorphism* $\varphi : I \longrightarrow I'$ *is entirely determined by its restriction to* $E_1(C) \cup \mathcal{A}$, *and Theorem 4.1.2. of loc. cit., explains how such an isomorphism may be induced:*
Given $C \in \mathrm{Ch}\, I$ *and* $C' \in \mathrm{Ch}\, I'$, *then every adjacence preserving bijection* $\varphi : E_2(C) \longrightarrow E_2(C')$ *extends to an isomorphism* $\varphi : I \longrightarrow I'$.

On the other hand, it is known that given two k-reductive groups G and G', each endowed with a frame, and a q-morphism $h : \mathcal{R}(G') \longrightarrow \mathcal{R}(G)$ of the corresponding root data, there exists a unique morphism $f : G \longrightarrow G'$ of k-reductive groups endowed with a frame inducing h (cf. [23], Exp. XXV, Théorème 1.1). In fact this Theorem results from a characterization of a morphism $f : G \longrightarrow G'$ in terms of generators and relations of the groups G and G' (cf. [23], Exp. XXIII, Théorème 2.3).

By comparing these two sets of results one obtains the table of generators and relations defining G may be interpreted in the building setting and that all building calculations may be translated into calculations of generators and relations.

The following comparison suggests how the geometrical constructions involving varieties of configurations (resp. of generalized galleries) which are carried out in buildings, lead to algebraic geometrical constructions, by comparing the incidence and the generators and relations description of a reductive group.

It is supposed that G is endowed with the frame $(T, M, R, R_0, (X_\alpha)_{\alpha \in R_0})$. Denote by $A(R)$ the apartment associated to R and by $C = C_{R_+}$ the chamber of $A(R)$ defined by the system of positive roots $R_+ = \mathbb{N} R_0 \cap R$.

Let $(F_\alpha)_{\alpha \in R_0}$ $\left(\text{resp. } (F_{\alpha\beta})_{(\alpha,\beta) \in R_0 \times R_0} \right)$ the set of codimension 1 (resp. codimension 2) facets of C and T_α (resp. $T_{\alpha\beta}$) the maximal torus contained in

$$\operatorname{Ker} \alpha \quad (\text{resp. } \operatorname{Ker} \alpha \cap \operatorname{Ker} \beta) .$$

With the notation of [23], Exp. XXIII, 1.7. Let

$$Z_\alpha = \underline{\operatorname{Cent}}_G(Z_\alpha) \quad (\text{resp. } Z_{\alpha\beta} = \underline{\operatorname{Cent}}_G(Z_{\alpha\beta})) ;$$

Z_α (resp. $Z_{\alpha\beta}$) is a k-reductive subgroup of G with radical subgroup T_α (resp. $T_{\alpha\beta}$). Write:

$$R_\alpha = \mathbb{Z} \cdot \{\alpha\} \cap R \quad (\text{resp. } R_{\alpha\beta} = \mathbb{Z} \cdot \{\alpha, \beta\} \cap R) ,$$

and one endows Z_α (resp. $Z_{\alpha\beta}$) with the canonical frame

$$(T, M, R_\alpha, \{\alpha\}, (X_\alpha))$$

$$(\text{resp. } (T, M, R_\alpha, \{\alpha, \beta\}, (X_\alpha, X_\beta)))$$

(cf. loc. cit., Exp. XXIII, 1.7). Let H be an S-sheaf in groups for the fffp (faithfully flat finite presentation) topology.

Theorem 2.3 of [23], Exp. XXIII asserts that the data

1. an S-group morphism $f_T : T \longrightarrow H$;

2. morphisms $f_\alpha : P_\alpha \longrightarrow H \quad (\alpha \in R_0)$;

3. sections $(h_\alpha)_{\alpha \in R_0}$ of H over S,

subjected to the following conditions:

a. there exists an S-group morphism $f_{N(T)} : N(T) \longrightarrow H$ extending f_T;

b. there exist morphisms $\overline{f}_\alpha : Z_\alpha \longrightarrow H \quad (\alpha \in R_0)$ extending f_α, such that $\overline{f}_\alpha(w_\alpha) = h_\alpha$ (Write $w_\alpha = \exp(X_\alpha)\exp(-X_{-\alpha})\exp(X_\alpha)$, where $X_{-\alpha}$ denotes the dual section of X_α in $\operatorname{Lie}(Z_\alpha)$);

c. there are morphisms $f_{\alpha\beta} : Z_{\alpha\beta} \longrightarrow H$ so that $f_{\alpha\beta} \,|\, P_\alpha = \overline{f}_\alpha$, $f_{\alpha\beta} \,|\, P_\beta = \overline{f}_\beta$, $\overline{f}_{\alpha\beta}(w_\alpha) = h_\alpha$, $\overline{f}_{\alpha\beta}(w_\beta) = h_\beta$.

Conditions a., b., and c. may be translated in terms of relations of the "generators" $T, (P_\alpha)_{\alpha \in R_0}, (w_\alpha)_{\alpha \in R_0}$ of G and $f_T, (f_\alpha)_{\alpha \in R_0}$. Denote by $(n_{\alpha\beta})$ the Coxeter matrix of R, and write

$$t_{\alpha\beta} = (w_\alpha w_\beta)^{n_{\alpha\beta}} \in T(S)$$

$$\left(\text{resp. } t_\alpha = t_{\alpha\alpha} = w_\alpha^2 = \alpha^\vee(-1) \in T(S)\right).$$

The automorphism of T defined by w_α is given by:

$$\text{int}(w_\alpha)(t) = t \, \alpha^\vee(\alpha(t))^{-1}.$$

One introduces conditions $A\_$, $B\_$, and $C\_$, corresponding to a., b., and c.

$$A\_ \ (i). \ h_\alpha f_T(t) h_\alpha^{-1} = f_T\left(\text{int}(w_\alpha)(t)\right);$$

$$(ii). \ f_T(t_{\alpha\beta}) = (h_\alpha h_\beta)^{n_{\alpha\beta}}.$$

Observe that $A\_$ allows to define a group homomorphism for every $\gamma \in R$

$$f_\gamma : P_\gamma \longrightarrow H.$$

For every root $\gamma \in R$, there is a section n of $N(T)$ so that its image in $N(T)/T$ gives rise to an element $\bar{n}$ of W with $\bar{n}(\gamma) = \alpha \in R_0$.
Condition $A\_$ also implies that there is a morphism

$$f_{N(T)} : N(T) \longrightarrow H$$

extending f_T. Write $h = f_{N(T)}(n)$ and given a section x of P_γ define

$$f_\gamma(x) = \text{int}(h) \left[f_\alpha\left(\text{int}(n^{-1})(x)\right)\right].$$

From loc. cit., Exp. XXIII, Lemme 2.3.5 it results that $f_\gamma : P_\gamma \longrightarrow H$ is a well defined group homomorphism.

$B\_ \quad (h_\alpha f_\alpha\left(\exp(X_\alpha)\right))^3 = e \, (= \text{the section of } H \text{ corresponding to the unity of } H).$

By definition a Chevalley system $(X_\alpha)_{\alpha \in R}$ of a splitted S-group (G, T, M, R) is a family of sections of $\text{Lie}(G)$ satisfying $X_\alpha \in \Gamma(S, \mathcal{G}^\alpha)^*$, i.e. X_α is a basis of $\mathcal{G}^\alpha$ satisfying the following condition:
for every $\alpha, \beta \in R$ one has

$$\text{ad}(w_\alpha(X_\alpha))X_\beta = \pm X_{s_\alpha}(\beta),$$

where $w_\alpha(X_\alpha) = \exp(X_\alpha \exp(-X_\alpha^{-1})\exp(X_\alpha)$, and s_α denotes the reflexion given by α.
Let $\alpha, \beta, \alpha + \beta \in R$, then one has

$$[X_\alpha, X_\beta] = \pm p \, X_{\alpha+\beta},$$

where $p > 0$ is the smallest integer so that $\beta - p\alpha$ is not a root (cf. [23], Exp. XXIII, Definition 6.1). The following result holds:

"Let G be an S-group endowed with a frame $(T, M, R, R_0, (X'_\alpha)_{\alpha \in R_0})$. Then there exists a Chevalley system $(X_\alpha)_{\alpha \in R}$ of G so that

$$\forall \alpha \in R_0 \ \ X_\alpha = X'_\alpha \text{"}.$$

Let $U = U_{R_+}$ be the unipotent radical of B_{R_+}, one defines $U_{\alpha\beta} = Z_{\alpha\beta} \cap U$. The unipotent group $U_{\alpha\beta}$ is generated by the set of root subgroups $(P_\gamma)_{\gamma \in R_{\alpha\beta} \cap R_+}$, satisfying the following commutation relations. Given $\gamma, \delta \in R_{\alpha\beta} \cap R_+$ there exists a unique set of constants $(C_{ij\gamma\delta})_{\{(i,j) \in \mathbb{N}^* \times \mathbb{N}^* \mid i\gamma + j\delta \in R_{\alpha\beta}\}} \subset \mathbb{Z}$ satisfying:

$$\text{for all } x, y \in \underline{G}_a(S') \ (S' \text{ an } S - \text{scheme})$$

there is:

$$\text{Com)} \ \exp(xX_\gamma) \exp(yX_\delta) \exp(X_\gamma)^{-1} = \exp(\gamma X_\delta)$$

$$\prod_{\substack{(i,j) \in \mathbb{N}^* \times \mathbb{N}^* \\ i\gamma + j\delta \in R}} \exp_{i\gamma + j\delta} \left(C_{ij\gamma\delta} \, x^i \, y^j \, X_{i\gamma + j\delta} \right),$$

where it is supposed that the factors are arranged according to some fixed order of $R_{\alpha\beta}$.

  $C\_$ (i) For every couple $(\alpha, \beta) \in R_0 \times R_0$ and every $n \in \underline{\mathcal{N}}\text{orm}_{Z_{\alpha\beta}}(T)(S)$ so that $\text{int}(n)(P_\alpha) = P_\alpha$ (resp. $\text{int}(n)(P_\alpha) = P_\beta$) one has: for every $x \in P_\alpha(S')$ $(S' \text{ an } S - \text{scheme})$

$$\text{int}\left(f_{N(T)}(n) \right) \ f_\alpha(x) = f_\alpha \left(\text{int}(n) \, x \right)$$

$$\left(\text{resp. int}\left(f_{N(T)}(n) \right) \ f_\alpha(x) = f_\beta \left(\text{int}(n) \, x \right) \right).$$

(ii) Let $f_{\alpha\beta} : U_{\alpha\beta} = \prod_{\gamma \in R_{\alpha\beta} \cap R_+} P_\gamma \longrightarrow H$ be defined by the set $(f_\gamma)_{\gamma \in R_{\alpha\beta} \cap R_+}$ and the composition morphism of H. Then $f_{\alpha\beta}$ must satisfy the set of relations $(Com)'$ obtained as the image by $f_{\alpha\beta}$ of the set of relations (Com).

  Thus one obtains conditions "a.", "b.", and "c." in terms of generators and relations as follows. One has the following implications:

  "A_" $\Rightarrow$ "a." (resp. "A_ and B_" $\Rightarrow$ "b.", "A_, B_, and C_" $\Rightarrow$ "c.").

Conditions (i) and (ii) of $C\_$ are made explicit for each type of rank 2 group in [23], Exp. XXIII (cf. sections 3, 3.1, 3.2, and 3.4 respectively for groups of type $A_1 + A_1$, A_2, B_2, and G_2.

  If S is given by the spectrum of an algebraically closed field k, i.e. $S = \text{Spec}(k)$, then the above defining relations for a morphism $f : G \longrightarrow H$, give rise to an explicit description of the group in terms of generators and relations (cf. [23], Exp. XXIII, 3.5.3).

The isomorphism $\mathrm{Aut}(I(G)) \simeq G^{\mathrm{ad}}$ generalizes the fundamental theorem of projective geometry (cf. [50]). The proof of this isomorphism is obtained by reducing the general case to the case of a rank 2 building (cf. [50], Th. 4.1.1, and Th. 4.1.2).

Observe that $E_2(C) = \bigcup \mathrm{Ch}\,\mathrm{St}_{\mathrm{F}_{\alpha\beta}}$, and that $\mathrm{St}_{\mathrm{F}_{\alpha\beta}}$ may be identified with the building $I(Z_{\alpha\beta})$ of $Z_{\alpha\beta}$.

As $Z_{\alpha\beta}(k)$ is a rank 2 group, one may suppose that $\mathrm{Aut}(I(Z_{\alpha\beta}))$ realizes as $Z_{\alpha\beta}^{ad}(k)$. Theorem 4.1.1 of [50] essentially states that the groups $Z_\alpha(k)$ and T generate $G(k)$. The following consequence of the incidence preserving hypothesis of Th. 4.1.2 of [50] suggests that incidence implicitly corresponds to the above defining relations of $G(k)$:

Let $\mathcal{A} \subset I$ (resp. $\mathcal{A}' \subset I'$) be an apartment of the building I (resp. I'), and $C \in \mathrm{Ch}\,\mathcal{A}$ (resp. $C' \in \mathrm{Ch}\,\mathcal{A}'$). Given an isomorphism $\alpha : \mathcal{A} \longrightarrow \mathcal{A}'$, and $\gamma : E_2(C) \longrightarrow E_2(C')$ an adjacence preserving mapping which coincides with α on $\mathcal{A} \cap E_2(C)$. Then α and γ are restrictions to $\mathcal{A}$ and $E_2(C)$ of an isomorphism of I onto I'.

References

[1] E. Cartan, Sur la Structure des groupes de transformations finis et continus (These) (Paris, Nony 1891).

[2] E. Cartan, Quelques remarques sur les 28 bitangentes d'une quartique plane et les 27 droites d'une surface cubique, Bull. Sci. Math., (2)**70**: 42–46 (1946).

[3] E. Cartan, Sur la Reduction a sa Forme Canonique de la Structure d'un Groupe de Transformations Fini et Continu, American Journal of Mathematics, Vol. **18**, No. 1 (Jan. 1896), pp. 1–61.

[4] N. Bourbaki, Groupes et Algèbres de Lie, Chap. IV, V, VI (Hermann 1968).

[5] N. Bourbaki, Groupes et algèbres de Lie, Chap. VII, VIII (Hermann 1975).

[6] A. Borel, Linear Algebraic Groups, Springer Verlag (1991).

[7] R. Bott, H. Samelson, Applications of the Theory of Morse to Symmetric Spaces, American Journal of Mathematics (1958).

[8] Fr. Bouhekenhout, A.M. Cohen, Diagram Geometry (Classical Groups, and Buildings), Springer (2013).

[9] Cl. Chevalley, Classification des Groupes Algébriques Semi-simples (Collected Works V3), Springer-Verlag, Germany Heidelberg 2005.

[10] Cl. Chevalley, Sur Certains Groupes Simples, Tohoku 1955.

[11] Wei-Liang Chow, On the Geometry of Algebraic Homogeneous Spaces, Annals of Mathematics, Vol. **50**, No. 1 (Jan. 1949).

[12] C. Contou-Carrère, Sur l'évaluation des nombres caractéristiques d'une sous-variété plongée dans $\mathbb{R}^{n+N}$, C.R.A.S., **281**, Série A-711 (1975).

[13] C. Contou-Carrère, Un modèle lisse d'un cycle de Schubert rendant localement libre l'image de son faisceau conormal dans le module des différentielles, C.R.A.S., **284**, Série A-171 (1977).

[14] C. Contou-Carrère, Immeuble de Tits d'un groupe réductif et construction d'une famille de modèles lisses pour les adhérences des cellules de Schubert généralisées, C.R.A.S., **288**, Série A-797 (1979).

[15] C. Contou-Carrère, Le Lieu singulier des variétés de Schubert, Advances in Mathematics, **71**, No. 2 (1988).

[16] C. Contou-Carrère, Géométrie des groupes semi-simples, résolutions équivariantes, et Lieu singulier de leurs variétés de Schubert, Thèse, Montpellier 1981.

[17] H.S.M. Coxeter, The polytope 2_{21}, whose twenty-seven vertices correspond to the lines on the general cubic surface, Amer. J. Math., **62**: 457–486 (1940).

[18] H.S.M. Coxeter, W.O.J. Moser, Generators and relations for discrete groups, Springer-Verlag (1965).

[19] M. Davis, The Geometry and Topology of Coxeter groups, Princeton University Press (2008).

[20] P. Deligne, Les immeubles de groupes de tresses généralisés, Inventiones Math., **17**: 273–302 (1972).

[21] M. Demazure, Désingularisation des variétés de Schubert généralisées, A.E.N.S., **7**: 53–88 (1974).

[22] M. Demazure, A. Grothendieck, Schémas en groupes (Vol. 1), LN in Mathematics **151**, Springer-Verlag.

[23] M. Demazure, A. Grothendieck, Schémas en groupes (Vol. 3), LN in Mathematics **153**, Springer-Verlag.

[24] J. Dieudonné, A. Grothendieck, Eléments de géométrie algébrique, Springer-Verlag (1971).

[25] J. Dieudonné, A. Grothendieck, Eléments de géométrie algébrique II, Publications de l'I.H.E.S., **8**.

[26] J. Dieudonné, A. Grothendieck, Eléments de géométrie algébrique IV (deuxième partie), Publications de l'I.H.E.S., **24** (1965).

[27] J. Dieudonné, A. Grothendieck, Eléments de géométrie algébrique IV (troisième partie), Publications de l'I.H.E.S., **28** (1966).

[28] I.V. Dolgachev, Classic Algebraic Geometry, a modern view, Cambridge University Press (2011).

[29] Ch. Ehresmann, Sur la Topologie de Certains Espaces Homogènes, Annals of Mathematics, Vol. **35**, No. 2 (1934).

[30] H. Freudenthal, H. de Vries, Linear Lie Groups, Academic Press, New York-London.

[31] S. Gaussent, Etude de la resolution de Bott-Samelson (Thèse Montpellier 2001).

[32] J. Harris, Galois Groups of Ennumerative Problems, Duke Mathematical Journal, Vol. **64**, No. 4 (Dec. 1979).

[33] R. Hartshorne, Algebraic Geometry, Graduate Texts in Mathematics, Vol. **52**, Springer (1997).

[34] H. Hironaka, Resolution of Singularities of an Algebraic Variety of Characteristic Zero: I, Annals of Mathematics , Vol. **79**, No. 1 (Jan. 1964).

[35] H. Hiller, Geometry of Coxeter Groups, Pitman, Boston-London-Melbourne 1982.

[36] W.V.D. Hodge, D. Pedoe, Methods of Algebraic Geometry, Vol. **2**, Cambridge University Press (1952).

[37] Huang, Yongdong, A Uniform Description of Riemannian Symmetric Spaces as Grassmannians using Magic Square (Thesis 2007), The Chinese University of Hong-Kong.

[38] A. Iliev, L. Manivel, The Chow ring of the Cayley plane, Compositio Math., **141**: 146–160 (2005).

[39] N. Jacobson, Lie Algebras, Interscience, New-York-London (1962).

[40] J.C. Jantzen, Representations of Algebraic groups (Second Edition), American Mathematical Society (2003).

[41] M. Morse, The Calculus of Variations in the Large, Colloquium Publications AMS, Vol. **18** (1934).

[42] D. Mumford, Geometric Invariant Theory, Springer-Verlag, Berlin-Heidelberg-New-York 1965.

[43] D. Mumford, Toroidal Embeddings LN **339**, Springer 1973.

[44] D. Mumford, Smooth Compactification of Locally Symmetric Varieties, Math Sci Press, Brookline, Massachusetts 1975.

[45] M. Pankov, Grassmannians of classical buildings, World Scientific (2010).

[46] B. Rosenfeld, Geometry of Lie Groups, Kluwer, Dordrech/Boston/London 1997.

[47] H. Salzmann, D. Betten, Th. Grundhöfer, H. Hähl, R. Löwen, M. Stroppel, Compact Projective Planes (An Introduction to Octonion Geometry), Walter de Gruyter (1995).

[48] J.P. Serre, Algebre Locale, Multiplicites, LN **11**, Springer, Berlin-Heidelberg-New-York (1965).

[49] R. Thom, Les singularités des applications différentiables, Annales de l'Institut Fourier, Vol. **4** (1955).

[50] J. Tits, Buildings of spherical type, Lecture Notes **386**, Springer Verlag, Berlin.

INDEX

Printed and bound by CPI Group (UK) Ltd, Croydon, CR0 4YY

01/11/2024

01782621-0013